The complex internal structure of the Sun can now be studied in detail through helioseismology and neutrino astronomy. The VI Canary Islands Winter School of Astrophysics was dedicated to examining these powerful new techniques. Based on this meeting, seven specially written chapters by world experts renowned for their teaching skills are presented in this timely volume.

With a clear and pedagogical style we are shown how the internal composition (density, He abundance, etc.) and dynamical structure (rotation, subsurface velocity fields, etc.) of the Sun can be deduced throught helioseismology; and how the central temperature can be inferred from measurements of the flux of solar neutrinos.

This volume provides an excellent introduction for graduate students and an up-to-date overview for researchers working on the Sun, neutrino astronomy and helio- and asteroseismology

THE STRUCTURE OF THE SUN

VI Canary Islands Winter School of Astrophysics

THE STRUCTURE OF THE SUN

VI Canary Islands Winter School of Astrophysics

Edited by

T. Roca Cortés, University of La Laguna, Tenerife, Spain

F. Sánchez, Instituto de Astrofísica de Canarias, Tenerife, Spain

CAMBRIDGE UNIVERSITY PRESS
Cambridge, New York, Melbourne, Madrid, Cape Town,
Singapore, São Paulo, Delhi, Tokyo, Mexico City

Cambridge University Press
The Edinburgh Building, Cambridge CB2 8RU, UK

Published in the United States of America by
Cambridge University Press, New York

www.cambridge.org
Information on this title: www.cambridge.org/9780521563079

First published 1996

A catalogue record for this publication is available from the British Library

ISBN 978-0-521-56307-9 Hardback

Contents

LIST OF PARTICIPANTS VI C.I.W.S.

AJUKOV, Sergey V.	Sternberg Astronomical Institute, Moscow State. RUSSIA
ANDRETTA, Vincenzo	Armagh Observatory. IRELAND
ARHONTIS, Basilis	Aristotelian University of Thessaloniki. GREECE
AUDARD, Nathalie	Observatoire de la Cote d'Azur.FRANCE
BABAEV, E.S.	IZMIRAN. RUSSIA.
BAGALA, Liria G.	Instituto de Astronomía y Física del Espacio. ARGENTINA
BALA, B.	Indian Institute of Astrophysics. INDIA
BANERJEE, Dipankar	Indian Institute of Astrophysics. INDIA
BARANYI, Tünde	Heliophysical Observatory. HUNGARY
BASU, Sarbani	Institute of Physics and Astronomy. DENMARK
BATOURINE, V.A.	Queen Mary and Westfield College. U.K.
BECK, John G.	National Solar Observatory. USA
BELLOT RUBIO, Luis R.	I.A.C. SPAIN
BIESECKER, Douglas	University of New Hampshire. USA.
BRAJSA, Roman	HVAR Observatory. CROACIA.
CANULLO, M. Victoria	Instituto de Astronomía y Física del Espacio. ARGENTINA
CHANG, Heon-Young	Institute of Astronomy. University of Cambridge. U.K.
DANIELL, Mark	Mathematical Sciences. University of St. Andrews. U.K.
DISTEL, James R.	University of Pennsylvania. USA
DZITKO, Hervé	Service d'Astrophysique. DAPNIA. FRANCE
EFF-DARWICH, Antonio M.	I.A.C. SPAIN
ERDELYI, Robert	Center for Plasma Astrophysics. BELGIUM
GARCIA BUSTINDUY, R. A.	I.A.C. SPAIN
GEORGOBIANI, Dali	National Solar Observatory. USA
GIBSON, Sarah	University of Colorado. USA
GONZALEZ HDEZ. , Irene E.	I.A.C. SPAIN
GRABOWSKI, Udo	Kiepenheuer Institut für Sonnenphysik. GERMANY
HAARDT, Francesco	Goteborg University. SWEDEN
HERNANDEZ C., Mario	I.A.C. SPAIN
HIREMATH, K.M.	Indian Institute of Astrophysics. INDIA
HIRZBERGAR, Johann	Institüt fur Astronomie. AUSTRIA
HOUDEK, Guenter	Institut for Fysik og Astronomi. DENMARK
JAIN, Rekha	University of St. Andrews. U.K.
KIEFER, Michael	Kiepenheuer Institut für Sonnenphysik.GERMANY
KIM, Yong-Cheol	Center for Solar and Space Research. Yale University. USA
LAPTHORN, Barry Thomas	Queen Mary Westfield College. U.K.
MADJARSKA, M. Sotirova	Institute of Astronomy. BULGARIA
MARTIN, Isabel	I.A.C. SPAIN
MATIAS, José	Observatoire de Paris - Meudon. FRANCE
MONTAGNE, Marc	Observatoire Midi-Pyrénées. FRANCE
MUGLACH, Karin	Kiepenheuer Institut für Sonnenphysik. GERMANY
OLIVER, Ramon	Universitat de les Illes Balears. SPAIN
OZISIK, Tuncay	Istanbul University Observatory. TURQUIA
PEREZ PEREZ, M. Elena	I.A.C. SPAIN.
P. F. GARCIA MONTEIRO, Mario	Queen Mary and Westfield College. U.K.
RABELLO SOARES, Cristina de A.	I.A.C. SPAIN
ROGL, Jadzia	University of Vienna. AUSTRIA
ROSENTHAL, Colin S.	JILA. University of Colorado. USA

RYBAK, Jan	Astronomical Institute. SLOVAK REPUBLIC
SANIGA, Metod	Astronomical Institute. SLOVAK REPUBLIC
SHIGUEOKA, Hisataki	Universidade Federal Fluminense. I. de Fisica. BRASIL
SHOUMKO, S.M.	Crimean Astrophysical Obs.. UKRANIA
STENVIT, Hilde	Center for Plasma Astrophysics. BELGIUM
SUTTERLIN, Peter	Kiepenheuer Institut für Sonnenphysik. GERMANY
SZAKALY, Gergely	Dept. of Astronomy. L. Eötvös University. HUNGARY
TRAMPEDACH, Regner	University of Aarhus. DENMARK
VISKUM, Michael	IFA, Aarhus University. DENMARK
WESTENDORP PLAZA, Carlos	I.A.C. SPAIN,

the members of IAC's scientific staff

Baudin, Frédéric
Jiménez, Antonio J.
Moreno Insertís, Fernando
Pallé, Pere L.
Patrón, Jesús
Pérez Hernández, Fernando
Régulo, Clara
Roca Cortés, Teodoro
del Toro, José C.
Vázquez Abeledo, Manuel,

the members of IAC's administrative staff
González, Lourdes
García, Nati
López, Begoña
del Puerto, Carmen
Campbell Warden,

and the lecturers.

PREFACE

Today, the study of *The Structure of the Sun* is one of the most exciting and rapidly evolving fields in physics. Helioseismology has provided us with a new tool to measure the physical state of the interior of a star, our Sun. This technique is successful to a depth of 0.7 $R_\odot$ (i.e. 0.3 $R_\odot$ from the centre). Deeper than this, observational data has been scarce. However, data are now becoming available from Earth-bound helioseismic networks (GONG, TON, IRIS, BISON,...) and from experiments on board SOHO (GOLF, MDI, VIRGO). These should allow the spectrum of gravity modes for the Sun to be determined, and thus the physical state of the solar core.

This book provides an up-to-date and comprehensive review of our current understanding of the Sun. Each chapter is written by a world expert. They are based on lectures given at the *VIth Canary Islands Winter School on Astrophysics.* This timely conference brought together leading scientists in the field, postgraduates and recent postdocs students. The aim was to take stock of the new understanding of the Sun and to focus on avenues for fruitful future research. Eight lecturers, around 60 students, and staff from the IAC met in the Hotel Gran Tinerfe in Playa de las Américas (Adeje, Tenerife) from the 5th to the 16th of December, 1994. It was a fortnight of intense and enjoyable scientific work.

At the meeting, oustanding lectures were given by Professors John Bahcall, Tim Brown, Jorgen Christensen-Dalsgaard, Douglas Gough, Jeff Kuhn, John Leibacher, Gene Parker and Yutaka Uchida. The students also presented their work in the form of poster papers which were discussed in special sessions. We are thankful to them all for making the meeting so scientifically profitable. Moreover, we benefited from the excellent work performed by our secretariat staff. Lourdes Gonzalez took care of all pre- and post-School organization; Nati García and Begoña López were on the everyday problems during the School; Carmen del Puerto and Begoña López surprised us with an excellent special issue of *Noticias* at the beginning of the School and Campbell Warden was there when we needed him. Nati García helped me in editing and assembling this manuscript. Without their help the School would not have been as profitable as it proved to be.

T. Roca Cortés and F. Sánchez, Editors.
Instituto de Astrofísica de Canarias , 38205 La Laguna, Tenerife, Spain.

Techniques for Observing Solar Oscillations

Timothy M. Brown

High Altitude Observatory/National Center for Atmospheric Research*
P.O. Box 3000
Boulder, CO 80307
USA

*The National Center for Atmospheric Research is sponsored by the National Science Foundation.

1 ANALYSIS TOOLS AND THE SOLAR NOISE BACKGROUND

When we observe solar oscillations, we are concerned with measuring perturbations on the Sun that are almost periodic in space and time. The periodic waves that interest us are, however, embedded in a background of broadband noise from convection and other solar processes, which tend to obscure and confuse the information we want. Also (and worse), the "almost-periodic" nature of the waves leads to problems in the interpretation of the time series that we measure. Much of the subject of observational helioseismology is thus concerned with ways to minimize these difficulties.

1.1 Fourier Transforms and Statistics

A common thread runs through all of the analysis tricks that one plays when looking at solar oscillations data, and indeed through many of the purely instrumental concerns as well: this thread is the Fourier transform. The reason for this commonality is, of course, that we are dealing with (almost) periodic phenomena – either the acoustic-gravity waves themselves, or the light waves that bring us news of them. Since many of the same notions will recur repeatedly, it is worth taking a little time (and boring the cognoscente) to review some of the most useful properties of Fourier transforms and power spectra. In what follows, I shall simply state results and indicate some of the more useful consequences. We shall see below that even when the Big Theorems of Fourier transforms do not apply, (as with Legendre transforms, for instance), analogous things happen, so that the Fourier example is a helpful guide to the kind of problems we may have. Those who want proofs and details can find them in Bracewell's (1965) classic book.

1.1.1 Shift and Convolution Theorems

Suppose we have a (possibly complex) function $f(x)$. Its Fourier transform $F(k)$ is defined as

$$F(k) = \int_{-\infty}^{\infty} f(x) exp(-2\pi ikx)\, dx \ . \tag{1}$$

To invert the transform, just invert the sign in the exponential:

$$f(x) = \int_{-\infty}^{\infty} F(k) exp(+2\pi ikx)\, dx \ . \tag{2}$$

A useful notation is to denote transforms (either forward or inverse) by overbars on the function being transformed:

$$\overline{f(x)} \;=\; F(k) \;=\; \int_{-\infty}^{\infty} f(x) exp(-2\pi ikx)\, dx \quad . \tag{3}$$

Aside from a sign change in the exponential, the Fourier transform is its own inverse, so that (with the looseness provided by the overbar notation) we can write

$$\overline{\overline{f(x)}} \;=\; \overline{F(k)} \;=\; f(x) \; . \tag{4}$$

This feature is extremely helpful, since it means that many of the following relations between a function and its transform may be applied in either direction. Another important property of Fourier transforms is their scaling property:

$$\overline{f(ax)} \;=\; \frac{1}{a} F(\frac{k}{a}) \; . \tag{5}$$

A few special cases are worth noting: If $f(x)$ is real, then $F(k)$ is conjugate-symmetric in k, that is, $F(-k) \;=\; F^{*}(k)$. If $f(x)$ is real and symmetric in x, then $F(k)$ is purely real. If $f(x)$ is real and antisymmetric in x, then $F(k)$ is purely imaginary.

Straightforward computation shows that $F(k)$ obeys the *shift theorem*, namely that if $f(x)$ has the Fourier transform $F(k)$, then $f(x-a)$ has the transform $exp(-2\pi iak)F(k)$.

The shift theorem may in turn be used to prove the single most useful result about Fourier transforms, the *convolution theorem.* A convolution $h(x)$ is defined as

$$h(x) \;=\; f \otimes g \;=\; \int_{-\infty}^{\infty} f(u) g(x-u) du \quad . \tag{6}$$

If the Fourier transforms of f and g are $\overline{f}$ and $\overline{g}$, then the convolution theorem says that

$$\overline{h} \;=\; \overline{f \otimes g} \;=\; \overline{f}\,\overline{g}. \tag{7}$$

That is, the Fourier transform of a convolution of two functions is the product of their Fourier transforms. Because if the self-inverse nature of the transform, it also follows that the transform of a product of two functions is the convolution of their transforms:

$$\overline{fg} \;=\; \overline{f} \otimes \overline{g} \; . \tag{8}$$

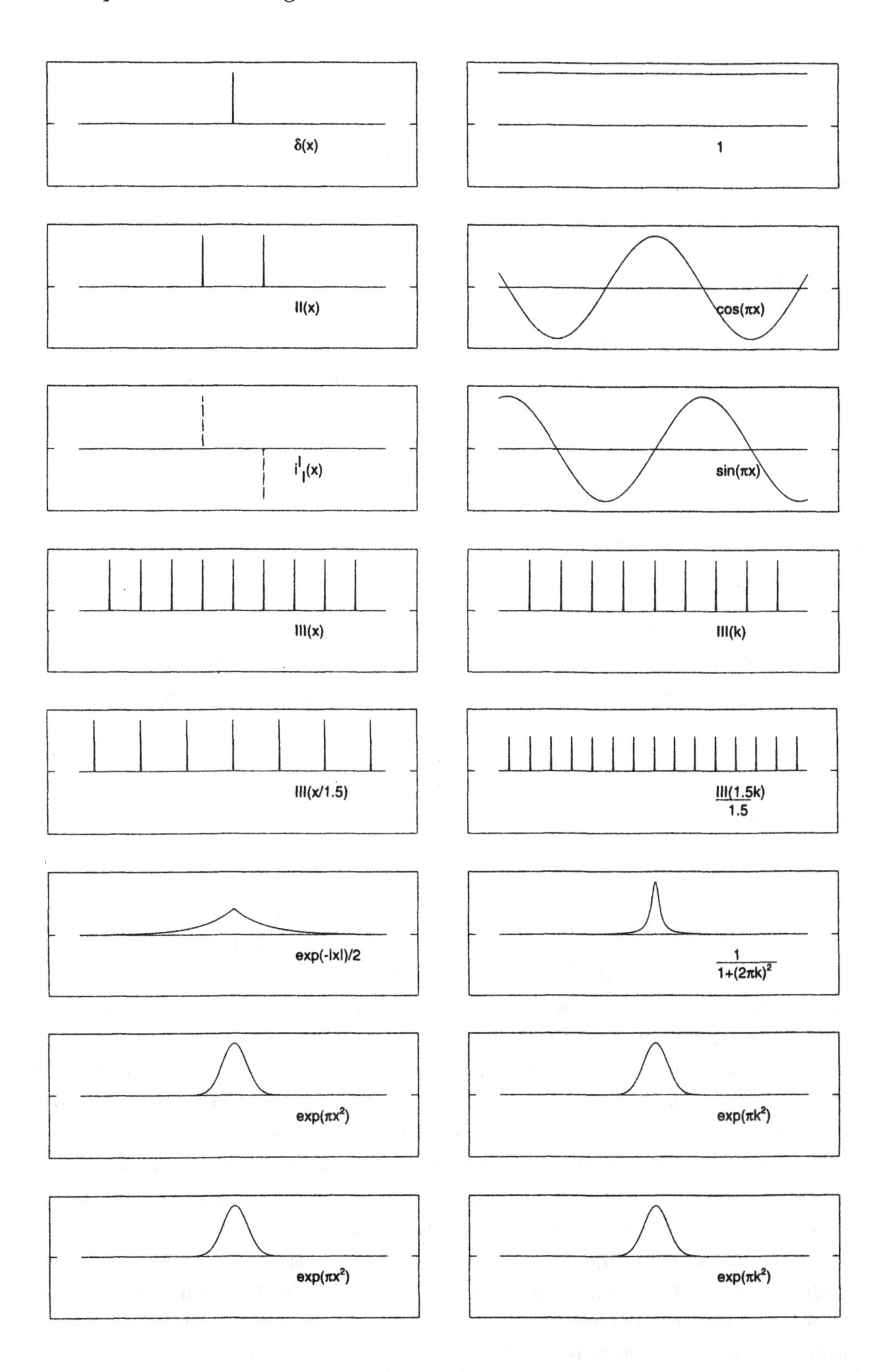

Figure 1: *Useful Fourier transform pairs (see Bracewell, 1965).*

1.1.2 Transforms of Useful Functions

At this point, it is useful to display a few commonly-encountered Fourier transform pairs. Those functions illustrated in fig. 1 will serve our needs for the rest of this discussion, but more may be found at the end of Bracewell's (1965) book. The one function illustrated here that may be unfamiliar is what Bracewell terms the "shah" function $III(x)$. It consists of an infinite string of equal delta functions with unit separation between them. This function is handy in many ways, not least being that it is its own Fourier transform. Notice that because of the scaling relation for Fourier transforms, spreading the delta functions farther apart in x causes the delta functions in k-space to grow closer together.

1.1.3 A Physical Example

To illustrate the use of some of these notions, let us consider a Fabry-Perot interferometer. The Fabry-Perot is doubly instructive in our context: it has been used to observe solar p-modes, and moreover it is a rough analogue for the cavity in which the p-modes propagate.

A Fabry-Perot interferometer consists of two plane mirrors lying parallel to one another, separated by a distance s. I will suppose that the mirrors reflect most of the light incident on them (independent of wavelength), and that the rest is transmitted. One way to understand how such an interferometer behaves is to imagine that a light source to the left of the interferometer emits a very short pulse of light in the form of a plane wave traveling toward the mirrors, with its propagation vector inclined at an angle θ to the normal to the mirrors. When this pulse hits the first mirror, some of the light is transmitted into the cavity between the mirrors. Part of this light is transmitted through the second mirror, and appears to an outside observer as an impulse at a time t_0. Of the light reflected from the second mirror, most is reflected again from the first mirror and arrives back at the second mirror after a time $\delta t = s/\cos(\theta)c$. Part of this delayed pulse is again transmitted by the second mirror, giving rise to a second output pulse at $t_0 + \delta t$. Subsequent reflections lead to a long string of pulses spaced δt apart in time, *i.e.*, a shah function in time. The frequency content of this output signal is given by the Fourier transform, namely a shah function in frequency, with peaks that are equally separated in frequency, with a spacing equal to $1/\delta t$ (see fig. 2a and 2b). These frequencies are the *resonant frequencies* of the interferometer; they evidently depend on the spacing s between the plates, the speed of wave propagation c, and on the propagation direction θ.

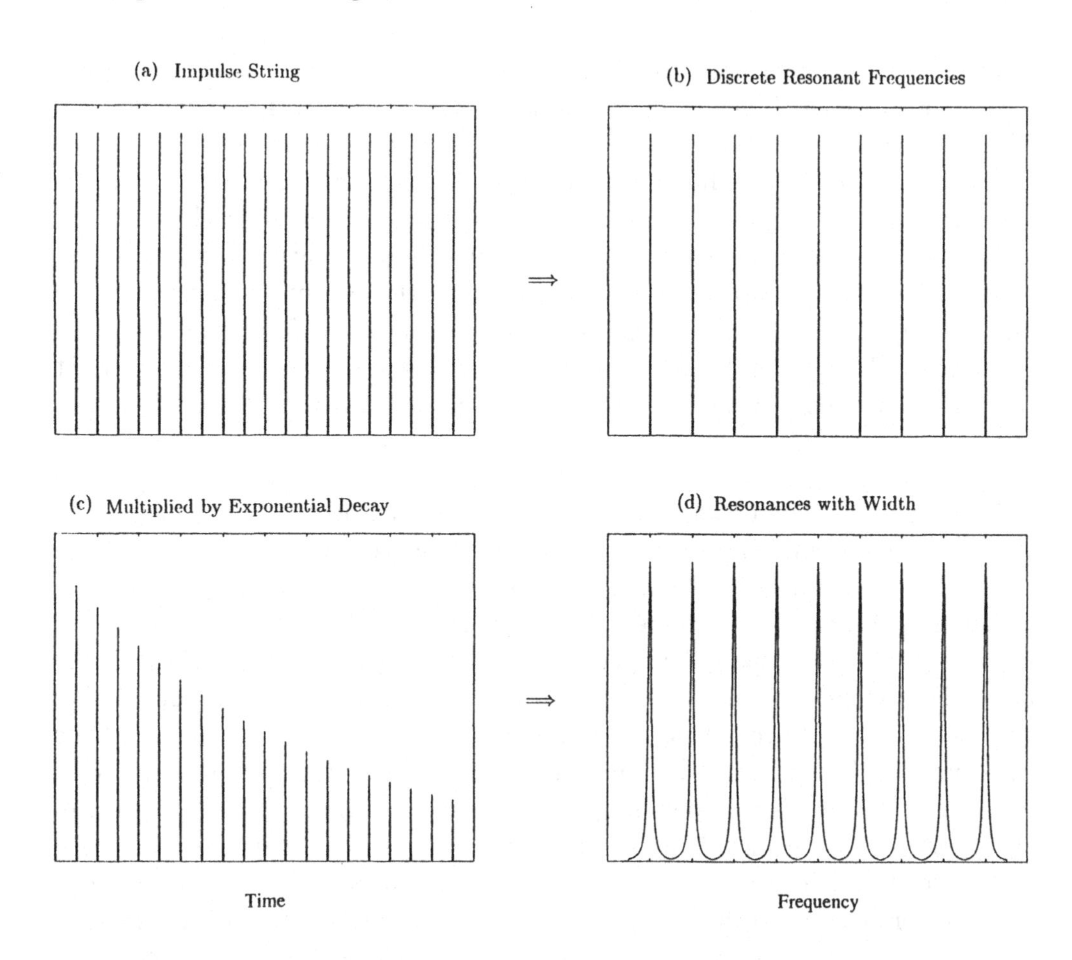

Figure 2: *(a) and (c): Time series of impulses emerging from a Fabry-Perot interferometer. (b) and (d): Corresponding Fourier transforms.*

An alternative way to analyze the Fabry-Perot interferometer is to insist that for resonant frequencies, an integral number of wavelengths of light must fit into one forward-and-back pass through the interferometer (with due allowance for phase shifts upon reflection, if any). This approach is equivalent to the travel-time argument just given; both are useful, with the clearest approach being dictated by the details of the situation. Here I emphasize the travel-time description mostly because it illustrates a general property of linear systems: The frequency response may be determined from the response of the system to a delta-function excitation.

Many physical systems (including the transmission of interferometers and the small-amplitude pulsations of stars) behave like this:

$$\text{Observed Quantity} = \Lambda(\text{Forcing Function}) , \tag{9}$$

where Λ is some linear operator that gives zero if the forcing function is always zero. (For instance, Λ may contain scaling factors, derivatives, or time delays.) Then by virtue of linearity and because of the orthogonality of the sines and cosines, the frequency components of any arbitrary forcing function may be considered one at a time; the effect of Λ is simply to multiply each frequency component by a (frequency dependent) complex factor. Since a δ function contains all frequency components in equal measure, the output resulting from a δ function input (the so-called *impulse response*) contains information about all frequencies. This is the "police brutality" view of systems analysis: if you want to find out all about a linear system, hit it once, and listen to what it has to say.

Now the Fabry-Perot analysis so far is unrealistic, because I have supposed that the string of pulses resulting from an impulse excitation will go on forever. In real life, the transmitted pulses bleed energy out of the cavity, so that the output impulse amplitude decreases exponentially with time (fig. 2c). What is the effect of this upon the spectrum? The new output signal is now the original shah function multiplied by an exponential envelope. By the convolution theorem, its transform is therefore the convolution of the transforms of the shah function (which is another shah function) and the transform of a decaying exponential (which is a Lorentzian). The effect is to change the resonances of the interferometer from infinitely sharp resonances to Lorentzian-shaped lines, whose widths are inversely proportional to the exponential decay time (fig. 2d). This sort of behavior turns out to be a pretty fair model of real interferometers, and is at least evocative of the processes that characterize the Sun's p-modes.

1.1.4 Stochastic Excitation

Suppose one has a system (like the Fabry-Perot) that is characterized by the spectrum of its impulse response, but one excites this resonator with a much more complicated driving function (a flock of impulses, say, with random amplitudes and times of occurrence). What is the resulting Fourier transform? Because of the system's linearity, the output signal (in the time domain) is the convolution of the system's impulse response with the driving function. The transform is then the product of the transforms of the impulse response and of the driving function. Many kinds of random excitation function have transforms whose expectation values are slowly-changing functions of frequency. However, a single realization of such a process usually looks very noisy (for the probability distributions related to such spectra, see the next section). In these cases, the emergent spectrum looks like noise that has been multiplied by the spectrum of the impulse response. In this sense, linear resonant systems act like filters, which (in the frequency domain) multiply but do not mix the frequency components of their driving function.

1.1.5 Probability Distributions for Power Spectra

For very many naturally-occurring time series, it turns out that different Fourier transform components behave as independent random variables, with the real and imaginary parts each being normally distributed with identical variance (Groth, 1975). The power is the sum of the squares of the real and imaginary parts of each frequency component, so its probability density is not Gaussian, but rather exponential (also called a χ^2 distribution with 2 degrees of freedom):

$$P(s) = e^{-s} . \tag{10}$$

An important property of this probability distribution is that the standard deviation associated with it is equal to its expectation value. Thus, the difference between the typical power at a given frequency and the value actually measured in a single realization is usually about as large as the typical power itself. Near resonances where the mean power is large, the measurement errors will be large. Far from resonances, where little power is expected, the distribution of measured power values is relatively narrow. Moreover, the high-power wing of the distribution in eq. (10) extends much further than does a Gaussian with the same variance. This means that one sees large-power events much more often than would naively be expected.

These peculiarities of the probability distribution for power spectra carry with them a severe danger of over-interpretation. It is all too easy to see a big feature in a power spectrum and think that it must be real, when in fact it is merely a (not very unlikely) statistical fluke. This sort of thing is illustrated in fig. 3, which shows a number of power spectra of a stochastically-excited damped oscillator. All of the spectra are for the same oscillator; all that is different is the random forcing. In all cases the limit spectrum (the spectrum that would be obtained by averaging many realizations) is shown as a dashed line, and the actual spectrum for each realization is solid. Fig. 3a shows the average of 60 spectra, which does indeed approach the limit spectrum. Among the individual realizations, almost any desired pathology may be found: narrow lines, wide lines, lines displaced either left or right from the average line center, as well as clear and symmetric doublets and triplets. Obviously great care must be taken when interpreting such spectra if one is to avoid erroneous conclusions.

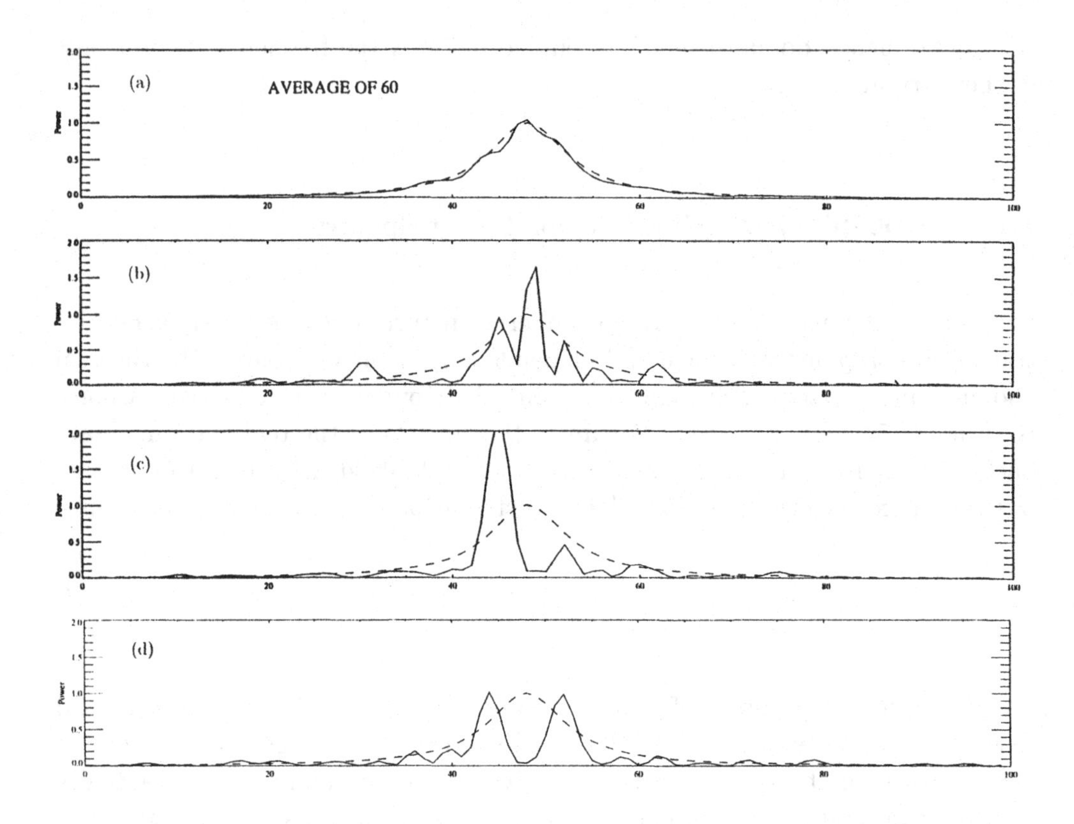

Figure 3: *Power spectra of a stochastically-driven damped oscillator. (a) shows the average of 60 realizations with different random driving. (b-d) show three single realizations, illustrating the wide range of behavior that characterizes such oscillators. In all cases, the limit spectrum is given by the dashed line.*

1.2 The Solar Noise Background

In addition to confusion resulting from random forcing, measurements of oscillating systems may be degraded by background noise that arises from other physical processes. A number of solar processes contribute to the noise polluting measurements of solar p-modes. The properties of these noise sources have implications for the design of instruments and of analysis methods.

Solar noise in p-mode measurements comes mostly from magnetic activity (at long timescales) and from convective motions (at shorter timescales). In the velocity signal, models of the noise roughly follow a power law, with the power spectral density P (in m^2/s^2 per Hz) estimated by Jiménez *et al.* (1988) given approximately by

$$P(\nu) = 10^7 \nu^{-1.6} . \tag{11}$$

(Note that a typical p-mode with amplitude 10 cm/s and linewidth 1 μHz has a power spectral density of about 10^4 m^2/s^2 per Hz. At p-mode frequencies, this is 2 or 3 orders of magnitude larger than the solar background.) The contrast between p-modes and the solar background is smaller in intensity than in velocity measurements, for two reasons. First, the amplitudes of the p-modes are relatively small in intensity because the radiative cooling time in the photosphere is short compared to the p-mode periods. Thus, though the p-modes are almost adiabatic when considered globally, in the solar atmosphere (where they are observed), they are almost isothermal, with only small associated intensity changes. Second, the solar background at p-mode frequencies comes mostly from thermal convection, the essential feature of which is temperature variation. The p-modes in irradiance data therefore appear superposed on a larger solar background than in velocity data, but they are nevertheless clearly visible.

1.3 Limits to Observable Accuracy

The noise background and the statistical nature of the p-mode driving processes set definite limits on one's ability to measure the properties of the p-modes. The things one would like to measure regarding an individual p-mode are its amplitude A, its frequency ν_0, and its frequency HWHM Γ. Assuming the mode spectrum to be represented by a Lorentzian line profile with additive noise, the expectation value of the observed power is

$$\langle P(\nu) \rangle = \epsilon + \frac{A}{[1 + (\nu - \nu_0)^2/\Gamma^2)]} . \tag{12}$$

The most interesting of the parameters is perhaps the line center frequency ν_0. For observing times T that are short compared with the mode decay time τ, the frequency precision that may be attained is roughly the reciprocal of the length of the observing interval:

$$(\delta\nu_0)_{rms} = \frac{1.}{T} . \tag{13}$$

This just reflects the scaling property of Fourier transforms (eq. 5), a property that is sometimes dignified with names like "the uncertainty principle." In this case, and as long as the power in the line is large enough so that the resonance is clearly visible above the background noise, neither the noise level nor the statistical variations in power due to stochastic excitation make much difference to the expected precision.

A more common case when analyzing solar p-modes is that in which $T \geq \tau$. In this case the resonance line is fully resolved, and the stochastic nature of the excitation does play a role. A good approximation to the rms uncertainty in $(\delta\nu_0)_{rms}$ is (Libbrecht 1992):

$$(\delta\nu_0)_{rms} \cong \left[f(\beta)\frac{\Gamma}{2\pi T}\right]^{1/2}, \tag{14}$$

where $\beta = \epsilon/A$ is the inverse of the signal-to-noise ratio, and

$$f = (1+\beta)^{1/2}[(1+\beta)^{1/2} + \beta^{1/2}]^3 .$$

As a typical numerical example, we might consider a solar mode with frequency near 3000 μHz, where the line HWHM is typically 0.5 μHz. At this frequency, the background noise can usually be ignored. For an observing run lasting 100 days (8.6×10^6 s), the frequency precision with which a single line can be measured is about 0.1 μHz. This frequency precision sounds pretty good (and it is), but it is not outstanding when compared to some of the frequency differences one wishes to measure. For instance, the activity-related frequency shifts, the frequency changes associated with time-dependent asphericity (these two are, in a sense, the same thing), and the departure from a linear m-dependence for rotational frequency splittings (which tell us about variations in rotation with latitude) at $l = 10$ are all on the order of a few tenths of a μHz. This merely emphasizes the importance of measuring many modes at once, so that usably precise average values may be obtained.

2 INSTRUMENTATION FOR MEASURING SOLAR p-MODES

The p-modes are acoustic waves; their passage both displaces and compresses the gas in the solar atmosphere. The displacements are observable principally because of the corresponding velocity signal, which may be detected by measuring the Doppler shift of absorption lines in the solar spectrum. The compressions may be detected because the gas tries to heat adiabatically as it is compressed ("Tries to," because thermal radiation tends to return the temperature to its unperturbed value. Any gas that can see the solar surface suffers to some extent from this radiative damping process.), and hence radiates more brightly than it would otherwise do.

For any individual solar p-mode, both the velocity and the brightness signals are quite small, by usual astronomical standards. Velocity amplitudes are usually no more than 15 cm/s, even at the peak of the p-mode band, near 3 mHz. The continuum brightness fluctuations for the same modes are roughly $\delta I/I = 3 \times 10^{-6}$, corresponding to temperature fluctuations of perhaps 0.005 K. That such small signals can be detected at all is largely attributable to the Sun's

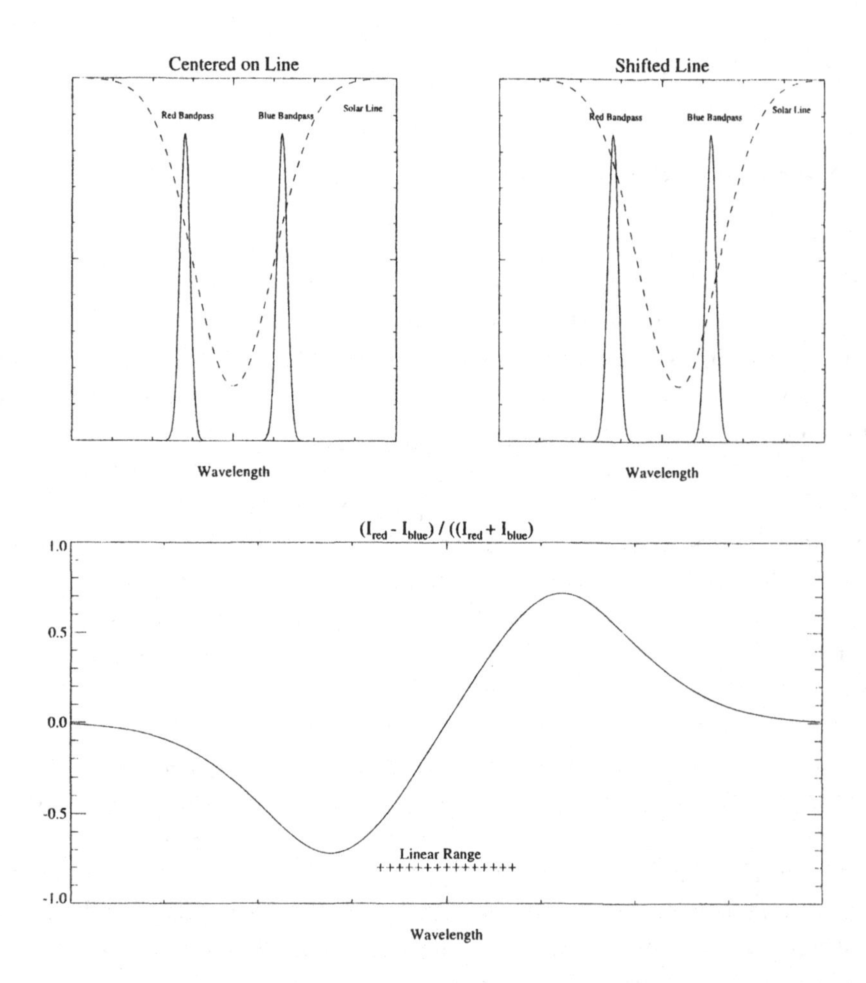

Figure 4: *The 2-point Doppler measurement technique. The top two panels show a solar absorption line (dashed) and two narrow-band filter transmission profiles, one in the red and one in the blue line wings. The bottom panel illustrates the nonlinearity associated with the normalized difference* $(I_{red} - I_{blue})/(I_{red} + I_{blue})$.

great brightness, which allows us to make very low-noise measurements. Since modern detector technology is equal to the challenge of detecting the p-modes, the real question when choosing an instrument becomes one of optimization for a particular application. Different approaches to the detection problem work best in different frequency and wavenumber ranges, offer different levels of convenience to the observer, and suffer from different kinds and amounts of solar background noise. The choice of instrumental strategies thus depends in some detail on the circumstances of the desired observation.

2.1 Velocity Measurements

2.1.1 Narrow-Band Filters

The most straightforward means of measuring a Doppler shift is to use a narrow-band filter (fig. 4). One measures the intensity in two narrow frequency bands, one in the red wing of the line and one in the blue. A shift of the line to the red (say) reduces the intensity in the red passband while increasing it in the blue passband. The ratio $(I_{red} - I_{blue})/(I_{red} + I_{blue})$ is then a measure of the displacement of the line from its nominal position. The reason to form the ratio is to reduce sensitivity to changing transmission of the atmosphere or instrument. Two-point Doppler methods always suffer from some degree of nonlinearity, because the wings of real line profiles do not go on forever with constant slope $dI/d\lambda$. For line shifts that are comparable to the line half-width (a few km/s, for most solar lines), significant corrections are therefore necessary if accurate velocities are to be obtained.

Several techniques are available for obtaining the necessary narrow and stable bandpasses for the two-point velocity measurement. The most time-honored is to use a slit spectrograph with appropriate spectral resolution. This approach has the advantage that (with proper setup) the intensity can be measured in both wings simultaneously (reducing seeing problems). The principal disadvantage is that observing a 2-dimensional area on the Sun requires scanning the slit, which is time-consuming and wasteful of light.

A technique that exploits the natural stability of resonating atoms but that sacrifices spatial resolution is the atomic resonance cell. In this approach (see fig. 5), incoming sunlight passes through a switchable circular polarizer and a cell containing the vapor of an alkali metal (sodium or potassium), which is permeated by a magnetic field. The Zeeman splitting of the atomic energy levels causes the atoms to scatter radiation from opposite sides of the unperturbed line profile, depending on the direction of circular polarization of the incoming light. The amount of light scattered from the beam for each polarization is a measure of the light intensity in the corresponding red and blue wings of the line profile. Instruments based on this principle can be extremely sensitive and stable, since their bandpass frequencies are determined mostly by the properties of the atoms in the vapor cell. The bandpass shapes and center frequencies are somewhat sensitive to the cell temperature and pressure and to the strength of the applied magnetic field, but as these changes are almost symmetric about the line center wavelength, they have at most a small effect on the estimated Doppler shifts. An obvious limitation of the technique is that, because it depends

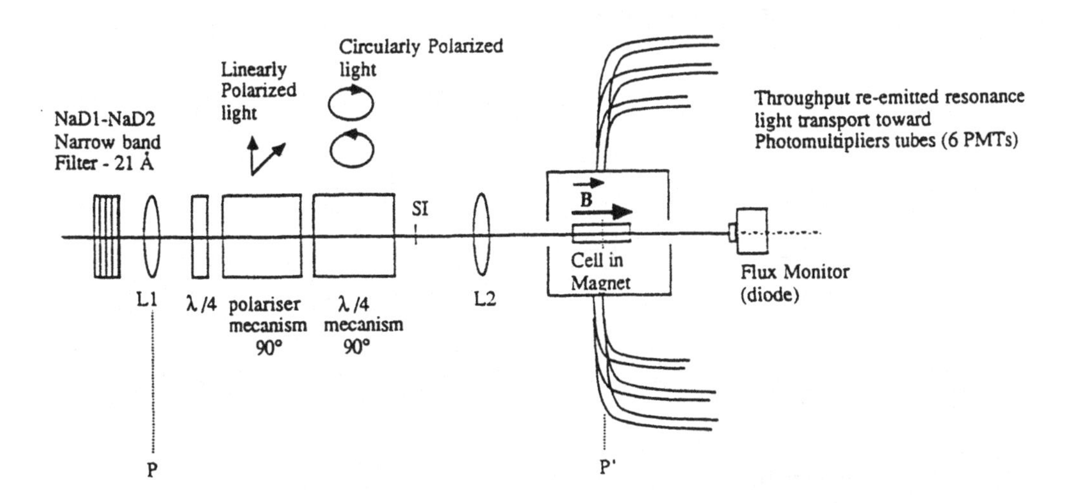

Figure 5: *A schematic illustration of the atomic scattering instrument (Global Oscillations at Low Frequencies, or GOLF) instrument, which is to fly on the SOHO spacecraft (figure from Damé 1988).*

on detecting scattered light, spatial resolution is possible only by scanning an aperture over the solar image before light enters the vapor cell. Also, the only solar absorption lines that one may observe are the resonance lines of alkali atoms – in practice, the Na D lines and the potassium line at 769.9 nm. This may sometimes restrict one to observing higher in the solar atmosphere than desired. The simplicity and stability of the atomic resonance technique are often sufficient to outweigh these limitations, however. As a result, this technique is used in a great number of contemporary instruments, including the IRIS (Fossat 1988) and BISON (Aindow *et al.* 1988) networks, and (with some modifications) in the GOLF (Damé 1988) instrument on the SOHO spacecraft.

The Magneto-Optical Filter (MOF; Rhodes *et al.* 1986, Tomczyk *et al.* 1994) is an atomic resonance two-point Doppler system that (at the cost of some additional complication) achieves spatial resolution as well as excellent stability. An important advantage of filter instruments such as the MOF is that large areas of the Sun (up to and including the whole solar disk) may be passed through the filter simultaneously, avoiding the need for spatial scanning. One realization of the MOF is illustrated in fig. 6. Unpolarized light from the Sun passes through a linear polarizer and then through an atomic vapor cell, which is placed in a magnetic field. The effect of the cell is to leave the polarization state of continuum light unchanged, but to convert light in the red and blue wings of the line to oppositely-directed circular polarizations. By passing the light through a polarizer that is crossed with the first one, the continuum light is rejected, and the

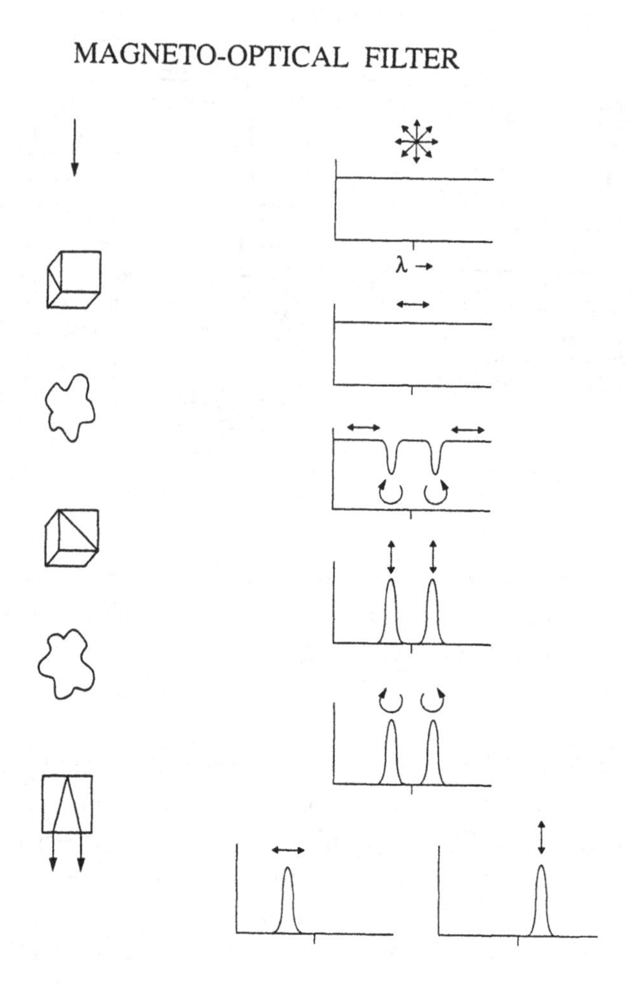

Figure 6: *Schematic showing the operating principle of the Magneto-Optical Filter (MOF), from Tomczyk* et al. *1994. See the text for a description.*

light in the transmitted bandpasses is converted to linear polarization. Passage through a second vapor cell returns the transmitted bandpasses to their state of opposite circular polarization, and a subsequent quarter-wave plate (not shown) and linear polarizer separates the beams from the red and blue bandpasses for separate detection. This implementation of the filter is attractive because it involves no moving parts, although it does need separate detectors for the red and blue images. MOF instruments suffer from the same line-choice restrictions as do the atomic scattering instruments, but they are capable of high spatial resolution. Various MOF instruments have been operated on Mt. Wilson for many years (Rhodes *et al.* 1986, Korzennik 1990), and an MOF is employed in the new LOWL instrument deployed by HAO on Mauna Loa in Hawaii (Tomczyk *et al.* 1994.)

The Fabry-Perot interferometer was discussed in Section 1, in another context. Such interferometers can be used as filters producing extremely narrow transmission bandpasses, and suitable designs can maintain the separation between the interferometer mirrors (and hence the wavelength of the bandpasses) within close tolerances. Generating 2 (or more) bandpasses that are closely spaced within a solar line profile is achieved by arranging for minute changes in the optical path length between the mirrors. Such filters are more flexible than those based on atomic resonance, because they can be tuned to use any of a wide selection of absorption lines. On the other hand, considerable complication is usually required to control the plate spacing with the desired precision, and to exclude other interferometer orders than the one at the desired wavelength. Also, unlike the MOF, the wavelength transmitted by a Fabry-Perot is a fairly sensitive function of the angle at which light passes through it. This means that it must be used in a nearly parallel beam of light, and to get adequate transmission, comparatively large interferometer elements must be employed. These practical considerations have militated against the use of Fabry-Perots in solar oscillations instruments, although useful observations have been obtained with the ingenious lithium niobate interferometers constructed by CSIRO and described by Rust *et al.* (1986).

Compared to a Fabry-Perot, a much greater acceptance angle and substantially decreased temperature sensitivity may be obtained by using Michelson interferometers as wavelength-selecting elements. The essential idea is to replace one or two of the largest path difference elements in a Lyot birefringent filter with equivalent elements involving polarizing Michelson interferometers. This approach is used in the Michelson Doppler Imager (MDI) instrument (Scherrer *et al.* 1989), which is to fly on the SOHO spacecraft. The penalty that one pays for the Michelson's advantages is additional complication, since more than one Michelson filter element is required to attain the desired spectral resolution, and since the construction of Michelson elements of the necessary quality is a demanding task.

The MDI uses a telescope aperture of about 7.5 cm to image the entire Sun onto a 1K x 1K CCD detector, giving a projected pixel size of about 2 arcsec, and allowing observation of p-modes with $l \leq 750$, approximately. It also allows a high-resolution mode of operation, in which a part of the solar disk may be imaged at a scale of about 0.7 arcsec per pixel. A further refinement is the use of a Doppler measurement scheme that uses 4 measurements across the line profile (rather than 2); this approach nearly eliminates the nonlinear Doppler response associated with 2-point techniques, and also allows estimates of quantities such as continuum intensity and line center depth.

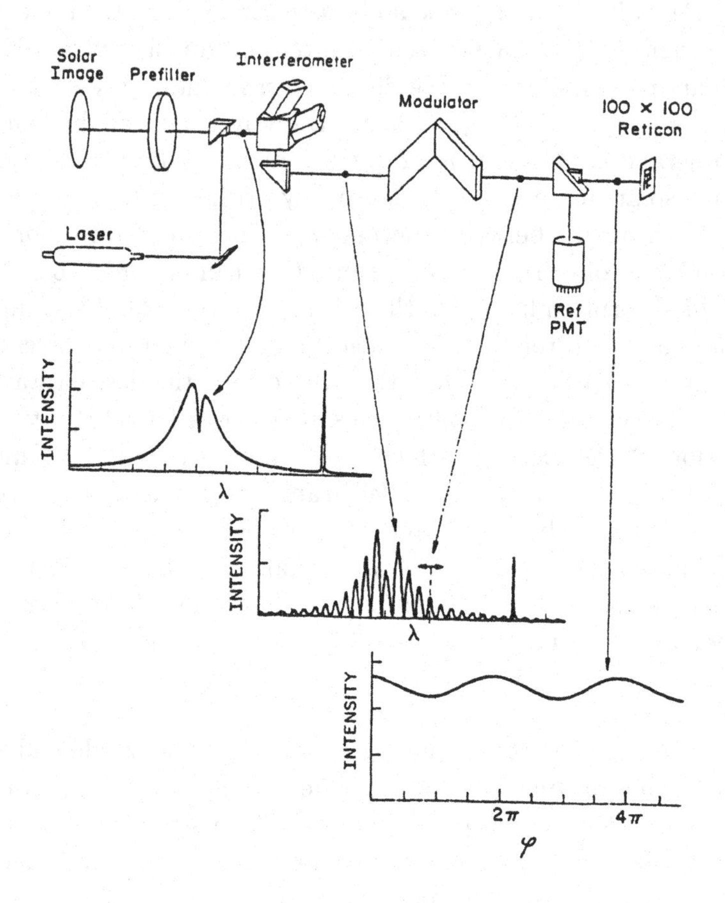

Figure 7: *Schematic of the optical operation of a Fourier tachometer, illustrating the dependence of intensity upon wavelength and modulation phase at several points along the optical path.*

2.1.2 Fourier Tachometers

Fourier tachometers are special-purpose versions of Fourier transform spectrographs. From a performance point of view, their most noteworthy characteristic is that they measure the line wavelength in a way that does not have the linearity problems of 2-point Doppler measurements, and that is largely unaffected by the line shape variations seen on the Sun. Practically speaking, the most important thing about Fourier tachometers is that they are the Doppler analyzers for the GONG network.

The basics of Fourier tachometer operation are illustrated in fig. 7, and explained in some detail in Brown (1980). Incoming sunlight passes through a prefilter with transmission function $T_P(\lambda)$, which isolates the desired solar absorption line. The light then passes through a single tunable Michelson interferometer, whose transmission function $T_M(\lambda)$ in the neighborhood of the prefilter bandpass is adequately approximated by

$$T_M(\lambda) = \frac{1}{2}[1 + cos(R\lambda + \phi)] , \tag{15}$$

where R is a constant proportional to the interferometer's mean optical path difference, and ϕ is a phase determined by the tuning of the interferometer. Subsequent optics image the Sun onto a detector, where each pixel sees an intensity given by

$$I(\phi) = \int I_S T_P[1 + \cos(R\lambda + \phi)]d\lambda . \tag{16}$$

By comparison with eq. (1), and recalling that $\cos(A+\phi) = \cos(A)\cos(phi) - \sin(A)\sin(\phi)$, this integral can be seen to be a sum of 3 terms: a constant related to the mean intensity, and one component each (corresponding to the wavenumber R) of the sine and cosine transforms of the filtered solar intensity $I_S T_P$. By measuring $I(\phi)$ at 3 different values of ϕ, one can obtain estimates for all 3 terms. The phase of the sinusoidal intensity variation $I(\phi)$ may be defined as $\psi = \text{atan}(S/C)$, where S and C are the amplitudes of the sine and cosine terms. Now if T_P is chosen to be wide enough that it contains no significant power at wavenumber R, the only contribution to the sine and cosine transforms must come from the solar line. (In practice, one chooses R so that the width of the line is about half of one cycle of the cosine function in eq. (15); this way the signal from the solar line is large.) Then the phase ψ is (by virtue of the shift theorem) a measure of the central wavelength of the solar absorption line, while the amplitude $(S^2 + C^2)^{1/2}$ is a measure of its strength. The third measurable term, the mean intensity, is mostly a measure of the brightness of the continuum near the chosen line. These three combinations of the observed intensities, termed Velocity, Modulation, and Intensity (V,M,I), are the three outputs of a Fourier tachometer.

GONG is a program to place six identical Fourier tachometers at sites spread about the globe, so that the sun may be observed nearly continuously (fig. 8). The GONG instruments will image the Sun at a resolution of roughly 8 arcsec per pixel onto CCD detectors that are currently sized at 256 x 243 pixels. Images in V, M, and I will be obtained every 60s by all stations, whenever they can see the Sun, throughout the life of the project (which is expected to be at least 3 years). Substantial preliminary data processing (along the lines described in the next section) will be undertaken by the project, so that optimally useful information may be provided to participating scientists.

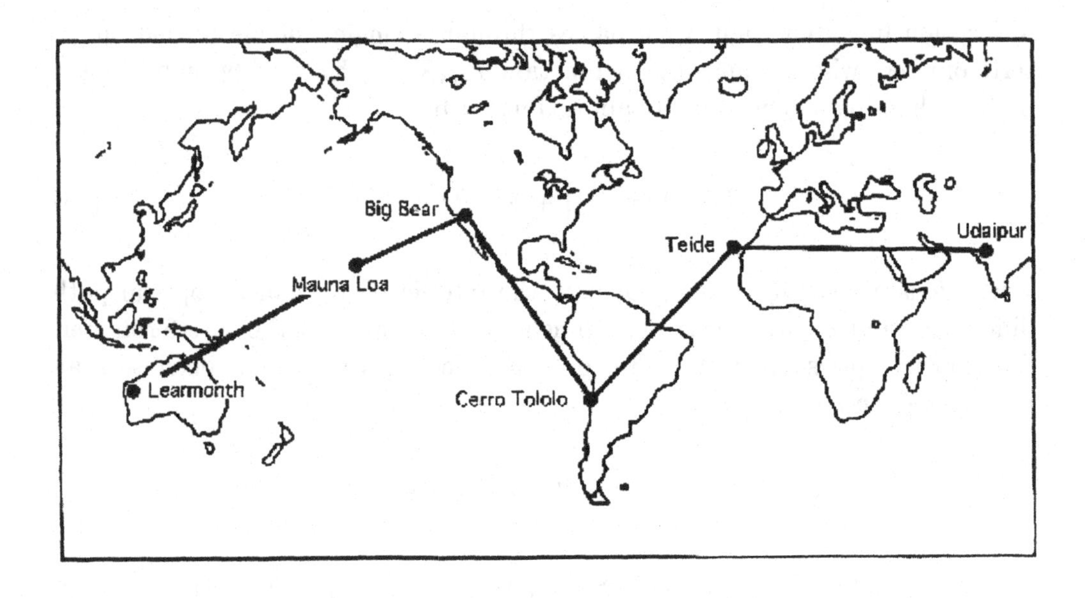

Figure 8: *Locations of the six GONG network sites.*

2.2 Intensity Measurements

Intensity measurements are the simplest kind to make, since the only equipment required is a suitable photodetector and (usually) a filter to isolate the desired spectral bandpass. They do, however, require precise detection techniques, because the mode amplitudes are close to the intrinsic noise limits of many detector arrangements; this is particularly true for continuum intensities, where the mode amplitudes are very small because of the short radiative equilibration time in the upper photosphere. Moreover, the fluctuating p-mode intensities are seen against a background of fluctuating convection intensities – the solar granulation. The granulation is a thermally driven convective process, so the fluid perturbations have large $\delta T/\delta v$ ratios, compared to the acoustic modes. This means that the solar background noise is more than an order of magnitude higher for continuum intensity measurements than it is for velocity. Both the small mode amplitude and the large granulation background can be mitigated somewhat by observing radiation from high up in the solar atmosphere – in the core of the Ca II K line, for instance. Mode velocity amplitudes increase with increasing height (even for evanescent waves), as does the thermal relaxation time. The granular temperature contrast, on the other hand, decreases above the photosphere, so the S/N ratio is improved by a large factor in going from the continuum to the K line. Most current intensity instruments (especially those with spatial resolution) now employ Ca II filters, to exploit this advantage.

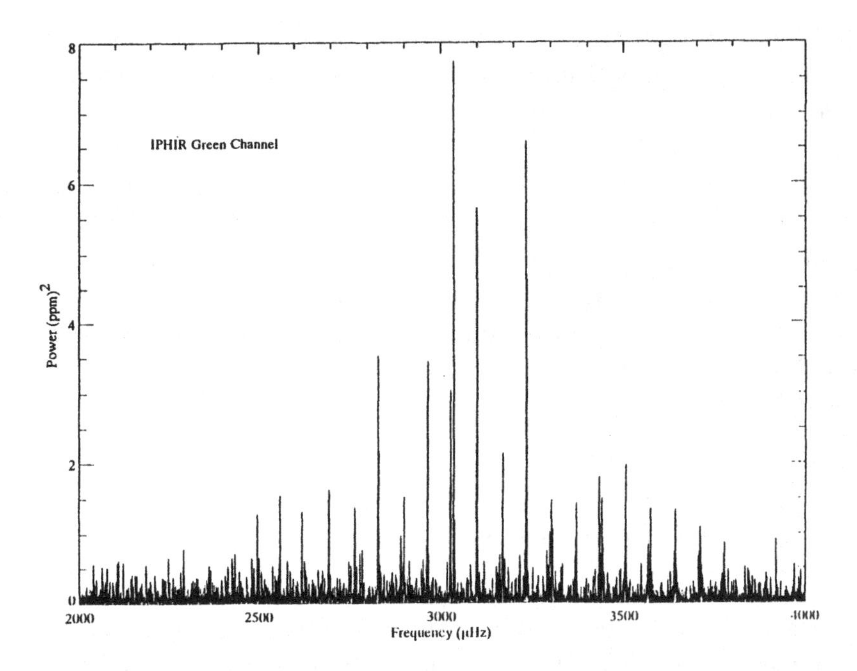

Figure 9: *A power spectrum of low-degree solar oscillations obtained with the green channel of the IPHIR instrument, which measured the intensity of broad-band sunlight (see Toutain & Fröhlich 1992). From data kindly provided by C. Fröhlich and T. Toutain.*

2.2.1 Broad-Band Intensity

Most measurements of the continuum intensity treat the Sun as an unresolved (or barely resolved) star and are conducted outside the Earth's atmosphere, to eliminate noise from fluctuating atmospheric transmission. The most famous of such instruments is surely the Active Cavity Radiometer (ACRIM, Willson et al., 1984), versions of which have flown on several spacecraft of the last decade or so, and continue to be placed on current missions.

Although the ACRIM instrument was capable of observing solar brightness oscillations, it was not intended to do so, and was far from optimally designed for this purpose. The first spacecraft photometer designed to observe solar oscillations was the IPHIR instrument, which operated during the interplanetary transit phase of the two Soviet Phobos Mars missions (Toutain & Fröhlich 1992). IPHIR used silicon detectors and filters to measure the intensity of disk-integrated sunlight in several color bands. Because it operated from a spacecraft on an interplanetary trajectory, IPHIR was able to obtain long uninterrupted time series of solar data. The power spectrum of one month of data from the green channel is shown in fig. 9. It is evident that, concerns about higher solar background noise notwithstanding, high-quality data may be obtained from measurements of the continuum intensity.

Perhaps more important than providing a different way of monitoring p-modes, continuum intensity measurements are interesting because they may be superior to velocity measurements for detecting g-modes. Based partly on this rationale, the VIRGO package on SOHO (Fröhlich *et al.* 1989) consists of a set of radiometers and broad-band photometers, including one photometer with a specially-fabricated detector to allow identification of oscillation modes with l up to about 7.

2.2.2 Ca II Line-Core Intensity

An advantage of spatially-resolved intensity measurements is that, unlike velocity measurements, the apparent amplitude of p-modes seen in intensity remains large even at the solar limb. The reason for this is that the velocities associated with p-modes are almost vertical, and hence are perpendicular to our line of sight (and unobservable) at the limb. Intensities, on the other hand, depend upon temperature fluctuations, which may be viewed equally easily from any angle. An implication of this difference is that a larger fraction of the Sun's surface area may be observed with intensity than with velocity, making modes with similar l and m values easier to distinguish. This point will be discussed at more length in Section 3.

To gain a more easily measured signal than is available in the continuum intensity, many observers have resorted to observations taken in the core of the Ca II K line. Instruments designed for such observations are simple and robust, consisting of little more than a telescope, a suitably controlled filter (typically with a bandpass width between .1 and 1 nm), a CCD detector, and a data acquisition system. One of the first, longest-running, and most successful observation programs using this approach is that pursued at the South Pole by workers from Bartol and from the National Solar Observatory (Duvall *et al.* 1986, Harvey *et al.* 1982). Observing from this location during the austral summer allows occasional almost-uninterrupted data runs of many days' duration. This relative absence of gaps in the time coverage (especially ones that recur at regular intervals) makes the South Pole observations particularly valuable, and has contributed greatly to their utility and importance.

A number of other instruments working on the same principle are now in operation. Among them are the High Resolution Helioseismometer (HRH) operated at Kitt Peak by the National Solar Observatory, and the p-mode Oscillations Instrument (POI) operated by the University of Hawaii and sited at Haleakala on the island of Maui. The Taiwan Oscillation Network (TON) is a recently-conceived network of Ca II K-line instruments that combine simple design with high spatial resolution (Chou *et al.* 1995). The first sites of this network have already been installed (including one on Tenerife), and observations are being obtained.

2.3 Summary

The next few years will be exciting ones for helioseismology, since data of unprecedented quality and quantity will soon be appearing from several sources. In upcoming years, we can expect nearly-continuous Doppler observations of the unresolved Sun from the atomic scattering networks IRIS and BISON, and (once SOHO is launched) from the GOLF experiment. Similar measurements of broad-band intensity will also accompany the SOHO launch, provided by the VIRGO package. There will soon be nearly-continuous Doppler observations at moderate spatial resolution provided by the GONG network, as well as (again associated with SOHO) MDI data at fairly high spatial resolution, covering intervals ranging from about 8 hours up to several months. The TON should also yield a near-continuous record of Ca II images with resolution comparable to that from MDI. In addition to all of these resources that are explicitly directed at p-modes, there are a number of general-purpose facilities around the world that can offer extremely high spatial resolution (adequate to resolve some of the structure in granulation, for instance), or the ability to measure vector magnetic fields, or can perform other unique campaign-style observations. Combining the information from all of these sources will prove to be a challenge both to observers and to theorists, but the combined data sets should make it possible to attack the problems of solar structure, cycle, and dynamics in ways that have never before been possible.

3 ANALYSIS TOOLS FOR HELIOSEISMOLOGY

Instruments and observations are only half of the story where p-modes are concerned. Unlike many astronomical situations, a single good image tells one nothing, and there are no gains to be had by selecting particularly interesting events. Instead, the raw observations must be Fourier transformed at least once to obtain any parameter that can be compared with theory. Thus, one simply has to reduce all of the available data, and then see what emerges. The ultimate objective is to process the raw data so as to isolate the information about individual modes, allowing more or less straightforward estimation of the modal properties.

Solar p-modes are characterized by the 3-dimensional shapes of their eigenfunctions (described by the radial order n, the angular degree l, and the azimuthal order m); each mode also has a characteristic frequency $\nu(n, l, m)$. The radial structure described by n is inaccessible to observation, so the characteristics that are used to isolate modes from one another are ordinarily their angular structure (described by l and m), and their frequency ν. In practice, this is done by first isolating that part of the observed surface variation resulting from modes with similar angular structure, and then Fourier transforming in time, so that modes with different n values may be distinguished by their different ν values. From the resulting transforms or power spectra, one estimates the mode parameters that are of physical interest – the mode amplitudes, frequencies, and lifetimes. As will become clear, all of these reduction processes are hindered by missing pieces of the raw data.

3.1 Spatial Analysis

The object of any spatial analysis in helioseismology is to separate the observable surface effects (velocity, intensity, or whatever) into components that have similar spatial characteristics. Exactly what one means by "spatial characteristics" depends somewhat on the circumstances, but most often these characteristics are related to the horizontal wavelength of the acoustic waves, and to their direction of travel. Since the waves are almost periodic in the horizontal spatial dimensions, the natural tools to accomplish the necessary separation are Fourier and related transforms. Complications arise because the surface of the Sun is a sphere, and because we cannot see all of its surface, and because many of the waves are indeed only *almost* periodic.

The idea underlying most spatial analysis schemes is the same as that underlying the Fourier transforms itself: one chooses a set of basis functions $u_i(\theta, \phi)$ that represent the acoustic waves in whatever geometry is appropriate, and that are at least roughly orthogonal on the observed domain, *i.e.*, $\int u_i u_j dA \cong \delta_{ij}$, where the integration is over the observed area of the Sun. The observed quantity $V(\theta, \phi)$ is expressed as a linear combination of different u_i:

$$V(\theta, \phi) = \sum A_i u_i(\theta, \phi) . \tag{17}$$

Then the amplitudes A_i in the expansion may be estimated by simple projection of the observed V onto the basis functions:

$$A_i \cong \int V(\theta,\phi)u_i(\theta,\phi)dA \,, \tag{18}$$

where the "$\cong$" is necessary because the u_i are only nearly orthogonal on the observed domain. (One may, and sometimes does, choose functions that are honestly orthogonal on the observed domain, but since the actual modes know about the whole Sun, the approximation in eq. (18) does not really go away.)

As a practical matter, the integral in eq. (18) can be very expensive to compute. If one has N^2 pixels in an image (where N is the edge length of one's detector), then N^2 coefficients will be needed to represent the image, and the total operations count to do the necessary sums is of order N^4. A lot of effort has therefore gone into finding efficient ways to approximate the desired integrals. The most typical approach is to use interpolation to remap $V(\theta,\phi)$ onto some coordinate system in which the integration over at least one dimension may be represented as a Fourier transform. In this case FFT methods may be used, and the operations count scales as $N^3 \ln N$. In the following subsections, I shall give a brief account of some of the ways in which this kind of program has been carried out.

3.1.1 2-D Fourier Transform

In cases where only a very small area of the Sun is being observed, the surface may reasonably be taken as flat, and the solution of the acoustic wave equations may be represented in terms of plane waves. This fact suggests the simplest (also the most efficient) form of spatial analysis, which is simply to Fourier transform the image in 2 dimensions. In this case the spatial frequencies corresponding to the individual plane waves are genuine wavenumbers k_x, k_y, with dimensions of an inverse length. A rough connection may be made with the spherical harmonic l and m values that I have used up to this point:

$$|k| \cong (l+\frac{1}{2})/R\,; \qquad m/l \cong k_y/|k|\,, \tag{19}$$

where R is the solar radius and I have assumed that the observations are taken near the solar equator and that the x direction is parallel to the equator. Recall, however, that this kind of analysis only makes sense for regions whose diameter D is significantly smaller than R. The resolution in wavenumber is only $2\pi/D$; since $D \leq R$, each resolution element in k corresponds to a range of l values – anything from half a dozen upward, depending upon D.

As an aside, the early history of observational helioseismology was dominated by an even simpler approach, in which all modes with nonzero k_y were filtered

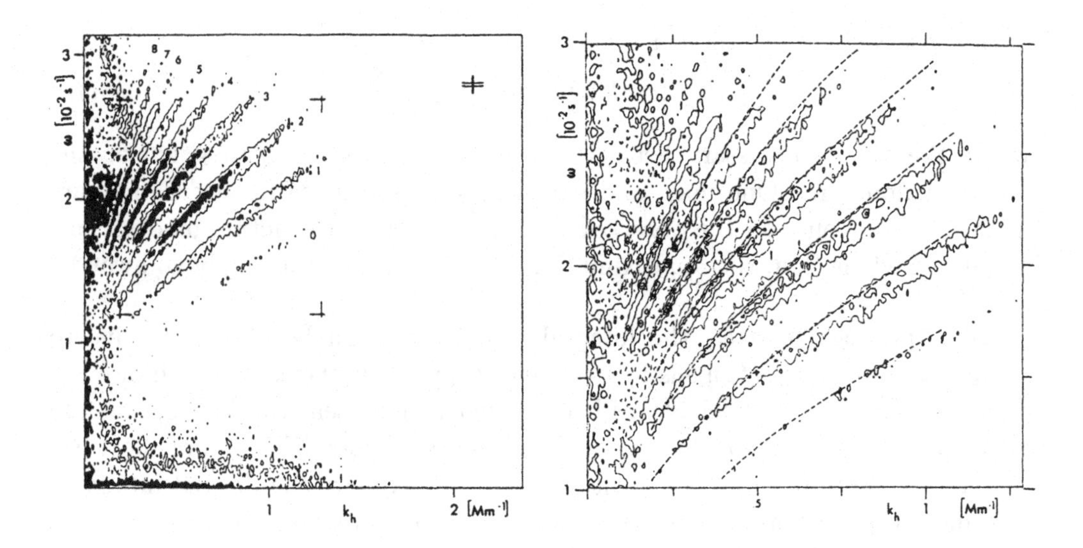

Figure 10: *A $k-\omega$ diagram formed by 2-D Fourier transformation of a relatively early data set due to F. Deubner (figure from Deubner* et al. *1979). The abscissa is k in* Mm^{-1}*; the ordinate is $\omega = 2\pi\nu$. The right panel shows an expanded view of a portion of the left panel. Each ridge f power contours corresponds to a set of p-modes with the same value of the radial order n but with different angular degree l.*

out by integration over the y direction (*i.e.*, over latitude). The resulting data array contained information only about the behavior of the velocity field in x and in time, leading to the early $k-\omega$ diagrams published by Deubner (1975) and others (fig. 10). Both the 2-dimensional and 1-dimensional versions of the Fourier transform analysis are notable in that they involve no interpolation of the data before the transforms are computed.

3.1.2 Interpolation

If one wishes to deal with global p-mode data, the functions u_i are most naturally chosen to be the spherical harmonics $Y_l^m(\theta, \phi) = P_l^m(\theta)\exp(im\phi)$, where θ is the colatitude, ϕ is the longitude, and P_l^m is an associated Legendre function. Since the ϕ part of these functions consists simply of sines and cosines, the most efficient procedure is to interpolate the observations from their native plane-of-the-sky coordinates onto a grid that is equally spaced in solar longitude, so that FFT methods may be used in the ϕ direction. It is essential that this interpolation be accurate (in the sense that the geometry of the Sun is correctly represented, so that image points are mapped into the correct longitudes and latitudes). If this is not done, systematic errors occur in the spatial analysis, complicating and possibly corrupting later results. Moreover, all interpolation processes introduce noise and add smoothing to the data; these effects tend to vitiate the instrument's hard-won capacity for producing pure data with high resolution, so they should be minimized. Several real-world details make it difficult to achieve either of these goals as well as one would like.

First, the solar image actually recorded by an instrument is not even completely circular. Differential refraction in the Earth's atmosphere as well as distortion and other flaws in the instrument lead to images that may be oblate or worse. One essential function of the interpolation procedure is to measure the size and nature of the image distortion, so that the mapping between image and heliographic coordinates can be accurate. One must also account for the orientation of the Sun's image; the orientation of the rotation axis in the plane of the image must be accurately known, and the effect of the B angle (the latitude of the sub-Earth point) must be included in the mapping. Finally, the image radius must be accurately measured, since an error in the assumed radius leads directly to errors in l (or wavenumber). It is conceptually simplest to correct for each of these effects with a separate interpolation process. This is unwise, however, since the noise and smoothing added by interpolation accumulates as successive interpolations are done. The best procedure is therefore to estimate all of the parameters that come into the geometric transformation between the image and heliographic coordinates, and then do one interpolation that incorporates them all.

The choice of a particular interpolation formula is influenced (but not driven) by a trade-off between accuracy and computation time. Solar p-mode data ordinarily contain a lot of power at scales that are similar to the pixel spacing. All interpolation methods work badly on such data; the question is really how much one is willing to pay to get more-or-less equally unsatisfactory results. This being the case, the usual approach is to rely on fairly simple interpolation schemes.

3.1.3 Spherical Harmonic Transform (Global)

When the aim is to decompose $V(\theta, \phi)$ in terms of spherical harmonics, the interpolated data are first Fourier transformed in the ϕ dimension (see fig. 11). The result is a set of complex amplitudes $W_m(\theta)$ describing V in terms of colatitude and azimuthal wavenumber. In existing routines, ϕ is usually defined so that $\phi = 0$ corresponds to the Sun's central meridian as seen from Earth. With this definition, the real components of W_m correspond to that portion of V that is

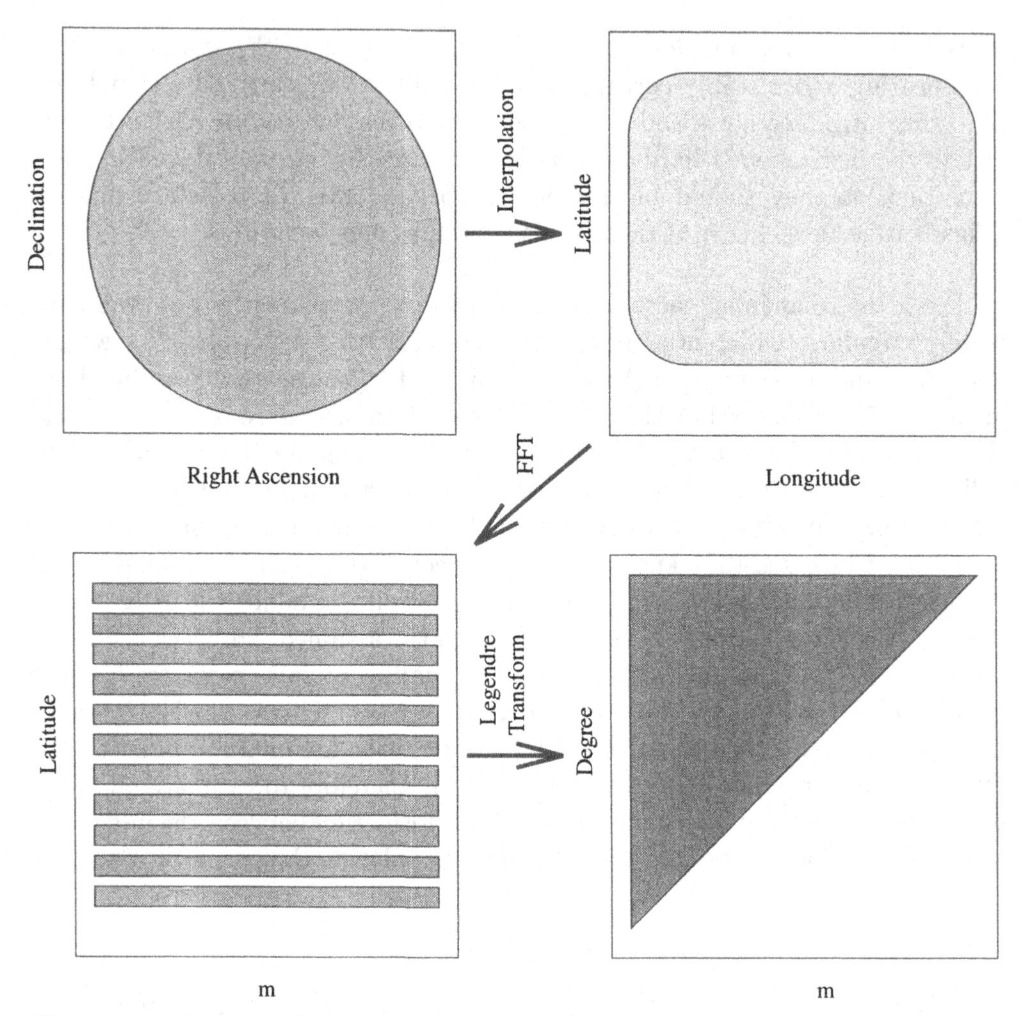

Figure 11: *Graphic flowchart of the major steps in a spherical harmonic decomposition of an observed solar image. See the text for a description of the process.*

symmetric about the central meridian, while the imaginary parts of W_m are the antisymmetric parts. Because V is real, the W_m are conjugate-symmetric in m, *i.e.*, $W_m = W^*_{-m}$. At this point, what we refer to as m is actually the absolute value of the spherical harmonic m; it is not possible at this stage to distinguish prograde from retrograde waves.

For each m ($m = m_0$, say), $W_{m_0}(\theta)$ is next expanded as a sum over the $P_l^{m_0}(\theta)$, with l running over all possible values, but with m restricted to the (positive) value m_0. The expansion coefficients $A_l^{m_0}$ are estimated simply by computing (using some numerical quadrature) the integral

$$A_l^{m_0} = \int W_{m_0}(\theta) P_l^{m_0}(\theta) \sin\theta d\theta \ . \tag{20}$$

The real and imaginary parts of the complex coefficient A_l^m are estimates of the amplitudes of the functions $P_l^m(\theta)\cos\phi$ and $P_l^m(\theta)\sin\phi$, respectively, evaluated at the time of the original image. It is not until the time analysis is done (so that one can see the prograde or retrograde direction of wave propagation) that one can meaningfully distinguish between $+m$ and $-m$. Moreover, the estimate A_l^m is an imperfect one. The Y_l^m are orthogonal when integrated over a sphere, but the available data cover something smaller than a hemisphere. The available data may be thought of as the "true" data (which cover the whole sphere) multiplied by a mask M that is nonzero where good data are available, and zero elsewhere. The convolution theorem (which does not apply to P_l^ms) nevertheless leads us to suspect that the spectrum A_l^m that we estimate will be the true spectrum, but smeared by some sort of a blurring function that grows larger and fuzzier as the nonzero area of M grows smaller. The smearing can be fairly large, and the shape of the fuzzy smearing function is quite different for modes with $l \cong 0$ than for modes with $l \cong |m|$ (this is an expression of the failure of the convolution theorem for Legendre transforms).

3.1.4 Transforms with an Axis in the Image

In some cases (sunspots, for instance) it is advantageous to choose different coordinates than those defined by the Sun's rotation axis. When investigating the interaction between p-modes and sunspots, for instance, it makes sense to use cylindrical coordinates (or something similar) with the axis placed in the center of the spot.

The simplest version of such an analysis uses the "flat Sun" approximation, and treats the observations as if they represent a portion of a plane, with distance from the coordinate axis denoted by r and azimuthal angle from some arbitrary direction denoted by η. In this coordinate system the solution to the wave equation is expressed in terms of Hankel functions, which are merely linear combinations of Bessel functions, with a sinusoidal dependence in the azimuthal direction:

$$H^{(1)}(kr,\eta) = [J_m(kr) + iY_m(kr)]\exp(im\eta) , \qquad (21)$$

where J_m and Y_m are Bessel functions of order m and of the first and second kinds, respectively. When performing this sort of analysis, one proceeds by first interpolating onto a grid with equal increments in r and η, then Fourier transforming in η for each r, and finally projecting onto the appropriate Bessel functions. Note that the functions $Y_m(kr)$ become infinite at $r = 0$, so that this analysis is only practical if some region surrounding the axis is excluded. See Braun *et al.* (1988) for details of this procedure. If curvature effects are important, the appropriate functions to use are again the spherical harmonics, but this time with the pole of the coordinate system coincident with the desired axis (Bogdan *et al.* 1993).

3.2 Transposition

When the spatial analysis is complete, one has complex amplitudes for all of the desired expansion functions and for every point in time (*i.e.*, for every image). In what follows I will suppose that the analysis is in terms of the global wavefunctions Y_l^m; analogous descriptions would apply to the other possibilities. The next desired step is to Fourier transform in time, to obtain the frequency dependence for each l and m. This is conceptually straightforward but can have its practical difficulties. The problem is that, after the spatial analysis, the data are ordered in the computer as $A(m, l, t)$; for purposes of transforming in time, one needs $A(t, m, l)$. The data sets involved are often quite large: a month of MDI data involves 5×10^4 time steps, each containing roughly 10^6 spatial coefficients. Transposing a matrix of this size is feasible, but doing it efficiently requires some attention.

3.3 Time Series Analysis & Peak Bagging

In the time-series analysis, one computes the temporal Fourier transforms of the various A_l^m, and somehow uses these transforms to estimate the frequencies, amplitudes, and linewidths of the multitude of modes that show up in each time series. The problem can be difficult, as suggested in fig. 12, because gaps in the data combine with incomplete spatial coverage to turn a relatively simple theoretical spectrum into a complicated real-world mess. Extensive simulations by Schou (1992) show that a large variety of innocuous-sounding problems in the upstream analysis can appear as systematic errors in the derived mode properties. What I will describe below are methods that are believed to work pretty well, but there is certainly a great deal of room for improvement in this area of helioseismic analysis.

3.3.1 Data Conditioning, Temporal FFT, Power Spectrum

When the data have been transposed into the correct order, one typically finds that some data points are missing, and some are corrupted.

Missing data are inevitable with single-site observations, because of the di-

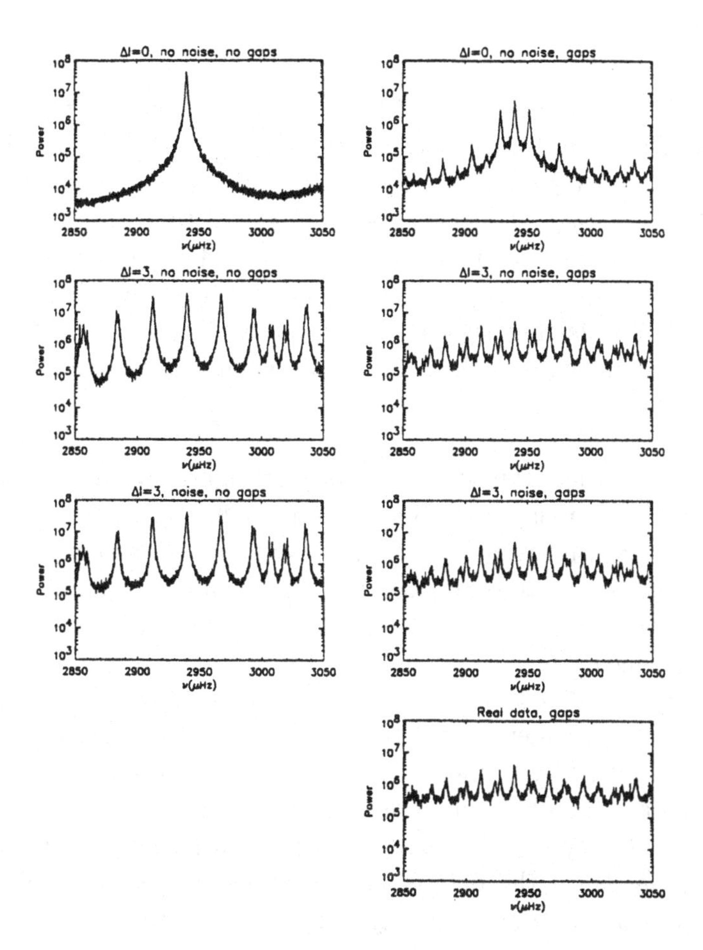

Figure 12: *Power spectra of artificial data, showing the influence of temporal gaps (right-hand set of panels) and incomplete spatial coverage (that is, cross-talk in l), from Schou 1992.*

urnal cycle or (even if the site is the South Pole) weather. Even with a fairly extensive network, missing data points will occur because of coincidences of bad weather among several sites, instrument failures, and time taken off for calibrations. For example, the expected duty cycle for the full 6-site GONG network is expected to be at most about 95%. For long intervals of missing data (more than a few oscillation cycles, say), the usual approach is simply to set the missing data to zero. There are differences of opinion about the advisability of multiplying the data adjacent to the gaps by a function that goes smoothly from unity to zero (a "tapering" function), to decrease the far sidelobes of the effective window function. For short data gaps (particularly those that span only a few minutes), it proves convenient and effective to replace the missing data with an extrapolation of real data from both ends into the gap. Presently, the favored extrapolation technique is based on an autoregressive model of the time series

(Anderson *et al.* 1993).

Corrupted data points are a fairly common feature in time series of Y_l^m coefficients. They may occur for a wide variety of reasons, ranging from marginal weather to momentary instrument failures to ill-placed moths, spiders, or A-10 aircraft. Most are easily detected by scanning each time series and searching for extreme statistical outliers; if the data for many A_l^m are all found to be bad at the same time, then one may exclude all of the coefficients for that particular time, on the grounds that they are all suspect. The typical action taken with corrupt data points is to set their time series values to zero, and then treat them as missing data.

The actual Fourier transform process is straightforward, the only complication being that, for reasons of computational efficiency, one likes to transform time series whose lengths are the products of small prime numbers (ideally but not necessarily 2 raised to some power). This often means padding the available time series with zeros in order to stretch it to the next convenient size larger than its true length.

One last rearrangement is needed after the Fourier transform is computed. Modes that propagate in opposite directions around the Sun (*i.e.*, modes with $+m$ and $-m$) are (finally!) separated. They appear after the Fourier transform with temporal frequencies of opposite sign; which sign of ν goes with which sign of m depends on the details of how wavefunctions are defined and how the transforms are computed. In any case, many routines fold the transformed data so that all the ν are positive, and reorder the data so that m increases monotonically.

Some algorithms for mode identification and parameter estimation work on power spectra (the GONG routines, for instance), while others (MDI) operate directly on the Fourier transforms. In the former case, power and phase spectra are computed from the Fourier transforms, and are stored as intermediate data products.

3.3.2 Maximum Likelihood & Lorentzian Fitting

Determining the frequencies, amplitudes, and linewidths for all the oscillation modes represented in a large data set is a complicated and computation-intensive task. Rather than attempt to describe the process in detail, I shall cover a few general points and refer the reader to work by Anderson *et al.* (1990) and Schou (1992) for a more complete discussion.

To date, efforts to determine p-mode parameters have depended on modeling the observed transforms or power spectra in terms of a collection of damped oscillators superposed on some sort of background noise spectrum. Since the transform of an exponentially damped sinusoid is a Lorentzian, the problem amounts to one of estimating the amplitudes, widths, and center frequencies of a substantial number of overlapping Lorentzians. This problem is ordinarily approached as a maximum-likelihood fit. That is, given the observations and the (presumed known) statistical properties of the data, find the set of parameters that is most probable. Because of the χ^2 probability distribution of stochastic power spectra (*c.f.* Section 1.1.5), a least-squares fit is inappropriate to this problem, and the fitting procedures used are nonlinear and iterative. The actual likelihood maximization techniques used vary, depending on what is assumed to be known about the correlations between neighboring frequency points (because of gaps in the time series) and between neighboring l and m values (because of limited spatial coverage on the Sun). Also, the organization of such codes varies considerably. For instance, the GONG procedure (Anderson *et al.* 1990) treats the convolution in frequency resulting from temporal gaps, but not the cross-talk between modes with the same l and different m. The MDI code takes the opposite approach. So far, application of both techniques to the same artificial data sets have revealed no compelling reason to choose one approach or the other. In fact, we are still in the process of understanding which factors really make a difference in this kind of analysis, and which ones only seem as if they ought to.

One final complication to the data fitting process arises from the observation by Duvall *et al.* (1993a) that, particularly at low frequencies, p-mode line shapes are significantly asymmetric. This effect is interpreted in terms of a combination of acoustic sources that lie outside the region of p-mode propagation and of phase shifts (possibly arising from non-adiabatic processes in the upper solar envelope) affecting the coupling of the sources to the cavity. In any event, it is clear that fits to symmetric Lorentzian profiles will give incorrect results if applied to asymmetric profiles, so allowance for the asymmetry should be made in the analysis routines. So far as I am aware, no commonly-used routines do this.

4 LOCAL ANALYSIS METHODS

Most helioseismic work to date has focused on global modes, because they carry information about interesting large-scale properties of the Sun (such as its spherical structure and the depth and latitude dependence of its angular velocity), and because the mathematical framework for dealing with the global modes is sophisticated and well understood. This approach is not effective for all purposes, however. If one is interested in probing the structure of small-scale features in the Sun (sunspots, for instance, or moderate-scale non-axisymmetric flows), then it is more attractive to treat the acoustic waves as local phenomena that do not know or care about the Sun as a whole. Indeed, waves with high frequencies or l-values or both do not survive long enough to propagate all the way around the Sun; for such waves a local description is the only one that makes sense.

4.1 Frequencies *vs.* Eigenfunctions

It is worth noting that there is a large variety of possible perturbations to the solar structure that do not perturb the mode frequencies (to first order), but that do cause first-order changes in the eigenfunctions. Any perturbation that is antisymmetric about the equator fits this description, for instance. This fact provides a motivation for two methods of local analysis described in the next two sections.

4.1.1 Rings and Trumpets

To understand the "ring and trumpet" analysis (Hill 1988), first consider only wave propagation in the presence of a mean flow field. The phase speed of a wave, seen in the frame moving with the fluid, is

$$v_{\phi 0} \equiv \lambda\nu = 2\pi\nu/k \ , \tag{22}$$

so that the frequency ν is given by

$$\nu = v_{\phi 0}k/2\pi \ , \tag{23}$$

where k is the magnitude of the wavenumber. From the point of view of a fixed observer, the phase velocity (which I shall now treat as a vector) is augmented by the flow velocity, with a corresponding change in the frequency:

$$v_\phi = v_{\phi 0} + v_{flow} \quad ; \quad \nu = (\vec{v}_{\phi 0} + \vec{v}_{flow}) \cdot \vec{k}/2\pi \ . \tag{24}$$

The observed frequency of a wave is thus perturbed by an amount

$$\delta\nu = \vec{v}_{flow} \cdot \vec{k}/2\pi \ . \tag{25}$$

Now in a local analysis in the rest frame of the fluid, sound waves do not know anything about their horizontal direction of propagation, so the resonant

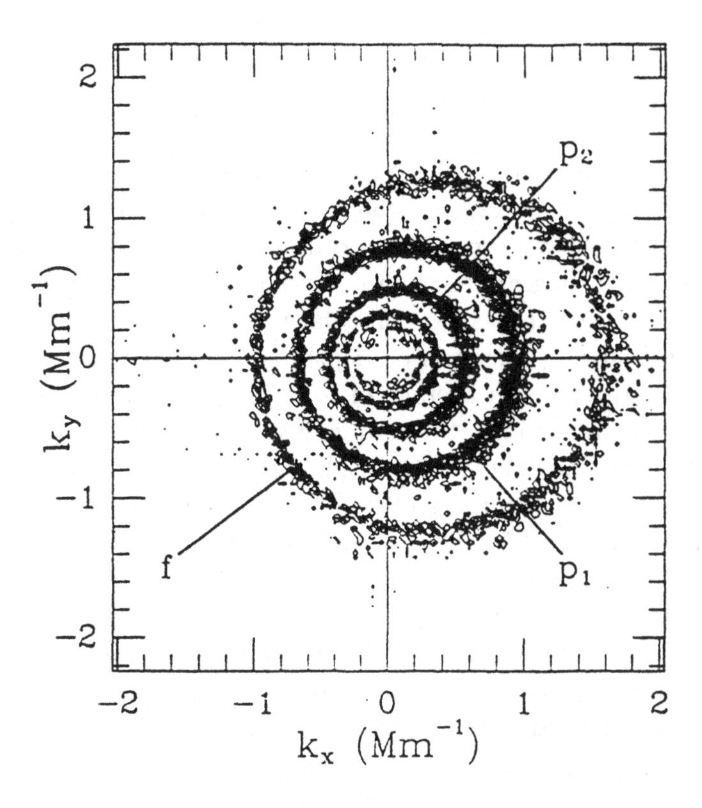

Figure 13: *Ring diagram in the presence of a large transverse flow, from Hill (1988). This diagram corresponds to* ν = 3010 μ*Hz; the* f, p_1, *and* p_2 *modes are marked.*

frequencies are independent of the azimuthal angle $\arctan(k_y/k_x)$. Thus one could obtain the 3-dimensional version of fig. 10 simply by rotating that figure about the frequency axis. One would then find significant power not along 1-dimensional "ridges" in $k - \nu$ space, but rather along 2-dimensional flaring surfaces in $k_x - k_y - \nu$ space. These surfaces are the "trumpets" of the ring and trumpet analysis. To obtain the rings, one simply slices this 3-dimensional trumpet diagram with a plane at constant ν; the trumpet cross-sections then appear as circles centered on the k_x, k_y origin.

What happens to the nice, symmetric ring picture if we now impose a bulk flow? From eq. (25), the frequencies are changed by amounts that depend on $\vec{k}$ and on $\vec{v}_{flow}$. These frequency changes displace the rings (which are measured at constant frequency) in the direction of $\vec{k}$ by an amount

$$\delta k = \frac{\delta\nu}{d\nu/dk} . \tag{26}$$

For p-modes, we have approximately $\nu \propto k^{1/2}$, so $d\nu/dk \propto -(k^{-1/2})$. Combined with eqs. (25-26), one finds that, to lowest order,

$$\delta\vec{k} \propto -\vec{k} \cdot v_{flow}|k|^{1/2}\hat{v}_{flow} \quad . \tag{27}$$

Thus in the first approximation, the observed rings remain circular, but are displaced in a direction opposite to the flow vector by an amount that is proportional to $k^{3/2}$. A relatively gross form of this behavior is illustrated in fig. 13, which shows a ring diagram for a case in which the observed field of view was fixed relative to the solar disk, so that solar rotation provided a v_{flow} of approximately 2 km/s. If $\vec{v}_{flow}$ depends on depth, then the displacement of the rings depends on whatever depth average of $\vec{v}_{flow}$ is appropriate to modes with the given (unperturbed) k and ν. Since different $k-\nu$ combinations have different depth sensitivities, one may use the displacements measured for different k and ν to infer the depth dependence of $\vec{v}_{flow}$. Since one need not use all (or even most) of the observable solar hemisphere for computing the ring diagrams, one can gain some information about longitude and latitude dependence of subsurface flows by repeating the analysis for subrasters of the solar image, centered at different locations. The details of the inversion procedure, and of the method for using 3-dimensional power spectra to estimate ring displacements, are described by Hill 1988, and by Patrón *et al.* (1993).

4.1.2 Hilbert Transforms

The Hilbert transform analysis is another way in which one might hope to extract information from p-modes about flow fields or thermodynamic perturbations to the solar background state (Gough *et al.* 1993, Julien *et al.* 1995). The method depends on the fact that nonuniformities in the Sun's background state must impose changes on the amplitude and (especially) the phase of sound waves moving through it – waves that would otherwise appear as simple plane waves. In the simplest 1-dimensional example, one describes the wave propagation with the Helmholtz equation

$$\frac{d^2\Psi}{dx^2} = -\kappa^2(x)\Psi \quad , \tag{28}$$

where x is the horizontal coordinate and κ describes the local wave propagation properties of the medium, about which one would like to learn. If Ψ is represented as a single wave $\Psi = A(x)\cos[\phi(x)]$, then one has approximately

$$\kappa(x) \cong \frac{d\phi}{dx} \quad . \tag{29}$$

To estimate $\frac{d\phi}{dx}$, one may use a Hilbert transform (see Bracewell 1965), which, combined with the original signal, gives an estimate of the instantaneous amplitude and phase of the wave.

A major complication is that one actually uses a group of modes with similar ν to perform such an analysis, and beating between the modes with different wavenumber in this group corrupts the phase measurement. This problem can be dealt with by using the fact that a wave packet travels with its group velocity, whereas the phase contamination resulting from beating is fixed in space. The two effects may therefore be separated and the analysis completed, in spite of wave beating problems.

In a real 2-dimensional setting, one creates pseudo 1-dimensional problems by averaging the observed wave field along some direction, thereby filtering out all waves except those traveling perpendicular to the chosen line. The Hilbert transform method then yields an average of κ, formed in the same way as the average of Ψ. If this process is repeated for many different averaging directions, the results may be combined with a tomographic procedure to recover the 2-dimensional structure of κ. By using wave packets that penetrate to different depths, inversion for 3-dimensional properties should be possible.

The Hilbert transform technique is extremely promising, but it is still in the early stages of development, and has not been tried on solar data. Also it has a few limitations, which should not be unduly restrictive for many purposes. For the mathematical approximations of the method to be valid, the background structure must vary on spatial scales that are large compared to the horizontal wavelengths of the waves employed. Moreover, the structures being measured must survive essentially unchanged for a time that is long enough for the group velocity to carry the wave packet across the observed area.

4.1.3 Time-Distance Methods

Time-distance helioseismology was first described by Duvall *et al.* (1993b). The underlying idea is to use a ray formulation of sound propagation in the Sun to arrive at a picture of the p-mode phenomenon that is similar to traditional approaches in terrestrial seismology. In the simplest application of time-distance methods, one considers an arbitrary point on the solar surface to be a source of acoustic waves. (Even if the waves do not originate at the chosen point, Huygens' principle assures that subsequent wave evolution will be the same as if waves passing through the point at time $t = t_0$ actually started there.) A wave with a particular frequency leaving the origin in a particular direction follows some trajectory as shown in fig. 14, eventually returning to the surface a distance D away, where it contributes to the local surface displacement. If there were no other sources of acoustic waves (this is pretty much the case in the Earth, where the origin may be considered the location of an earthquake), then the

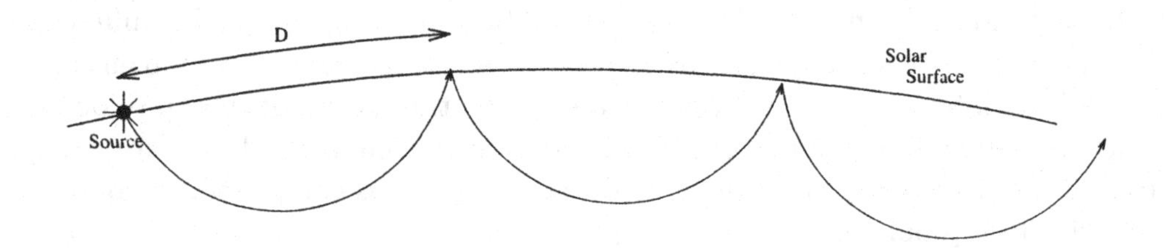

Figure 14: *The geometry of a time-distance analysis. Acoustic waves emitted from (or merely passing through) the source follow paths below the surface, emerging after a time τ at a distance D from the source. Successive bounces cause correlated signals at integer multiples of τ and D.*

wave signal at D would look just like the signal at the origin, but delayed by a time

$$\tau(D) = \int_{path} \frac{ds}{c} , \tag{30}$$

where c is the local sound speed and the integration is taken along the path followed by the ray. The waves arriving at distances similar to D travel along slightly different paths, sample different parts of the Sun, and hence contain different information. By mapping out the complete dependence $\tau(D)$ for all possible distances up to half a solar circumference, one samples all depths in the Sun, and may estimate (say) the sound speed everywhere. As an added attraction, one may watch waves bounce more than once (if that is what they do). Aside from losses from scattering and nonadiabatic mechanisms, one expects that a wave that appears at distance D and time τ will reappear at distance qD and time $q\tau$, with q being the number of internal bounces the wave has undergone since leaving the origin.

There is an obvious technical difficulty with the procedure as just described, namely that there are very many sources of acoustic noise on the Sun at any given time, so that to measure $\tau(D)$, one must somehow isolate the desired signal from all of the unwanted ones. A direct approach for doing this is to first form an average over a ring with radius D centered on the desired origin. Signals from the origin should add in phase for points on this ring, while those from other places should tend to cancel out. Then one cross-correlates the signal at the origin with the average over the ring; if long enough time series are analyzed, the correlated part of the signals prevails over the noise, and a clean correlation peak emerges at the desired time delay τ.

An improvement on the averaging method may be used if one assumes that

$\tau(D)$ is independent of the choice of origin (*i.e.*, if one is uninterested in variations in the time-distance relation from one place to another). In this case, one could average the correlation obtained above over all possible locations for the origin. This average is actually the azimuthally-averaged cross-correlation function for the observed wave field; it is efficiently determined simply by transforming the azimuthally averaged power spectrum

Really, measuring $\tau(D)$ is (once again) the notion of the impulse response: to learn about a system's resonances, thump it once and see what happens. It therefore should not be surprising that the information in the $\tau(D)$ relation is the same as that in the $k - \omega$ diagram (the power spectrum). Some interpretational points are, however, easier to see when presented as time-distance plots than when illustrated with the power spectrum. An example of this effect is the non-reflective character of acoustic waves with ν greater than the photospheric maximum in the acoustic cutoff frequency (Duvall *et al.* 1993b, also see below). Moreover, recent work with time-distance methods (*e.g.* Braun 1995, d'Silva 1994) use phase or reflection information in the time-distance relation to infer things about sunspots that are, at best, quite difficult to extract from the power spectrum. Although still under development, it seems likely that the time-distance approach will prove to be a useful one for unraveling the nature of subsurface flows and other inhomogeneities within the Sun.

4.2 Scattering, Absorption, and Lifetimes

To date, most analyses of p-modes treat the acoustic waves as if they propagate adiabatically and do not couple to one another. Braun *et al.* (1988) were the first to notice that, in regions where strong magnetic fields exist, these assumptions are not satisfied even approximately. This area of helioseismology is making rapid progress, though to date it remains largely phenomenological – many of the physical process at work are not understood.

4.2.1 Sunspot Absorption of p-Modes

To approach the problem of how acoustic waves propagate in and near sunspots, one must first deal with the problem that, for a variety of reasons, such waves are quite difficult to observe within the actual boundaries of a spot. The approach taken by Braun *et al.* (1988) and in most subsequent investigations is to ignore the sunspot proper, and concern oneself only with an annular region surrounding the spot (fig. 15). The data analysis problem now comes down to representing

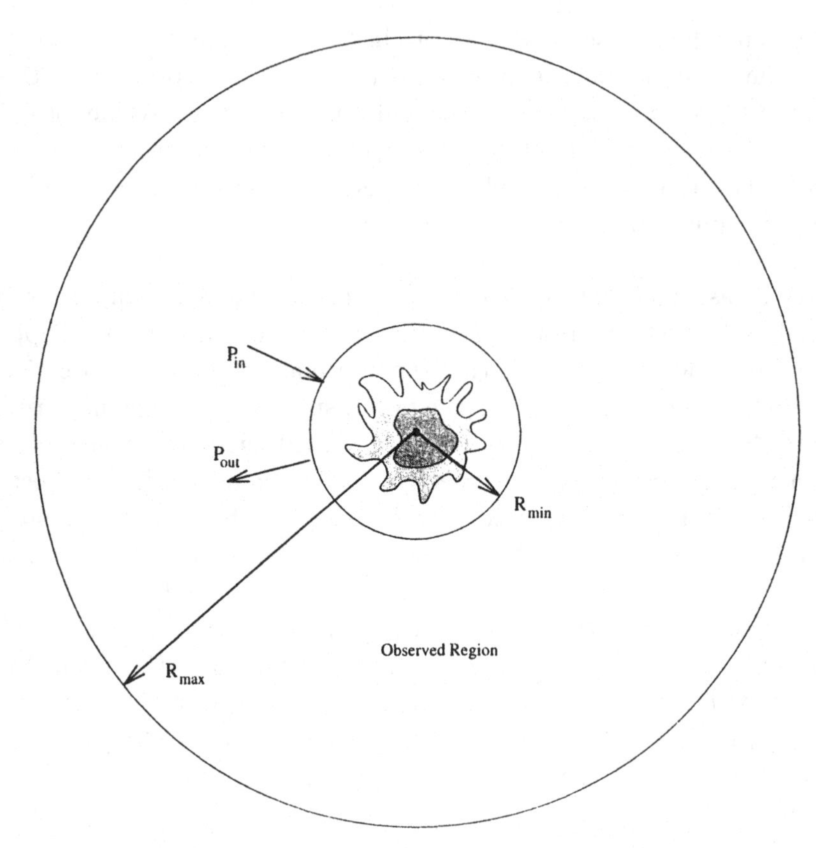

Figure 15: *Geometry of the p-mode analysis for studies of absorption by sunspots. The spot is placed in the central circular area of an annulus with inner radius* R_{min} *and outer radius* R_{max}*. The wave field outside the annular region is ignored, for the purposes of this analysis. Absorption of p-modes is detected by measuring the difference* $P_{in} - P_{out}$ *between the ingoing and outgoing energy fluxes.*

the observed wave field in the annulus in terms of inward- and outward-traveling waves. This is done in terms of either Hankel functions or spherical harmonics, as outlined in section 3.1.4.

If the sunspot contained within the central circle of the annulus had no effect on wave propagation, one would expect that all of the inward-traveling energy crossing the inner boundary would reappear as outward-traveling waves. Thus, the power in the inward-going wave field should be the same as that in the outward-going field. Braun *et al.* (1988) observed something quite different from this. They found that there is a deficit of outward-going power, and that this deficit grows with increasing wavenumber. This deficit is generally described in terms of a power absorption coefficient

$$\alpha \equiv \frac{P_{in} - P_{out}}{P_{in}} , \tag{31}$$

where P_{in} and P_{out} are the ingoing and outgoing powers, respectively. For wavenumbers k that are large enough (roughly $kL \geq 2\pi$, where L is the size of the spot, *i.e.*, waves whose horizontal scale is comparable to or smaller than the spot size), α grows as large as 0.5, *i.e.*,, roughly half of the incoming power disappears within the spot. This behavior is common to all of the spots so far studied, but is not seen in regions of quiet Sun that are observed as control areas. Something about the size of the p-mode absorbing region can be learned by examining the way in which the absorption depends on the azimuthal quantum number m in eq. (21). The value of m may be thought of as relating to the angular momentum of the corresponding wave, measured about the origin of the coordinate system. Waves with $m = 0$ pass straight through the origin, while those with nonzero m pass by the origin with an impact parameter given approximately by $kr_{min} = m$. The observed dependence of α on m at constant k, ν suggests that most of the absorption occurs in the spot itself, although a noticeable fraction of it takes place in the plage areas that surround the spot.

4.2.2 Scattering by Convection

When Bogdan *et al.* (1993) used high-resolution observations to push the analysis the analysis of absorption by sunspots up to wavenumbers corresponding to $l \geq 1000$, a curious feature appeared (fig. 16). After rising to a maximum of about 0.5 at $l \simeq 500$, α decreased to values near zero for $l \geq 1200$. The interpretation of this curious behavior is that the waves with high l simply do not live long enough to propagate all the way across the observed annulus. Instead, these waves are scattered (or perhaps absorbed and re-emitted, though this seems less likely) before they can make the journey from the edge of the annulus to the spot and back again. The situation is analogous to trying to observe a black feature on a white wall in foggy air. As the fog grows thicker (or the distance to the wall increases), the contrast of the feature decreases. The practical importance of this phenomenon is that the presence of the partially-absorbing sunspot causes an anisotropy in the acoustic flux that one may use to investigate phenomena outside the spot – in this case, scattering processes in the quiet Sun. It is thought (based on global mode lifetimes) that most of the scattering results from the entropy and velocity fluctuations associated with convection in the Sun's upper envelope. To my knowledge, no attempt to check the consistency of this conclusion using sunspot absorption has yet been made, but it probably should be.

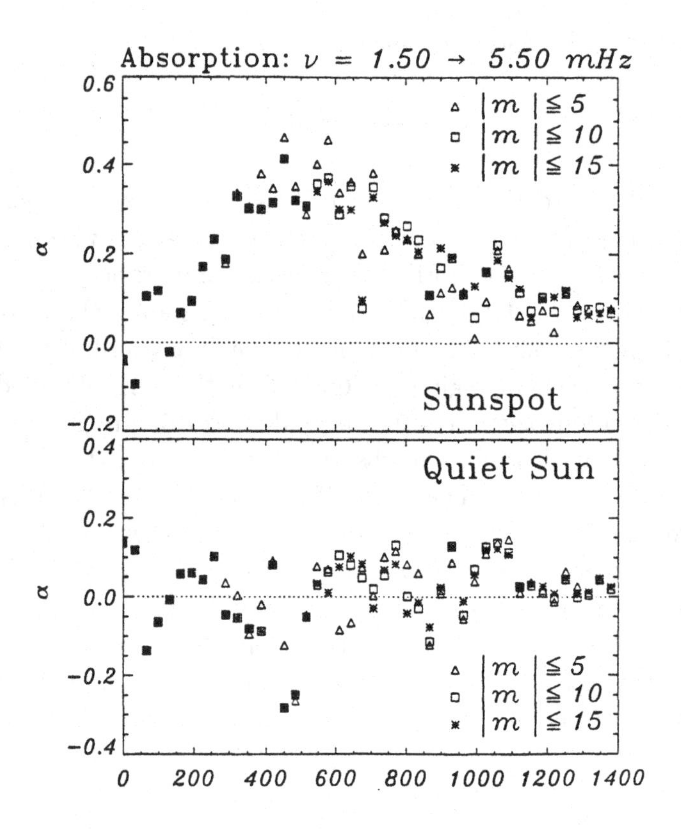

Figure 16: *Absorption of p-mode energy plotted against angular degree l, from Bogdan* et al. *(1993). The top panel shows results when a sunspot occupies the center of the observed annulus; the bottom panel shows results for the quiet Sun. See the text for further explanation.*

4.2.3 Waves above the Acoustic Cutoff

According to the simple Fabry-Perot-like picture of acoustic resonance in the solar envelope, one does not immediately expect resonant behavior for frequencies above the photospheric acoustic cutoff frequency, which is to say, above roughly $\nu = 5.5$ mHz. At such frequencies there should be no top boundary to region of wave propagation, so that a resonant cavity does not really exist. One does, however, see ridges in the $l - \nu$ diagram at frequencies that are much higher than 5.5 mHz. The observed ridges seem to be smooth continuations of the lower-frequency p-mode ridges, though they are wider and have lower contrast. They are more easily seen in lines that are formed high in the solar atmosphere than in lines that are formed close to the photosphere.

Two explanations have been offered for this phenomenon. The first, which

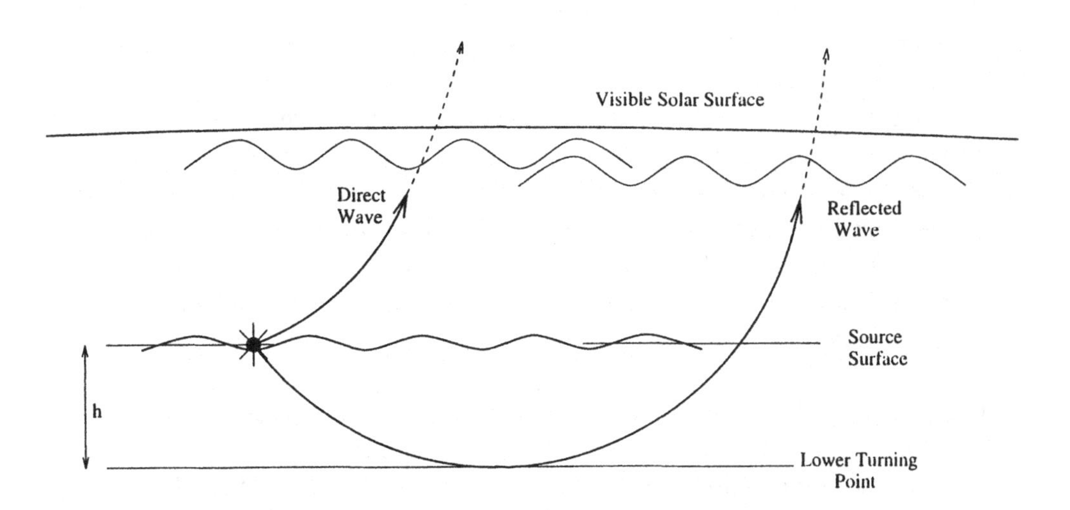

Figure 17: *Geometry resulting in resonance-like features in the $k - \omega$ diagram at frequencies above the photospheric acoustic cutoff frequency. Waves with a given ν and horizontal wavenumber k traveling from the source to the visible surface may proceed directly, or may reflect once at the lower turning point. Interference between waves taking these two different paths is responsible for the apparent resonance.*

has little observational support, is that there actually is a top to the cavity, formed perhaps by the abrupt change in atmospheric properties in the chromosphere-corona transition region (Balmforth & Gough 1990). The second explanation, which reproduces the observations pretty well, is that the observed quasi-resonance is produced by the presence of the bottom boundary combined with a fairly localized source of acoustic waves. The mechanism involves interference between waves that travel directly from the source to the surface and those that come to the surface via one reflection from the lower boundary, as illustrated in fig. 17. Let us first consider only one Fourier transform component of the source, which oscillates with a horizontal wavenumber k and frequency ν, and is located a height h above the lower turning point for waves of the given k and ν. This source produces two waves: one moving up and one moving down, both at an angle θ to the horizontal that is determined by ν/k. Both of these waves emerge at the solar surface with the same frequency and wavenumber, but the wave that initially propagated downward is delayed relative to the upward-propagating wave by the time τ_{turn} required for it to travel down to the lower turning point and return to the source depth. The surface amplitude that is observed depends on the phase difference $2\pi\nu\tau_{turn}$ between the direct and the reflected waves; it may be large or small, depending on how the waves interfere.

The most important feature distinguishing the high-frequency ridges from their lower-frequency p-mode cousins is that, for the high-frequency waves, the loci of large amplitude in frequency-wavenumber space depend on the depth of the source surface. The high-frequency ridges therefore carry information about the location (and indeed the depth distribution) of the source region, and hence about the properties of turbulence in the upper convection zone. Current estimates based on this kind of analysis place the acoustic source surface at a depth of 140 ± 60 km below the photosphere (Kumar 1994). The fit between observed and calculated power spectra computed from this model is not perfect, but it is quite good.

Observations of the high-frequency ridges have mostly been conducted with data having moderate spatial resolution and relating to lines formed in the high photosphere to middle chromosphere. The canonical sampling time for p-mode observations is one image per minute, yielding a temporal Nyquist frequency of 8.33 mHz. For many kinds of observations dealing with high-frequency waves this sampling is inadequate, since there are measurable ridges in some data sets reaching up to frequencies of 10 mHz or more.

5 REFERENCES

Aindow, A., Elsworth, Y.P., Isaak, G.R., McLeod, C.P., New, R., van der Raay, H.B.: 1988, in E.J. Rolfe (ed) *Seismology of the Sun and Sun-Like Stars*, (ESA), p. 157.

Anderson, E.R.: 1993, in T.M. Brown (ed.), *GONG 1992: Seismic Investigation of the Sun and Stars*, (Astronomical Society of the Pacific), p. 445.

Anderson, E.R., Duvall, T.L. Jr., Jefferies, S.M.: 1990, *Astrophys. J.* **364**, 699.

Balmforth, N., Gough, D.O.: 1990, *Astrophys. J.* **362**, 256.

Bogdan, T.J., Brown, T.M., Lites, B.W., Thomas, J.H.: 1993, *Astrophys. J.* **406**, 723.

Bracewell, R.: 1965, *The Fourier Transform and its Applications*, McGraw-Hill.

Braun, D.C.: 1995, in R. Ulrich (ed.), *GONG 1994*, (Astronomical Society of the Pacific) (in press).

Braun, D.C., Duvall, T.L. Jr., LaBonte B.J.: 1988, *Astrophys. J.* **335**, 1015.

Brown, T.M.: 1980, in R.B. Dunn (ed.), *Solar Instrumentation: What's Next?*, (Sacramento Peak Observatory), p. 151.

Chou, D.Y. *et al.*: 1995, in R. Ulrich (ed.), *GONG 1994*, (Astronomical Society of the Pacific) (in press).

Damé, L.: 1988, in E.J. Rolfe (ed) *Seismology of the Sun and Sun-Like Stars*, (ESA), p. 367.

Deubner, F.-L.: 1975, *Astron. Astrophys.* **44**, 371.

Deubner, F.-L., Ulrich, R.K., Rhodes, E.J., Jr.: 1979, *Astron. Astrophys.* **72**, 177.

d'Silva, S.: 1994 *Astrophys. J.* **435**, 881.

Duvall, T.L. Jr., Harvey, J.W., Pomerantz, M.A.: 1986, *Nature* **321**, 500.

Duvall, T.L. Jr., Jefferies, S.M., Harvey, J.W., Osaki, Y., Pomerantz, M.A.: 1993a, *Astrophys. J.* **410**, 829.

Duvall, T.L. Jr., Jefferies, S.M., Harvey, J.W., Pomerantz, M.A.: 1993b, *Nature* **362** 430.

Fossat, E.: 1988, in E.J. Rolfe (ed.), *Seismology of the Sun & Sun-Like Stars*, (ESA), p. 161.

Fröhlich, C., *et al.*: 1989, in ESA SP-1104, p. 19.

Gough, D.O., Merryfield, W.J., and Toomre, J.: 1993, in T.M. Brown (ed.), *GONG 1992: Seismic Investigation of the Sun and Stars*, (Astronomical Society of the Pacific), p. 257.

Groth, E.J.: 1975, *Astrophys J. Suppl.* **29**, 286.

Harvey, J., Pomerantz, M., Duvall, T.: 1982, *Sky & Telescope* **64**, 520.

Hill, F.: 1988, *Astrophys. J.* **333**, 996.

Jiménez, A., Pallé, P.L., Péres Hernandez, F., Régulo, C., Roca Cortes, T.: 1988, *Astron. Astrophys.* **192**, L7.

Julien, K., Gough, D., Toomre, J.: 1995, in R. Ulrich (ed.), *GONG 1994*, (Astronomical Society of the Pacific) (in press).

Korzennik, S.G.: 1990, Ph.D. Thesis, University of California.

Kumar, P.: 1994, *Astrophys. J.* **428**, 827.

Libbrecht, K.G.: 1992, *Astrophys. J.* **387**, 712.

Patrón, J., Hill, F., Rhodes, E.J., Korzennik, S.G., Cacciani, A., Brown, T.M.: 1993, in T.M. Brown (ed.), *GONG 1992: Seismic Investigation of the Sun and Stars*, (Astronomical Society of the Pacific), p. 437.

Rhodes, E.J. Jr., Cacciani, A., Tomczyk, S., Ulrich, R.K.: 1986, in D.O. Gough (ed.), *Seismology of the Sun and the Distant Stars*, (D. Reidel), p. 309.

Rust, D.M., Burton, C.H., Leistner, A.J.: 1986, in D.L. Crawford (ed.), *Instrumentation in Astronomy VI*, (SPIE) **627**, 39.

Scherrer, P.H., Hoeksema, J.T., Bogart, R.S., Walker, A.B.C., Title, A.M. *et al.*: 1989, in ESA SP-1104, 25.

Schou, J.: 1992, Ph.D. Thesis, Århus University.

Tomczyk, S., Streander, K., Card, G., Elmore, D., Cacciani, A.: 1994, *Solar Phys.* (in press).

Toutain, T., Fröhlich, C.: 1992, *Astron. Astrophys.* **257**, 287.

Willson, R.C.: 1984, *Space Science Reviews* **38**, 203.

Testing a solar model: the forward problem

Jørgen Christensen-Dalsgaard

Teoretisk Astrofysik Center, Danmarks Grundforskningsfond, and
Institut for Fysik og Astronomi
Aarhus Universitet
DK 8000 Aarhus C, Denmark

1 INTRODUCTION

The present chapter addresses *the forward problem*, *i.e.*, the relation between the structure of a solar model and the corresponding frequencies. As important, however, is the extent to which the frequencies reflect the physics and other assumptions underlying the model calculation. Thus in Section 2 I consider some aspects of solar model computation. In addition, the understanding of the diagnostic potential of the frequencies requires information about the properties of the oscillations, which is provided in Section 3. Section 4 investigates the relation between the properties of solar structure and the oscillations by considering several examples of modifications to the solar models and their effects on the frequencies, while Section 5 considers further analyses of the observed frequencies. Finally, the prospects of extending this type of work to other stars are addressed in Section 6.

A more detailed background on the theory of solar oscillations was given, for example, by Christensen-Dalsgaard & Berthomieu (1991), Gough (1993), and Christensen-Dalsgaard (1994). For other general presentations of the properties of solar and stellar oscillations see, *e.g.*, Unno *et al.* (1989) and Gough & Toomre (1991).

1.1 A little history

The realization that observed frequencies of solar oscillation might provide information about the solar interior goes back at least two decades. Observations of fluctuations in the solar limb intensity (Hill & Stebbins 1975; Hill, Stebbins & Brown 1976), and the claimed detection of a Doppler velocity oscillation with a period close to 160 minutes (Brookes, Isaak & van der Raay 1976; Severny, Kotov & Tsap 1976) provided early indications that global solar oscillations might be detectable and led to the first comparisons of the reported frequencies with those of solar models (*e.g.* Scuflaire *et al.* 1975; Christensen-Dalsgaard & Gough 1976; Iben & Mahaffy 1976; Rouse 1977). Although the reality of these early claims is questionable, they undoubtedly provided an important starting point for this type of work.

At about the same time the first detailed observations of the five-minute oscillations of high degree (Deubner 1975; Rhodes, Ulrich & Simon 1976) confirmed their nature as trapped acoustic modes of oscillation in the outermost parts of the Sun, previously inferred by Ulrich (1970) and Leibacher & Stein (1971). More detailed computations of frequencies for solar envelope models yielded results in overall agreement with the observations and suggested that the observed modes were overstable (*e.g.* Ando & Osaki 1975, 1977; it should be noted that later stability analyses, taking into account the interaction with convection, indicate that the modes are stable; see Balmforth 1992a). Further comparisons of the observed and computed frequencies indicated that the solar convection zone was deeper than previously assumed (Gough 1977a; Ulrich & Rhodes 1977) and showed that the frequencies were sensitive to details of the equation of state of matter in the Sun (*e.g.* Berthomieu *et al.* 1980; Lubow, Rhodes & Ulrich 1980).

The first definite detection of modes extending through most of the Sun resulted from whole-disk Doppler measurements which clearly showed oscillations in the five-minute region (Fossat & Ricort 1975; Claverie *et al.* 1979); particularly important was the identification in the latter data of an approximately uniformly spaced set of peaks in the power spectrum, corresponding to the asymptotically predicted behaviour of the frequencies of low-degree acoustic modes. A much more detailed spectrum, resolving individual modes, was obtained in almost continuous observations over several days from the geographical South Pole (Grec, Fossat & Pomerantz 1980).

The connection between these low-degree modes and the high-degree oscillations mentioned above was established by observations by Harvey & Duvall (1984). This enabled an unambiguous identification of the radial orders of the low-degree modes and provided extensive data on the structure of the solar interior. Further observations using a variety of techniques have since then dramatically increased the number of identified modes and the accuracy with which the frequencies have been determined, providing the current basis for helioseismic investigations.

1.2 Definition of the forward problem

Given a set of observed frequencies, probably the most immediate and obvious method of analysis is to compare them with frequencies computed for a solar model. This essentially defines the forward problem, as a test of a solar model. As such, the oscillation frequencies have several major advantages over other measurements that might relate to the structure of the solar interior: they can be determined observationally with great accuracy; different modes probe very different aspects of the structure; and, given a solar model, the frequencies can be computed with substantial precision. As discussed in more detail below the last statement must be qualified: aspects of the superficial region of the Sun introduce uncertainties in the computed frequencies which must be kept in mind when carrying out the comparison with the observations. Nonetheless, the frequencies remain very powerful diagnostics of the solar interior.

In some sense the solar model in itself is uninteresting: what requires testing are the assumptions and physical properties that underly the calculation of the model, thus improving the basis for general stellar-evolution calculations. Furthermore, the accuracy and extent of the solar data allow the properties of matter in the Sun to be probed in considerable detail, thus providing information that is totally inaccessible in laboratory studies. In this sense, therefore, the forward problem links the physics to the observed frequencies. This point of view will be employed extensively in the following.

Faced with the inevitable discrepancies between the computed and observed frequencies, what does one do? The obvious goal is to correct the model, or more fundamentally the physics, in such a way as to reduce the discrepancies. One approach is to compute several models and frequencies, adopting that model which best fits the data. At the opposite extreme, techniques for inverse analysis (see the chapter by Gough) offer systematic ways of determining the corrections. In between is a grey area of least-squares fits to the data, varying small sets of suitably chosen parameters. Such procedures were employed extensively in the analysis of the early, limited helioseismic data (*e.g.* Christensen-Dalsgaard & Gough 1980; Gabriel, Scuflaire & Noels 1982). Although the current wealth of data makes inverse techniques attractive, parameter fitting still has an important rôle to play for more specialized applications, where the data can be combined in such a way as to isolate specific aspects of the solar interior. An example of this will be considered in Section 5.2.

2 PHYSICS OF SOLAR MODELS

2.1 Introduction

The computed solar models, and hence their frequencies, depend on assumptions about the physical properties of matter in stars, in particular the equation of state, the opacity and the rates of nuclear reactions; these aspects of the calculation might be called the *microphysics*. Furthermore, the computations involve a number of simplifying assumptions, often covering much complex physics which might be called the *macrophysics*:

- The treatment of convection is approximated by mixing-length theory which provides a parametrization of the structure of the uppermost part of the convection zone in terms of the mixing-length parameter α_c.
- The dynamical effects of convection (the so-called turbulent pressure) are ignored.

- It is assumed that there is no mixing outside convectively unstable regions.
- Effects of magnetic fields are ignored.

Similarly, the calculations of oscillation frequencies are often done in the adiabatic approximation. Even when nonadiabatic effects are taken into account, their treatment is uncertain, since there is no definite theory for the perturbation in the convective flux, induced by the oscillations. Also, the perturbations in the turbulent pressure are usually neglected.

The goals of the analysis of observed frequencies are evidently to test both the microphysics and the simplifying assumptions. This is complicated by the fact that a given region of the model in general is affected by several aspects of the microphysics, *e.g.* both the opacity and the equation of state; under these circumstances it may evidently be difficult or impossible to isolate the cause of discrepancies between observations and models.

The computation of solar models requires the specification of a number of parameters. The age of the Sun can be estimated from ages determined for meteorites (*e.g.* Guenther 1989; Appendix by G. Wasserburg in the paper by Bahcall & Pinsonneault 1995). The present ratio Z/X between the abundances of heavy elements and hydrogen on the solar surface is approximately known from spectroscopy (*e.g.* Anders & Grevesse 1989; Grevesse & Noels 1993). Also, the computed models must match the photospheric radius and surface luminosity of the present Sun. This is achieved by adjusting the initial abundance Y_0 of helium and a parameter characterizing convective energy transport. The latter parameter (in mixing-length theory taken to be the ratio α_c between the mixing length and the pressure scale height) serves to fix the value s of the specific entropy in the bulk of the convection zone, where the temperature stratification is essentially adiabatic and where s is therefore nearly constant.

In the following I provide a brief overview of the some aspects of the physics of particular relevance to the analysis of the oscillation frequencies.

2.2 Microphysics

2.2.1 Equation of state

As mentioned in Section 1.1 the potential for using the observed frequencies to test the equation of state was recognized quite early. A detailed analysis of the treatment of the thermodynamical properties of solar matter and its effect on the frequencies was given by Christensen-Dalsgaard & Däppen (1992).

The theoretical description of the solar plasma is complicated by the interactions between its constituents, which strongly affect, *e.g.*, the degree of partial ionization in the outer parts of the Sun. A specific problem is to ensure that matter becomes fully ionized in the deep solar interior; this is normally achieved by including some formulation for the so-called "pressure ionization", taking effect at high density. We are still far removed from a definitive treatment of these processes. However, an essential feature of any treatment is that it be thermodynamically consistent. If this is not the case, the results obtained depend on the details of how, for example, the oscillation equations are formulated, potentially leading to unpredictable results.

Here I consider the following equations of state:

- The Eggleton, Faulkner & Flannery (1973) formulation (in the following EFF). This uses the basic Saha equation, assuming all atoms or ions to be in the ground state, and includes a simple but thermodynamically consistent treatment of pressure ionization.

- The CEFF formulation. This corresponds to EFF, but with the addition of Coulomb effects in the Debye-Hückel approximation. It should be noted that these effects result in corrections to pressure and internal energy of the gas, as well as to the chemical potentials and hence ionization balance. The effects on ionization, which dominate in the outer parts of the model, have sometimes been ignored, leading to thermodynamic inconsistencies.

- The so-called MHD formulation (Hummer & Mihalas 1988; Mihalas, Däppen & Hummer 1988; Mihalas *et al.* 1990). This provides a detailed treatment of the interactions within the gas, involving a probabilistic description of the level populations.

These formulations are all based on what has been called the *chemical picture* (*e.g.* Däppen 1992), where the constituents of the gas are regarded as atoms, ions and electrons. However, equally detailed descriptions exist in the *physical picture*, where the properties of the gas are described directly in terms of the interactions between fundamental particles, handled through many-body activity expansion. For practical applications the most important example is the Livermore equation of state (*e.g.* Rogers, Swenson & Iglesias 1995), which forms the basis for the OPAL opacities discussed in the following section.

2.2.2 Opacity

Uncertainties in the opacity have substantial effects on the solar models and frequencies. Early inversions for the sound speed (Christensen-Dalsgaard *et al.* 1985) indicated that the the opacity should be increased in the solar interior, relative to the then used tables. Such increases have in fact been found in recent opacity calculations, substantially reducing the discrepancies in the structure of the radiative interior between the models and the Sun. Nonetheless, it is likely that some of the remaining difference may still be caused by errors, at a level of a few per cent, in the opacities. A particular uncertainty concerns the opacity in the solar atmosphere: much of the difference between the observed and computed frequencies can be eliminated through a substantial increase in the low-temperature opacity (*cf.* Section 5.3), although the reality of such large opacities is doubtful.

Of particular importance in the recent revisions of the opacity calculations has been the inclusion of large numbers of spectral lines, which have been found to dominate the opacity in extensive regions of the density-temperature plane. The result has been increases in opacity of up to factors of 2 – 3, although the changes under conditions relevant to the Sun have been somewhat smaller. Two independent calculations, using rather different techniques, have been carried out: by the Opacity Project (OP; *e.g.* Seaton *et al.* 1994); and by the Livermore group (*e.g.* Iglesias *et al.* 1992) resulting in the so-called OPAL table. The results of these two sets of calculations agree to within a few per cent.

The models discussed in this chapter were computed with the OPAL tables; the low-temperature opacities were obtained from Kurucz (1991). To illustrate the sensitivity to the opacity I also consider several cases of modifying artificially the opacity in restricted temperature ranges.

2.2.3 Nuclear reactions

The nuclear reactions responsible for the Sun's energy output are discussed in the chapter by Bahcall. The details of the reaction networks are of crucial importance to the computed neutrino fluxes. In contrast, the oscillation frequencies are relatively insensitive to the reaction parameters. This is to some extent caused by the calibration of the initial composition to obtain the correct luminosity for the model of the present Sun: a change in the reaction parameters is compensated by a change in the composition, leaving the structure of the model largely unchanged. Dziembowski *et al.* (1994) found that a 3 per cent change in the basic $p-p$ reaction rate caused changes in sound speed and density which are barely detectable with the present helioseismic accuracy. They also considered an increase in the ^{3}He+^{3}He rate of a sufficient magnitude to bring the computed capture rate in the ^{37}Cl radiochemical neutrino experiment into agreement with the measurements; interestingly, this resulted in changes in sound speed and density which might well be detectable in the oscillation frequencies, at least if other uncertainties in the structure of the model could be eliminated.

Here I largely employ nuclear parameters from Parker (1986).

2.2.4 Microscopic diffusion and gravitational settling

Microscopic diffusion and gravitational settling are not normally considered part of a "standard model calculation". Indeed, crude estimates (*e.g.* Eddington 1926) suggest that the average diffusive time scale far exceeds the typical lifetime of stars. However, more careful calculations show that the effect is significant in solar evolution (*e.g.* Noerdlinger 1977; Wambsganss 1988), particularly in view of the high precision with which solar models can be tested.

An initial study of the effects of diffusion and settling on solar oscillations was carried out by Demarque & Guenther (1988). Cox, Guzik & Kidman (1989) considered detailed effects on solar models and oscillation frequencies. The dominant effect is the settling of helium out of the convection zone. Solar models with diffusion were also computed by Proffitt & Michaud (1991), while Christensen-Dalsgaard, Proffitt & Thompson (1993) showed that helium settling, causing a steep gradient in the helium abundance at the base of the convection zone, very substantially improved the agreement between the sound speed of the model and the solar sound speed as inferred from helioseismic inversion. In addition, there was a significant effect of the accumulation of helium in the core. These results are discussed in Sections 4.3 and 5.1.

Settling similarly causes the heavy elements to sink, slightly depleting the convection zone and the outer parts of the radiative interior and enriching the core. This mainly affects the structure through a modification of the opacity in the radiative interior. Also, the current photospheric value of Z/X is smaller than the initial value. Proffitt (1994) found that when this effect was taken into account, the resulting sound-speed profile below the convection zone, but excluding the core, was rather similar to the result obtained with helium settling and diffusion alone. It should be noted, however, that different heavy elements diffuse at different rates. With diffusion and settling the relative mixture of the heavy elements therefore varies with position. This causes very substantial complications, which so far have not been dealt with consistently, in the interpolation in the opacity tables.

2.3 Macrophysics

2.3.1 Outer convection zone

There seems little doubt that most of the solar convection zone is very nearly adiabatically stratified: here a minute superadiabaticity is sufficient to drive the convective motion required for the energy transport. The calibration of solar models to the correct radius essentially fixes the structure of this region, and hence the depth of the convection zone (Gough & Weiss 1976). In the uppermost part of the convection zone, however, the density is so low that energy transport requires a substantial superadiabatic gradient. Only here do different treatments of convection result in significant differences in solar structure.

In general, prescriptions for convection include free parameters, such as the mixing-length parameter α_c in mixing-length theory. One such parameter is required for the calibration, controlling the change in specific entropy s from the bottom of the atmosphere (where s is essentially determined by atmospheric structure, as fixed by theoretical or semi-empirical atmospheric models) to the nearly adiabatic interior of the convection zone. This calibration fixes the integral of the superadiabatic gradient $\nabla - \nabla_{\rm ad}$, where $\nabla = {\rm d}\ln T / {\rm d}\ln p$ and $\nabla_{\rm ad}$ is its adiabatic value. However, the detailed behaviour of $\nabla - \nabla_{\rm ad}$ differs between different treatments of convection. Canuto & Mazzitelli (1991, 1992) developed descriptions based on assuming a full turbulent spectrum, and using as mixing length the distance to the top of the convection zone; this formulation resulted in a substantially higher and sharper $\nabla - \nabla_{\rm ad}$ in the superadiabatic region than for traditional mixing-length models. Lydon *et al.* (1992) based their treatment on correlations obtained from hydrodynamical simulations of turbulent convection, finding $\nabla - \nabla_{\rm ad}$ similar to, although slightly lower and broader than, the results of mixing-length theory. In this treatment, as well as the in the work of Canuto & Mazzitelli, adjustable free parameters were not explicitly used. Monteiro, Christensen-Dalsgaard & Thompson (1995ab) made an extensive survey of various types of convection formulations and their effects on the computed oscillation frequencies.

These convection treatments are all *local*, in that they assume that the convective flux at a given point is determined by the conditions, including the superadiabatic gradient, at that point. In reality convective eddies must sample a range of positions within the Sun, and the energy transport is determined by eddies originating over a range of levels. Procedures for taking such effects into account were developed by Spiegel (1963) and Gough (1976), although the application to solar modelling has been slow. Exceptions are the work by Xiong & Chen (1992) and by Balmforth (1992) who in addition considered effects of a time-dependent formulation of non-local mixing-length theory on solar oscillations.

In the strongly superadiabatic region the convective velocities reach a substantial fraction of the sound speed. As a result, the convective momentum transport makes a significant contribution to hydrostatic balance in the average model. Nonetheless, this so-called turbulent pressure is usually ignored in computations of solar models. Indeed, a consistent treatment of turbulent pressure in local mixing-length models is plagued by mathematical difficulties at the boundaries of the convectively unstable region (Gough 1977b). These problems are avoided in non-local treatments where there are no such sharp boundaries.

The near-surface part of the convection zone has also been modelled through detailed hydrodynamical simulations (*e.g.* Stein & Nordlund 1989). These show a strong asymmetry of the flow: rapid radiative cooling at the surface causes strong narrow cold downdrafts which persist throughout the computational domain, with slow upwelling in between. From such simulations mean models can be constructed through horizontal and temporal averaging. These models confirm that the stratification becomes nearly adiabatic at depth, with an adiabat which is relatively close to that obtained in calibrated mixing-length models. Furthermore, the models clearly demonstrate the importance of turbulent pressure.

Conditions at the base of the convection zone remain uncertain. It seems inevitable that motion extends beyond the unstable region. Simple models (Shaviv & Salpeter 1973; Schmitt, Rosner & Bohn 1984; Pidatella & Stix 1986; Zahn 1991) predict a nearly adiabatic extension of the convection zone with relatively vigorous motion, followed by an abrupt transition to the radiative gradient. As discussed in more detail in Section 4.5 the resulting near-discontinuity in the sound-speed gradient might have observable effects on the oscillation frequencies.

2.3.2 Mixing in the solar interior?

A potentially serious uncertainty in solar modelling concerns mixing caused by material motion in the solar interior. It is plausible that penetration beyond the convection zone induces weaker motion, possibly in the form of internal gravity waves, in the radiative interior. Such motion could have substantial effects on the composition profile resulting from microscopic diffusion and gravitational settling. Indeed, the reduction of the solar photospheric lithium abundance by roughly a factor 100 relative to the solar-system abundance indicates that mixing well beneath the convection zone must have occurred at some phase of solar evolution (*e.g.* Christensen-Dalsgaard, Gough & Thompson 1992); lithium depletion in the pre-main-sequence phase may also have been an important factor (*e.g.* Ahrens, Stix & Thorn 1992; Swenson *et al.* 1994). Schatzman *et al.* (1981) considered the effects on solar evolution of substantial turbulent diffusion, finding that this might lead to a significant reduction in the predicted neutrino flux; the physical basis for the assumed diffusion coefficient was somewhat questionable, however.

The solar spin-down from a normally assumed initial rapid rotation is likely to have caused regions of strong shear in the solar radiative interior which could have led to instabilities and hence mixing. In an ambitious effort, Pinsonneault *et al.* (1989) have computed models of such effects, with approximate expressions for the transport resulting from the instabilities. The results showed substantial effects on the photospheric lithium abundance, reproducing the observed depletion, while the effect on the central hydrogen abundance was small. Detailed analysis of models including rotational instabilities indicated no significant changes in the oscillation frequencies (Chaboyer *et al.* 1995). It should be noted, however, that the computed rotation profiles seem to be inconsistent with helioseismic determinations of solar internal rotation (*e.g.* Duvall *et al.* 1984; Tomczyk, Schou & Thompson 1995). Zahn (1992) developed a procedure to evaluate mixing due to turbulence induced by meridional circulation. Detailed results for the Sun have apparently so far not been obtained, however.

2.4 Structure of the solar convection zone

The fact that the bulk of the convection zone is nearly adiabatically stratified considerably simplifies the analysis of solar oscillation data. Here to a very good approximation pressure p and density ρ are related by

$$\frac{\mathrm{d}\ln\rho}{\mathrm{d}\ln p} \simeq \frac{1}{\Gamma_1} , \tag{1}$$

where $\Gamma_1 = (\partial \ln p / \partial \ln \rho)_{\mathrm{ad}}$. As a result, the structure of the convection zone is independent of opacity: it is determined by the value of the specific entropy s, by the composition (uniform, because of the very efficient mixing) and the equation of state. Consequently, modes of oscillation that are trapped in the convection zone are ideally suited to test the properties of the equation of state and to determine the solar composition (see also Christensen-Dalsgaard & Däppen 1992).

From equation (1) we may easily obtain an approximate expression for the adiabatic sound speed c. Introducing $u = p/\rho$ we have from the equation for hydrostatic support that

$$\frac{\mathrm{d}u}{\mathrm{d}r} = \frac{1}{\rho}\frac{\mathrm{d}p}{\mathrm{d}r}\left(1 - \frac{\mathrm{d}\ln\rho}{\mathrm{d}\ln p}\right) \simeq -\frac{Gm}{r^2}\left(1 - \frac{1}{\Gamma_1}\right), \tag{2}$$

where G is the gravitational constant and m is the mass interior to radius r. In equation (2) we may take $m \simeq M$, the total mass of the Sun, since the convection zone contains only about 2 per cent of the solar mass. Outside the dominant ionization zones of hydrogen and helium, which are confined to the outer 2 – 3 per cent of the radius, we can furthermore assume Γ_1 to be constant. Within this rather crude approximation we can therefore integrate equation (2), to obtain

$$u \simeq GM\left(1 - \frac{1}{\Gamma_1}\right)\left(\frac{1}{r} - \frac{1}{R^*}\right), \tag{3}$$

or

$$c^2 = \Gamma_1 u \simeq GM(\Gamma_1 - 1)\left(\frac{1}{r} - \frac{1}{R^*}\right); \tag{4}$$

here R^*, which serves as a constant of integration, is approximately equal to the photospheric radius R of the star. Thus in the deeper parts of the convection zone $c(r)$ is essentially determined by the total mass and radius of the Sun, as well as by (a suitable average of) Γ_1. It should also be noted that equation (2) relates the derivative of u, which can be determined from the observed oscillation frequencies, to the thermodynamic properties of the gas as described by Γ_1. This relation may therefore be used to determine the helium abundance of the convection zone (Däppen & Gough 1984, 1986) or to test the equation of state (Dziembowski, Pamyatnykh & Sienkiewicz 1992).

2.5 Overview of solar structure

Figure 1 provides a summary of the physics and uncertainties of solar internal structure. In the radiative interior the situation is complicated by the simultaneous dependence on the equation of state, opacity, energy generation rate and composition profile. On the other hand, since matter is essentially fully ionized in this region the uncertainties introduced by the equation of state are substantially reduced. Also, I indicated in Section 2.2.3 that the details of nuclear reactions have modest effects on the oscillation frequencies. Thus in this part of the Sun we expect that the oscillations will be used predominantly to measure aspects of the opacity and the composition profile. As argued in Section 2.4 the convection zone, on the other hand, depends on composition, specific entropy and the equation of state. Furthermore, the composition can be assumed to be independent of position in the convection zone; this offers some hope that uncertainties in the composition and the equation of state can be partially separated

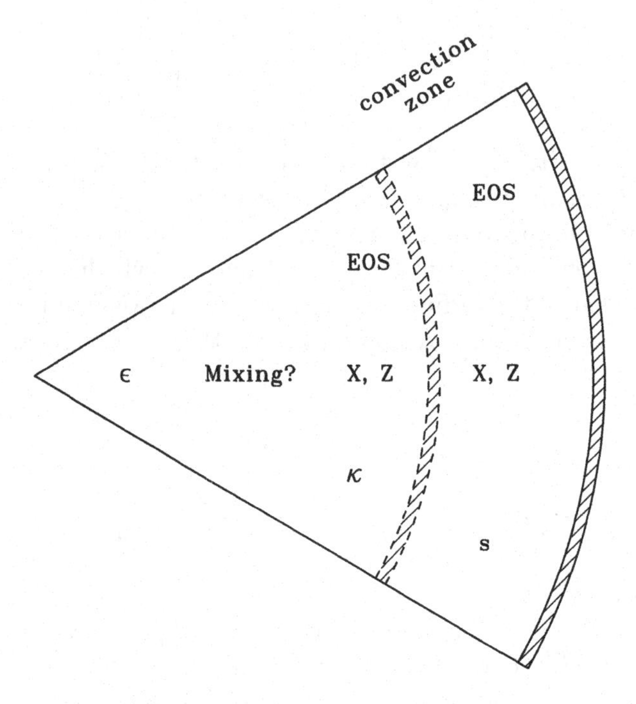

Figure 1: *Schematic representation of solar structure. The thin hashed area near the surface indicates the region where the physics is uncertain, because of effects of convection, nonadiabaticity, etc. At the base of the convection zone, convective overshoot and diffusion introduce additional uncertainty. The structure of the adiabatic part of the convection zone is determined by the equation of state (EOS), and the constant values of specific entropy s, and composition (given by the abundances X and Z of hydrogen and heavy elements). Beneath the convection zone the structure also depends on opacity κ and the energy generation rate ϵ.*

(Gough 1984a). It is obvious, however, that one cannot exclude errors in the equation of state which essentially mimic a change in composition.

Much of the uncertainty in the physics is concentrated very near the surface. This is true of the dynamical effects of convection, since convective velocities are likely to be very small elsewhere, and of the details of convective energy transport; furthermore, unless the interior field is much stronger than the observed photospheric field, effects of magnetic fields on the average solar structure are similarly concentrated. Also, potential errors in the treatment of the oscillations, associated with nonadiabatic effects and fluctuations in the turbulent pressure, are concentrated in this region; in contrast, the oscillations are expected to be adiabatic to very high precision in the solar interior. These near-surface uncertainties have a substantial effect on the oscillation frequencies which may influence the analysis of the observations. Fortunately, as argued in Section 3.4.4 it is to a large extent possible to separate the effects of the superficial errors, hence isolating those aspects of the frequencies which provide reliable information about the solar interior.

3 PROPERTIES OF ADIABATIC OSCILLATIONS

3.1 Introduction

For general properties of solar oscillations reference may be made, for example, to Christensen-Dalsgaard & Berthomieu (1991). However, for completeness the present chapter contains certain basic definitions. Also, in the following analysis of the sensitivity of the oscillation frequencies to aspects of structure we shall need insight on several specific points. Much of the material presented here has been adapted from Christensen-Dalsgaard (1994).

Although claims for detection of g modes have been made, the existing identified observations of solar oscillation are confined to p and f modes. Thus I shall restrict myself to these modes. Furthermore, I generally assume the oscillations to be adiabatic. In fact, nonadiabatic effects on the frequencies are important very near the surface; however, the physical nature of these effects is still badly understood, and so they must at present be regarded as a source of uncertainty in the analysis of the observed frequencies.

As usual, the modes are described in terms of spherical harmonics Y_l^m; the *degree* l is related to the the length $k_{\rm h}$ of the horizontal component of the wavenumber, at distance r from the centre, by

$$k_{\rm h} = \frac{L}{r} , \qquad \text{with} \quad L^2 = l(l+1) . \tag{5}$$

We neglect rotation and other departures from spherical symmetry. Then the angular frequencies ω_{nl} depend only on the degree and the radial order n of the modes. In addition to ω_{nl}, we shall often consider the cyclic frequencies $\nu_{nl} = \omega_{nl}/2\pi$.

The frequencies of a solar model are complicated functions of the structure of the model. However, all available evidence indicates that our present solar models are fairly close to the actual solar structure. This motivates analyzing the observed frequencies in terms of *departures* from the frequencies of some reference model. If the differences between the Sun and the model are sufficiently small, there is an approximately linear relation between the model and the frequency differences (see also Section 3.4). Similarly, the effects of changing some aspect of the model physics is conveniently described in terms of the corresponding model and frequency differences. Several examples of this are given in Section 4.

Computation of adiabatic frequencies for a given model is relatively straightforward. However, the results are not immediately easy to interpret. Insight into the relation between the frequencies and the structure of the model can be obtained from the asymptotic properties of the modes. Most of the observed frequencies correspond to acoustic modes of relatively high radial order or degree, for which the asymptotic description is fairly accurate. Although the asymptotic theory cannot replace precise numerical computation, it may still be used to obtain quantitative results. Examples of this are given in Section 5.1.

The f-mode frequencies depend only weakly on solar structure. To leading order, they satisfy

$$\omega^2 \simeq g_{\rm s} k_{\rm h} \, , \tag{6}$$

where $g_{\rm s}$ is the surface gravitational acceleration. To this accuracy the information content in the f-mode frequencies is clearly limited. The observed frequencies show departures from equation (6) which are substantially larger than the corresponding departures for computed frequencies for normal solar models; the origin of these departures is so far uncertain (*e.g.* Rosenthal & Gough 1994; Rosenthal & Christensen-Dalsgaard 1995; Rosenthal *et al.* 1995a). Thus in the following I shall predominantly consider p modes.

3.2 What do the oscillation frequencies depend on?

The adiabatic oscillation equations obviously depend on the structure of the equilibrium model. A closer inspection reveals that the coefficients are determined solely by the set of equilibrium variables

$$\rho, \; p, \; \Gamma_1, \; g \, , \tag{7}$$

as functions of r. However, the equilibrium model satisfies the stellar structure equations; in addition it may be assumed to have a given mass and radius, which at least in the case of the Sun are known with high precision. If $\rho(r)$ is given, the interior mass and hence $g(r)$ can be determined from simple integration; given g, the equation of hydrostatic support may be integrated from the surface to provide $p(r)$ (the surface pressure is known from empirical models of the solar atmosphere). Thus of the set (7) only the two functions $\rho(r)$ and $\Gamma_1(r)$ are independent, and the adiabatic oscillation frequencies are determined solely by these two functions. Conversely, if no other constraints are imposed, the observed frequencies give direct independent information only about ρ and Γ_1; from ρ, p may then be determined from the constraint of hydrostatic support.

The equation of state relates Γ_1 to p, ρ and the chemical composition. To a fair degree of approximation the chemical composition can be specified by a single parameter, such as the abundance Y by mass of helium. In this approximation just three quantities should suffice to specify fully the thermodynamic state. Given that ρ, p and Γ_1 can be obtained from the oscillation observations, knowledge of the equation of state should enable the determination of any other thermodynamic variable, including Y, from these observed quantities. Outside the major ionization zones Γ_1 is very nearly constant and hence gives a poor determination of the thermodynamic state; however, it varies sufficiently in the helium ionization zones to allow a determination of Y, provided that the properties of the equation of state are known with sufficient accuracy. Indeed, procedures for such a determination of the helium abundance of the Sun have been proposed (*e.g.* Däppen & Gough 1984, 1986; Vorontsov, Baturin & Pamyatnykh 1991; Kosovichev *et al.* 1992). Alternatively, such analyses provide a test of the equation of state, if the helium abundance is otherwise constrained. As discussed in Section 5.2, an apparently quite efficient technique can be obtained on the basis of the asymptotic behaviour of the oscillations in the helium ionization zone.

The preceding discussion was made in terms of the pair (ρ, Γ_1). However, any other independent pair of model variables, related directly to ρ and Γ_1, may be used instead. Since most of the observed solar oscillations have essentially the nature of standing acoustic waves, their frequencies are largely determined by the behaviour of sound speed c; hence it is natural to use c as one of the variables, combined with, *e.g.*, ρ or Γ_1. As discussed in Section 5.1 below, the observations are sufficiently rich that the observed frequencies may be inverted to obtain an estimate of the sound speed in most of the Sun. It follows from the equation of state, approximated by the ideal gas law, that this essentially provides a measure of T/μ, T being temperature and μ the mean molecular weight. However, it is important to note that measurements of adiabatic oscillation frequencies do not by themselves allow a determination of the temperature in a star. Only if the mean molecular weight can be otherwise constrained (*e.g.* by demanding that its variation in the stellar interior results from normal stellar evolution) is it possible to estimate the stellar interior temperature. This limitation is of obvious importance for the use of observed solar oscillation frequencies to throw a light on the apparent deficit of observed solar neutrinos (*e.g.* Christensen-Dalsgaard 1991).

3.3 Simple p-mode asymptotics

Much insight into the information content in solar oscillation frequencies can be obtained from the asymptotic properties the modes. Here I present a few simple aspects of the asymptotic description, for use later in the present chapter.

3.3.1 Simple derivation of Duvall law

A relation central to the understanding of p-mode behaviour is the *Duvall law*. This can be derived very simply from the dispersion relation for plane sound waves:

$$\omega^2 = c^2 |\mathbf{k}|^2 , \tag{8}$$

where $\mathbf{k}$ is the wave number. I write $|\mathbf{k}|^2 = k_r^2 + k_\mathrm{h}^2$, where k_h is given by equation (5). Thus

$$k_r^2 = \frac{\omega^2}{c^2} - \frac{L^2}{r^2} . \tag{9}$$

This equation describes the geometry of the ray along which the wave propagates; propagation, with real k_r, is confined to the region outside the *turning point* $r = r_\mathrm{t}$, where r_t satisfies

$$\frac{c(r_\mathrm{t})}{r_\mathrm{t}} = \frac{\omega}{L} . \tag{10}$$

A more careful analysis (briefly summarized in Section 3.3.5) shows that the waves have an outer turning point R_t located just below the photosphere. To obtain a standing wave (*i.e.*, a mode of oscillation) we must require, roughly, an integral number of oscillations in the radial direction between the lower turning point and the photosphere, although taking into account the phase shifts at the extremes of the propagating region; thus

$$\int_{r_\mathrm{t}}^{R} k_r \mathrm{d}r = n(\pi + \alpha) , \tag{11}$$

where α takes care of the behaviour near the turning points. From equation (9) we therefore obtain

$$\int_{r_\mathrm{t}}^{R} \left(1 - \frac{L^2 c^2}{\omega^2 r^2}\right)^{1/2} \frac{\mathrm{d}r}{c} = \frac{[n + \alpha(\omega)]\pi}{\omega} , \tag{12}$$

which is the Duvall law.

It might be noted that analysis of the asymptotics of low-degree modes indicates that in the asymptotic relations L should be defined as

$$L = l + 1/2 , \tag{13}$$

instead of the definition given in equation (5). (Note that the distinction between these two definitions is only significant for low-degree modes.) In the following applications of the asymptotic theory equation (13) will be used.

3.3.2 Effects of changing the dispersion relation

To investigate the effects of changes to the model in the asymptotic approximation, we replace equation (8) by

$$\omega^2 = c^2|\mathbf{k}|^2 + \delta f(r) \,, \tag{14}$$

where δf is a small modification which, as indicated, is assumed to be given as a function of r. Instead of equation (9) we therefore obtain, to leading order in δf,

$$\begin{aligned} k_r &= \left(\frac{\omega^2}{c^2} - \frac{L^2}{r^2} - \frac{1}{c^2}\delta f\right)^{1/2} \\ &\simeq \frac{\omega}{c}\left[\left(1 - \frac{L^2c^2}{\omega^2 r^2}\right)^{1/2} - \frac{1}{2\omega^2}\left(1 - \frac{L^2c^2}{\omega^2 r^2}\right)^{-1/2}\delta f\right] . \end{aligned} \tag{15}$$

By substituting equation (15) into the condition (11) for a standing wave we obtain

$$\frac{(n+\alpha)\pi}{\omega} \simeq \int_{r_t}^{R}\left(1 - \frac{L^2c^2}{\omega^2 r^2}\right)^{1/2}\frac{\mathrm{d}r}{c} - \frac{1}{2\omega^2}\int_{r_t}^{R}\left(1 - \frac{L^2c^2}{\omega^2 r^2}\right)^{-1/2}\delta f\frac{\mathrm{d}r}{c} . \tag{16}$$

Here the last term shows the effect of the perturbation on the Duvall law.

We can now find the effect on the oscillation frequencies of the perturbation. We assume that the result is to change the frequency from ω to $\omega + \delta\omega$. Also it should be recalled that $\alpha = \alpha(\omega)$ in general depends on ω. Finally, we note that the perturbation, in addition to δf, may involve a change $\delta\alpha$ to the phase function α. Multiplying equation (16) by ω and linearizing in $\delta\omega$ and $\delta\alpha$ yields

$$\begin{aligned} \pi\frac{\mathrm{d}\alpha}{\mathrm{d}\omega}\delta\omega &= \delta\omega\int_{r_t}^{R}\left(1 - \frac{L^2c^2}{\omega^2 r^2}\right)^{1/2}\frac{\mathrm{d}r}{c} + \omega\int_{r_t}^{R}\left(1 - \frac{L^2c^2}{\omega^2 r^2}\right)^{-1/2}\frac{L^2c^2}{\omega^2 r^2}\frac{\delta\omega}{\omega}\frac{\mathrm{d}r}{c} \\ &\quad - \frac{1}{2\omega}\int_{r_t}^{R}\left(1 - \frac{L^2c^2}{\omega^2 r^2}\right)^{-1/2}\delta f\frac{\mathrm{d}r}{c} - \pi\delta\alpha . \end{aligned} \tag{17}$$

From this we finally obtain

$$S\frac{\delta\omega}{\omega} \simeq \frac{1}{2\omega^2}\int_{r_t}^{R}\left(1 - \frac{L^2c^2}{\omega^2 r^2}\right)^{-1/2}\delta f\frac{\mathrm{d}r}{c} + \pi\frac{\delta\alpha}{\omega} , \tag{18}$$

where

$$S = \int_{r_t}^{R} \left(1 - \frac{L^2 c^2}{\omega^2 r^2}\right)^{-1/2} \frac{\mathrm{d}r}{c} - \pi \frac{\mathrm{d}\alpha}{\mathrm{d}\omega} . \tag{19}$$

This is the desired general expression.

If the terms in $\mathrm{d}\alpha/\mathrm{d}\omega$ and $\delta\alpha$ are neglected, equation (18) has a very simple physical interpretation: the equation shows that the relative change in ω^2 is just a weighted average of $\delta f/\omega^2$, with the weight function

$$\mathcal{W}(r) = \frac{1}{c} \left(1 - \frac{L^2 c^2}{\omega^2 r^2}\right)^{-1/2} . \tag{20}$$

It is easily seen that $\mathcal{W}(r)\mathrm{d}r$ is just the sound travel time, corresponding to the radial distance $\mathrm{d}r$, along the ray describing the mode. Hence the weight in the average simply gives the time that the mode, regarded as a superposition of plane waves, spends in a given region of the star.

3.3.3 Asymptotic effects of changes in structure

To investigate the effects of modifications in solar structure, we consider a change in the sound speed from c to $c + \delta c$; as a result, the dispersion relation for sound waves is changed to

$$\omega^2 = c^2 |\mathbf{k}|^2 + 2c\delta c |\mathbf{k}|^2 = c^2 |\mathbf{k}|^2 + 2\omega^2 \frac{\delta c}{c} . \tag{21}$$

This is of the form given in equation (14), with $\delta f = 2\omega^2 \delta c/c$. It follows from equation (18) that the frequency change is given by

$$S \frac{\delta\omega}{\omega} \simeq \int_{r_t}^{R} \left(1 - \frac{L^2 c^2}{\omega^2 r^2}\right)^{-1/2} \frac{\delta c}{c} \frac{\mathrm{d}r}{c} + \pi \frac{\delta\alpha}{\omega} . \tag{22}$$

Note that equation (22) is of the form

$$S \frac{\delta\omega}{\omega} \simeq \mathcal{H}_1 \left(\frac{\omega}{L}\right) + \mathcal{H}_2(\omega) , \tag{23}$$

where

$$\mathcal{H}_1(w) = \int_{r_t}^{R} \left(1 - \frac{c^2}{r^2 w^2}\right)^{-1/2} \frac{\delta c}{c} \frac{dr}{c} , \tag{24}$$

and

$$\mathcal{H}_2(\omega) = \frac{\pi}{\omega} \delta\alpha(\omega) . \tag{25}$$

Some properties of this equation were discussed by Christensen-Dalsgaard, Gough & Pérez Hernández (1988) and Christensen-Dalsgaard, Gough & Thompson (1989). As pointed out in the latter reference, $\mathcal{H}_1(\omega/L)$ and $\mathcal{H}_2(\omega)$ can be obtained separately, to within a constant, by means of a double-spline fit of the expression (23) to p-mode frequency differences. The dependence of $\mathcal{H}_1$ on ω/L is determined by the sound-speed difference throughout the star, whereas $\mathcal{H}_2(\omega)$ depends on differences in the upper layers of the models.

3.3.4 Effect of perturbation in gravitational potential

The derivation of the Duvall law implicitly made the Cowling approximation, neglecting the perturbation Φ' in the gravitational potential. It is very simple to estimate the effect on the frequencies of including Φ'. For plane acoustic waves, including the effect of self-gravity, the dispersion relation is

$$\omega^2 = c^2 \, |\mathbf{k}|^2 - 4\pi G\rho \tag{26}$$

(Jeans 1929). This is again of the form (14), with $\delta f = -4\pi G\rho$. It follows from equation (16) that the modified Duvall law is

$$\frac{\pi(n+\alpha)}{\omega} \simeq \int_{r_t}^{R} \left(1 - \frac{L^2c^2}{\omega^2 r^2}\right)^{1/2} \frac{\mathrm{d}r}{c} + \frac{2\pi G}{\omega^2} \int_{r_t}^{R} \rho \left(1 - \frac{L^2c^2}{\omega^2 r^2}\right)^{-1/2} \frac{\mathrm{d}r}{c} . \tag{27}$$

Note that, as r_t is a function of ω/L, equation (27) may be written as

$$\frac{\pi(n+\alpha)}{\omega} = F\left(\frac{\omega}{L}\right) + \frac{1}{\omega^2} F_\Phi\left(\frac{\omega}{L}\right) , \tag{28}$$

where the functions $F(w)$ and $F_\Phi(w)$ are defined by equation (27). An expression of this form, with additional correction terms, was obtained by Vorontsov (1991).

From equations (18) and (19) we obtain an approximate expression for the difference $\delta\omega^{(\Phi)} = \omega^{(F)} - \omega^{(C)}$ between the frequency $\omega^{(F)}$ obtained taking the perturbation in the gravitational potential into account, and the frequency $\omega^{(C)}$ obtained in the Cowling approximation. The result is

$$\delta\omega^{(\Phi)} \simeq -\frac{1}{\omega} \frac{2\pi G \displaystyle\int_{r_t}^{R} \rho \left(1 - \frac{L^2c^2}{\omega^2 r^2}\right)^{-1/2} \frac{\mathrm{d}r}{c}}{\displaystyle\int_{r_t}^{R} \left(1 - \frac{L^2c^2}{\omega^2 r^2}\right)^{-1/2} \frac{\mathrm{d}r}{c}} . \tag{29}$$

Thus the frequency change induced by the gravitational potential perturbation depends on an average of the density structure of the equilibrium model, over the region where the mode is trapped.

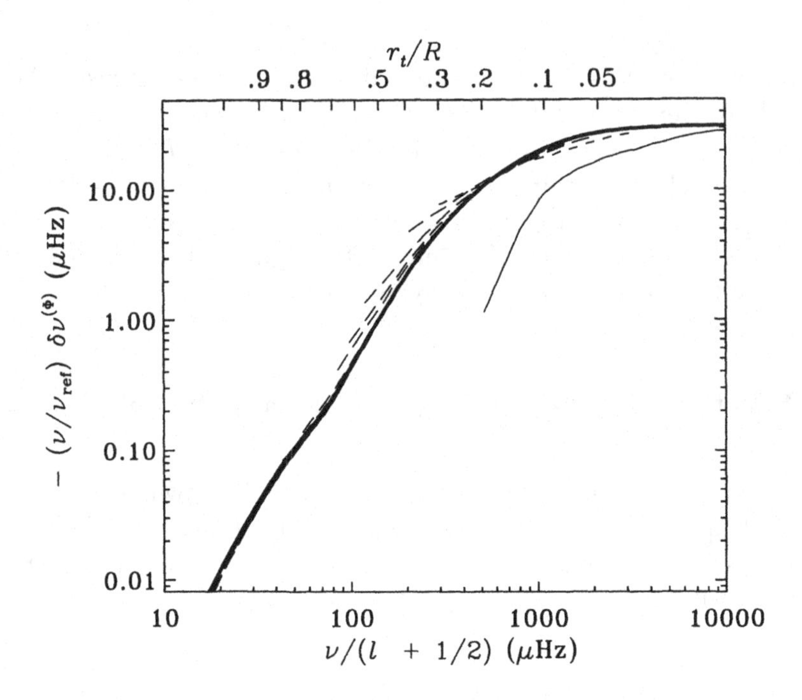

Figure 2: *The thin lines show scaled frequency corrections $\delta\nu^{(\Phi)}$ resulting from the inclusion of the effect of the perturbation in the gravitational potential, for a normal solar model. Points corresponding to a given degree have been connected with a continuous line ($l = 0$), short dashes ($l = 1$) or long dashes ($l > 1$). In accordance with the asymptotic expression (29) $\delta\nu^{(\Phi)}$ has been scaled by $-\nu/\nu_{\rm ref}$, using a reference frequency $\nu_{\rm ref} = 3000\,\mu{\rm Hz}$, and the results have been plotted against $\nu/(l+1/2)$ (lower abscissa) and r_t/R [cf. equation (10); upper abscissa]. The heavy continuous line corresponds to the asymptotic frequency difference $\delta\nu^{(\Phi)} = \delta\omega^{(\Phi)}/2\pi$ computed from equation (29).*

It might be noticed that equation (29) directly relates the frequencies to the density structure of the model. However, the effect is evidently strongly diminished for modes of moderate or high degree, which do not penetrate into the high-density region near the core. This is illustrated in Figure 2 which shows the asymptotic and computed frequency corrections resulting from including the perturbation in the gravitational potential.

3.3.5 Refinements of asymptotic theory

The preceding asymptotic analysis was highly simplified, although sufficient for much of the following. For completeness, I here note some extensions.

It was shown by Gough (*cf.* Deubner & Gough 1984) that a more accurate asymptotic representation, valid both for p and g modes, is

$$\omega \int_{r_1}^{r_2} \left[1 - \frac{\omega_c^2}{\omega^2} - \frac{L^2 c^2}{r^2 \omega^2} \left(1 - \frac{N^2}{\omega^2} \right) \right]^{1/2} \frac{\mathrm{d}r}{c} \simeq \pi(n - 1/2) , \tag{30}$$

where r_1 and r_2 are adjacent zeros of the integrand. Here the characteristic *acoustical cut-off frequency* ω_c is defined by

$$\omega_c^2 = \frac{c^2}{4H^2} \left(1 - 2\frac{\mathrm{d}H}{\mathrm{d}r} \right) , \tag{31}$$

where we have introduced the density scale height H by

$$H^{-1} = -\frac{\mathrm{d}\ln\rho}{\mathrm{d}r} ; \tag{32}$$

also, N is the buoyancy frequency. Near the surface, the term in ω_c dominates and defines the point $r = R_t$, with $\omega_c(R_t) \simeq \omega$, where the mode is reflected: this occurs essentially at the photosphere although for frequencies below about $2000\,\mu$Hz the reflection point shifts deeper with decreasing frequency. Waves with frequencies exceeding the (nearly constant) value of ω_c in the atmosphere (corresponding to a cyclic frequency $\nu_{ac} \simeq 5300\,\mu$Hz) are essentially free to travel out into the solar atmosphere, and hence do not form normal modes (see, however, the chapter by Brown).

One may obtain the original form of the Duvall law, equation (12), from equation (30) by neglecting the term in N^2 and expanding the bracket in the integral on the left-hand side to take out the dependence on ω_c^2 (*e.g.* Christensen-Dalsgaard & Pérez Hernández 1992); the contribution from ω_c^2 is essentially absorbed in $\alpha(\omega)$. A more careful analysis (Brodsky & Vorontsov 1993; Gough & Vorontsov 1995) results in a further refinement of the Duvall law. The result may be written as

$$F\left(\frac{\omega}{L}\right) + \frac{1}{\omega^2} F_2\left(\frac{\omega}{L}\right) = \pi\omega^{-1} \left[n + \alpha(\omega) + \left(\frac{L}{\omega}\right)^2 \alpha_2(\omega) \right] . \tag{33}$$

Here $F(w)$ is defined by equations (27) and (28), while $F_2(w)$ is analogous to $F_\Phi(w)$ introduced in those equations, but includes also a contribution from N^2 as well as other terms. The term in $\alpha_2(\omega)$ on the right-hand side arises because the waves propagate obliquely at the upper turning point. It becomes of substantial importance for modes whose degree exceeds a few hundred.

3.3.6 Asymptotics of low-degree modes

For low-degree modes one may show, by expanding the Duvall law, that

$$\nu_{nl} \simeq (n + \frac{l}{2} + \frac{1}{4} + \alpha)\Delta\nu - (AL^2 - \delta)\frac{\Delta\nu^2}{\nu_{nl}} , \tag{34}$$

where

$$\Delta\nu = \left[2 \int_0^R \frac{\mathrm{d}r}{c} \right]^{-1} \tag{35}$$

is the inverse of twice the sound travel time between the centre and the photosphere, and

$$A = \frac{1}{4\pi^2 \Delta\nu} \left[\frac{c(R)}{R} - \int_0^R \frac{\mathrm{d}c}{\mathrm{d}r} \frac{\mathrm{d}r}{r} \right] \tag{36}$$

(*e.g.* Tassoul 1980; Gough 1986). Neglecting the term in A, equation (34) predicts a uniform spacing $\Delta\nu$ in n of the frequencies of low-degree modes. Also, modes with the same value of $n + l/2$ are predicted to be almost degenerate,

$$\nu_{nl} \simeq \nu_{n-1,l+2} \ . \tag{37}$$

This frequency pattern has been observed for the solar five-minute modes of low degree and may be used in the search for stellar oscillations of solar type. The *deviations* from the simple relation (37) have considerable diagnostic potential. They may be expressed in terms of

$$\delta_{nl} \equiv \nu_{nl} - \nu_{n-1,l+2} \simeq -(4l + 6) \frac{\Delta\nu}{4\pi^2 \nu_{nl}} \int_0^R \frac{\mathrm{d}c}{\mathrm{d}r} \frac{\mathrm{d}r}{r} \ , \tag{38}$$

where we neglected the term in the photospheric sound speed $c(R)$. This equation indicates that δ_{nl} is predominantly determined by conditions in the solar core. It should be noted, however, that the accuracy of equation (38) is questionable: it appears to agree fortuitously with frequencies computed for models of the present Sun, whereas it is less successful for models of different ages or masses (Gabriel 1989; Christensen-Dalsgaard 1991). In fact, the normal asymptotic approximation breaks down in the core where conditions vary rapidly, and the derivation of equation (34) neglected the perturbation in the gravitational potential which, as we have seen, is important for low-degree modes. Nonetheless it remains true that δ_{nl} is a useful diagnostics of the structure of the core.

3.4 Perturbation analysis of effects of model changes

So far I have considered effects of solar structure only in terms of the asymptotic description of acoustic modes. To obtain more precise results, which in addition are valid for f and g modes, we must return to the original equations of adiabatic oscillations. In this way we obtain relations, precisely valid in the linear approximation, between changes in the structure and the corresponding frequency changes. In addition, the analysis provides further insight into aspects of the oscillations.

3.4.1 Reformulation of the oscillation equations

A great deal of insight into the properties of adiabatic oscillations can be obtained by regarding the equations as an eigenvalue problem in a Hilbert space (Eisenfeld 1969; Dyson & Schutz 1979; Christensen-Dalsgaard 1981). The starting point is the perturbed equations of motion (see, for example, Christensen-Dalsgaard & Berthomieu 1991). After separation of the time dependence as $\exp(-i\omega t)$, these can be written as

$$\omega^2 \boldsymbol{\delta}\mathbf{r} = \mathcal{F}(\boldsymbol{\delta}\mathbf{r}) , \tag{39}$$

where

$$\mathcal{F}(\boldsymbol{\delta}\mathbf{r}) = \frac{1}{\rho}\nabla p' - \nabla\Phi' - \frac{\rho'}{\rho}\mathbf{g} . \tag{40}$$

Here $\boldsymbol{\delta}\mathbf{r}$ is the displacement vector, p', ρ' and Φ' are Eulerian perturbations of pressure, density and gravitational potential and $\mathbf{g}$ is the equilibrium gravitational acceleration. As indicated, $\mathcal{F}$ is a linear functional of $\boldsymbol{\delta}\mathbf{r}$. To see this, note that from the continuity equation,

$$\rho' + \operatorname{div}(\rho\boldsymbol{\delta}\mathbf{r}) = 0 , \tag{41}$$

ρ' is a linear functional of $\boldsymbol{\delta}\mathbf{r}$. The gravitational potential perturbation Φ' may then be obtained by integrating the perturbed Poisson's equation. In the adiabatic case p' can be obtained directly from ρ' and $\boldsymbol{\delta}\mathbf{r}$. This defines *the adiabatic operator* $\mathcal{F}_a$. The nonadiabatic case is more complicated, but here also it is possible to obtain p' as a linear functional of $\boldsymbol{\delta}\mathbf{r}$ (see Christensen-Dalsgaard 1981).

I now introduce a space $\mathcal{H}$ of vector functions of position in the star, with suitable regularity properties, and define an inner product on $\mathcal{H}$ by

$$< \boldsymbol{\xi}, \boldsymbol{\eta} > = \int_V \rho \boldsymbol{\xi}^* \cdot \boldsymbol{\eta} \, \mathrm{d}V , \tag{42}$$

for $\boldsymbol{\xi}, \boldsymbol{\eta}$ in $\mathcal{H}$; here "*" denotes the complex conjugate, and the integration is over the volume V of the star. I also introduce the domain $\mathcal{D}(\mathcal{F})$ of the operator $\mathcal{F}$ as those vectors in $\mathcal{H}$ satisfying the surface boundary condition that the Lagrangian pressure perturbation vanish. The central result is now that, as shown by Lynden-Bell & Ostriker (1967), the operator $\mathcal{F}_a$ corresponding to equation (40) for adiabatic oscillations is symmetric, in the sense that

$$< \boldsymbol{\xi}, \mathcal{F}_a(\boldsymbol{\eta}) > = < \mathcal{F}_a(\boldsymbol{\xi}), \boldsymbol{\eta} > , \qquad \text{for } \boldsymbol{\xi}, \boldsymbol{\eta} \in \mathcal{D}(\mathcal{F}) . \tag{43}$$

From equation (43) a number of useful properties of $\mathcal{F}_a$ follow immediately. The simplest result is that the squared eigenfrequencies are real. I introduce the functional Σ on $\mathcal{D}(\mathcal{F})$ by

$$\Sigma(\boldsymbol{\xi}) = \frac{< \boldsymbol{\xi}, \mathcal{F}_a(\boldsymbol{\xi}) >}{< \boldsymbol{\xi}, \boldsymbol{\xi} >} ; \tag{44}$$

it follows from equation (43) that $\Sigma(\boldsymbol{\xi})$ is real. If ω_0^2 is an eigenvalue of the problem with eigenvector $\boldsymbol{\xi}_0$, *i.e.*,

$$\mathcal{F}_a(\boldsymbol{\xi}_0) = \omega_0^2 \boldsymbol{\xi}_0 , \tag{45}$$

then

$$\Sigma(\boldsymbol{\xi}_0) = \omega_0^2 , \tag{46}$$

and hence ω_0^2 is real. Also, the eigenfunctions may be chosen to be real at all r.

As is well known, a second property of a symmetric operator is that eigenvectors corresponding to different eigenvalues are orthogonal. Thus if

$$\mathcal{F}_a(\boldsymbol{\xi}_1) = \omega_1^2 \boldsymbol{\xi}_1; \qquad \mathcal{F}_a(\boldsymbol{\xi}_2) = \omega_2^2 \boldsymbol{\xi}_2; \qquad \omega_1^2 \neq \omega_2^2 , \tag{47}$$

then

$$< \boldsymbol{\xi}_1, \boldsymbol{\xi}_2 >= 0 . \tag{48}$$

A very important result concerns the effect of a small perturbation to the oscillation equations. This perturbation could result from a small change to the equilibrium model, to the inclusion of nonadiabatic effects (Christensen-Dalsgaard 1981) or to the inclusion of the effect of large-scale velocity fields, such as rotation, in the model. I characterize the perturbation by a change $\delta\mathcal{F}$ in the operator defining the oscillation equations. If $\boldsymbol{\delta r}_0$ and ω_0 are solutions to the adiabatic oscillation equations,

$$\omega_0^2 \boldsymbol{\delta r}_0 = \mathcal{F}_a(\boldsymbol{\delta r}_0) , \tag{49}$$

the change in ω^2 caused by the perturbation $\delta\mathcal{F}$ can be obtained from first order perturbation analysis (*e.g.* Schiff 1949) as

$$\delta\omega^2 \simeq \frac{< \boldsymbol{\delta r}_0, \delta\mathcal{F}(\boldsymbol{\delta r}_0) >}{< \boldsymbol{\delta r}_0, \boldsymbol{\delta r}_0 >} . \tag{50}$$

Thus the frequency change can be computed from the unperturbed eigenvector. Some consequences of this relation are discussed in the following sections.

3.4.2 Effects of changes in solar structure

As an example of the use of equation (50), we consider in more detail changes in the frequencies caused by changes in the equilibrium model. The effects on the oscillations of the change in the structure can be described as a perturbation $\delta\mathcal{F}_a$ in the operator characterizing adiabatic oscillations. According to equation (50), the corresponding relative frequency change, for a mode (n, l) with eigenvector $\boldsymbol{\xi}_{nl}$ in $\mathcal{H}$, is then

$$\frac{\delta\omega_{nl}}{\omega_{nl}} = \frac{1}{2}\frac{\delta\omega_{nl}^2}{\omega_{nl}^2} = \frac{< \boldsymbol{\xi}_{nl}, \delta\mathcal{F}_a(\boldsymbol{\xi}_{nl}) >}{2\omega_{nl}^2 < \boldsymbol{\xi}_{nl}, \boldsymbol{\xi}_{nl} >} . \tag{51}$$

It is convenient in the following to express $\boldsymbol{\xi}_{nl}$ in terms of its radial and horizontal amplitudes $\xi_{r,nl}(r)$ and $\xi_{\mathrm{h},nl}(r)$ (assumed to be real) in a spherical-harmonic decomposition, chosen such that

$$< \boldsymbol{\xi}_{nl}, \boldsymbol{\xi}_{nl} >= 4\pi \int_0^R [\xi_{r,nl}(r)^2 + \xi_{\mathrm{h},nl}(r)^2] r^2 \rho \, \mathrm{d}r . \tag{52}$$

We can express this quantity in terms of a normalized inertia E_{nl}, defined by

$$E_{nl} = \frac{4\pi \int_0^R [\xi_{r,nl}(r)^2 + \xi_{\mathrm{h},nl}(r)^2] \rho_0 r^2 \mathrm{d}r}{M\,[\xi_{r,nl}(R)^2 + \xi_{\mathrm{h},nl}(R)^2]} = \frac{M_{\mathrm{mode}}}{M} , \tag{53}$$

where M is the total mass of the star, and M_{mode} is the so-called modal mass. Also, we similarly represent $\delta\mathcal{F}_a$ on component form as

$$\delta\mathcal{F}_a(\boldsymbol{\xi}_{nl}) = (\phi_r[\boldsymbol{\xi}_{nl}], \phi_{\mathrm{h}}[\boldsymbol{\xi}_{nl}]) , \tag{54}$$

where $\phi_r[\boldsymbol{\xi}_{nl}](r)$ and $\phi_{\mathrm{h}}[\boldsymbol{\xi}_{nl}](r)$ are functions of r. Then we can write equation (51) as

$$E_{nl} \frac{\delta\omega_{nl}}{\omega_{nl}} = I_{nl} , \tag{55}$$

where

$$I_{nl} = \frac{2\pi \int_0^R [\xi_{r,nl}(r)\phi_r[\boldsymbol{\xi}_{nl}](r) + \xi_{\mathrm{h},nl}(r)\phi_{\mathrm{h}}[\boldsymbol{\xi}_{nl}](r)] \, \rho r^2 \mathrm{d}r}{M\omega_{nl}^2 [\xi_{r,nl}^2(R) + \xi_{\mathrm{h},nl}^2(R)]} . \tag{56}$$

Thus I_{nl} gives the integrated effect of the perturbation, normalized to the total photospheric displacement.

There is a close analogy between the exact equations (55) and (56) and the asymptotic expression (22). In both cases the factor multiplying $\delta\omega$ represents the fact that modes with larger inertia are more difficult to perturb. The effect on the frequencies of the change in the model is described by the right hand sides of equations (22) and (56), and depends on the overlap between the eigenfunction and the change in the model.

Equations (55) and (56) provide a somewhat formal linear relation between the change in the model and the change in the frequency. It follows from the discussion in Section 3.2 that the changes in the coefficients of the oscillation equations, and hence the changes $\phi_r[\boldsymbol{\xi}_{nl}](r)$ and $\phi_{\mathrm{h}}[\boldsymbol{\xi}_{nl}](r)$ in the components of $\delta\mathcal{F}_a$, can be expressed in terms of changes in two suitably chosen model variables, for example density and sound speed. For simplicity, I assume that the change in the model occurs without a change in its radius (this would in general be the case for models of the Sun, where the radius is known with high accuracy) and let $\delta\rho$ and δc denote the differences in ρ and c between the equilibrium models, at fixed r. Then equations (55) and (56) can be expressed as

$$\frac{\delta\omega_{nl}}{\omega_{nl}} = \int_0^R \left[K_{nl}^{(c,\rho)}(r)\frac{\delta c}{c}(r) + K_{nl}^{(\rho,c)}(r)\frac{\delta\rho}{\rho}(r) \right] \mathrm{d}r \tag{57}$$

(*e.g.* Gough & Thompson 1991), where the kernels $K_{nl}^{(c,\rho)}$ and $K_{nl}^{(\rho,c)}$ are computed from the eigenfunctions. Examples of such kernels are shown in Figure 3.

As implied by Section 3.2, the pair (c, ρ) in equation (57) can be replaced by other suitable pairs, such as for example (c, Γ_1) or (ρ, Γ_1). Further transformations are possible if the equation of state and the heavy-element abundances are assumed to be known. Then Γ_1 is determined by p, ρ and the helium abundance Y. Also, since p can be obtained from ρ under the assumption of hydrostatic equilibrium, $\delta\Gamma_1$ can be expressed in terms of just $\delta\rho$ and δY. Here ρ may be replaced by suitable quantities obtained from ρ, p or their derivatives. A convenient parameter is $u = p/\rho$, related to the sound speed by $c^2 = \Gamma_1 u$. Thus equation (57) becomes

$$\frac{\delta\omega_{nl}}{\omega_{nl}} = \int_0^R \left[K_{nl}^{(u,Y)}(r)\frac{\delta u}{u}(r) + K_{nl}^{(Y,u)}(r)\frac{\delta Y}{Y}(r) \right] \mathrm{d}r \; ; \tag{58}$$

here the kernels $K_{nl}^{(Y,u)}(r)$ are only of substantial magnitude in the regions where Γ_1 varies significantly, *i.e.*, in the main ionization zones.

3.4.3 Effects of sharply localized feature

Equation (57) provides the general relation between differences between solar models and the corresponding frequency differences. We now assume that the Sun contains a sharply localized feature at some radius $r = r_0$; examples of such features are the sudden change in the slope of the sound-speed gradient at the base of the convection zone, or the rapid variation of Γ_1 in the second helium ionization zone. The effect of the feature on the frequencies can be estimated by introducing a suitably smoothed model which only differs from the actual model in the close vicinity of the feature (Monteiro, Christensen-Dalsgaard & Thompson 1994); δc and $\delta\rho$ are then taken to be the differences between the actual and the smoothed structure, such that they would be non-zero only near r_0. Hence the effects of the feature on the frequencies are essentially given by $K_{nl}^{(c,\rho)}(r_0)$ and $K_{nl}^{(\rho,c)}(r_0)$.

The behaviour of the frequency effects can be understood by noting that the kernels are, very roughly, proportional to the square of the eigenfunction.

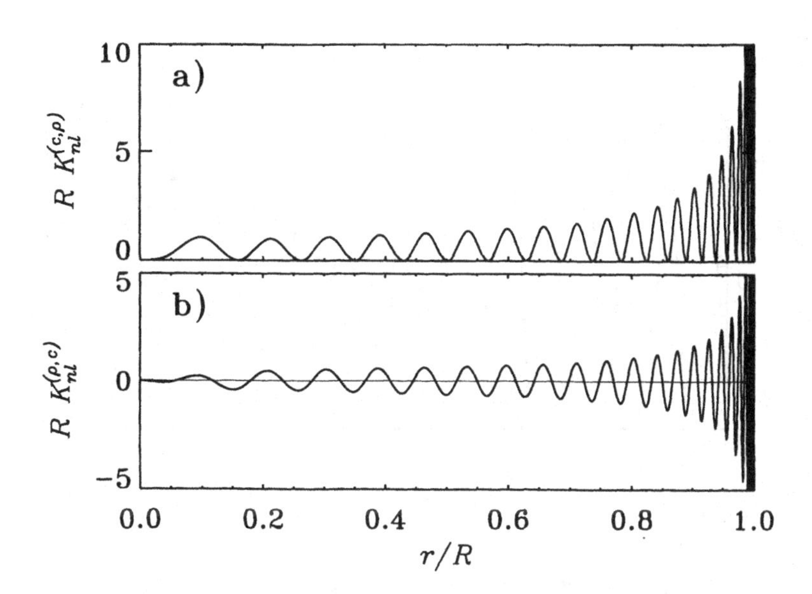

Figure 3: *Kernels* $RK_{nl}^{(c,\rho)}$ *(panel* a*) and* $RK_{nl}^{(\rho,c)}$ *(panel* b*) for an* $l = 1$, $n = 21$ *mode (with frequency* $\nu = 3.1\,\mathrm{mHz}$*) of a model of the present Sun. For clarity the plots have been truncated: the maximum value of* $RK_{nl}^{(c,\rho)}$ *is* 45*, while the extreme values of* $RK_{nl}^{(\rho,c)}$ *are* −24 *and* 16.

For high-order acoustic modes one finds that the amplitude ξ_r of the radial displacement depends on r and ω roughly as

$$\xi_r \sim \cos[\omega\tau - (\alpha + \frac{1}{4})\pi] \,, \tag{59}$$

where

$$\tau = \int_r^R \frac{\mathrm{d}r}{c} \tag{60}$$

is the acoustical depth, and α is the phase function introduced in equation (12). Thus the kernels behave roughly as $\cos(2\omega\tau + \phi)$ for some phase ϕ. It follows that the frequency changes induced by a sharp feature oscillate as a function of frequency, with a "period" determined by the depth τ at which the feature is located: features localized near the surface give rise to frequency changes varying slowly with frequency, whereas deeper features cause more rapid variations (*e.g.* Thompson 1988; Vorontsov 1988; Gough 1990). This is illustrated in Figure 4, where kernels of low-degree modes have been plotted against frequency at fixed locations corresponding approximately to the base of the convection zone (Figure 4*a*), the second helium ionization zone (Figure 4*b*) and the photosphere (Figure 4*c*).

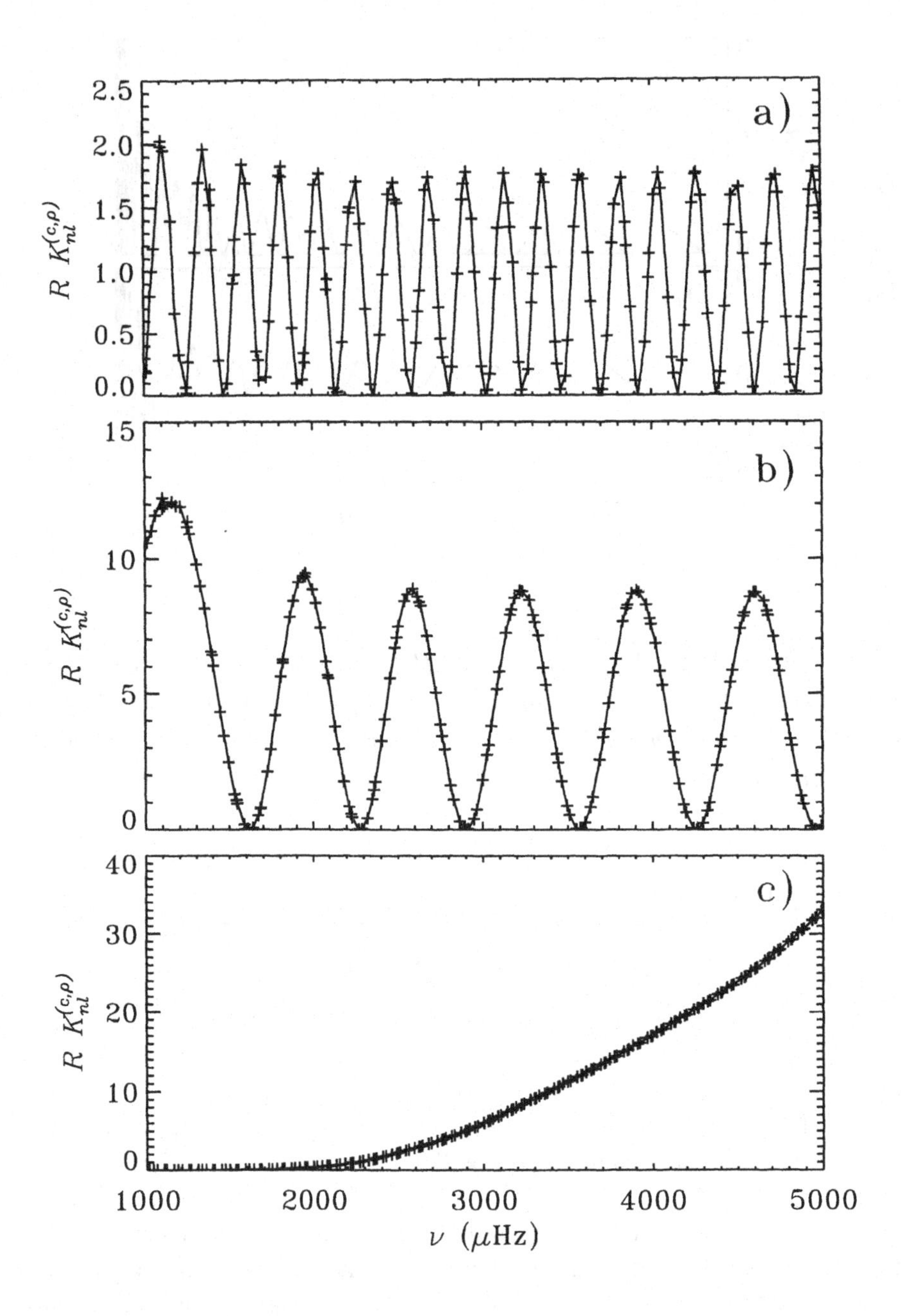

Figure 4: *Kernels* $K_{nl}^{(c,\rho)}(r_0)$ *at* $r_0 = 0.72R$ *(panel* a*)* $r_0 = 0.98R$ *(panel* b*) and* $r_0 = R$ *(panel* c*), for modes of degree* $l \leq 5$*, plotted against frequency. For clarity the points have been connected.*

3.4.4 Effects of near-surface modifications

It was argued in Section 2.5 that the near-surface region introduces substantial uncertainty in the computation of oscillation frequencies. The effects of these errors on the frequencies can be analyzed by means of an analogue of equations (55) and (56), where now the perturbation $\delta\mathcal{F}$ represents incorrectly treated features in the model and the physics of the oscillations. These may include errors in the hydrostatic structure, such as would be introduced by the neglect of turbulent pressure, as well as errors in the physics of the oscillations, *e.g.* the assumption of adiabaticity. I assume that $\delta\mathcal{F}$ is localized near the solar surface, in the sense that

$$\phi_r[\xi](r) \simeq 0, \qquad \phi_{\rm h}[\xi](r) \simeq 0 \quad \text{for} \quad R - r > \delta \, , \tag{61}$$

for some small δ. Thus in equation (56) for I_{nl} the integration extends essentially only over the region $[R-\delta, R]$. Modes extending substantially more deeply, *i.e.*, with $R - r_{\rm t} \gg \delta$, correspond to waves which propagate almost vertically in the region of modification; thus at a given frequency the eigenfunctions are essentially independent of l in this region: as a result I_{nl} depends little on l at fixed ω. The same is therefore true of $E_{nl}\delta\omega_{nl}$. To get a more convenient representation of this property, we introduce

$$Q_{nl} = \frac{E_{nl}}{\overline{E}_0(\omega_{nl})} \, , \tag{62}$$

where $\overline{E}_l(\omega)$ is obtained by interpolating to ω in E_{nl} at fixed l. Then $Q_{nl}\delta\omega_{nl}$ is independent of l, at fixed ω, for modes such that $R - r_{\rm t} \gg \delta$. This behaviour may be used to identify, and eliminate, the effects of the near-surface uncertainties. Q_{nl} has been plotted in Figure 5 for selected values of l. Its variation with l is largely determined by the change in the penetration depth. Modes with higher degree penetrate less deeply and hence have a smaller inertia at given surface displacement. As a consequence of this their frequencies are more susceptible to changes in the model. Thus unscaled frequency differences resulting from near-surface effects are expected to show strong dependence on l. An example of this is discussed in Section 4.1.

It should also be noticed that, according to trapping of the modes near the surface as determined by the behaviour of $\omega_{\rm c}$ [*cf.* equations (30) and (31)] low-frequency modes are evanescent in the uncertain region, with much smaller amplitudes than in the interior. Thus for such modes the near-surface effects are expected to yield very small frequency changes. Finally, the discussion in Section 3.4.3 indicates that near-surface effects are likely to give rise to frequency changes varying slowly with frequency (see also Figure 4), unless, of course, the effects are themselves rapidly-varying functions of frequency.

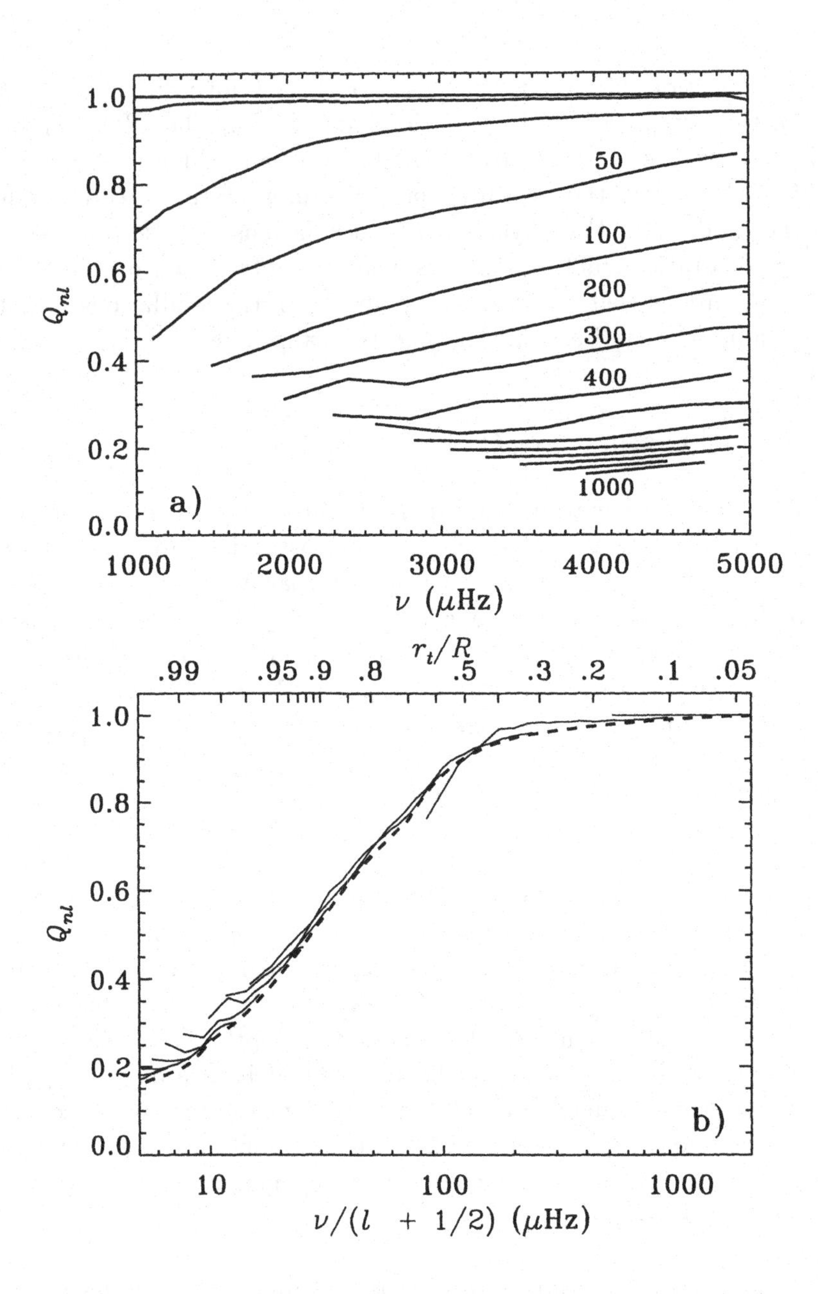

Figure 5: *The solid lines show the inertia ratio Q_{nl}, defined in equation (62), for p modes in a normal solar model. Each curve corresponds to a given degree l, In panel* (a) *Q_{nl} is shown against frequency ν, and selected values of l are indicated. In panel* (b) *the lower abscissa is ν/L, related to the turning-point radius r_t shown as the upper abscissa through equation (10). Here the heavy dashed curve shows the asymptotic scaling $\tilde{S}_{nl}/\tau_0$ (see text).*

It follows from the preceding discussion that equation (57) should be replaced by

$$\frac{\delta\omega_{nl}}{\omega_{nl}} = \int_0^R \left[K_{nl}^{(c,\rho)}(r)\frac{\delta c}{c}(r) + K_{nl}^{(\rho,c)}(r)\frac{\delta\rho}{\rho}(r) \right] dr + Q_{nl}^{-1}\mathcal{G}(\omega_{nl}) \,, \qquad (63)$$

where the function $\mathcal{G}(\omega)$ accounts for the effect of the near-surface uncertainties. $\mathcal{G}$ must then be determined as part of the analysis of the frequency differences, or suppressed by means of suitable filtering of the data. This equation is closely equivalent to equation (23), where the term $\mathcal{H}_2(\omega_{nl})$ contains the contributions corresponding to $\mathcal{G}(\omega)$ Furthermore, Q_{nl} corresponds asymptotically to S_{nl}, apart from a constant scaling factor. Indeed, in Figure 5(*b*) the dashed line shows $\tilde{S}_{nl}/\tau_0$; here $\tilde{S}_{nl}$ is defined as in equation (19), but excluding the term in $\mathrm{d}\alpha/\mathrm{d}\omega$, and τ_0 is the limit of S_{nl} for $r_\mathrm{t} \to 0$, corresponding to the acoustical radius of the star evaluated from equation (60), with $r = 0$.

The arguments presented here assumed that the eigenfunctions of the modes were essentially independent of degree in the region of modification. This ceases to be true at sufficiently high l, leading to departures from the simple frequency dependence of the scaled frequency differences. Antia (1995) showed that even for moderate degree the resulting l-dependence may be comparable with the observational errors so that it might affect the results of inversion. Procedures to handle such effects have been developed on the basis of higher-order asymptotic treatments, such as equation (33) (*e.g.* Gough & Vorontsov 1995).

4 APPLICATION TO MODELS AND OBSERVED FREQUENCIES

It was argued in Section 1.2 that the principal interest in the forward problem lies in the connection between the physics of solar models and their frequencies. To explore this connection, the present section considers various examples of physics modifications and their effects on the models and frequencies. The results are interpreted in terms of the properties of the oscillations presented in Section 3, particularly the asymptotic frequency changes discussed in Section 3.3.3. Furthermore, I briefly consider some aspects of the observed frequencies. More extensive results from the observations are presented in Section 5.

The results discussed here are largely based on complete solar models, obtained from evolution calculations starting at chemically homogeneous zero-age main-sequence models. These have been calibrated to have solar radius and luminosity, by adjusting the composition, characterized by the initial helium abundance Y_0, and a parameter describing convection (*cf.* Section 2.2). To save

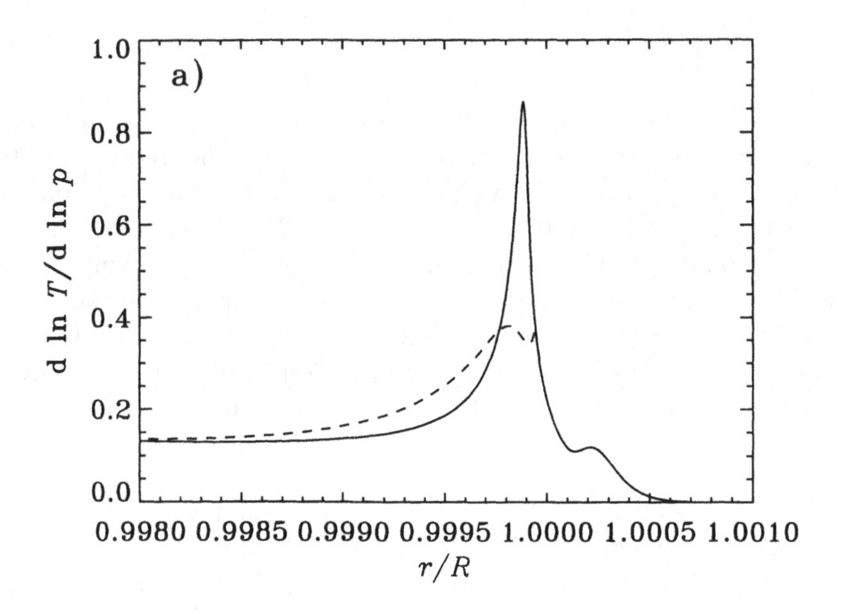

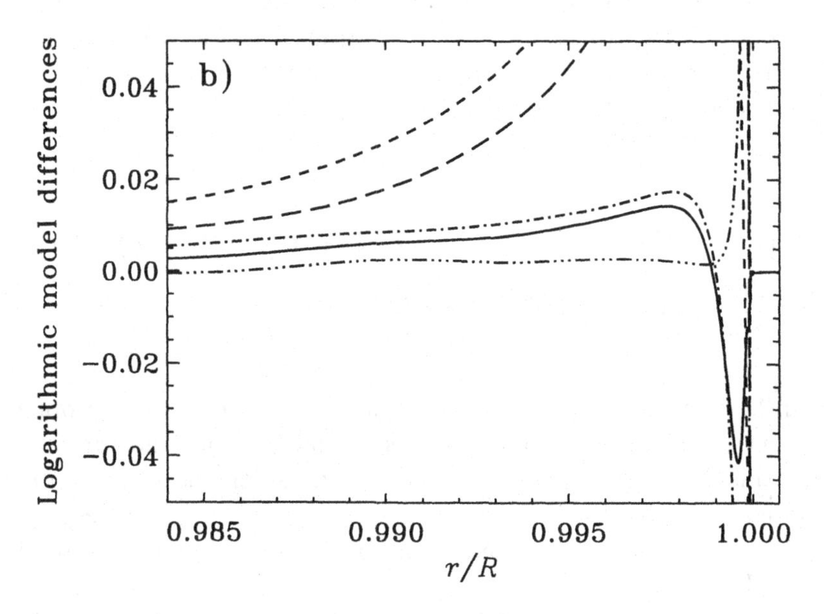

Figure 6: (a) $\nabla = \mathrm{d}\ln T/\mathrm{d}\ln p$ *for a mixing-length model (solid line) and a modified model (dashed line).* (b) *Logarithmic differences at fixed radius r between the modified model and the mixing-length model, in the sense (modified) – (mixing length). The following line styles have been used:* $\delta\ln c$*: ————; $\delta\ln p$: - - - - - - - - - - - -; $\delta\ln\rho$: — — — — — —; $\delta\ln T$: —·—·—; $\delta\ln\Gamma_1$: —···—···—···.*

computational effort, an alternative is to consider static models of the present Sun: here the abundance profile is based on scaling the hydrogen-abundance profile $X(m)$ for a given reference model, as a function of mass m, by a factor χ such as to obtain the correct luminosity. This is often adequate to obtain insight into effects of modifications to the physics, as long as these occur outside the energy-generating core. Finally, when studying the properties of the convection zone, it is convenient to consider models of the solar envelope alone. In such models, the surface luminosity and radius are chosen to have solar values and the composition is typically assumed to be constant. Also, the convection parameter is typically adjusted such that the model has a prescribed depth of convection zone. For physics modifications that are largely confined to the convection zone, such as those resulting from modifying the equation of state, this ensures that the radiative interior is approximately unchanged.

4.1 Effects of changes in the superadiabatic region of the convection zone

To illustrate the effects of near-surface uncertainties, I first consider a model where the treatment of the superadiabatic gradient has been artificially modified [see Christensen-Dalsgaard (1986) for details]. In Figure 6, panel (a) compares the resulting ∇ with the result of using mixing-length theory, whereas panel (b) shows differences, at fixed r, between various quantities in the modified and the reference models. Here static models of the present Sun were used. The effects are confined to the outermost parts of the convection zone; the deeper parts of the convection zone and the radiative interior are virtually unchanged.

Frequency differences between the modified and the reference model are illustrated in Figure 7. Panel (a) shows the original differences, which clearly depend strongly on both degree and frequency. It was argued in Section 3.4.4 that the l-dependence should be mainly associated with the variation of the mode inertia. This is confirmed by panel (b) where the frequency differences have been scaled by the normalized inertia Q_{nl} defined in equation (62). The scaled differences are virtually independent of degree for $l \lesssim 300$, corresponding to modes that propagate essentially vertically in the region where the model is modified. It should also be noticed that, in accordance with the discussion in Section 3.4.4, the differences are very small at low frequency and vary slowly with frequency.

This example illustrates the usefulness of scaling frequency differences to highlight effects of near-surface errors in the models. Thus in the following I shall almost exclusively consider differences scaled in this manner.

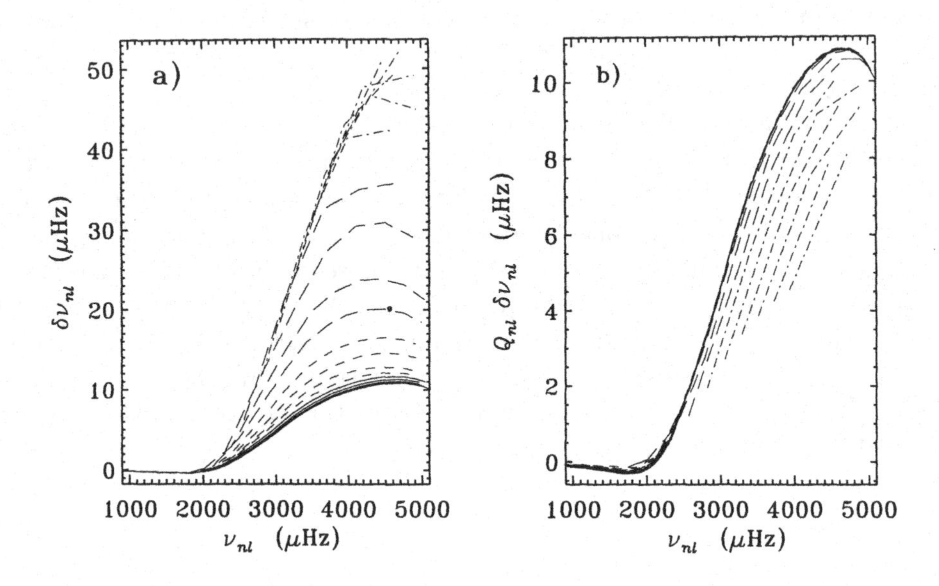

Figure 7: *Unscaled (panel* a*) and scaled (panel* b*) frequency differences corresponding to the model differences illustrated in Figure 6*(b). *Modes of the same degree l have been connected, according to the following line styles:* $l = 0 - 30$: ————; $l = 40 - 100$: ------------; $l = 150 - 400$: – – – – – – – –; $l = 500 - 1100$: —·—·—·—.

The behaviour of $\nabla - \nabla_{\rm ad}$ illustrated in Figure 6 was chosen with no physical basis. However, as noted in Section 2.3.1 refined versions of mixing-length theory have suggested that $\nabla - \nabla_{\rm ad}$ might be sharper and higher than in mixing-length models. Figure 8 illustrates an example of such a model and the resulting scaled frequency differences; it is based on a parametrization by Monteiro *et al.* (1995b) of the convection treatment developed by Canuto & Mazzitelli (1991). Within the range of degrees considered, between 20 and 300, the scaled differences are essentially independent of degree, as shown by the small scatter of the points in the plot. Also, as discussed in Section 5.3, the frequency differences for $\log_{10}(\beta_c) = 3$ bear a striking resemblance to the differences between the observed frequencies and those of a model computed with the normal mixing-length theory.

4.2 Opacity increase near the base of the convection zone

The effects of modifications of the physics of the radiative interior may be illustrated by considering changes in the opacity. Indeed, as mentioned in Section 2.5, the opacity is likely to be a dominant source of uncertainty in the deep interior of the model. Although the actual error in the opacity is likely to be

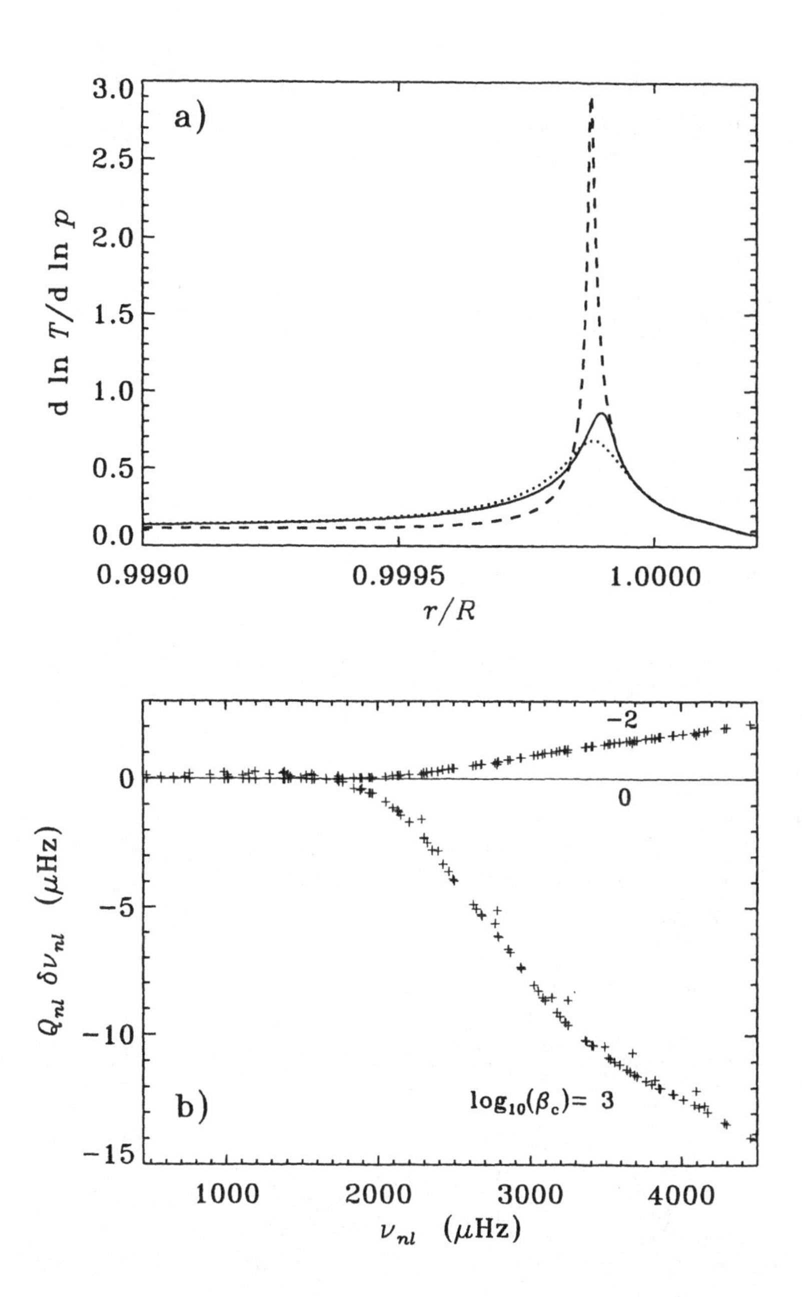

Figure 8: (a) $\nabla = \mathrm{d}\ln T/\mathrm{d}\ln p$ *for three different treatments of convection, characterized by the parameter* β_c. *Results are shown for a mixing-length model* ($\beta_\mathrm{c} = 1$*; solid line), a model approximating the Canuto & Mazzitelli (1991) formulation* ($\beta_\mathrm{c} = 10^3$*; dashed line), and a model with lower and broader* $\nabla - \nabla_\mathrm{ad}$ ($\beta_\mathrm{c} = 10^{-2}$*; dotted line).* (b) *Scaled frequency differences, relative to the mixing-length model, for* $l = 20, 30, 40, 50, 100, 200, 300$ *for the models shown in panel* (a). *The points have been identified by the value of* $\log_{10}\beta_\mathrm{c}$. *(Adapted from Monteiro et al. 1995a).*

a complicated function of temperature and density, the effects are most easily understood by analyzing the consequences of a localized increase. The analysis is based on static models with scaled hydrogen abundance.

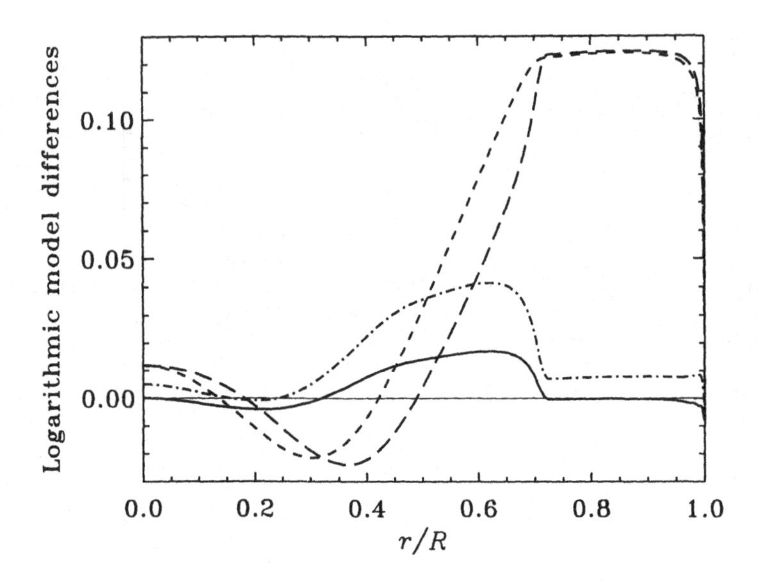

Figure 9: *Logarithmic differences at fixed radius r between the model with modified opacity and the reference model, in the sense (modified) – (reference). The following line styles have been used:* $\delta \ln c$: ————; $\delta \ln p$: - - - - - - - -; $\delta \ln \rho$: — — — — ; $\delta \ln T$: —·—·—·— .

I consider an increase in the Rosseland mean opacity κ defined as a function of temperature by

$$\delta \log \kappa = A_\kappa \exp[-(y/\Delta \log T)^2] , \tag{64}$$

where

$$y = \begin{cases} x - \log T_1 & \text{for } x < \log T_1 \\ 0 & \text{for } \log T_1 \leq x \leq \log T_2 \\ x - \log T_2 & \text{for } x > \log T_2 , \end{cases} \tag{65}$$

log being the logarithm to base 10. I confine the change near the base of the convection zone, by choosing $\log T_1 = 6$, $\log T_2 = 6.6$ and $\Delta \log T = 0.15$, and take the maximum change to be $A_\kappa = 0.1$. The resulting changes in sound speed, pressure and density are shown in Figure 9. The corresponding frequency changes are largely determined by the the change in the sound speed. This is dominated by the increased depth of the convection zone in the modified model, resulting directly from the increase in the opacity: since the gradients of temperature and sound speed are steeper in the convection zone than in the radiative region below, there is a region where the sound speed increases more rapidly with depth in the modified model; therefore, the sound speed is sharply higher in the modified model just beneath the convection zone, as seen in the figure. In contrast, the sound-speed difference is very small in the bulk of the convection zone; this is consistent with equation (4) according to which the sound speed in this region, at given r, depends little on the details of the structure. The only visible exception is in the ionization zones near the surface, where the changes in composition and mixing length required to calibrate the modified model cause small additional differences in the sound speed.

The corresponding scaled frequency differences are shown in Figure 10(a). They clearly reflect the behaviour of the sound-speed difference and the region to which the modes are confined, as determined by the location of the turning point r_t (*cf.* equation 10). For low-degree modes which penetrate well beyond the base of the convection zone the frequencies are increased by the sound-speed increase in the outer parts of the radiative region. In contrast, high-degree modes are trapped in the convection zone and are dominated by the small negative sound-speed differences in the ionization zones. For $l = 20 - 50$ the behaviour depends strongly on frequency: higher-frequency modes penetrate more deeply, according to equation (10), and hence sense the positive sound-speed difference below the convection zone, whereas low-frequency modes are largely confined to the convection zone.

The preceding discussion indicates the close link between the frequency differences and the location of the turning point. This is clearly in accordance with the asymptotic relation, equation (23), between the sound-speed difference and the frequency differences. It becomes obvious when, as in Figure 10(b), scaled relative differences are plotted against ν/L which according to equation (10) determines r_t. It is obvious that the general behaviour of the frequency differences is indeed dominated by $\mathcal{H}_1(\omega/L)$, with a sharp transition where the turning-point position r_t of the modes coincides with the base of the convection zone, at $r \simeq 0.72R$. For modes trapped in the convection zone, the frequency-dependent term $\mathcal{H}_2(\omega)$ dominates; this comes predominantly from the negative sound-speed difference (hardly visible in Figure 9) in the hydrogen ionization zone, and hence varies slowly with frequency. A similar contribution is visible for the deeply-penetrating modes; here an additional rapid variation with frequency is induced by the sharp difference at the base of the convection zone (*cf.* Section 3.4.3).

4.3 Diffusion and settling of helium

Helium diffusion and settling cause significant changes in the abundance profile of models of the present Sun and hence in the structure and frequencies of the models. Figure 11 shows differences between a model with diffusion and a normal

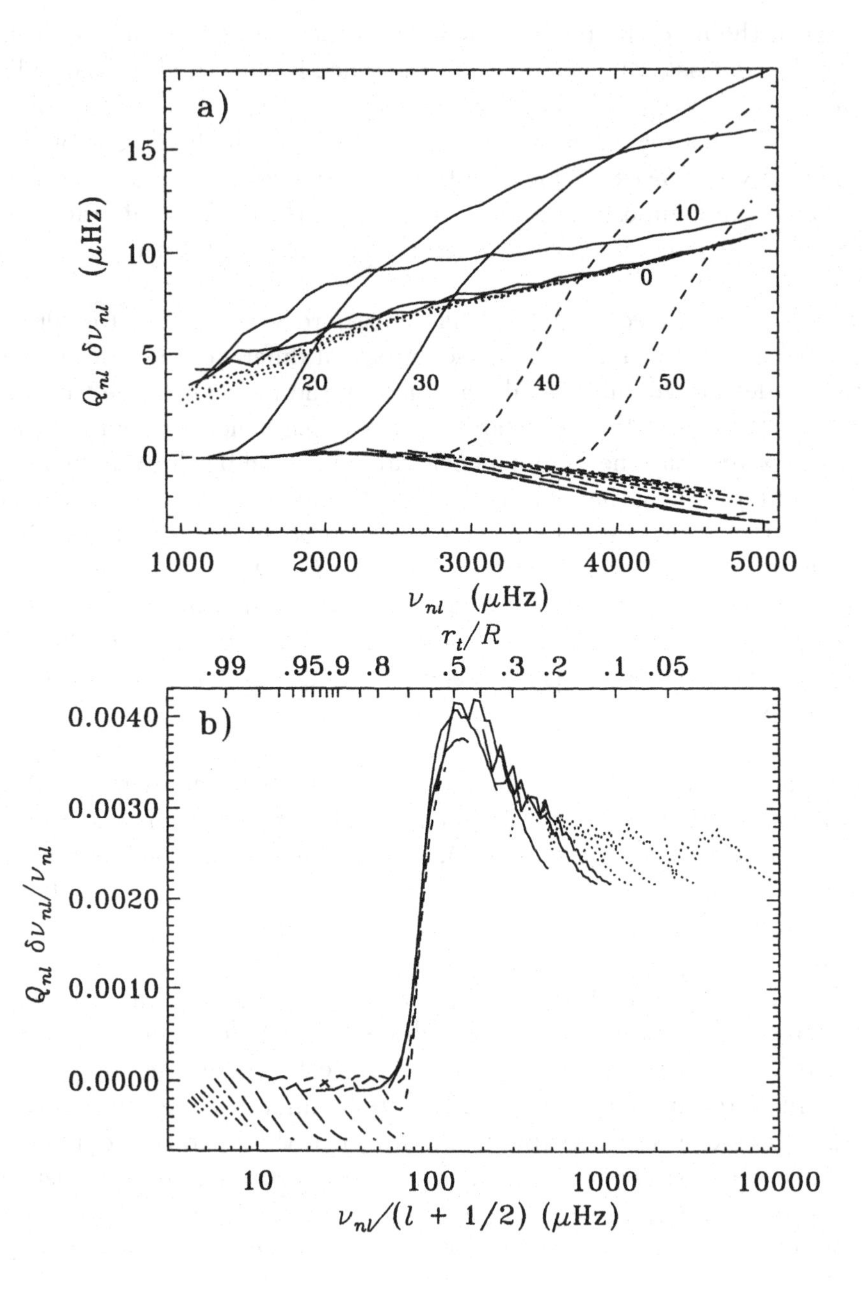

Figure 10: *Scaled frequency differences corresponding to the model differences, resulting from opacity increase near the base of the convection zone, illustrated in Figure 9. Modes of the same degree l have been connected, according to the following line styles: $l = 0-3$: ················; $l = 4-30$: ————————; $l = 40-100$: ------------; $l = 150-400$: – – – – – – –; $l = 500-1100$: —·—·—·—.* (a) *Absolute differences, plotted against frequency. Selected values of l have been indicated.* (b) *Relative differences, plotted against ν/L (lower abscissa) and the corresponding turning-point position [cf. equation (10); upper abscissa].*

solar model (Christensen-Dalsgaard *et al.* 1993). Settling of helium causes an increase in the hydrogen abundance by about 0.03 in the convection zone, with a sharp gradient at its base. In the outer parts of the radiative interior X is still somewhat higher than in the normal model, while the central abundance is reduced, as a result of the accumulation of helium. The increase in X near the base of the convection zone leads to an increase in the depth of the convection zone, and hence a substantial increase in sound speed in this region, just as in the case of the opacity increase discussed in Section 4.2. As in that case $\delta c/c$ is small in most of the convection zone, whereas the change in composition results in a considerable change in c in the ionization zones of hydrogen and helium, due to the change in Γ_1.

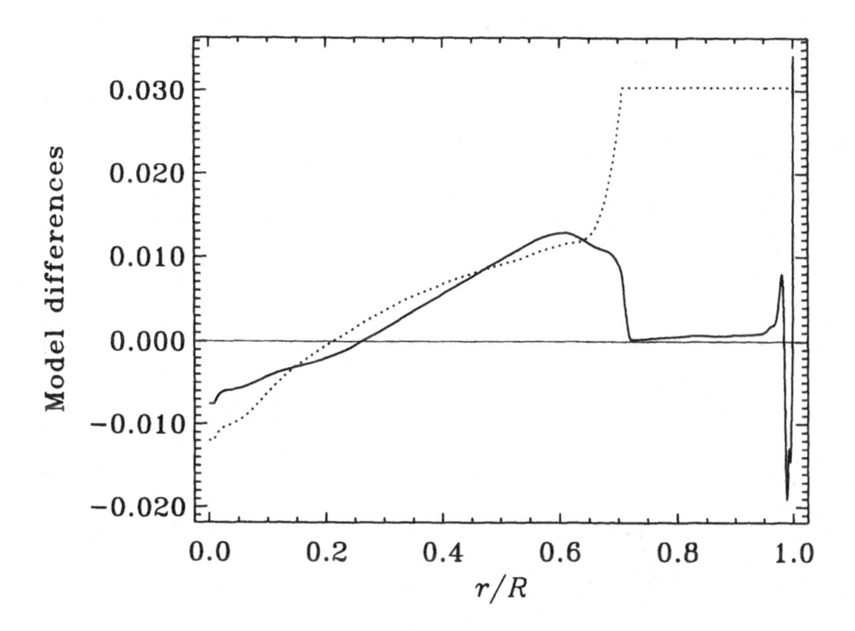

Figure 11: *Differences, at fixed r, between a model of the present Sun including diffusion and gravitational settling of helium and a normal model without diffusion. The dotted line shows the difference δX in the hydrogen abundance, and the continuous line shows the fractional difference $\delta c^2/c^2$ in squared sound speed. Adapted from Christensen-Dalsgaard et al. (1993).*

As in the case of the opacity increase discussed in Section 4.2 the change in the frequencies is largely controlled by the location of the lower turning point. Thus Figure 12(a) shows scaled frequency differences, at selected values of l, between the diffusive and non-diffusive models, plotted against ν/L (with $L = l + 1/2$). Here the scaling has been done in terms of the asymptotic factor S_{nl}, normalized by τ_0, such that it tends to unity at low degree (note that, as indicated by Figure 5, S_{nl}/τ_0 is closely equivalent to Q_{nl}). Hence the

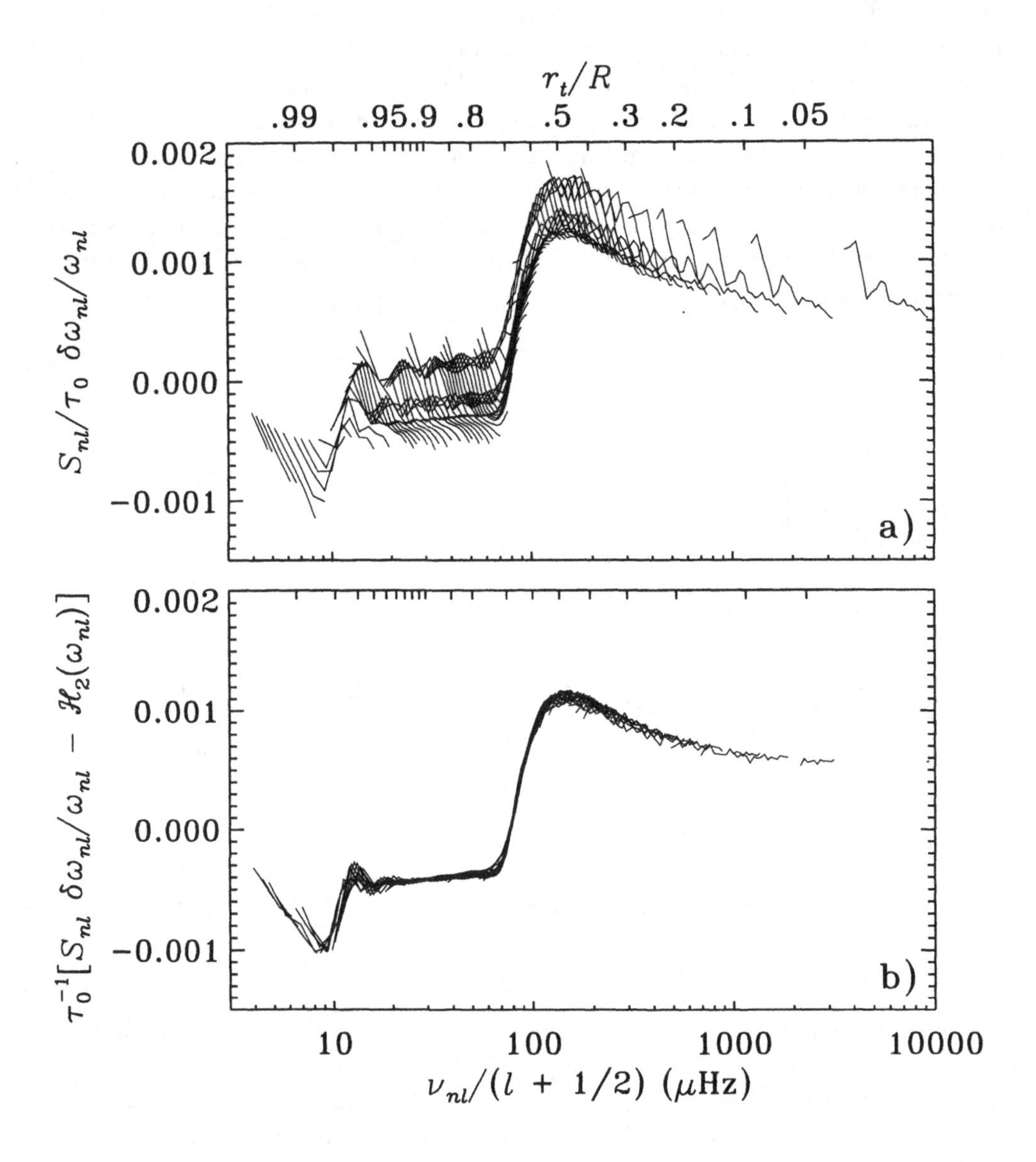

Figure 12: *Scaled frequency differences corresponding to the model differences shown in Figure 11, plotted against* $\nu/(l+1/2)$. *The upper abscissa shows the location of the lower turning point, which is related to* $\nu/(l+1/2)$ *through equation (10). Points corresponding to fixed* l *have been connected.* (a) *Original scaled frequency differences.* (b) *Scaled differences, after subtraction of the function* $\mathcal{H}_2(\omega)$ *obtained from the spline fit.*

scaled frequency differences correspond in magnitude to the differences for low-degree modes. As in Figure 10(*b*) the dependence of $S\delta\nu/\nu$ on ν/L is dominated by a substantial positive sound-speed difference at the base of the convection zone: modes with $\nu/L > 100\,\mu$Hz sense this feature and hence display a positive frequency difference; in contrast, for $\nu/L \lesssim 100\,\mu$Hz the modes are entirely trapped in the convection zone, and the frequency difference corresponds to the term $\mathcal{H}_2(\nu)$ arising from differences near the surface, particularly the difference in X.

This qualitative description suggests that the frequency differences may be analyzed in detail in terms of equation (23). To do so, I have determined the functions $\mathcal{H}_1$ and $\mathcal{H}_2$ by means of the spline fit of Christensen-Dalsgaard *et al.* (1989), where details about the fitting method may be found. Briefly, the procedure is to approximate $\mathcal{H}_1(\omega/L)$ and $\mathcal{H}_2(\omega)$ by splines, the coefficients of which are determined through a least-squares fit to the scaled frequency differences. The knots of the splines in $w \equiv \omega/L$ are distributed uniformly in $\log w$ over the range considered, whereas the knots for the ω-splines are uniform in ω. I used 28 knots in w and 20 knots in ω. Figure 12(*b*) shows the result of subtracting the function $\mathcal{H}_2(\omega)$ so obtained from the scaled frequency differences. It is evident that what remains is in fact very nearly a function of ω/L alone, directly reflecting the behaviour of $\delta c/c$, as discussed above.

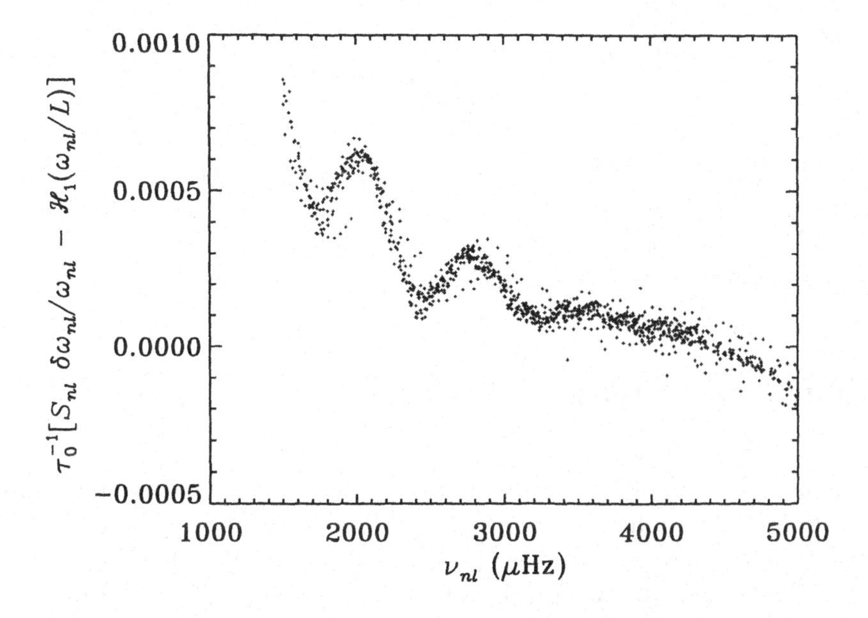

Figure 13: *Scaled frequency differences corresponding to the model differences shown in Figure 11, after subtraction of the function $\mathcal{H}_1(\omega/L)$ resulting from the spline fit.*

The residual scaled frequency differences after subtraction of the term in $\mathcal{H}_1(\omega/L)$ are shown in Figure 13. These are clearly largely a function of frequency, although with some scatter. The behaviour is dominated, at low frequency, by an oscillation with a 'period' of around 800 μHz. According to Figure 4 this corresponds to the effect of a sharp feature located around $r \simeq 0.98R$, *i.e.*, at the second helium ionization zone: it is caused by differences in the ionization zone resulting from the difference in helium abundance. The remaining slow trend is associated with changes in the hydrogen ionization zone and the atmosphere.

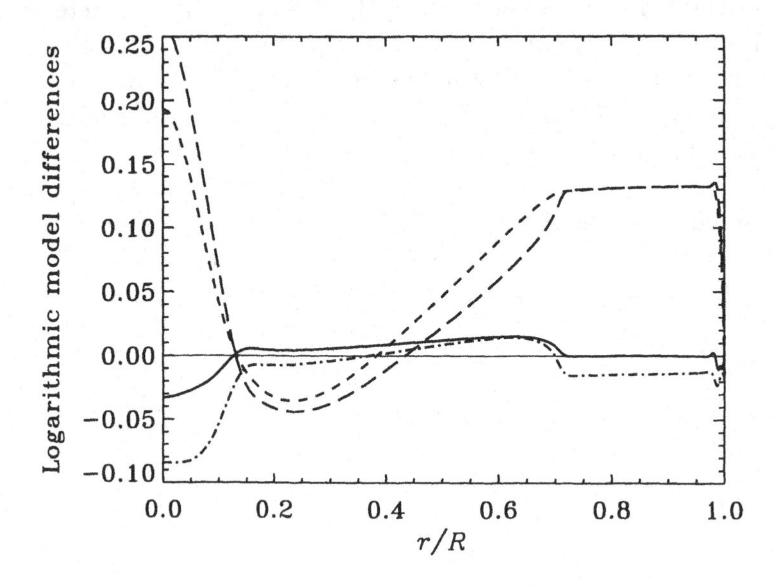

Figure 14: *Logarithmic differences at fixed radius r between the model with reduced central opacity and the reference model, in the sense (modified) – (reference). The following line styles have been used:* $\delta \ln c$: ———————; $\delta \ln p$: - - - - - - - - - - - -; $\delta \ln \rho$: — — — — — — —; $\delta \ln T$: — · — · — · — .

4.4 Opacity decrease in the core

The solar neutrino problem (see the chapter by Bahcall) has motivated a number of suggestions for modifications to solar models designed to reduce the flux of high-energy neutrinos, by reducing the core temperature of the Sun. One such suggestion involved postulating the presence in the Sun of Weakly Interacting Massive Particles (WIMPs) whose motion was assumed to contribute to the energy transport. In this way the temperature gradient required for radiative transport, and hence the central temperature, could be reduced (Steigman *et al.* 1978; Spergel & Press 1985; Faulkner & Gilliland 1985).

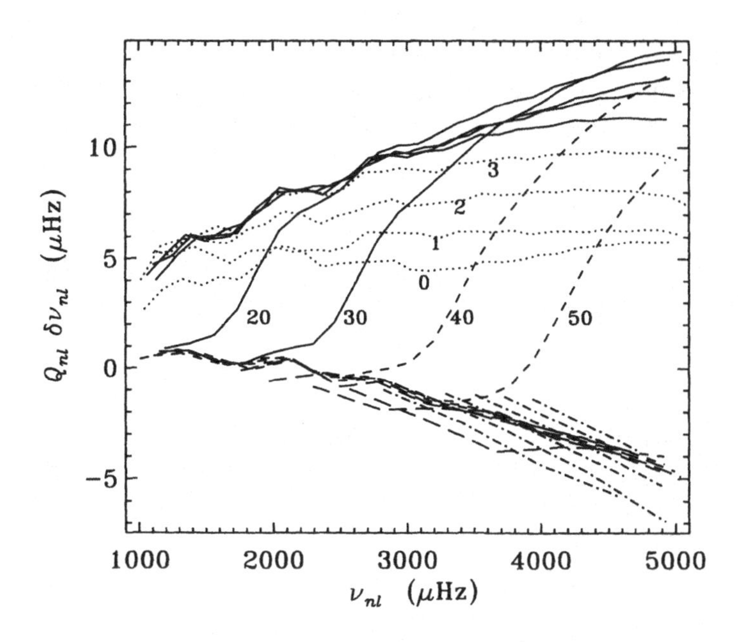

Figure 15: *Scaled frequency differences corresponding to the model differences, resulting from opacity decrease in the core, illustrated in Figure 14. Modes of the same degree l have been connected, according to the following line styles: $l = 0 - 3$: ················; $l = 4 - 30$: ————————; $l = 40 - 100$: ------------; $l = 150 - 400$: – – – – – – – –; $l = 500 - 1100$: —·—·—·—. Selected values of l have been indicated.*

Here I model this effect through a reduction in the core opacity (see also Christensen-Dalsgaard 1992). Specifically, the opacity was modified as in equations (64) and (65), but with $A_\kappa = -0.4$, $\log T_1 = 7.1$, $\log T_2 = 7.5$ and $\Delta \log T = 0.04$. The resulting model differences, based on scaled static models, are shown in Figure 14. Clearly the central temperature has been reduced, leading to a decrease in the neutrino flux. This is accompanied by a reduction in the central sound speed and a dramatic increase in the core pressure and density. The changes in composition and mixing length required to obtain the correct luminosity and radius induce additional modifications in the outer parts of the model, including a modest increase in the depth of the convection zone, visible in the sound-speed difference.

Scaled frequency differences between the models are shown in Figure 15. The positive sound-speed difference just below the convection zone leads the the now familiar variations for the modes penetrating beyond the base of the convection zone. However, in addition the strong variations in the core cause a substantial dependence of the frequency differences on l, amongst the low-degree modes. This arises both from the depression of the sound speed and the increase in density; according to equation (29) the latter modification increases the effect of the perturbation in the gravitational potential and hence contributes to decreasing the frequencies. The resulting negative contributions to the frequencies are largest for the lowest-degree modes, leading to the variation in $\delta\nu$ with l.

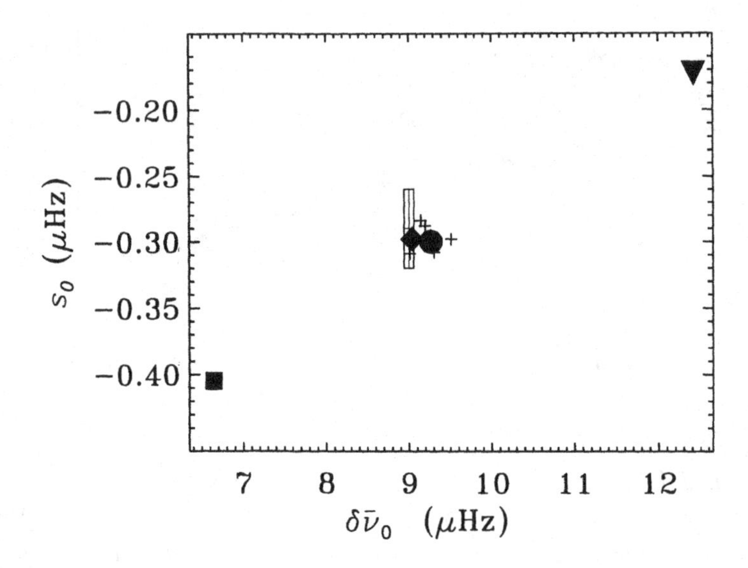

Figure 16: *Parameters in least-square fit to the small frequency separation (cf. equation 66), using a reference order* $n_0 = 21$. *The small crosses show various normal solar models using varying physics. The filled circle and diamond are models with up-to-date physics, without and with helium settling, respectively. The filled square shows results for the model with reduced core opacity, illustrated in Figs 14 and 15, while the filled triangle is a model with a partially mixed core. The error box shows observed values from Elsworth et al. (1990).*

As a result of this l-dependence the small frequency separation δ_{nl} (*cf.* equation 38) is reduced substantially by the reduction in the core opacity. To illustrate the effect, and compare it with unmodified solar models and the observed values, I follow Elsworth *et al.* (1990), approximating δ_{nl} as

$$\delta_{nl} \simeq \bar{\delta}_l + s_l(n - n_0) \; , \tag{66}$$

where n_0 is a suitable reference order, and the coefficients $\bar{\delta}_l$ and s_l are determined through a least-squares fit. The results are shown in Figure 16. Normal solar models, particularly with the inclusion of helium settling in the core, are in good agreement with the observations. In contrast, the model with reduced core opacity is clearly inconsistent with the observed values. The figure also shows results for a model with partial mixing of the core, based on the hydrogen profile of Schatzman *et al.* (1981). This has a reduced neutrino flux relative to normal models, but is again inconsistent with the observed frequency separation, although the computed value is now too large.

Results such as these clearly argue against an astrophysical solution to the solar neutrino problem: there is a strong tendency that proposed models with reduced neutrino fluxes are inconsistent with the helioseismic data. However, as mentioned in Section 3.2 the observed frequencies do not in themselves constrain the solar internal temperature and hence the neutrino flux. In fact, one might imagine constructing a model involving both reduced core opacity and partial mixing, choosing the magnitude of the effects such as to bring both oscillation frequencies and neutrino fluxes into agreement with the measured values. Such a model would clearly be somewhat contrived, and hardly plausible; but its possible existence highlights the need for further assumptions, if helioseismology is to provide constraints on the neutrino production of the solar core.

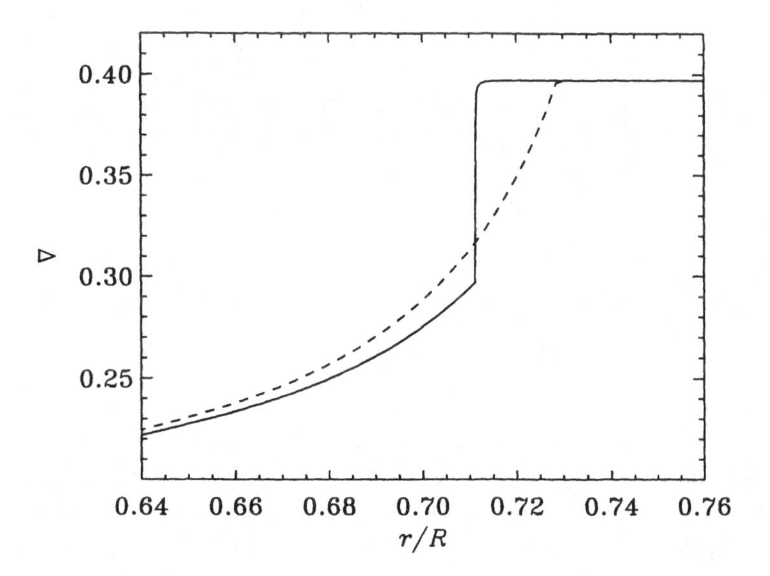

Figure 17: *Temperature gradient* $\nabla = \mathrm{d}\ln T/\mathrm{d}\ln p$ *in normal solar model (dashed line) and model with overshoot of* $0.21H_p$, H_p *being the pressure scale height. (Adapted from Monteiro et al. 1994.)*

4.5 The base of the convection zone

Normal solar models predict a sharp transition between the lower part of the convection zone, where the temperature gradient is very nearly adiabatic, and the radiative region below where the temperature gradient decreases sharply with the increasing temperature. This is illustrated in Figure 17. It is evident that the sound-speed gradient displays a very similar behaviour, resulting in a near-discontinuity in the second derivative of sound speed at the base of the convection zone. The resulting inflection was visible even in early determinations of the solar sound speed through asymptotic inversions of observed frequencies (Christensen-Dalsgaard *et al.* 1985). Christensen-Dalsgaard, Gough & Thompson (1991) carried out careful analyses to determine the location of the break in the gradient of the sound speed as inferred from inversion, and testing the methods on artificial data; in this way they determined the depth of the solar convection zone as $d_b = (0.287 \pm 0.003)R$.

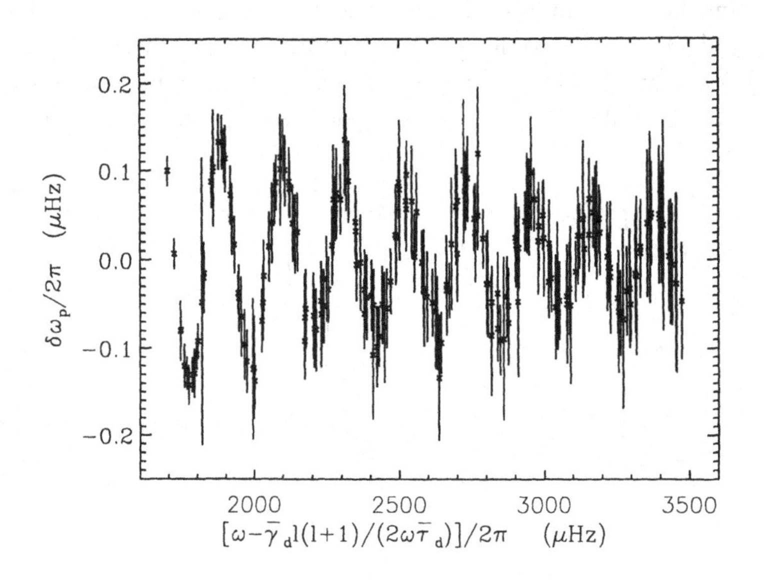

Figure 18: *Oscillatory signal corresponding to the base of the convection zone, for observed solar frequencies from Libbrecht, Woodard & Kaufman (1990), plotted against reduced frequency. The parameters* $\overline{\gamma}_d$ *and* $\overline{\tau}_d$ *(the latter being the acoustical depth of the discontinuity) have been obtained from a fit to the data. (From Monteiro et al. 1994.)*

The discontinuity in the second derivative is a sharp feature, in the sense introduced in Section 3.4.3, and hence may be expected to introduce an oscillatory signal in the oscillation frequencies. An even more extreme behaviour is predicted by simple models of convective overshoot (*cf.* Section 2.3.1). The resulting temperature gradient, also illustrated in Figure 17, is essentially discontinuous at the point where the motion stops. Thus at this point there is also a discontinuity in the sound-speed gradient, which again introduces oscillations in the frequencies.

The detailed effects on the oscillation frequencies of these properties were analyzed by Basu, Antia & Narasimha (1994), Monteiro *et al.* (1994), Roxburgh

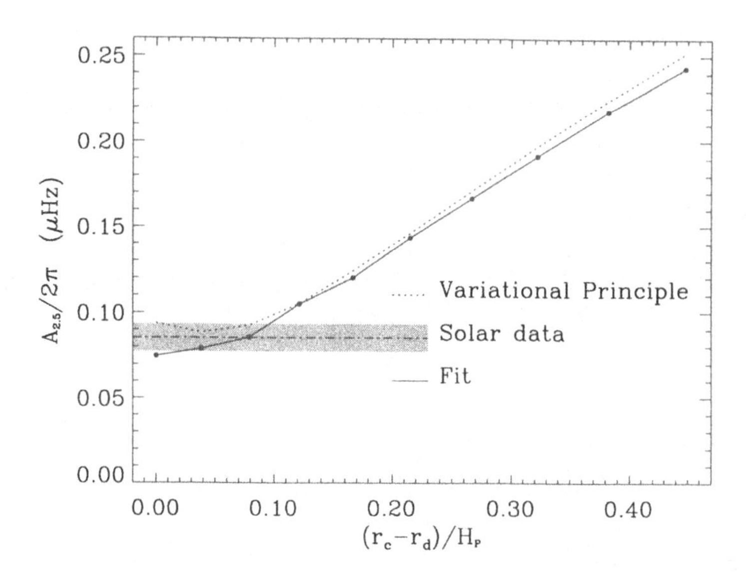

Figure 19: *Normalized amplitude of the oscillatory frequency signal induced by the base of the convection zone, as a function of overshoot distance in units of the pressure scale height H_p. The solid line shows a fit to computed frequencies, whereas the dotted line was obtained from an analytical approximation to the amplitude. The shaded area shows the amplitude inferred from the solar data (Libbrecht et al. 1990), with a width corresponding to an estimate of its error. (From Christensen-Dalsgaard et al. 1995a.)*

& Vorontsov (1994) and Christensen-Dalsgaard, Monteiro & Thompson (1995a). As shown by Monteiro *et al.* (1994) the signal in the frequencies is essentially a function of $\omega - \gamma_d L^2/(2\omega\tau_d)$, where τ_d is the acoustical depth of the discontinuity (in the first or second sound-speed derivative) and $\gamma_d = \int_{r_d}^{R}(c/r^2)\mathrm{d}r$, r_d being the radius of the discontinuity. In Figure 18 the oscillatory component of the observed frequencies have been plotted in this form. There is indeed a clear signal, with a 'frequency' which corresponds to the depth of the convection zone as inferred by Christensen-Dalsgaard *et al.* (1991).

Frequencies of solar models without and with overshoot show a similar behaviour. However, the amplitude of the oscillatory signal depends on the extent of overshoot. Figure 19 shows fitted amplitudes, normalized to a frequency of 2.5 mHz, as a function of the overshoot distance in units of the pressure scale height H_p, together with the similarly fitted amplitude for the solar frequencies. It is evident that in the Sun overshoot of this form can at most extend approximately $0.1 H_p$ (Basu *et al.* 1994; Monteiro *et al.* 1994).

Although these results place interesting constraints on the structure at the base of the solar convection zone, it must be realized that the proposed model for overshoot is highly simplified. In particular, it assumes that the effects on the frequencies can be characterized by a spherically symmetric and time-independent structure. In reality, overshoot must display substantial variations as a function of position and time. The oscillations sense an average of these variations; thus it is no surprise that the observed frequencies indicate a relatively smooth structure. More careful investigations, involving also hydrodynamical simulations of conditions at the base of the convection zone, will be required to obtain firmer helioseismic bounds on the extent of overshoot.

4.6 The equation of state

It was argued in Section 2.4 that the convection zone is well suited for studies of the properties of the equation of state of solar matter. Although conditions very near the surface are complicated by the uncertain structure of the strongly superadiabatic region, in the deeper parts of the convection zone the stratification is very nearly adiabatic, and the structure is largely controlled by the equation of state and the composition.

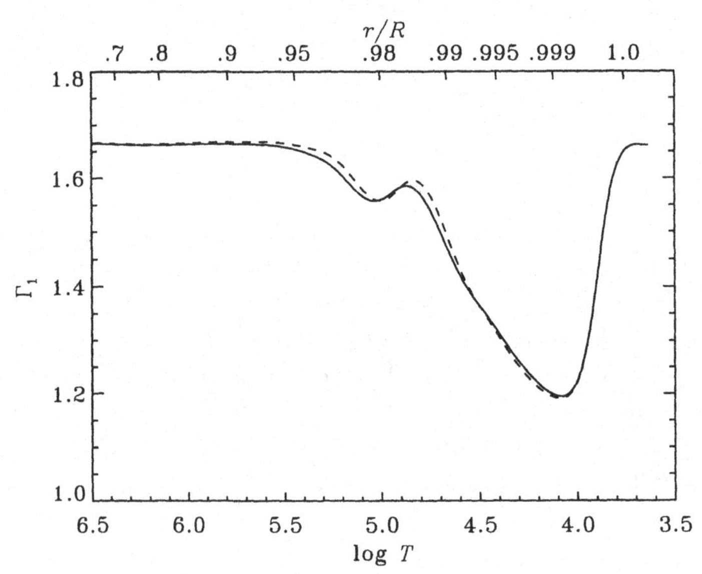

Figure 20: *Adiabatic exponent* $\Gamma_1 = (\partial \ln p / \partial \ln \rho)_s$*, plotted at the conditions* $(\ln \rho, \ln T)$ *in a normal solar model. The lower abscissa indicates* $\log T$ *(*log *being to base* 10*) while the upper abscissa shows the corresponding fractional radius* r/R*. The solid line was obtained using the EFF equation of state, while the dashed line is based on the CEFF formulation.*

Much of the uncertainty in the treatment of the thermodynamic properties of solar matter is related to the ionization processes, which depend crucially on the interactions between the constituents of the gas. The hydrogen and first helium ionization zones are situated so close to the surface that their structure may be affected by the uncertain physics of convection. However, at the second helium ionization zone these effects are probably sufficiently weak that analysis of the influence of this region on the frequencies may be used for tests of the equation of state, or for determinations of the solar envelope helium abundance.

In the present section I illustrate the sensitivity of the structure and the oscillation frequencies to the treatment of the thermodynamic properties, by considering formulations of increasing complexity [see Christensen-Dalsgaard & Däppen (1992) for a much more detailed treatment]. Comparisons with the observed frequencies are deferred to Section 5.2.

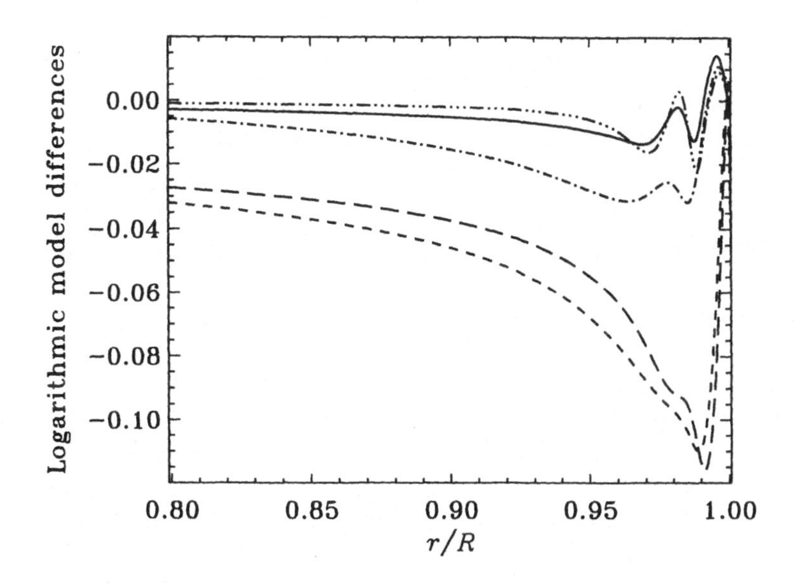

Figure 21: *Logarithmic differences at fixed radius r between models computed with the EFF and CEFF equations of state, in the sense (EFF) – (CEFF). The following line styles have been used:* $\delta \ln c$*: ———————; $\delta \ln p$: - - - - - - - - - - - -; $\delta \ln \rho$: – – – – – – –; $\delta \ln T$: —·—·—·—; $\delta \ln \Gamma_1$: —···—···—···.*

4.6.1 Comparison of EFF and CEFF formulations

The EFF and CEFF treatments (see Section 2.2.1 for details) differ in the inclusion in CEFF of Coulomb effects. These predominantly affect the structure of the convection zone through a change in the ionization balance. To illustrate this, Figure 20 shows Γ_1 computed with the EFF and CEFF equations of state, at conditions corresponding to a solar model. The ionization zones are reflected by dips in Γ_1: the dominant dip near the surface results from the combined effects of the hydrogen and first helium ionization zones, while the second helium ionization zone is visible as a separate dip, at $r \simeq 0.98R$, $\log T \simeq 5$. Inclusion of Coulomb effects, illustrated by the dashed line, causes a shift of the second helium zone towards lower temperature.

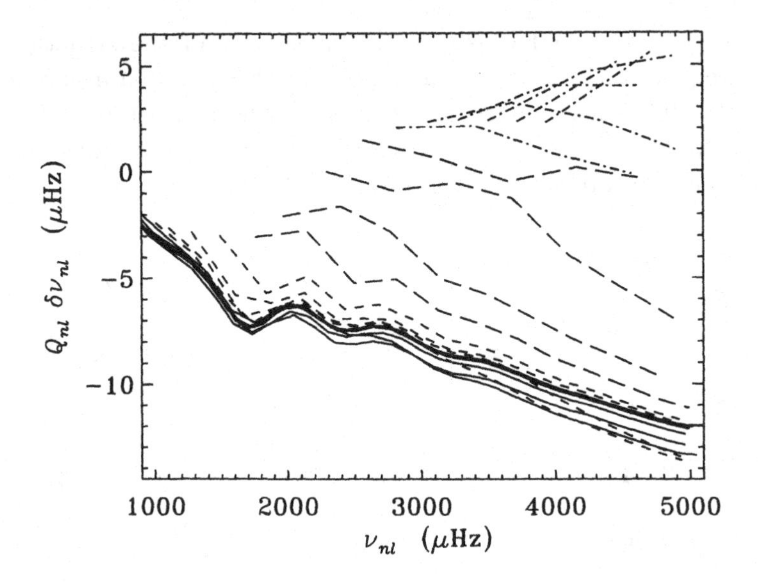

Figure 22: *Scaled frequency differences corresponding to the differences illustrated in Figure 21 between models computed with the EFF and CEFF equations of state. Modes of the same degree l have been connected, according to the following line styles:* $l = 0-30$: ————; $l = 40-100$: - - - - - - - - -; $l = 150-400$: — — — — —; $l = 500 - 1100$: —·—·—·—.

Differences between models computed with the EFF and CEFF equations of state are illustrated in Figure 21. The shift in the ionization causes an oscillatory feature in $\delta\Gamma_1$ which is reflected in the sound-speed difference. The effects in the inner parts of the model are considerably smaller. The corresponding scaled frequency differences are shown in Figure 22. These are dominated by the model changes in the hydrogen and helium ionization zones, leading the frequency differences which depend little on degree for $l \lesssim 100$; here the variation with frequency clearly displays an oscillation associated with the second helium ionization zone. For higher degree, the modes get trapped close to the surface and hence are affected predominantly by the positive sound-speed difference in the hydrogen and first helium ionization zones. It should be noticed that the frequency changes are comparatively large, up to about 10 μHz. Hence they are easily visible in the observed frequencies. Indeed, it was shown by Christensen-Dalsgaard, Däppen & Lebreton (1988) that the observations were

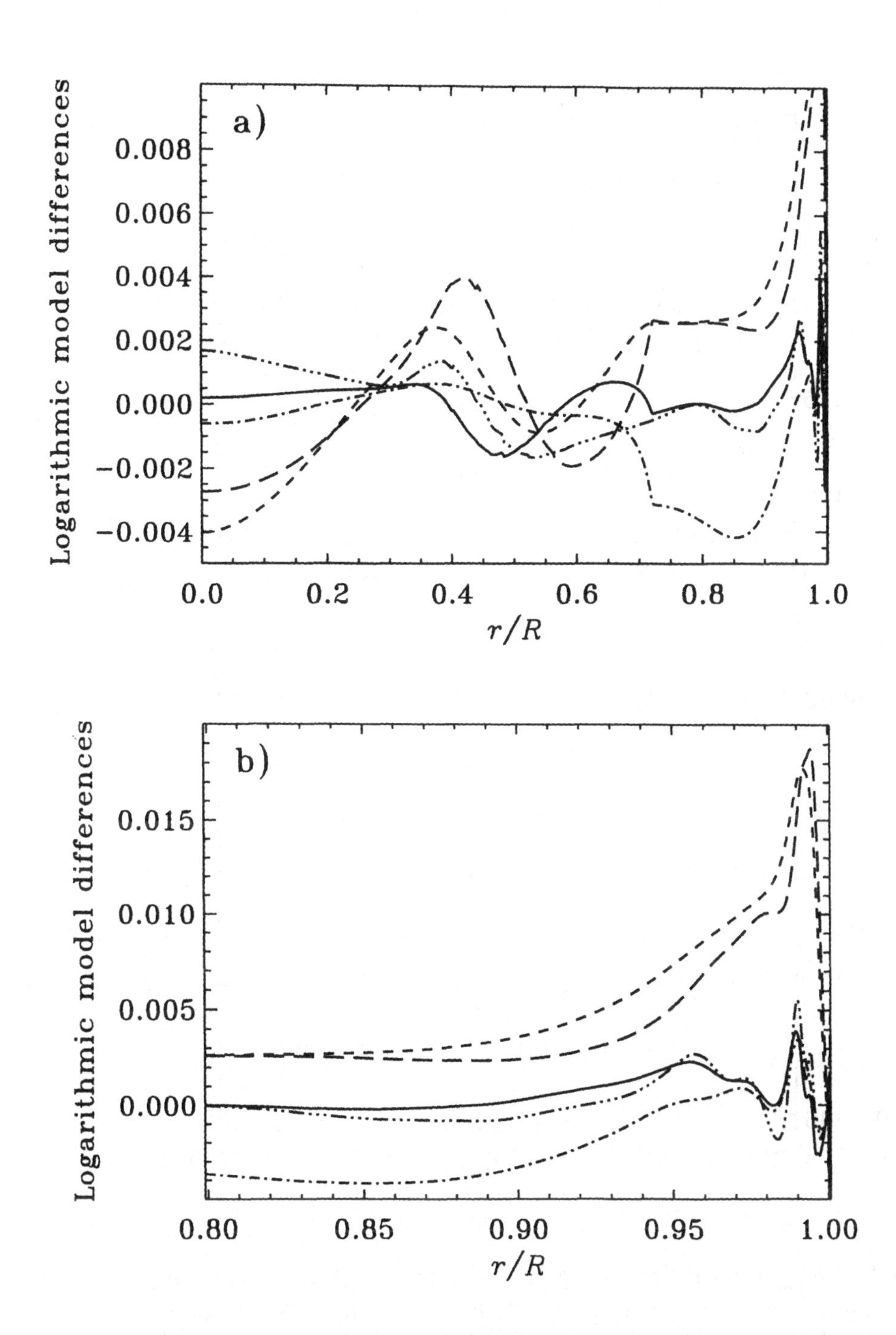

Figure 23: *Logarithmic differences at fixed radius r between models computed with the CEFF and MHD equations of state, in the sense (CEFF) − (MHD). The following line styles have been used:* $\delta \ln c$*: ————; $\delta \ln p$: - - - - - - - - - -; $\delta \ln \rho$: — — — — — —; $\delta \ln T$: —·—·—·—; $\delta \ln \Gamma_1$: —···—···—···. Panel* (a) *shows the entire model, panel* (b) *the outer parts of it.*

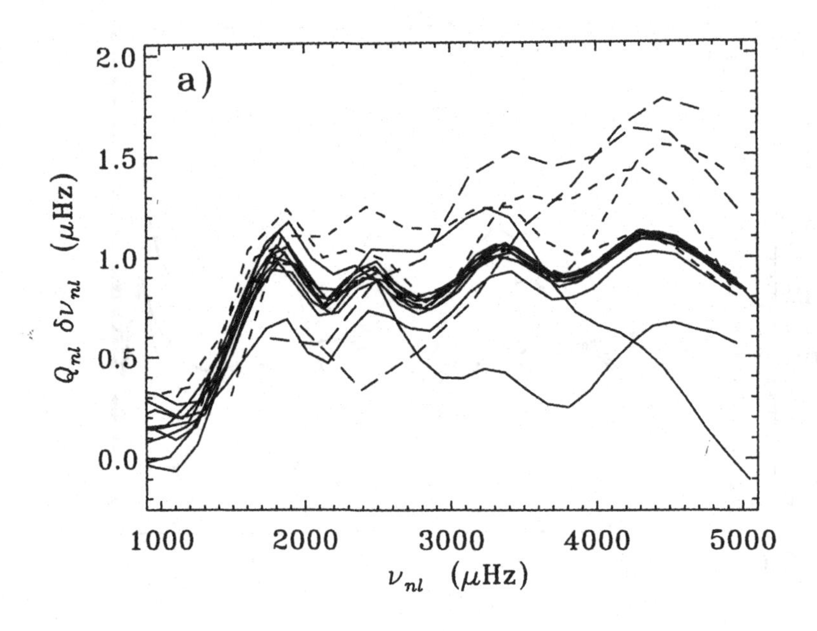

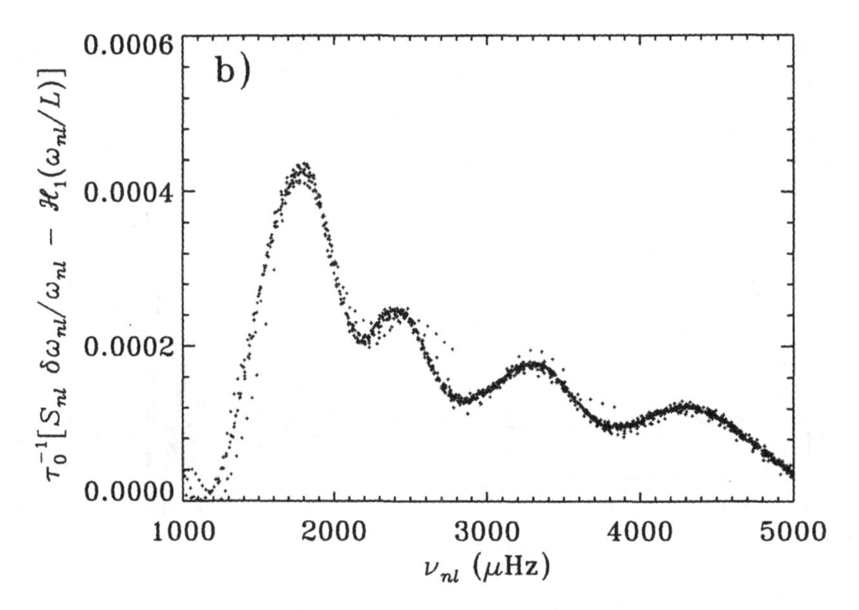

Figure 24: *Frequency differences corresponding to the differences illustrated in Figure 23 between models computed with the CEFF and MHD equations of state.* (a) *Original scaled frequency differences. Modes of the same degree l have been connected, according to the following line styles: $l = 0 - 30$: ————; $l = 40 - 100$: ------------; $l = 150 - 400$: — — — — —; $l = 500 - 1100$: —·—·—·—.* (b) *Scaled differences after subtraction of the function $\mathcal{H}_1(\omega/L)$ resulting from the spline fit in equation* (23).

clearly inconsistent with the EFF equation of state (see also Section 5.2 below).

4.6.2 Comparison of CEFF and MHD formulations

To illustrate the considerable sensitivity of the oscillation frequencies to the equation of state I next consider differences between models computed with the CEFF and MHD equations of state. These formulations differ mainly in the treatment of the partition function. The resulting model differences, shown in Figure 23, display a great deal of structure, much of it evidently still associated with the hydrogen and helium ionization zones. Also, the changes are roughly an order of magnitude smaller than those obtained between the EFF and CEFF models. The associated scaled frequency differences, shown in Figure 24(a), are correspondingly smaller and somewhat more complex than those obtained for the EFF – CEFF differences. Even so, the oscillatory signature of the second helium ionization zone is clearly visible. Also, it should be noticed that the magnitude of the differences is still much bigger than the observational error; thus it is plausible that the observations can distinguish between these two formulations. In Section 5.2 I show that this indeed the case.

The finer details in the frequency differences can be shown more clearly by carrying out a fit of the form given in equation (23). Here I concentrate on the function $\mathcal{H}_2(\omega)$, by showing in Figure 24(b) the residuals after subtraction of the fitted function $\mathcal{H}_1(\omega/L)$. These residuals are indeed predominantly a function of frequency and very clearly displays an oscillation, with a 'period' approximately corresponding to a feature at the location of the second helium ionization zone.

5 ANALYSIS OF OBSERVED FREQUENCIES

In the preceding section I considered a few aspects of the observed frequencies of solar oscillation; these indicated that the structure of the solar core is similar to that of normal solar models, and placed stringent limits of a possible adiabatic extension of the convective envelope through convective overshoot. Here I make a more detailed comparison of the observed frequencies with the models.

Two different sets of observed frequencies are used in the analysis. One is the compilation by Libbrecht *et al.* (1990) which combines Big Bear Solar Observatory data for $l \geq 3$ with low-degree data obtained from whole-disk observations (Jiménez *et al.* 1988). In the second (in the following BISON-BBSO)

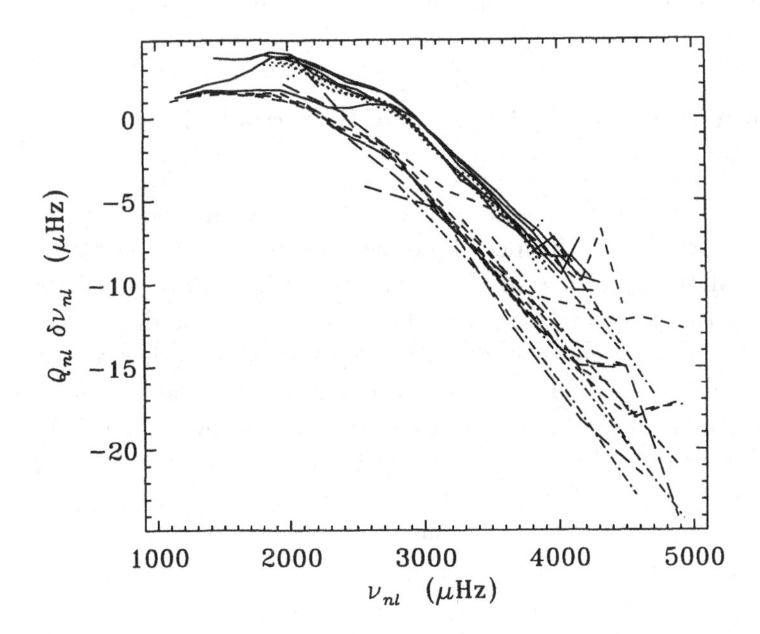

Figure 25: *Scaled frequency differences between the BISON-BBSO set of observed frequencies and the normal solar model of Christensen-Dalsgaard* et al. *(1993), in the sense (observations) – (model), plotted against frequency. Modes of the same degree l have been connected, according to the following line styles: $l = 0-3$: ················; $l = 4-30$: ————————; $l = 40-100$: ------------; $l = 150-400$: – – – – – – –; $l = 500-1100$: —·—·—·—.*

the modes with $l \leq 3$ were obtained from more recent data from the BISON network (Elsworth *et al.* 1994). The observations are compared with models differing predominantly in the equation of state, or in the inclusion or neglect of diffusion and gravitational settling of helium. Except where otherwise noted, the models were computed as done by Christensen-Dalsgaard *et al.* (1993). This includes the CEFF equation of state, OPAL opacities and the Parker (1986) nuclear reaction parameters. Calibration to the luminosity of the present Sun required an initial helium abundance $Y_0 \simeq 0.28$.

5.1 Asymptotic analysis of observed frequencies

A first step in the analysis is evidently to consider differences between observed and computed frequencies. The suspicion of problems in the superficial layers of the models motivates the inclusion of scaling by the normalized mode inertia Q_{nl} or the equivalent S_{nl}/τ_0 (*cf.* Section 3.4.4). Typical results, for a normal solar model with no diffusion, are shown in Figure 25. It is evident that the scaled differences do indeed depend mostly on frequency, as would be the case if the errors in the model were concentrated near the surface. The departures from this trend are so small as to be barely visible in a plot of this form. The dominant effect is a separation between modes of degree $l \leq 20$ and $l \geq 50$, with a transition for intermediate modes. This is strongly reminiscent of the behaviour found in Sections 4.2 and 4.3 to result from an increased depth of the convection zone, suggesting that the convection zone in the model is too shallow (this effect was first noticed by Christensen-Dalsgaard & Gough 1984). Nonetheless, it is clear that a more careful analysis is required to uncover the details of this and possible other problems in the interior of the model.

To isolate the more subtle features I employ the asymptotic analysis already used on computed differences in Section 4.3. Figure 26(*a*) shows the scaled differences (using now the asymptotic scaling) against ν/L and hence turning-point position. It is evident already from this raw difference plot that in this case the term in $\mathcal{H}_2$ dominates, as was also noted in Figure 25. However, there is also evidence for a contribution from $\mathcal{H}_1$. This becomes clearer if the spline fit is carried out and the contribution from $\mathcal{H}_2$ is subtracted from the scaled differences. The result is shown in Figure 26(*b*), together with the fitted function $\mathcal{H}_1(\omega/L)$. There is again a sharp step corresponding in position to $r_t \simeq 0.7R$, *i.e.*, the base of the convection zone. This confirms the evidence from the simple inspection of frequency differences in Figure 25 that the convection zone in the Sun is slightly deeper than in the model. Indeed, the convection-zone depth in the model is $d_b = 0.278R$, somewhat smaller than the solar value of $d_b = 0.287R$ inferred by Christensen-Dalsgaard *et al.* (1991).

It is evident that there is considerably more scatter in Figure 26(*b*) than in the corresponding Figure 12(*b*). This is due to observational errors, both random and systematic. In particular, it may be noticed that there is an apparent break at around $\nu/L \simeq 15\,\mu\text{Hz}$. In fact, the observed frequencies were obtained from two separate sets of observations, the merge taking place at $l = 400$; it has later been found that there were slight systematic errors in the high-degree set. Furthermore, there appear to be problems at low degree, corresponding to the highest values of ν/L. These difficulties are clearly reflected in the fitted $\mathcal{H}_1(\omega/L)$.

The residual after subtraction of the fitted $\mathcal{H}_1$ from the scaled differences, and the fitted $\mathcal{H}_2$, are shown in Figure 27. The residuals are indeed largely a function of frequency. They are dominated by a slowly varying trend which, as argued in Section 5.3 below, reflects errors in the near-surface region of the model. However, there is also a weak but clearly noticeable oscillatory signal. As discussed in Section 5.2 this probably reflects differences between the Sun and the model in the helium abundance and equation of state in the convective envelope.

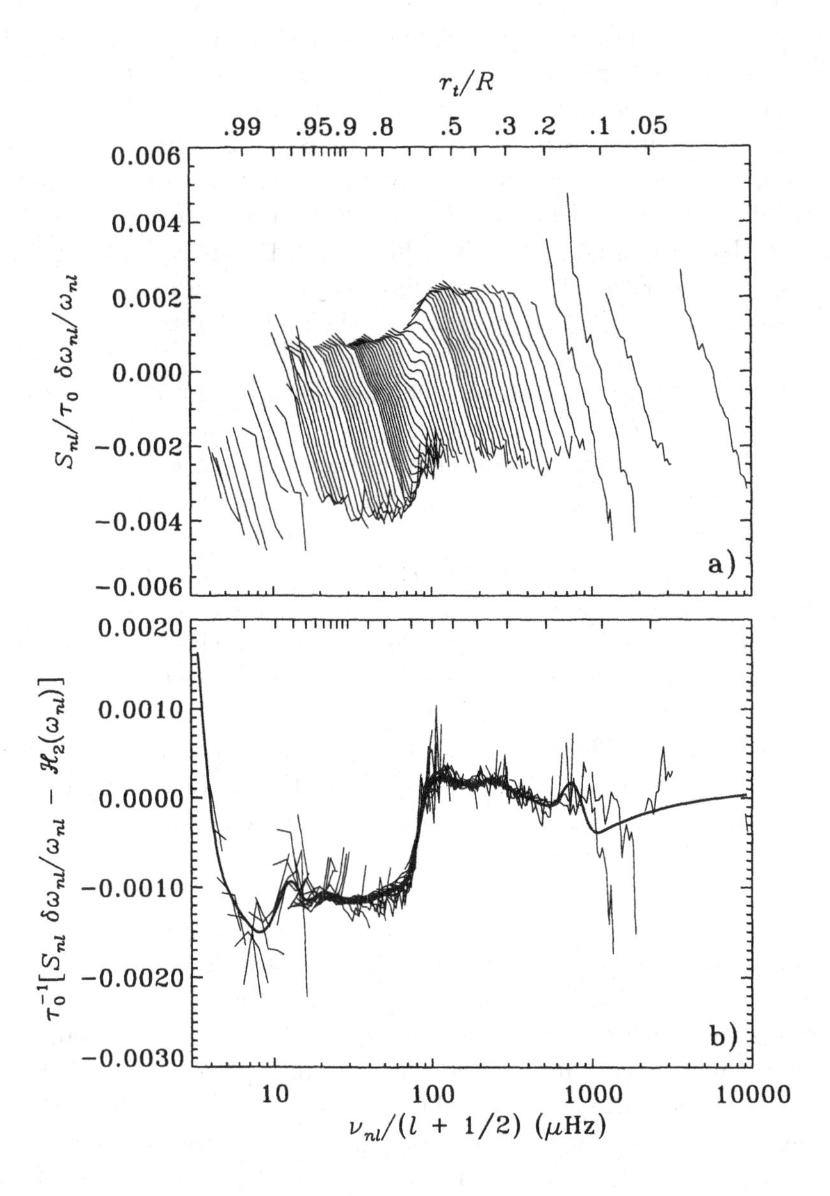

Figure 26: *Asymptotically scaled frequency differences between the observed frequencies of Libbrecht et al. (1990) and the normal solar model of Christensen-Dalsgaard et al. (1993), in the sense (observations) – (model), plotted against* $\nu/(l+1/2)$. *The upper abscissa shows the location of the lower turning point, which is related to* $\nu/(l+1/2)$ *through equation (10).* (a) *Original scaled frequency differences.* (b) *Scaled differences, after subtraction of the function* $\mathcal{H}_2(\omega)$ *obtained from the spline fit. The heavy solid line shows the fitted function* $\mathcal{H}_1(\omega/L)$.

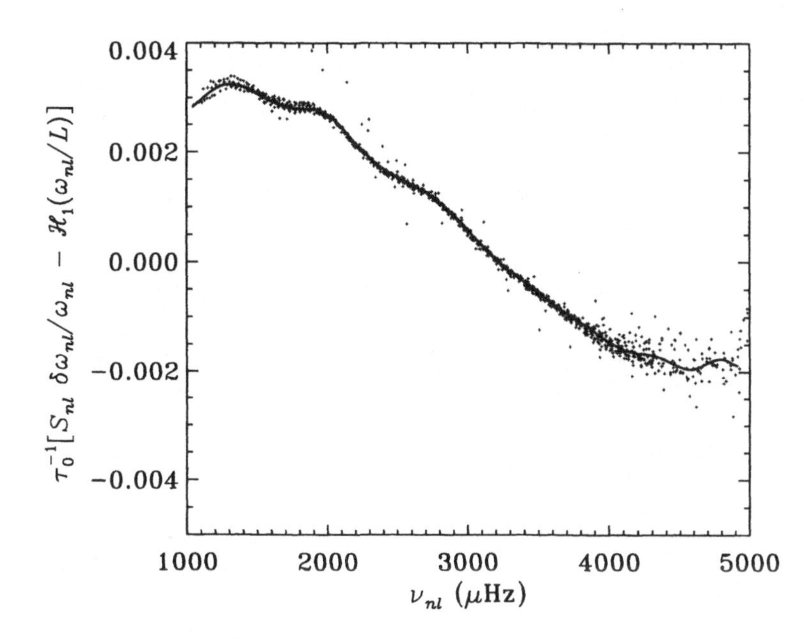

Figure 27: *Scaled frequency differences between observations and model, shown in Figure 26*(a), *after subtraction of the function $\mathcal{H}_1(\omega/L)$ resulting from the spline fit. The heavy solid line shows the fitted function $\mathcal{H}_2(\omega)$.*

The function $\mathcal{H}_1(\omega/L)$ is related to the sound-speed difference between two models, or between the Sun and a model, through equation (24). Given a determination of $\mathcal{H}_1$ this equation is an integral equation for $\delta c/c$, with the solution

$$\frac{\delta c}{c} = -\frac{2a}{\pi}\frac{\mathrm{d}}{\mathrm{d}\ln r}\int_{a_s}^{a}(a^2 - w^2)^{-1/2}\mathcal{H}_1(w)\mathrm{d}w \tag{67}$$

(Christensen-Dalsgaard *et al.* 1989), where $a = c/r$ and $a_s = a(R)$. This provides one of the simplest examples of an *inverse analysis* to infer properties of the solar interior from the observed frequencies (see the chapter by Gough). In fact, the asymptotic relation (12) and refinements of it also lead to *absolute* inversion methods whereby the solar sound speed is determined without reference to a solar model (*e.g.* Gough 1984b; Christensen-Dalsgaard *et al.* 1985; Vorontsov & Shibahashi 1991).

Tests of the inversion method given by equation (67) show that it provides reasonably accurate results in the range $0.2R < r < 0.95R$, where the asymptotic description is approximately valid (Christensen-Dalsgaard *et al.* 1989). Here I apply it to the differences between the solar and the computed frequencies shown in Figure 26. The results are shown as the dotted line in Figure 28. The thin lines illustrate the effects of the error estimates for the solar frequencies, as quoted by the observers. Evidently the formal error on the result is extremely

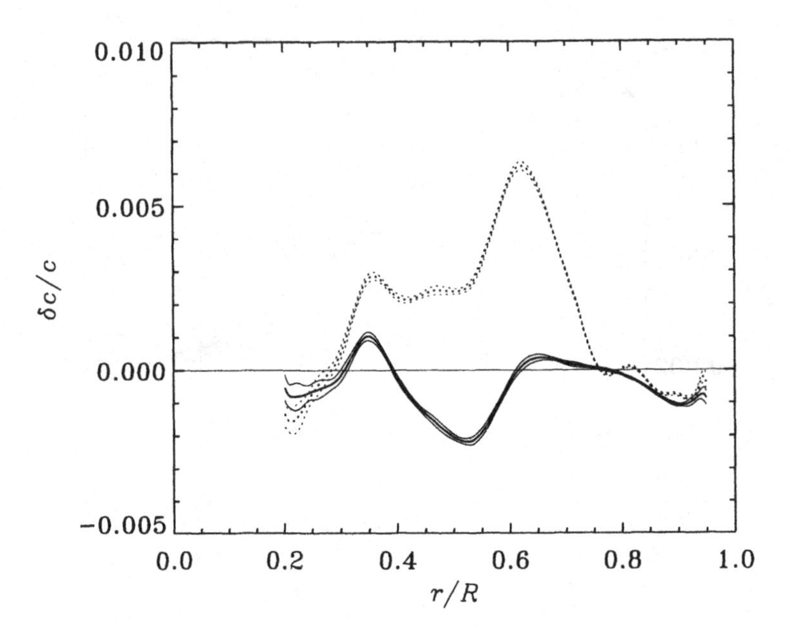

Figure 28: *The dotted line shows the sound-speed difference* $\delta c/c$ *between the Sun and a normal solar model, inferred by applying equation (67) to the function* $\mathcal{H}_1(\omega/L)$ *shown in Figure 26*(b)*, corresponding to differences between observed and model frequencies. The solid line shows* $\delta c/c$*, similarly obtained, between the Sun and a model including diffusion and gravitational settling of helium. The thin lines indicate* 1σ *error limits, based on the errors in the observed frequencies. Adapted from Christensen-Dalsgaard et al. (1993).*

small. Also, the sound-speed differences are small, corresponding to errors in T/μ in the models of generally less than 1 per cent. Nonetheless, the differences are clearly much larger than the errors resulting from the observations.

To illustrate the effects of improvements in the description of the solar interior the solid lines show the sound-speed difference obtained from the differences between the solar frequencies and those of a model including the effects of helium diffusion and settling (Christensen-Dalsgaard *et al.* 1993). It is interesting that the relatively subtle, and previously commonly neglected, effect of gravitational settling leads to a substantial improvement between the model and the observations, partly by increasing the depth of the convection zone in the model. This is a striking illustration of the power of helioseismology to probe the details of the physics of the solar interior. Furthermore, it should be stressed that the models were computed without any attempt to match the observed frequencies. It is remarkable that our relatively simple description of solar evolution, using physics based on laboratory experiments, permits us to reproduce the sound speed in the solar interior to within a fraction of a per cent. On the other hand, it must be pointed out that the model is not unique: it is likely that modest modifications in the opacity, well within the precision of current opacity tables, might introduce changes in the sound speed of similar magnitude. The separation of opacity uncertainties from effects of diffusion and settling is a major challenge, which will undoubtedly require better physical understanding of the processes involved.

The results of Christensen-Dalsgaard *et al.* (1993) are in apparent conflict with those obtained by Guenther, Pinsonneault & Bahcall (1993), who concluded that current frequencies do not permit a definite helioseismic test of the effects of diffusion. However, Guenther *et al.* (1993) based their analysis on simple differences between observed and computed frequencies. Thus their results were dominated by the the effects of the near-surface uncertainties in the model, causing frequency differences of order $10\,\mu$Hz, which apparently masked the rather smaller effects resulting from diffusion. This provides a clear illustration of the need for careful analysis to isolate the sometimes quite subtle features of the solar interior in the observed frequencies.

5.2 Test of the equation of state

In Section 4.6 I demonstrated the sensitivity of the computed frequencies to description of the thermodynamics of the solar interior. A closely related issue is the use of the observed frequencies to determine the helium abundance in the convection zone. Here I illustrate how comparisons with the observed frequencies may be used to test the equation of state.

An early indication of the power of the frequencies in this regard was obtained by Christensen-Dalsgaard, Däppen & Lebreton (1988) who compared observed frequencies with frequencies computed with the EFF and MHD equations of state (*cf.* Section 2.2.1). Figure 29 shows corresponding results, although comparing instead the EFF and CEFF formulations. In both cases, the differences are dominated by a frequency-dependent trend, clearly resulting from errors in the near-surface layers. However, it is evident that in the EFF model there is a very considerable dependence of the differences on degree at given frequency, indicating a dependence on the turning-point position and hence a sign of errors in the interior of the model. It should be noticed that this spread is found even amongst modes of degree $l \geq 50$ trapped in the convection zone, where the equation of state is the dominant source of uncertainty. Also, the differences are comparatively large at low frequency, again indicating a component of the error in the model outside the superficial layers. In contrast, the spread with l is much reduced in the CEFF model; also, much of it occurs for modes with turning points near the base of the convection zone (compare, for example, with

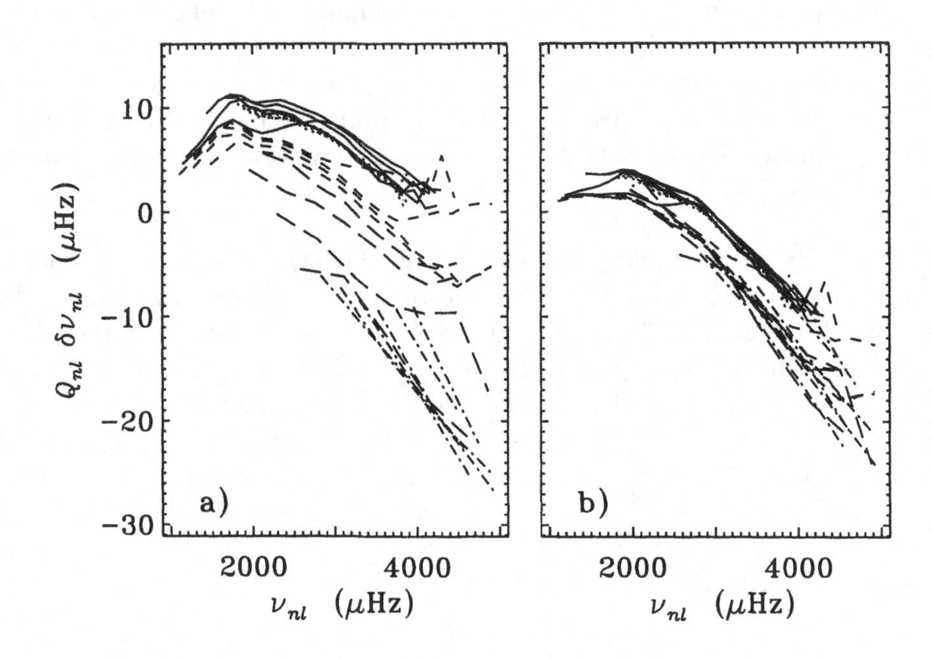

Figure 29: *Scaled frequency differences between the BISON-BBSO set of observed frequencies and frequencies of two solar models, in the sense (observations) – (model), plotted against frequency. Modes of the same degree l have been connected, according to the following line styles: $l = 0 - 3$:* ················*;* $l = 4 - 30$: ————*;* $l = 40 - 100$: - - - - - - - - -*;* $l = 150 - 400$: — — — — — — —*;* $l = 500 - 1100$: —·—·—·—. (a) *Model computed with the EFF equation of state.* (b) *Model computed with the CEFF equation of state (this uses the same data as Figure 25).*

Figure 10) and corresponds to the comparatively large sound-speed difference between the Sun and the model already inferred from the inversion in Section 5.1 (*cf.* Figure 28). Furthermore, the differences at low frequency are now quite small. Thus there are strong indications that the CEFF model provides a better representation of the solar convection zone than does the EFF model. The improvement in the models resulting from the consistent inclusion of Coulomb effects was also noted by Stix & Skaley (1990).

A more detailed comparison can be made, as usual, by carrying out a fit of the form given in equation (25) to the differences between the observed and computed frequencies. Figure 30 shows the resulting functions $\mathcal{H}_1(\omega/L)$ for models computed with the EFF, the CEFF and the MHD equations of state. The fit only determines $\mathcal{H}_1$ to within a constant: in fact, it follows from equation (67) that the sound-speed difference is essentially determined by the gradient of $\mathcal{H}_1$. Thus the figure clearly confirms the improved agreement resulting from using

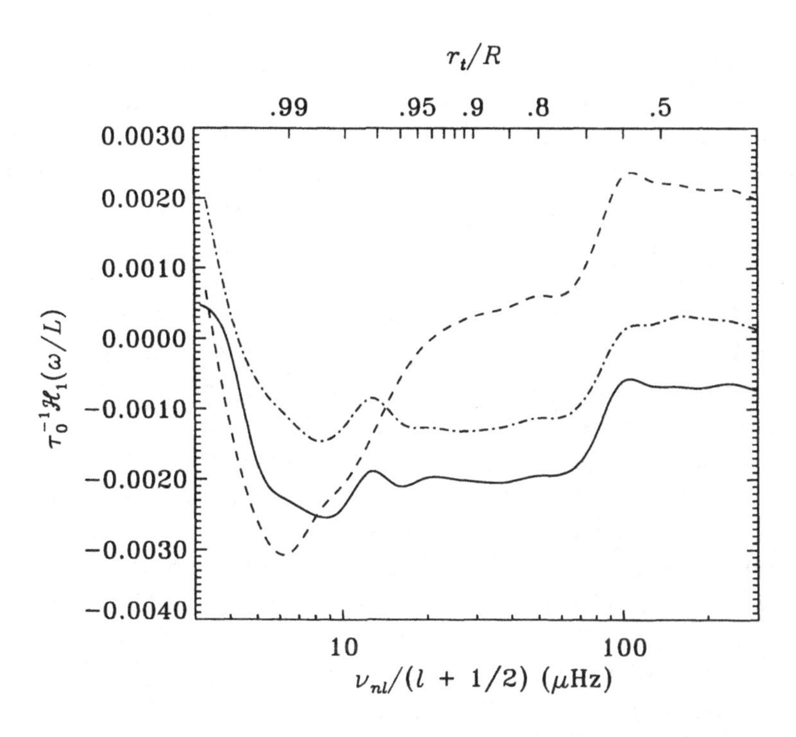

Figure 30: *Functions $\mathcal{H}_1(\omega/L)$ fitted to asymptotically scaled frequency differences between the BISON-BBSO set of observed frequencies, in the sense (observations) – (model), and three solar models: a model using the EFF equation of state (dashed line), a model using the CEFF equation of state (solid line), and a model using the MHD equation of state (dot-dashed line). The physics otherwise corresponds to the non-diffusive model of Christensen-Dalsgaard et al. (1993). The upper abscissa shows the location of the lower turning point, which is related to $\nu/(l + 1/2)$ through equation (10).*

the CEFF formulation, compared with EFF. On the other hand, there is no clear distinction in this type of analysis between the MHD and CEFF formulations (see also Christensen-Dalsgaard & Däppen 1992). The fairly steep rise in $\mathcal{H}_1(\omega)$ at low $\nu/(l + 1/2)$ may be associated with near-surface errors in the model and oscillation physics or with errors in the assumed asymptotic formulation. Also, the curves clearly show the increase in $\mathcal{H}_1$ near $\nu/(l+1/2) = 100\,\mu$Hz, associated with the difference in convection-zone depth between the models and the Sun. The effects of the equation of state on $\mathcal{H}_1(\omega/L)$ were also investigated by Antia & Basu (1994) and Basu & Antia (1995).

The phase function $\alpha(\omega)$ appearing in the Duvall law (equation 12), or the function $\mathcal{H}_2(\omega)$ obtained from the asymptotic fit, apparently provide even more sensitive discrimination between different equations of state and measures of the envelope helium abundance Y_e. Vorontsov *et al.* (1991) analyzed properties of $\alpha(\omega)$ to infer that $Y_e \simeq 0.25$. A similar value was obtained by Christensen-

Dalsgaard & Pérez Hernández (1991) from analysis of $\mathcal{H}_2(\omega)$. Also, Pamyatnykh, Vorontsov & Däppen (1991) investigated the sensitivity of functions related to $\alpha(\omega)$ to various aspects of the convective envelope. A serious problem in using these phase functions is the fact that they are generally dominated by contributions coming from the uncertain near-surface region. However, it was argued in Section 3.4.4 that these contributions are generally slowly varying functions of frequency, whereas contributions coming from somewhat deeper parts of the Sun oscillation with frequency (see also Figure 4 and Section 4.1). Vorontsov, Baturin & Pamyatnykh (1992) developed a polynomial fitting procedure which provided a separation between the slowly and the rapidly varying parts of the phase function. Working in terms of phase-function differences, Pérez Hernández & Christensen-Dalsgaard (1994a) introduced a filtered function $\mathcal{H}_2^{\rm f}(\omega)$, obtained by passing $\mathcal{H}_2(\omega)$ through a high-pass filter and hence suppressing the near-surface effects. Here I illustrate the use of such procedures by showing in Figure 31 the result of applying this filter to the $\mathcal{H}_2(\omega)$ resulting from differences between observed and computed frequencies, together with the original $\mathcal{H}_2(\omega)$. It is evident that the slowly-varying trend has indeed been eliminated, leaving an oscillatory function of frequency with a period corresponding roughly to the depth of the second helium ionization zone (*cf.* Section 3.4.3).

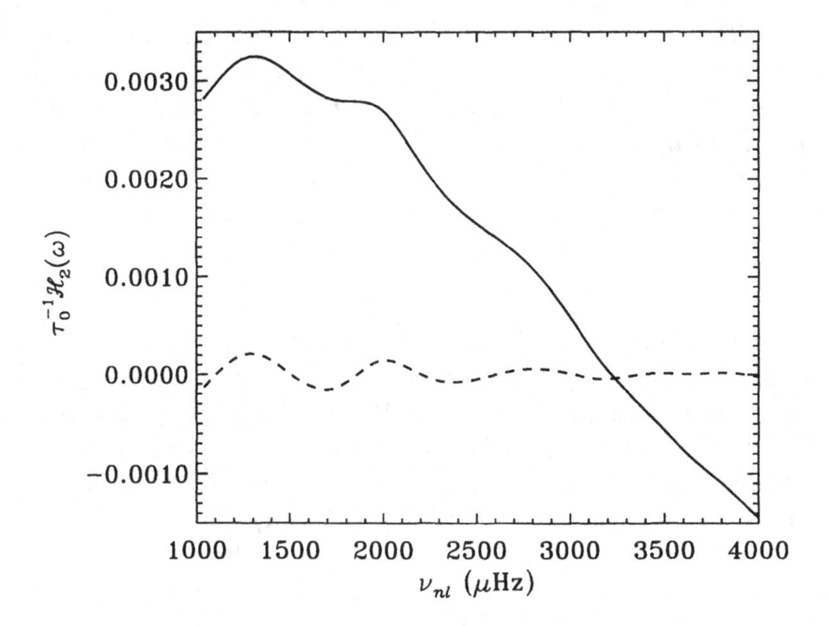

Figure 31: *The solid curve shows the function $\mathcal{H}_2(\omega)$ obtained in a fit to scaled frequency differences between observations and model (see Figure 27). The dashed curve shows the result $\mathcal{H}_2^{\rm f}(\omega)$ of applying the high-pass filter of Pérez Hernández & Christensen-Dalsgaard (1994a) to this function.*

The signal shown in Figure 31 contains contributions from several different

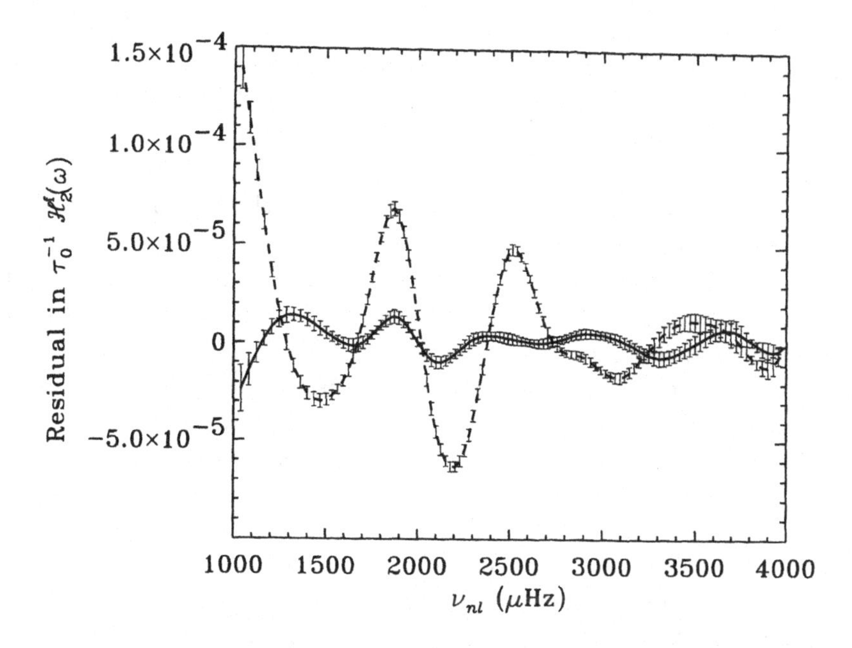

Figure 32: *Residuals, with 3σ error bars, from fits of the filtered phase functions $\mathcal{H}_2^{\rm f}(\omega)$ corresponding to differences between observed and model frequencies. The fits involve contributions from changes in the helium abundance, mixing-length parameter and near-surface structure. The solid curve corresponds to a reference model using the MHD equation of state, while the dashed curve corresponds to the CEFF equation of state. (From Pérez Hernández & Christensen-Dalsgaard 1994b.)*

sources, including errors in the equation of state, differences between the composition and the specific entropy of the solar and the model convection zones and residual effects of the near-surface errors. To separate these effects Pérez Hernández & Christensen-Dalsgaard (1994b) carried out a least-squares fit to the observed $\mathcal{H}_2^{\rm f}(\omega)$ of a linear combination of three contributions: a contribution from a change in the envelope helium abundance; a contribution of a change in the mixing-length parameter (and hence the specific entropy in the adiabatic part of the convection zone); and a contribution corresponding to a change in the atmospheric opacity and representing effects of near-surface errors. The effect of each individual parameter was represented by a function $\mathcal{H}_2^{\rm f}(\omega)$ obtained from differences between an envelope model incorporating a change in the given parameter and a reference model. Figure 32 shows the residuals of such fits to $\mathcal{H}_2^{\rm f}(\omega)$ obtained from differences between the observed frequencies and frequencies for two reference models: a model computed using the CEFF equation of state and a model computed with the MHD equation of state. It is evident that the MHD model provides a much closer fit to the observations, when analyzed in this way, than does the CEFF model. In fact, the residual for the CEFF case is of the same order of magnitude as the oscillatory component of $\mathcal{H}_2(\omega)$ for the differences between the CEFF and MHD equations of state (*cf.* Figure 24*b*); this suggests that there is relatively little cross-talk between the effects of using the CEFF equation of state and the remaining uncertainties in the model, in particular the helium abundance. It should be noted, however, than even for the MHD model the residuals are substantially larger than the 3σ error bars shown in the figure, indicating that the MHD model is still not consistent with the Sun at the level of the observational errors. Nonetheless, these results suggest that the MHD equation of state provides a better representation of the thermodynamic properties of solar matter than the CEFF formulation, at least in the second helium ionization zone which dominates the signal shown in Figure 31. This is hardly surprising: the CEFF treatment is relatively simple, compared with the complex handling by MHD of interactions between the constituents of the gas.

A similarly detailed test of the Livermore equation of state (*cf.* Section 2.2.1) has yet to be carried out. However, preliminary results indicate that the $\mathcal{H}_2(\omega)$ corresponding to the difference between models computed with the Livermore and the MHD equations of state differs in shape from the effect of a change in Y_e; this suggests that the Livermore formulation may be less successful in fitting the observations than MHD. On the other hand, it appears from sound-speed inversions there may be problems with the MHD formulation at temperatures somewhat exceeding that of the second helium ionization zone (Dziembowski *et al.* 1992).

As a result of their fit, Pérez Hernández & Christensen-Dalsgaard (1994b) estimated the envelope helium abundance as $Y_e \simeq 0.242$, largely consistent with the value of 0.25 quoted above. Similar values were also obtained by Basu & Antia (1995) from analyzing $\mathcal{H}_1$ and $\mathcal{H}_2$, using either the MHD or the Livermore equations of state. However, these results may still compromised by errors in the equation of state, and by possible systematic errors in the analysis procedure. Kosovichev *et al.* (1992), using non-asymptotic inversion techniques, made an extensive investigation of the uncertainties in the inferred value of Y_e resulting from differences in the equation of state, the choice of mode set and the inversion procedure; the results suggest that current estimates of Y_e must be viewed with some caution. Nevertheless, it is interesting to compare the results with those obtained from the calibration of solar models to the correct present luminosity, where a value of the initial helium abundance Y_0 of 0.27 – 0.28 is typically required. This is probably inconsistent with the helioseismic estimates, even given the uncertainty in the equation of state. However, the results of Section 4.3 on models with helium diffusion and settling (see in particular Figure 11) shows that these effects reduce Y_e by about 0.03 relative to the initial value. Thus for calibrated models that include helium settling the present value of Y_e is close to the values inferred from helioseismology.

Despite the problems of separating effects of composition and equation of state, the results shown here demonstrate that current observations of solar oscillations are sensitive to aspects of the equation of state beyond the inclusion of the Coulomb effects. This offers hope that these data, and the substantially more accurate data expected from new helioseismic experiments, will provide new physical insight into the properties of the thermodynamics of hot partially ionized gases.

5.3 Matching the near-surface effects

Figure 28 indicates that with the inclusion of helium diffusion and settling the interior structure of solar models is very close to the that of the Sun. Also, the results in Section 5.2 suggested that the MHD equation of state provides a reasonable representation of the thermodynamics of the solar convection zone. To illustrate how these properties are reflected in the comparison between observed and computed frequencies, Figure 33(*a*) shows scaled frequency differences between the Sun and a model including helium settling, computed with the MHD equation of state (Basu *et al.* 1995). It is obvious that there is little evidence here for errors in the interior of the model, as would have been indicated by an l-dependence of the scaled differences or a substantial difference at low frequency. The only significant exception is probably at very high degree where the assumption of vertical propagation in the near-surface layers breaks down (*e.g.* Antia 1995; see also equation 33). The remaining scatter is likely to be predominantly observational.

I argued in Section 2.5 that the frequency-dependent difference could derive from a number of errors in the model or frequency computations. As a simple illustration of the effects of near-surface modifications on the comparison between observations and models I here consider a model using the same physics as for Figure 33(*a*), except that the opacity has been increased by a factor 2.34 at temperatures below about 8 000 K. The resulting scaled frequency differences are shown in Figure 33(*b*). The change in atmospheric structure resulting from the opacity increase has clearly eliminated much of difference between observations and model; similar effects were also noted by Christensen-Dalsgaard (1990), Kim, Demarque & Guenther (1991) and Turck-Chièze & Lopes (1993). However, it is evident that there remain significant variations. Some of these are undoubtedly associated with the remaining differences between the interiors of the Sun and the model, illustrated in Figure 28. In addition, there is an indication of an oscillatory variation with frequency which might be related to errors in the equation of state or convection-zone helium abundance, as well as a contribution varying more slowly with frequency and hence probably concentrated very near the surface.

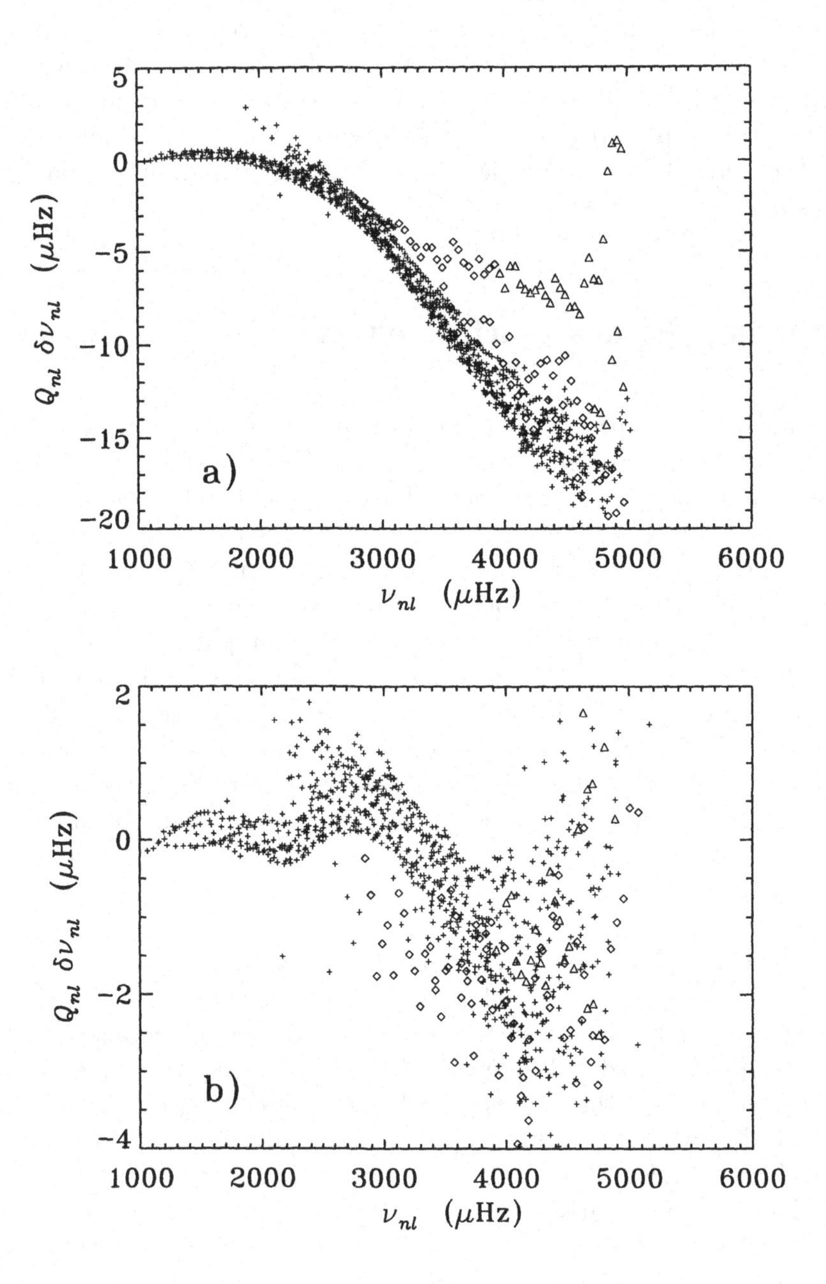

Figure 33: (a) *Scaled differences between observed frequencies and frequencies of a model including helium diffusion and settling, computed with the MHD equation of state. Crosses indicated modes with* $l \leq 500$, *diamonds are modes with* $500 < l \leq 1000$ *and triangles are modes with* $1000 < l$. (b) *As* (a), *but for a model where in addition the atmospheric opacity has been increased by a factor 2.34.*

It should be stressed that opacity errors of this magnitude in recent tables are quite unlikely. Thus this calculation cannot be regarded as a realistic attempt to explain the frequency-dependent part of the difference between observations and models. Instead, it is presented here as a simple example of the effects of near-surface changes. Similar effects arise from other types of modifications. It was shown in Section 4.1 that the observed behaviour can be mimicked by making the superadiabatic gradient steeper, as in, for example, the Canuto & Mazzitelli (1991) formulation (*e.g.* Physics Paternó *et al.* 1993; Monteiro *et al.* 1995ab; *cf.* Figure 8). Superficially similar effects arise when turbulent pressure is taken into account in the equilibrium model, or from effects of convective fluctuations (*e.g.* Rosenthal *et al.* 1995b), while the effects of nonadiabaticity or the perturbation in the turbulent pressure are still somewhat uncertain, due to the difficulties in modelling the effects of convection (*e.g.* Christensen-Dalsgaard & Frandsen 1983; Cox, Guzik & Kidman 1989; Balmforth 1992b; Guenther 1994; Rosenthal *et al.* 1995b).

The situation concerning the near-surface problems in the modelling is clearly rather unsatisfactory at present. However, hope is provided by the availability of increasingly realistic hydrodynamical simulations of this region, accompanied by a better physical understanding of the relevant processes. Also, the expected more accurate observations may permit us to distinguish between the different models proposed to account for the frequency behaviour. Finally, potentially very important information concerning convective effects might be obtained from observations of solar-like oscillations in other stars covering a range in effective temperature and surface gravity.

6 TOWARDS THE STARS?

It is evident that observations of solar oscillations are providing extremely detailed information about the properties of the solar interior. This gives a precise test of stellar evolution theory, including the physical information that enter into it, as applied to the Sun. However, the Sun is only a single specific example of a star, with a comparatively simple structure. It is obviously of great interest to obtain similar information about other stars.

This encounters two problems. Detection of small-amplitude oscillations, such as those observed in the Sun, is greatly complicated by the low light-level available for stars other than the Sun. And the apparent extent of such stars is so small that essentially no spatial resolution is possible; thus with few exceptions the observations are limited to modes of low degree, as in whole-disk observations of solar oscillations. As a result, the data for any individual remote star will always be much less extensive than the data available for the Sun. This, however, must be balanced against the possibility of studying stars of greatly varying parameters, such as mass and age, covering a corresponding range of physical properties and phenomena. A very important example are convective cores which are only found in stars more massive than the Sun: the properties of such convective cores and the associated mixing, which may include overshooting or weaker turbulence, are highly uncertain; yet such processes play a major rôle for the evolution of the stars on the main sequence and beyond.

The early analyses of this nature predates the first results of helioseismology. For a long time there appeared to be discrepancies between the observed periods of classical Cepheids and evolution models of these stars; the problem was particularly acute for double-mode Cepheids, as first pointed out by Petersen (1973). This and other discrepancies between pulsation observations and evolution calculations were reviewed by Cox (1980). The new computations of opacities, where a more careful treatment of lines has led to very substantial opacity increases (see Section 2.2.2), have largely solved these so-called "Cepheid mass problems" (*e.g.* Moskalik, Buchler & Marom 1992; Kanbur & Simon 1994; Christensen-Dalsgaard & Petersen 1995). It is interesting that the effects in the Cepheids are dominated by the opacity at temperatures between 10^5 and 10^6 K: this temperature range falls within the solar convection zone and the corresponding opacities have no effect on solar structure. Thus studies of the double-mode Cepheids complement the information that can be obtained about opacity from helioseismology.

Extensive data are now available for pulsating white dwarfs (for a review, see *e.g.* Winget 1991), providing precise measures of white-dwarf masses, information about the thickness of the outer hydrogen layer, constraints on the rotation rate and magnetic field and in some cases measurements of evolutionary effects, visible as frequency changes. On or just after the main sequence observations of β Cephei stars and other pulsating B stars are providing information about the properties of relatively massive stars (*e.g.* Gautschy 1990); the recent opacity revisions have provided a natural explanation for the excitation of oscillations in these stars (*e.g.* Cox *et al.* 1992; Kiriakidis, El Eid & Glatzel 1992; Moskalik & Dziembowski 1992). Also, extensive sets of frequencies, which may even include g modes, are becoming available for δ Scuti stars (*e.g.* Breger *et al.* 1993; Belmonte *et al.* 1994; Frandsen *et al.* 1995; for a review, see Matthews 1993). Both B stars and δ Scuti stars have convective cores and hence promise information that is unavailable for the Sun. Furthermore, by solar standards the amplitudes are large, with relative intensity variations of order 10^{-3}, making the oscillations relatively easy to detect; even so, the identification of the modes still give rise to considerable uncertainty. Finally, the rapidly oscillating Ap stars (*e.g.* Kurtz 1995) display frequency spectra with some superficial similarity to the solar oscillations of low degree, although strongly affected by the large-scale magnetic fields of these stars.

Even though a variety of stars are therefore good targets for asteroseismic investigations, it would clearly be of particular value to detect and study solar-like oscillations in stars other than the Sun. The diagnostic potential of such data is relatively well understood (Ulrich 1986; Christensen-Dalsgaard 1984, 1988, 1993; Gough & Novotny 1993; Brown *et al.* 1994; Audard, Provost & Christensen-Dalsgaard 1995). Furthermore, information on the dependence of the mode amplitudes and line widths on stellar parameters would provide important information about the excitation and damping processes, eventually perhaps leading to a better understanding of the properties of outer stellar convection zones. The detection of such oscillations has been elusive, however. So far a number of suggestive results have been obtained (*e.g.* Gelly, Grec & Fossat 1986; Innis *et al.* 1991; Brown *et al.* 1991; Pottasch, Butcher & van Hoesel 1992), although with no definitive identification of oscillations. Also, a very ambitious project involving a substantial number of large telescopes reached a very low detection threshold but still failed to make definitive detection (Gilliland *et al.* 1993). Basic problems in these attempts have been atmospheric noise in broad-band intensity measurements and photon noise and spectrograph stability in velocity measurements.

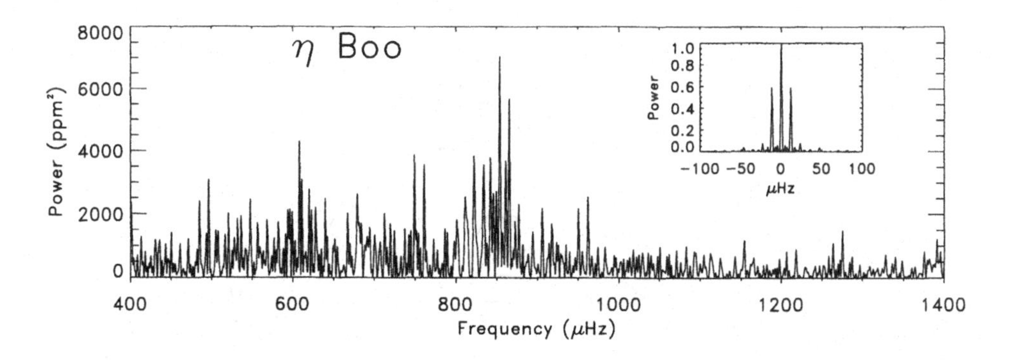

Figure 34: *Observed power spectrum of η Bootis. The inset shows the window function. (From Bedding & Kjeldsen 1995).*

6.1 η Bootis

The atmospheric effects can to a large extent be eliminated by making differential measurements, comparing the intensity in spectral lines with the intensity in the neighbouring continuum. Kjeldsen *et al.* (1995) showed that the integrated intensity in the Balmer lines, expressed in terms of the equivalent width, provides a sensitive measure of oscillations in stellar atmospheres. In this way they were able to detect probable solar-like oscillations in the sub-giant star η Bootis, from 6 nights of observations with the 2.5 m Nordic Optical Telescope. Figure 34 shows the relevant part of the resulting power spectrum. There are clear indications of excess power in the frequency range $700 - 950\,\mu$Hz, of a shape superficially similar to the amplitude distribution in the solar five-minute oscillations.

The interpretation of the spectrum is greatly complicated by the presence of strong daily side lobes. Through a correlation analysis Kjeldsen *et al.* determined the large frequency separation $\Delta\nu$ (*cf.* equation 34) as $\Delta\nu \simeq 40.3\,\mu$Hz. To determine individual frequencies they carried out a so-called CLEAN analysis on the spectrum. Such techniques are subject to considerable uncertainty in data with comparatively low duty cycle; in particular, there is some risk of mistaking a side lobe for the main peak. However, Kjeldsen *et al.* inferred frequencies of thirteen modes, including several closely-spaced pairs which they identified as having $l = 0$ and 2, in accordance with the asymptotic expression (34).

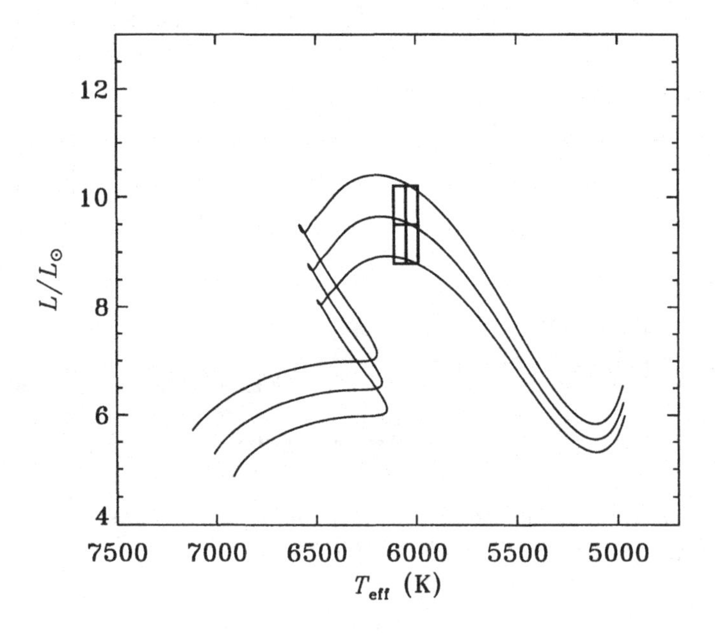

Figure 35: *Evolutionary tracks in the HR diagram, for models with* $Z = 0.03$, $X = 0.7$ *and a mixing-length parameter calibrated to obtain the proper solar radius. Models are shown with masses of* $1.6\,\mathrm{M}_\odot, 1.63\,\mathrm{M}_\odot$ *and* $1.66\,\mathrm{M}_\odot$. *The error box indicates the observed location of* η *Bootis. (Adapted from Christensen-Dalsgaard et al. 1995b).*

The interpretation of these data was considered by Christensen-Dalsgaard, Bedding & Kjeldsen (1995b). The star is sufficiently close that its distance is known with reasonable precision; from this its luminosity can be determined as $L = 9.5 \pm 0.7\,\mathrm{L}_\odot$. Also, spectroscopy shows that the effective temperature is $T_{\mathrm{eff}} = 6050 \pm 60\,\mathrm{K}$ and that the heavy-element abundance is somewhat higher than solar. Figure 35 shows the location of the star in a Hertzsprung-Russell diagram, together with evolutionary tracks for $Z = 0.03$ and three masses. These identify the star as being past the phase of central hydrogen burning, and with a mass of about $1.6\,\mathrm{M}_\odot$. Calculation of adiabatic frequencies shows that it is possible to find models in the error box with a $\Delta\nu$ which is consistent with the observed value. This provides an excellent test of the consistency of the frequency observations with the more classical stellar data: $\Delta\nu$ is essentially proportional to the the characteristic dynamical frequency $\omega_{\mathrm{dyn}} \equiv (GM/R^3)^{1/2}$ and hence is predominantly determined by the stellar radius; thus it is largely fixed by the location of the star in the HR diagram. On the other hand, this property also indicates that $\Delta\nu$ is relatively insensitive to the details of the stellar internal structure.

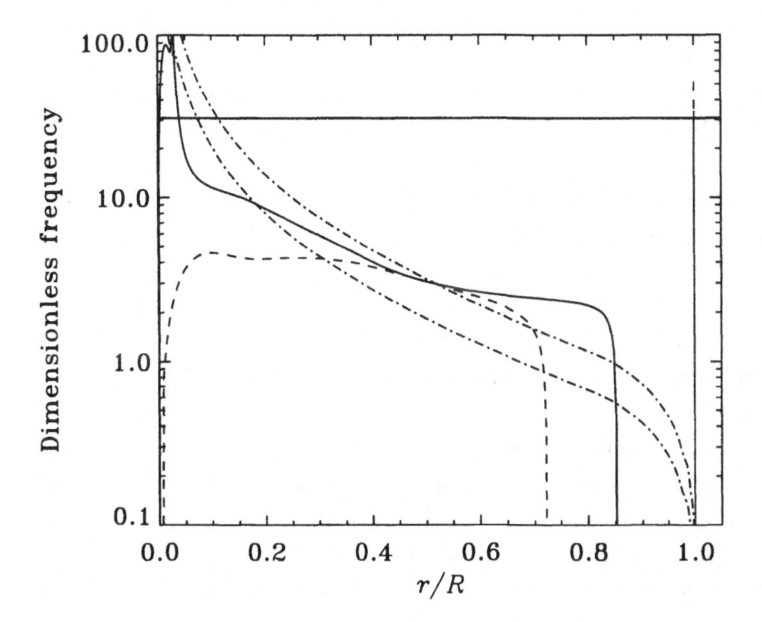

Figure 36: *Dimensionless buoyancy frequency* $\tilde{N} \equiv (GM/R^3)^{-1/2}N$ *plotted against fractional radius* r/R *for a model of the present Sun (dashed line) and a model of* η *Bootis (solid line). The dot-dashed lines show the dimensionless characteristic acoustic frequency* $\tilde{S}_l = (GM/R^3)^{-1/2}S_l$*, where* $S_l = cL/r$*, for* $l = 1$ *and* 2 *in* η *Bootis. The heavy horizontal line indicates the location of a mode in* η *Bootis of frequency* $850\,\mu\mathrm{Hz}$*, typical of the observed frequencies.*

To assist the understanding of the behaviour of the oscillations in η Bootis, Figure 36 shows the buoyancy frequency and characteristic acoustic frequencies in a model of η Bootis, in units of ω_{dyn}, and compare them with the buoyancy frequency in the present Sun. The dominant difference between the two models is the very large peak in $\tilde{N}$ near the centre of the η Bootis model. This is caused by two effects: during main-sequence evolution the retreating convective core leaves behind a steep gradient in the hydrogen abundance, leading to a highly stable stratification and hence contributing to a large value of N (*e.g.* Dziembowski & Pamyatnykh 1991; Audard, Provost & Christensen-Dalsgaard 1995); in addition, the increasing central condensation as the core contracts after hydrogen exhaustion drives up the gravitational acceleration in the core, further increasing N. As a result, the maximum value of N exceeds the acoustical cut-off frequency in the stellar atmosphere. Thus *all* trapped acoustic modes may in principle be affected by the buoyancy frequency, taking on g-mode character in the core.

It follows from the asymptotic properties of stellar oscillations (see, for example, Christensen-Dalsgaard & Berthomieu 1991) that a given mode behaves like a p mode where its frequency ω satisfies $\omega > N$ and $\omega > S_l$ and like a g mode where $\omega < N$ and $\omega < S_l$; in regions where $N < \omega < S_l$ the mode is evanescent, either decreasing or growing exponentially. Thus, at the frequencies characteristic for the observations of η Bootis, indicated by the horizontal line in Figure 36, the modes have extended p-mode regions in the outer parts of the star and a small g-mode region near the centre. The separation between these two regions is quite small for $l = 1$, leading to a substantial coupling between the two types of behaviour; with increasing l, the separation increases rapidly and the coupling becomes small.

The effects of this structure on the oscillations are illustrated in Figure 37. The frequencies of the radial modes, shown by dashed lines in panel (a), decrease approximately with $\omega_{\rm dyn}$ as a result of the increasing stellar radius. The same general trend is shared by the $l = 1$ modes when they behave like p modes. However, the figure shows the presence in addition of g-mode branches, with frequencies increasing with age as the maximum value of N increases. This leads to resonances where frequencies of modes of the same degree undergo avoided crossings instead of crossing (*e.g.* Aizenman, Smeyers & Weigert 1977); on the other hand, there is no interaction between modes of different degree. The effect on the mode inertia E normalized at the photospheric amplitude, defined in equation (53), is shown in panel (b); for clarity two modes with $l = 1$ have been indicated in both panels by triangles and diamonds, respectively, at the points corresponding to the models in the evolution sequence. Where the $l = 1$ modes behave as p modes, their inertia is very close to that of a radial mode of similar frequency. However, the g-mode behaviour corresponds to an increase in the amplitude in the interior and hence in E. At the avoided crossing there is an interchange of character between the two interacting modes. (It should be noted that the density of models in the sequence is insufficient to resolve fully the variations with age in E, leading to the somewhat irregular behaviour in

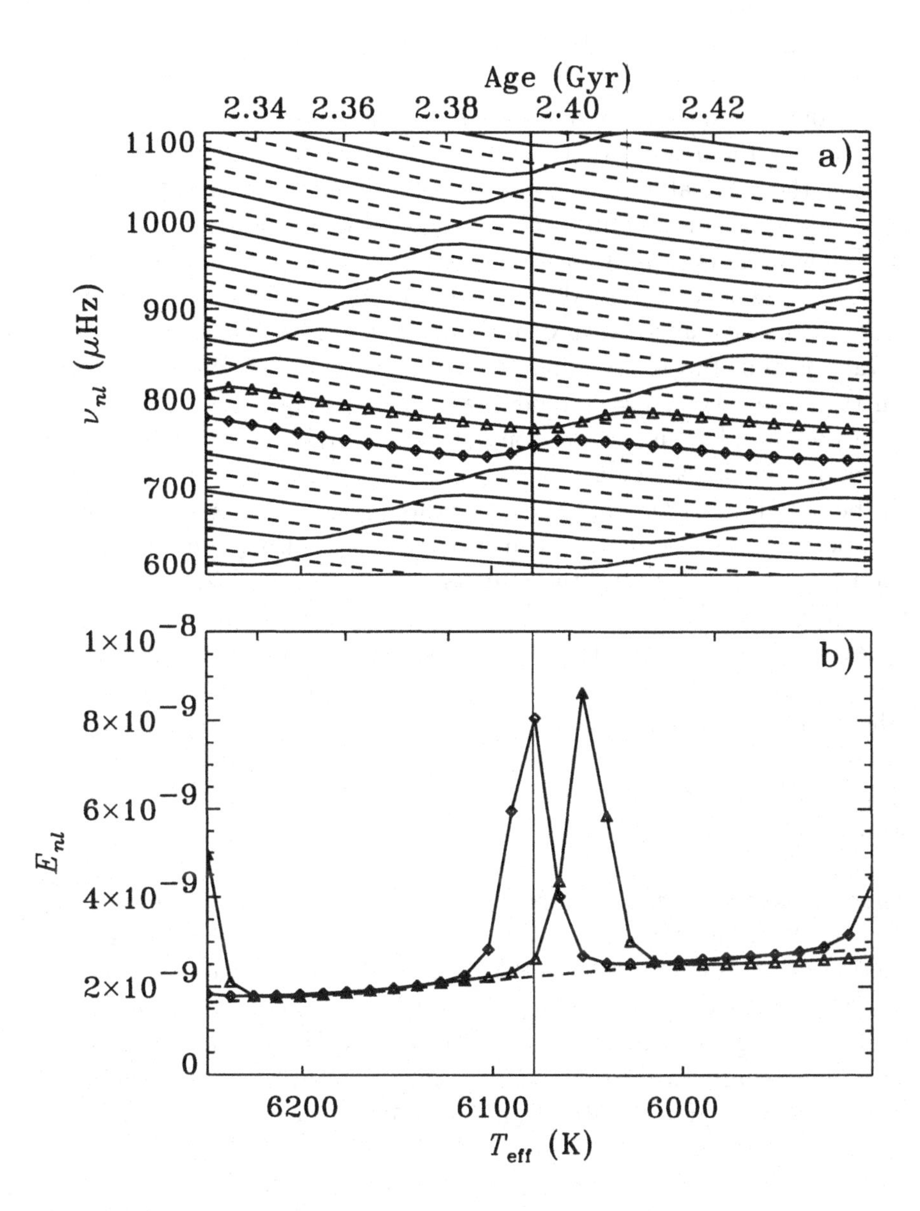

Figure 37: (a) *Evolution of adiabatic frequencies for model of mass* $1.60\,\mathrm{M}_\odot$. *The lower abscissa shows the effective temperature* $T_{\rm eff}$, *the upper abscissa the age of the model in Gyr. The dashed lines correspond to modes of degree* $l = 0$, *and the solid lines to* $l = 1$. *The vertical solid line indicates the location of the model whose frequencies are illustrated in Figure 39. (Adapted from Christensen-Dalsgaard et al. 1995b).* (b) *The change with age in the normalized mode inertia (cf. equation 53). The solid lines show modes with* $l = 1$, *each model being indicated by triangles or diamonds as in panel* (a), *whereas the dashed line shows the radial mode with approximately the same frequency.*

panel (*b*); however, the overall variation is clearly visible.)

The properties of the oscillations are further illustrated by the eigenfunctions shown in Figure 38, for the two modes with $l = 1$ undergoing avoided crossing at the vertical line in Figure 37 as well as for the neighbouring radial mode. The displacement amplitudes have been weighted by $\rho^{1/2} r$, so that they directly shows the contribution at a given radius to the mode inertia E_{nl} (*cf.* equation 53). The $l = 1$ mode in panel (*a*) is evidently very nearly a pure acoustic mode, with an vertical displacement behaving almost as for the radial mode, apart from the phase shift associated with the difference in frequency. In contrast, the second $l = 1$ mode has very substantial displacement amplitudes in the core, leading to the comparatively large normalized inertia shown in Figure 37; this is particularly visible in the enlarged view in Figure 38(*c*). It should be noted, however, that the separation between the g-mode and p-mode propagation regions is quite small in this case (see also Figure 36), leading to substantial coupling between the two regions and causing the large minimum separation in the avoided crossing and a maximum normalized inertia which is still relatively small, despite the g-mode like behaviour in the core. In contrast, for modes with $l \geq 2$ the separation between the propagation regions is larger and the coupling is much weaker; as a result, a frequency plot corresponding to Figure 37(*a*) shows two sets of frequencies apparently crossing with no avoidance, and the maximum inertia for, for example, $l = 2$ in the frequency region illustrated is around 3×10^{-7}.

The normalized inertia may provide a rough estimate of the likely surface amplitude of the modes, at least if the modes are excited stochastically by convection (*e.g.* Houdek *et al.* 1995): in that case the mode energy is likely to be independent of degree, at fixed frequency. It follows from equation (53) that kinetic energy of a mode can be expressed as $A^2 E_{nl}$, where A is the surface amplitude. Assuming that the energy is independent of degree, the amplitude A_{nl} of a mode of degree l, order n and normalized inertia E_{nl} satisfies

$$\frac{A_{nl}}{A_0(\nu_{nl})} \simeq \left[\frac{E_{nl}}{\overline{E}_0(\nu_{nl})} \right]^{-1/2} , \qquad (68)$$

where $A_0(\nu)$ and $\overline{E}_0(\nu)$ are obtained by interpolating to frequency ν in the results for radial modes. In particular, the modes with strong g-mode character in Figure 37 would be expected to have roughly half the surface amplitude of the pure acoustic modes.

To compare the fine structure in the observed and computed frequency spectra it is convenient to use an *echelle diagram* (*e.g.* Grec, Fossat & Pomerantz 1983). Here, the frequencies are reduced modulo $\Delta\nu$ by expressing them as

$$\nu_{nl} = \nu_0 + k\Delta\nu + \tilde{\nu}_{nl} \qquad (69)$$

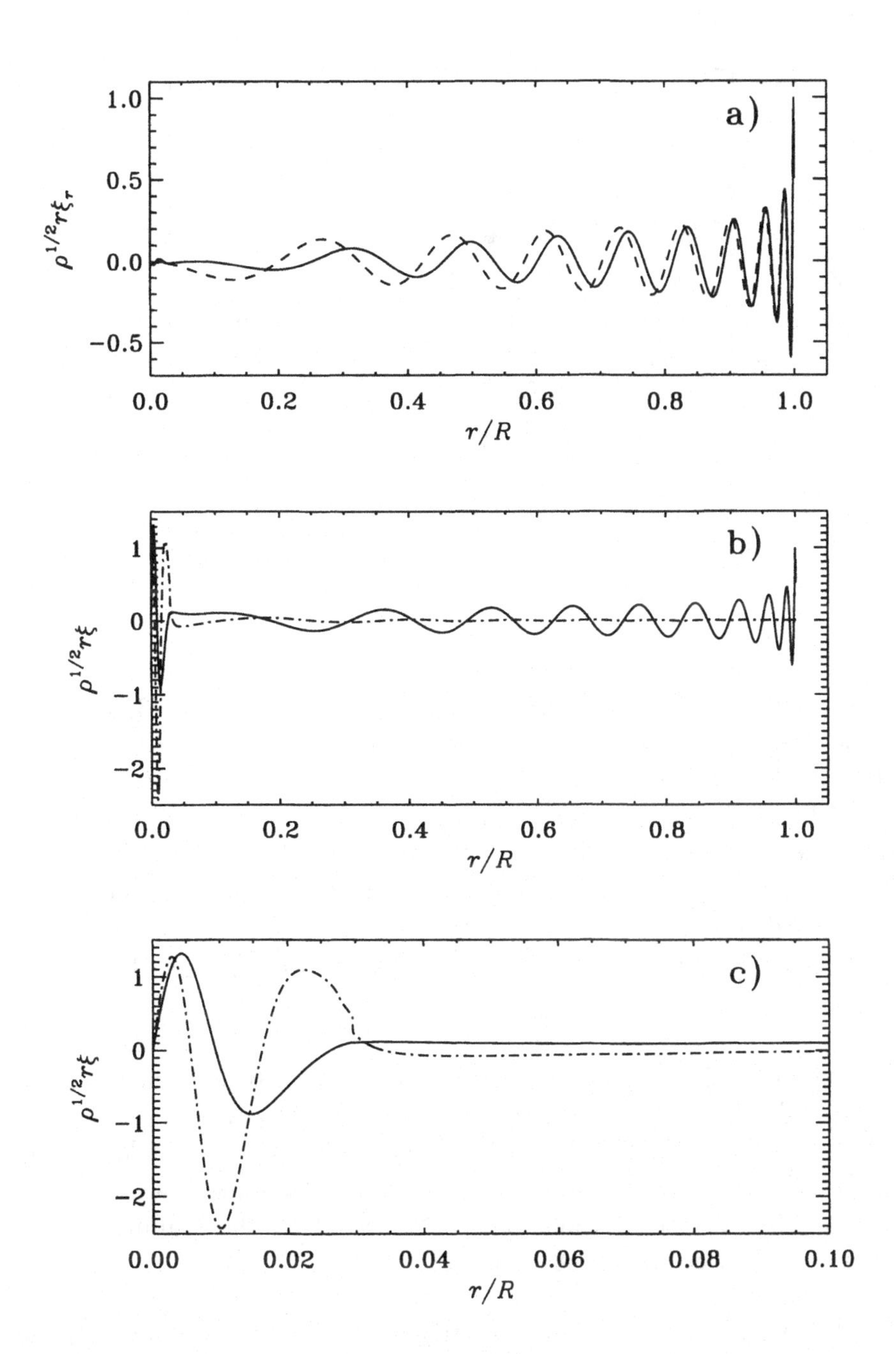

Figure 38: *Eigenfunctions of selected modes in the model indicated by a vertical line in Figure 37. In panel* (a) *the amplitudes of the vertical displacement are shown for the* $l = 1$ *mode indicated by triangles (solid line) and the neighbouring radial mode (dashed line). Panels* (b) *and* (c) *are for the* $l = 1$ *mode marked by diamonds: the solid and dot-dashed lines show the amplitudes of the vertical and horizontal displacement, respectively.*

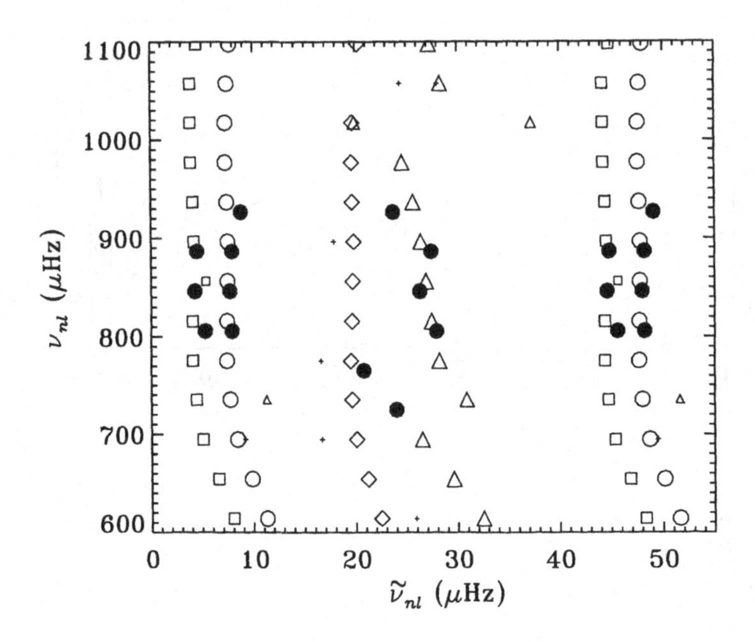

Figure 39: *Echelle diagram with a frequency separation of* $\Delta\nu = 40.3\,\mu\mathrm{Hz}$. *The open symbols show computed frequencies for a model with* $M = 1.60\,\mathrm{M}_\odot$ *and* $Z = 0.03$; *here the reference frequency was* $\nu_0^{(\mathrm{mod})} = 856\,\mu\mathrm{Hz}$. *Circles are used for modes with* $l = 0$, *triangles for* $l = 1$, *squares for* $l = 2$ *and diamonds for* $l = 3$. *The size of the symbols indicates the expected relative amplitude of the modes (see text); symbols that would otherwise be too small have been replaced by crosses. The filled circles show observed frequencies from Kjeldsen et al. (1995), plotted with the same* $\Delta\nu$ *but with a reference frequency of* $\nu_0^{(\mathrm{obs})} = 846\,\mu\mathrm{Hz}$.

(where ν_0 is a suitable reference frequency and k is an integer), such that $\tilde{\nu}_{nl}$ is between 0 and $\Delta\nu$. In the echelle diagram $\nu_0 + k\Delta\nu$ is plotted against $\tilde{\nu}_{nl}$. The result is shown in Figure 39. The open symbols are for a $1.6\,\mathrm{M}_\odot$ model that was chosen to have $\Delta\nu \simeq 40.3\,\mu\mathrm{Hz}$; the reference frequency was $\nu_0^{(\mathrm{mod})} = 856\,\mu\mathrm{Hz}$. The sizes of the symbols have been scaled by the amplitude ratio A_{nl}/A_0 determined by equation (68).

The model results for $l = 0$, 2 and 3 clearly reflect the behavior predicted by equation (34). In particular, the points for $l = 0$ and 2 run parallel, with a small separation $\delta_{n0} \simeq 3.3\,\mu\mathrm{Hz}$ resulting from the last term in that equation. For $l = 1$, equation (34) predicts an almost vertical series of points separated by roughly $\Delta\nu/2$ from those for $l = 0$. The model frequencies deviate from this. Comparison with Figure 37(a) (where the location of this model is marked by a vertical solid line) indicates that this behavior is associated with the avoided crossings, which change the frequency separation and therefore shift the frequencies relative to the location expected from p-mode asymptotics. As discussed above, even $l = 1$ modes behaving partly like g modes still have sufficiently small normalized inertia E_{nl} that their estimated amplitudes are close to those of the pure p modes. (The figure shows a single exception: a mode at $730\,\mu$Hz shifted almost to the $l = 0$ line, with somewhat reduced amplitude.) In contrast, since the g modes of degree 2 and 3 are trapped quite efficiently in the deep interior, their estimated amplitudes are so small as to make the points virtually invisible in Figure 39.

The filled circles in Figure 39 show the frequencies observed by Kjeldsen *et al.* (1995), again plotted with $\Delta\nu = 40.3\,\mu$Hz, but with the reference frequency $\nu_0^{(\mathrm{obs})} = 846\,\mu$Hz. We can immediately identify modes with degrees $l = 0$ and 2, and the small frequency separation found by Kjeldsen et al. ($\delta_{n0} = 3.1 \pm 0.3\,\mu$Hz) is in excellent agreement with the model value. The remaining six observed frequencies coincide quite well with $l = 1$ modes in the model and display an irregularity similar to the model frequencies (although differing in detail). This might indicate that the observations of η Bootis show evidence for avoided crossings involving g modes. Note, however, that some of the observed frequencies may arise from modes with $l = 3$.

The interpretation of the observations must clearly be regarded as preliminary, given the uncertainty in extracting individual frequencies in a single-site power spectrum complicated by side lobes. In particular, the indications in the echelle diagram of effects of g-mode trapping is clearly extremely tentative. On the other hand, the close agreement between the observed and computed value of the $l = 0 - 2$ frequency separation is suggestive. The difference of $10\,\mu$Hz in the reference frequency ν_0 required to obtain agreement between the location of the modes in Figure 39 is clearly a concern; however, it should be noted that this is of a similar magnitude to the differences observed in comparisons of solar observed and computed frequencies and attributed to errors in the treatment of the superficial layers. Similar effects might be expected for η Bootis (see also Christensen-Dalsgaard *et al.* 1995c).

6.2 Concluding remarks

The helioseismic investigations of the solar interior have undergone a dramatic development since the initial determination of individual p-mode frequencies of the Sun more than 15 years ago. Inverse analyses now let us determine density and sound speed in most of the solar interior with a relative precision exceeding 10^{-3}. Remarkably, the results are quite close to normal solar models, provided that gravitational settling and diffusion of helium are taken into account, although significant differences remain. The frequencies have been shown to be sensitive to fine details in the physics of the solar interior, particularly the thermodynamic properties and the opacity; this was used early to indicate the need for opacity increases, later confirmed, and is permitting tests of sophisticated formulations of the equation of state against the Sun.

The study of solar oscillations is on the threshold of a new era, with the deployment of the GONG project (*e.g.* Harvey *et al.* 1993; Leibacher *et al.* 1995) and the instruments on the SOHO satellite, expected to be launched towards the end of in 1995 (*e.g.* Scherrer *et al.* 1995; Appourchaux *et al.* 1995; Gabriel *et al.* 1991). Together with the existing BISON and IRIS networks (*e.g.* Elsworth *et al.* 1995; Fossat 1995) these projects will provide a major expansion of the helioseismic data. This should result in greatly improved information about the structure of the solar core, of obvious importance to the understanding of solar evolution and the neutrino problem, substantially better knowledge about conditions at the base of the convection zone, constraining possible overshooting and turbulent mixing in this region, and a much stronger basis for testing the equation of state and determining the envelope helium abundance. Furthermore, high-quality data on high-degree modes may improve our understanding of the properties of the near-surface region, including the effects of convection on the structure of the Sun and on the frequencies and excitation of the oscillations.

With the possible detection of modes in η Bootis, asteroseismology of solar-like stars may now be in a position similar to helioseismology 15 years ago, with a corresponding promise for the future development. The technique developed by Kjeldsen *et al.* (1995) is being applied to other bright stars; an important example where extensive observations have already been obtained is α Centauri. Also, it is obvious that further observations of η Bootis are required, to confirm and extend the initial results. Particularly important are multi-site observations, to reduce the complications of the side lobes in the spectrum. Also, the theoretical analysis in Section 6.1 indicates the substantial richness in the oscillations of sub-giant stars of somewhat more than solar mass, compared with stars in the phase of central hydrogen burning. Combined with their intrinsic brightness this makes subgiants attractive targets for asteroseismic investigations.

The effects of the Earth's atmosphere are likely to constrain ground-based investigations of solar-like oscillations to bright stars, observed with large telescopes. From space, however, the observations are limited essentially only by photon noise; here broad-band intensity measurements will allow the study of oscillations at solar amplitudes in fairly faint stars with modest-sized telescopes. An initial experiment of this nature, the EVRIS project (Baglin 1991), is scheduled for launch in November 1996 on the Russian Mars probe MARS96. It will use a 9 cm telescope to observe a limited number of rather bright stars during the 300-day cruise phase of the probe. Two more ambitious projects are currently under evaluation. The French COROT project (Catala *et al.* 1995) is aimed at providing extensive observations of a modest number of stars, to obtain highly accurate measurements of frequencies and frequency splittings. The STARS project, which is currently undergoing Phase A studies within ESA with a view towards possible selection in 1996, will use an 80 cm telescope with a CCD detector to make simultaneous observations of several stars (Fridlund *et al.* 1995). An important goal is to study solar-like oscillations of stars in open clusters: since stars in a cluster can be assumed to have approximately the same age, distance and chemical composition, measured frequencies for stars in a cluster provide far more stringent constraints on the internal structure of the stars than do frequencies of a single star, where the basic parameters are often rather uncertain.

Given the new helioseismic experiments, the continuing observations of "classical" pulsating stars and the growing potential for study of solar-like oscillations in other stars there seems little doubt that helio- and asteroseismology will remain a very active and fertile field of research for years to come.

Acknowledgements

I am very grateful to the organizers of the VI IAC Winter School for inviting me to participate in that splendid event. I thank the many colleagues who have contributed to the development of this field and my understanding of it, for enlightening discussions. The work presented here was supported in part by the Danish National Research Foundation through its establishment of the Theoretical Astrophysics Center.

7 REFERENCES

Ahrens, B., Stix, M. & Thorn, M.: 1992, "On the depletion of lithium in the Sun", *Astron. Astrophys.* **264**, 673 - 678.

Aizenman, M., Smeyers, P. & Weigert, A.: 1977, "Avoided crossing of modes of non-radial stellar oscillations", *Astron. Astrophys.* **58**, 41 - 46.

Anders, E. & Grevesse, N.: 1989, "Abundances of the elements: meteoritic and solar", *Geochim. Cosmochim. Acta* **53**, 197 - 214.

Ando, H. & Osaki, Y.: 1975, "Nonadiabatic nonradial oscillations: An application to the five-minute oscillation of the sun", *Publ. Astron. Soc. Japan* **27**, 581 - 603.

Ando, H. & Osaki, Y.: 1977, "The influence of the chromosphere and corona on the solar atmospheric oscillations", *Publ. Astron. Soc. Japan* **29**, 221 - 233.

Antia, H. M.: 1995, "Effects of surface layers on helioseismic inversion", *Mon. Not. R. astr. Soc.* **274**, 499 - 503.

Antia, H. M. & Basu, S.: 1994, "Measuring the helium abundance in the solar envelope: the role of the equation of state", *Astrophys. J.* **426**, 801 - 811.

Appourchaux, T., Domingo, V., Fröhlich, C., Romero, J., Wehrli, Ch., Andersen, B. N., Berthomieu, G., Delache, Ph., Crommelynck, D., Jiménez, A., Roca Cortés, T. & Jones, A. R.: 1995, "VIRGO - The solar monitor experiment on SOHO", in R. K. Ulrich, E. J. Rhodes Jr & W. Däppen (eds), *Proc. GONG'94: Helio- and Astero-seismology from Earth and Space*, ASP Conf. Ser. **76**, 408 - 415.

Audard, N., Provost, J. & Christensen-Dalsgaard, J.: 1995, "Seismological effects of convective-core overshooting in stars of intermediate mass", *Astron. Astrophys.* **297**, 427 - 440.

Baglin, A.: 1991, "Stellar seismology from space: the EVRIS experiment on board MARS 94", *Solar Phys.* **133**, 155 - 160.

Bahcall, J. N. & Pinsonneault, M. H.: 1995, "Solar models with helium and heavy element diffusion", *Rev. Mod. Phys.*, in the press.

Balmforth, N. J.: 1992a. "Solar pulsational stability. I: Pulsation-mode thermodynamics", *Mon. Not. R. astr. Soc.* **255**, 603 - 631.

Balmforth, N. J.: 1992b. "Solar pulsational stability. II: Pulsation frequencies", *Mon. Not. R. astr. Soc.* **255**, 632 - 638.

Basu, S. & Antia, H. M.: 1995, "Helium abundance in the solar envelope", *Mon. Not. R. astr. Soc.*, in the press.

Basu, S., Antia, H. M. & Narasimha, D.: 1994, "Helioseismic measurement of the extent of overshoot below the solar convection zone", *Mon. Not. R. astr. Soc.* **267**, 209 - 224.

Basu, S., Christensen-Dalsgaard, J., Schou, J., Thompson, J. & Tomczyk, S.: 1995, "Solar structure inversion with LOWL data", in V. Domingo *et al.* (eds), *Proc. 4th SOHO Workshop: Helioseismology*, ESA SP-376, ESTEC, Noordwijk, in the press.

Bedding, T.R. & Kjeldsen, H.: 1995, "More on solar-like oscillations in η Boo", in R. S. Stobie & P. A. Whitelock (eds), *IAU Colloquium 155: Astrophysical Applications of Stellar Pulsation,* ASP Conf. Ser., in the press.

Belmonte, J. A., Michel. E., Alvarez, M., Jiang, S. Y., Chevreton, M., Auvergne, M., Liu, Y. Y., Goupil, M. J., Baglin, A., Roca Cortés, T., Mangeney, A., Dolez, N., Valtier, J. C., Massacrier, G., Sareyan, J. P., Schmider, F. X. & Vidal, I.: 1994, "Time-resolved photometry of BN and BU Cancri, two δ Scuti stars in the Praesepe cluster. Fourth photometry campaign of the STEPHI network", *Astron. Astrophys.* **283**, 121 - 128.

Berthomieu, G., Cooper, A. J., Gough, D. O., Osaki, Y., Provost, J. & Rocca, A.: 1980, "Sensitivity of five minute eigenfrequencies to the structure of the Sun", in H. A. Hill & W. Dziembowski (eds), *Nonradial and nonlinear stellar pulsation, Lecture Notes in Physics* **125**, Springer-Verlag, Berlin, 307 - 312.

Breger, M., Stich, J., Garrido, R., Martin, B., Shi-yang, Jiang, Zhi-ping, Li, Hube, D. P., Ostermann, W., Paparo, M., Schreck, M.: 1993, "Nonradial pulsation of the δ Scuti star BU Cancri in the Praesepe cluster", *Astron. Astrophys.* **271**, 482 - 486.

Brodsky, M. A. & Vorontsov, S. V.: 1993, "Asymptotic theory of intermediate- and high-degree solar acoustic oscillations", *Astrophys. J.* **409**, 455 - 464.

Brookes, J. R., Isaak, G. R. & van der Raay, H. B.: 1976, "Observation of free oscillations of the Sun", *Nature* **259**, 92 - 95.

Brown, T. M., Christensen-Dalsgaard, J., Mihalas, B. & Gilliland, R. L.: 1994, "The effectiveness of oscillation frequencies in constraining stellar model parameters", *Astrophys. J.* **427**, 1013 - 1034.

Brown, T. M., Gilliland, R. L., Noyes, R. W. & Ramsey, L. W.: 1991, "Detection of possible p-mode oscillations of Procyon", *Astrophys. J.* **368**, 599 - 609.

Canuto, V. M. & Mazzitelli, I.: 1991, "Stellar turbulent convection: a new model and applications", *Astrophys. J.* **370**, 295 - 311.

Canuto, V. M. & Mazzitelli, I.: 1992, "Further improvements of a new model for turbulent convection in stars", *Astrophys. J.* **389**, 724 - 730.

Catala, C., Auvergne, M., Baglin, A., Bonneau, F., Magnan, A., Vuillemin, A., Goupil, M. J., Michel, E., Boumier, P., Dzitko, H., Gabriel, A., Gautier, D., Lemaire, P., Mangeney, A., Mosser, B., Turck-Chièze, S. & Zahn, J. P.: 1995, "COROT, a space project devoted to the study of convection and rotation in stars", in V. Domingo *et al.* (eds), *Proc. 4th SOHO Workshop: Helioseismology*, ESA SP-376, ESTEC, Noordwijk, in the press.

Chaboyer, B., Demarque, P., Guenther, D. B. & Pinsonneault, M. H.: 1995, "Rotation, diffusion and overshoot in the Sun: effects on the oscillation frequencies and the neutrino flux", *Astrophys. J.* **446**, 435 - 444.

Christensen-Dalsgaard, J.: 1981, "The effect of non-adiabaticity on avoided crossings of non-radial stellar oscillations", *Mon. Not. R. astr. Soc.* **194**, 229 - 250.

Christensen-Dalsgaard, J.: 1984, "What will asteroseismology teach us?", in F. Praderie (ed.), *Space Research Prospects in Stellar Activity and Variability*, Paris Observatory Press, 11 - 45.

Christensen-Dalsgaard, J.: 1986, "Theoretical aspects of helio- and asteroseismology", in D. O. Gough (ed.), *Seismology of the Sun and the distant Stars*, Reidel, Dordrecht, 23 - 53.

Christensen-Dalsgaard, J.: 1988, "A Hertzsprung-Russell diagram for stellar oscillations", in J. Christensen-Dalsgaard & S. Frandsen (eds), *Proc. IAU Symposium No 123, Advances in helio- and asteroseismology*, Reidel, Dordrecht, 295 - 298.

Christensen-Dalsgaard, J.: 1990, "Helioseismic investigation of solar internal structure", in G. Berthomieu & M. Cribier (eds), *Proc. IAU Colloquium No 121, Inside the Sun*, Kluwer, Dordrecht, 305 - 326.

Christensen-Dalsgaard, J.: 1991, "Some aspects of the theory of solar oscillations", *Geophys. Astrophys. Fluid Dynamics* **62**, 123 - 152.

Christensen-Dalsgaard, J.: 1992, "Solar models with enhanced energy transport in the core", *Astrophys. J.* **385**, 354 - 362.

Christensen-Dalsgaard, J.: 1993, "On the asteroseismic HR diagram", in T. M. Brown (ed.), *Proc. GONG 1992: Seismic investigation of the Sun and stars,* ASP Conf. Ser. **42**, 347 - 350.

Christensen-Dalsgaard, J.: 1994, *Lecture Notes on Stellar Oscillations (Third Edition)*, D.f.I. Print, Aarhus.

Christensen-Dalsgaard, J. & Berthomieu, G.: 1991, "Theory of solar oscillations", in A. N. Cox, W. C. Livingston & M. Matthews (eds), *Solar interior and atmosphere*, Space Science Series, University of Arizona Press, 401 - 478.

Christensen-Dalsgaard, J. & Däppen, W.: 1992, "Solar oscillations and the equation of state", *Astron. Astrophys. Rev.* **4**, 267 - 361.

Christensen-Dalsgaard, J. & Frandsen, S.: 1983, "Radiative transfer and solar oscillations", *Solar Phys.* **82**, 165 - 204.

Christensen-Dalsgaard, J. & Gough, D. O.: 1976, "Towards a heliological inverse problem", *Nature* **259**, 89 - 92.

Christensen-Dalsgaard, J. & Gough, D. O.: 1980, "Is the Sun helium-deficient?", *Nature* **288**, 544 - 547.

Christensen-Dalsgaard, J. & Gough, D. O.: 1984, "Implications of observed frequencies of solar p modes", in R. K. Ulrich, J. Harvey, E. J. Rhodes Jr & J. Toomre (eds), *Solar Seismology from Space*, NASA, JPL Publ. 84-84, 199 - 204,

Christensen-Dalsgaard, J. & Pérez Hernández, F.: 1991, "Influence of the upper layers of the Sun on the p-mode frequencies", in D. O. Gough & J. Toomre (eds), *Challenges to theories of the structure of moderate-mass stars, Lecture Notes in Physics* **388**, Springer, Heidelberg, 43 - 50.

Christensen-Dalsgaard, J. & Pérez Hernández, F.: 1992, "Phase-function differences for stellar acoustic oscillations - I. Theory", *Mon. Not. R. astr. Soc.* **257**, 62 - 88.

Christensen-Dalsgaard, J. & Petersen, J. O.: 1995, "Pulsation models of the double–mode Cepheids in the Large Magellanic Cloud", *Astron. Astrophys.*, in the press.

Christensen-Dalsgaard, J., Duvall, T. L., Gough, D. O., Harvey, J. W. & Rhodes Jr, E. J.: 1985, "Speed of sound in the solar interior", *Nature* **315**, 378 - 382.

Christensen-Dalsgaard, J., Däppen, W. & Lebreton, Y.: 1988, "Solar oscillation frequencies and the equation of state", *Nature* **336**, 634 - 638.

Christensen-Dalsgaard, J., Gough, D. O. & Pérez Hernández, F.: 1988, "Stellar disharmony", *Mon. Not. R. astr. Soc.* **235**, 875 - 880.

Christensen-Dalsgaard, J., Gough, D. O. & Thompson, M. J.: 1989, "Differential asymptotic sound-speed inversions", *Mon. Not. R. astr. Soc.* **238**, 481 - 502.

Christensen-Dalsgaard, J., Gough, D. O. & Thompson, M. J.: 1991, "The depth of the solar convection zone", *Astrophys. J.* **378**, 413 - 437.

Christensen-Dalsgaard, J., Gough, D. O. & Thompson, M. J.: 1992, "On the rate of destruction of lithium in late-type main-sequence stars", *Astron. Astrophys.* **264**, 518 - 528.

Christensen-Dalsgaard, J., Proffitt, C. R. & Thompson, M. J.: 1993, "Effects of diffusion on solar models and their oscillation frequencies", *Astrophys. J.* **403**, L75 - L78.

Christensen-Dalsgaard, J., Monteiro, M. J. P. F. G. & Thompson, M. J.: 1995a. "Helioseismic estimation of convective overshoot in the Sun", *Mon. Not. R. astr. Soc.*, in the press.

Christensen-Dalsgaard, J., Bedding, T. R. & Kjeldsen, H.: 1995b. "Modelling solar-like oscillations in η Bootis", *Astrophys. J.* **443**, L29 - L32.

Christensen-Dalsgaard, J., Bedding, T. R., Houdek, G., Kjeldsen, H., Rosenthal, C., Trampedach, R., Monteiro, M. J. P. F. G. & Nordlund, Å.: 1995c. "Near-surface effects in modelling oscillations of η Boo", in R. S. Stobie & P. A. Whitelock (eds), *IAU Colloquium 155: Astrophysical Applications of Stellar Pulsation,* ASP Conf. Ser., in the press.

Claverie, A., Isaak, G. R., McLeod, C. P., van der Raay, H. B. & Roca Cortes, T.: 1979, "Solar structure from global studies of the 5-minute oscillations", *Nature* **282**, 591 - 644.

Cox, A. N.: 1980, "The masses of Cepheids", *Ann. Rev. Astron. Astrophys.* **18**, 15 - 41.

Cox, A. N., Guzik, J. A. & Kidman, R. B.: 1989, "Oscillations of solar models with internal element diffusion", *Astrophys. J.* **342**, 1187 - 1206.

Cox, A. N., Morgan, S. M., Rogers, F. J. & Iglesias, C. A.: 1992, "An opacity mechanism for the pulsations of OB stars", *Astrophys. J.* **393**, 272 - 277.

Däppen, W.: 1992, "The equation of state for stellar envelopes: comparison of results from different formalisms", in C. Lynas-Gray, C. Mendoza & C. Zeippen (eds), *Proc. Workshop on Astrophysical Opacities, Revista Mexicana de Astronomia y Astrofisica* **23**, 141 - 149.

Däppen, W. & Gough, D. O.: 1984, "On the determination of the helium abundance of the solar convection zone", in *Theoretical Problems in Stellar Stability and Oscillations*, Institut d'Astrophysique, Liège, 264 - 269.

Däppen, W. & Gough, D. O.: 1986, "Progress report on helium abundance determination", in D. O. Gough (ed.), *Seismology of the Sun and the distant Stars*, Reidel, Dordrecht, 275 - 280.

Demarque, P. & Guenther, D. B.: 1988, "Sensitivity of solar p-modes to solar envelope structure", in J. Christensen-Dalsgaard & S. Frandsen (eds), *Proc. IAU Symposium No 123, Advances in helio- and asteroseismology*, Reidel, Dordrecht, 91 - 94.

Deubner, F.-L.: 1975, "Observations of low wavenumber nonradial eigenmodes of the Sun", *Astron. Astrophys.* **44**, 371 - 375.

Deubner, F.-L. & Gough, D. O.: 1984, "Helioseismology: Oscillations as a diagnostic of the solar interior", *Ann. Rev. Astron. Astrophys.* **22**, 593 -

619.

Duvall, T. L., Dziembowski, W. A., Goode, P. R., Gough, D. O., Harvey, J. W. & Leibacher, J. W.: 1984, "The internal rotation of the Sun", *Nature* **310**, 22 - 25.

Dyson, J. & Schutz, B. F.: 1979, "Perturbations and stability of rotating stars. I. Completeness of normal modes", *Proc. Roy. Soc. London* **A368**, 389 - 410.

Dziembowski, W. A. & Pamjatnykh, A. A.: 1991, "A potential asteroseismological test for convective overshooting theories", *Astron. Astrophys.* **248**, L11 - L14.

Dziembowski, W. A., Pamyatnykh, A. A. & Sienkiewicz, R.: 1992, "Seismological tests of standard solar models calculated with new opacities", *Acta Astron.* **42**, 5 - 15.

Dziembowski, W. A., Goode, P. R., Pamyatnykh, A. A. & Sienkiewicz, R.: 1994, "Seismic model of the sun's interior", *Astrophys. J.* **432**, 417 - 426.

Eddington, A. S.: 1926, *The internal constitution of the stars*, Cambridge University Press, Cambridge.

Eggleton, P. P., Faulkner, J. & Flannery, B. P.: 1973, "An approximate equation of state for stellar material", *Astron. Astrophys.* **23**, 325 - 330.

Eisenfeld, J.: 1969, "A completeness theorem for an integro-differential operator", *J. Math. Anal. Applic.* **26**, 357 - 375.

Elsworth, Y., Howe, R., Isaak, G. R., McLeod, C. P. & New, R.: 1990, "Evidence from solar seismology against non-standard solar-core models", *Nature* **347**, 536 - 539.

Elsworth, Y., Howe, R., Isaak, G. R., McLeod, C. P., Miller, B. A., New, R., Speake, C. C. & Wheeler, S. J.: 1994, "Solar *p*-mode frequencies and their dependence on solar activity: recent results from the BISON network", *Astrophys. J.* **434**, 801 - 806.

Elsworth, Y., Howe, R., Isaak, G. R., McLeod, C. P., Miller, B. A., van der Raay, H. B. & Wheeler, S. J.: 1995, "Performance of the BISON network 1981-present", in R. K. Ulrich, E. J. Rhodes Jr & W. Däppen (eds), *Proc. GONG'94: Helio- and Astero-seismology from Earth and Space*, ASP Conf. Ser. **76**, 392 - 397.

Faulkner, J. & Gilliland, R. L.: 1985, "Weakly interacting, massive particles and the solar neutrino flux", *Astrophys. J.* **299**, 994 - 1000.

Fossat, E.: 1995, "IRIS status report", in R. K. Ulrich, E. J. Rhodes Jr & W. Däppen (eds), *Proc. GONG'94: Helio- and Astero-seismology from Earth and Space*, ASP Conf. Ser. **76**, 387 - 391.

Fossat, E. & Ricort, G.: 1975, "Photospheric oscillations. I. Large scale observations by optical resonance method", *Astron. Astrophys.* **43**, 243 - 252.

Frandsen, S., Balona, L. A., Viskum, M., Koen, C. & Kjeldsen, H.: 1995, "Multisite CCD photometry of δ Scuti stars in the open cluster NGC6134 (the 1. STACC campaign)", submitted to *Astron. Astrophys.*.

Fridlund, M., Gough, D.O., Jones, A., Appourchaux, T., Badiali, M., Catala, C., Frandsen, S., Grec, G., Roca Cortés, T. & Schrijver, K.: 1995, "STARS: A proposal for a dedicated space mission to study stellar, structure and evolution", in R. K. Ulrich, E. J. Rhodes Jr & W. Däppen (eds), *Proc. GONG'94: Helio- and Astero-seismology from Earth and Space*, ASP Conf. Ser. **76**, 416 - 425.

Gabriel, A. H. and the GOLF team: 1991, "Global oscillations at low frequency from the SOHO mission (GOLF)", *Adv. Space Res.* **vol. 11, No. 4**, 103 - 112.

Gabriel, M.: 1989, "The D_{nl} values and the structure of the solar core", *Astron. Astrophys.* **226**, 278 - 283.

Gabriel, M., Scuflaire, R. & Noels: 1982, "The solar structure and the low l five-minute oscillation. I", *Astron. Astrophys.* **110**, 50 - 53.

Gautschy, A.: 1990, "On the pulsation - evolution connection of early-type stars", in D. Baade (ed.), *Proc. ESO Workshop on Rapid variability of OB-stars: nature and diagnostic value*, ESO Munich, 315 - 329.

Gelly, B., Grec, G. & Fossat, E.: 1986, "Evidence for global pressure oscillations in Procyon and α Centauri", *Astron. Astrophys.* **164**, 383 - 394.

Gilliland, R. L., Brown, T. M., Kjeldsen, H., McCarthy, J. K., Peri, M. L., Belmonte, J. A., Vidal, I., Cram, L, E., Palmer, J., Frandsen, S., Parthasarathy, M., Petro, L., Schneider, H., Stetson, P. B. & Weiss, W. W.: 1993, "A search for solar-like oscillations in the stars of M67 with CCD ensemble photometry on a network of 4 m telescopes", *Astron. J.* **106**, 2441 - 2476.

Gough, D. O.: 1976, "The current state of stellar mixing-length theory", in E. Spiegel & J.-P. Zahn (eds), *Problems of stellar convection, IAU Colloq. No. 38, Lecture Notes in Physics* **71**, Springer-Verlag, Berlin, 15 - 56.

Gough, D. O.: 1977a. "Random remarks on solar hydrodynamics", in R. M. Bonnet & Ph. Delache (eds), *Proc. IAU Colloq. No. 36: The energy balance and hydrodynamics of the solar chromosphere and corona*, G. de Bussac, Clairmont-Ferrand, 3 - 36.

Gough, D. O.: 1977b. "Mixing-length theory for pulsating stars", *Astrophys. J.* **214**, 196 - 213.

Gough, D. O.: 1984a. "On the rotation of the Sun", *Phil. Trans. R. Soc. London, Ser. A* **313**, 27 - 38.

Gough, D. O.: 1984b. "Towards a solar model", *Mem. Soc. Astron. Ital.* **55**, 13 - 35.

Gough, D. O.: 1986, "EBK quantization of stellar waves", in Y. Osaki (ed.), *Hydrodynamic and magnetohydrodynamic problems in the Sun and stars*, University of Tokyo Press, 117 - 143.

Gough, D. O.: 1990, "Comments on helioseismic inference", in Y. Osaki & H. Shibahashi (eds), *Progress of seismology of the sun and stars, Lecture Notes in Physics* **367**, Springer, Berlin, 283 - 318.

Gough, D. O.: 1993, "Course 7. Linear adiabatic stellar pulsation", in J.-P. Zahn & J. Zinn-Justin (eds), *Astrophysical fluid dynamics, Les Houches*

Session XLVII, Elsevier, Amsterdam, 399 – 560.
Gough, D. O. & Novotny, E.: 1993, "Asteroseismic calibration of stellar clusters", in T. M. Brown (ed.), *Proc. GONG 1992: Seismic investigation of the Sun and stars*, ASP Conf. Ser. **42**, 355 – 357.
Gough, D. O. & Thompson, M. J.: 1991, "The inversion problem", in A. N. Cox, W. C. Livingston & M. Matthews (eds), *Solar interior and atmosphere*, Space Science Series, University of Arizona Press, 519 – 561.
Gough, D. O. & Toomre, J.: 1991, "Seismic observations of the solar interior", *Ann. Rev. Astron. Astrophys.* **29**, 627 – 685.
Gough, D. O. & Vorontsov, S. V.: 1995, "Seismology of the solar envelope: measuring the acoustic phase shift generated in the outer layers", *Mon. Not. R. astr. Soc.* **273**, 573 – 582.
Gough, D. O. & Weiss, N. O.: 1976, "The calibration of stellar convection theories", *Mon. Not. R. astr. Soc.* **176**, 589 – 607.
Grec, G., Fossat, E. & Pomerantz, M.: 1980, "Solar oscillations: full disk observations from the geographic South Pole", *Nature* **288**, 541 – 544.
Grec, G., Fossat, E. & Pomerantz, M.: 1983, "Full-disk observations of solar oscillations from the geographic South Pole: latest results", *Solar Phys.* **82**, 55 – 66.
Grevesse, N. & Noels, A.: 1993, "Cosmic abundances of the elements", in N. Prantzos, E. Vangioni-Flam & M. Cassé (eds), *Origin and evolution of the Elements*, Cambridge Univ. Press, Cambridge, 15 – 25.
Guenther, D. B.: 1989, "Age of the Sun", *Astrophys. J.* **339**, 1156 – 1159.
Guenther, D. B.: 1994, "Nonadiabatic nonradial *p*-mode frequencies of the standard solar model, with and without helium diffusion", *Astrophys. J.* **422**, 400 – 411.
Guenther, D. B., Pinsonneault, M. H. & Bahcall, J. N.: 1993, "The effects of helium diffusion on solar *p*-mode frequencies", *Astrophys. J.* **418**, 469 – 475.
Harvey, J. W. & Duvall, T. L.: 1984, "Observations of intermediate-degree solar oscillations", in R. K. Ulrich, J. Harvey, E. J. Rhodes Jr & J. Toomre (eds), *Solar Seismology from Space*, NASA, JPL Publ. 84-84, 165 – 172.
Harvey, J. W., Hill, F., Kennedy, J. R. & Leibacher, J.: 1993, "GONG project update", in T. M. Brown (ed.), *Proc. GONG 1992: Seismic investigation of the Sun and stars*, ASP Conf. Ser. **42**, 397 – 409.
Hill, H. A. & Stebbins, R. T.: 1975, "The intrinsic visual oblateness of the Sun", *Astrophys. J.* **200**, 471 – 483.
Hill, H. A., Stebbins, R. T. & Brown, T. M.: 1976, "Recent oblateness observations: Data, interpretation and significance for earlier work", in J. H. Sanders & A. H. Wapstra (eds), *Atomic Masses and Fundamental Constants* **5**, Plenum Press, 622 – 628.
Houdek, G., Rogl, J., Balmforth, N. J. & Christensen-Dalsgaard, J.: 1995, "Excitation of solarlike oscillations in main–sequence stars", in R. K. Ulrich, E. J. Rhodes Jr & W. Däppen (eds), *Proc. GONG'94: Helio- and Astero-*

seismology from Earth and Space, ASP Conf. Ser. **76**, 528 - 531.

Hummer, D. G. & Mihalas, D.: 1988, "The equation of state for stellar envelopes. I. An occupation probability formalism for the truncation of internal partition functions", *Astrophys. J.* **331**, 794 - 814.

Iben, I. & Mahaffy, J.: 1976, "On the sun's acoustical spectrum", *Astrophys. J.* **209**, L39 - L43.

Iglesias, C. A., Rogers, F. J. & Wilson, B. G.: 1992, "Spin-orbit interaction effects on the Rosseland mean opacity", *Astrophys. J.* **397**, 717 - 728.

Innis, J. L., Isaak, G. R., Speake, C. C., Brazier, R. I. & Williams, H. K.: 1991, "High-precision velocity observations of Procyon A – I. Search for *p*-mode oscillations from 1988, 1989 and 1990 observations", *Mon. Not. R. astr. Soc.* **249**, 643 - 653.

Jeans, J. H.: 1929, *Astronomy and cosmogeny*, Cambridge University Press, Cambridge (1961: Dover Publications, New York).

Jiménez, A., Pallé, P. L., Pérez, J. C., Régulo, C., Roca Cortés, T., Isaak, G. R., McLeod, C. P. & van der Raay, H. B., 1988, "The solar oscillations spectrum and the solar cycle", in J. Christensen-Dalsgaard & S. Frandsen (eds), *Proc. IAU Symposium No 123, Advances in helio- and asteroseismology*, Reidel, Dordrecht, 205 - 209.

Kanbur, S. M. & Simon, N. R.: 1994, "Comparative pulsation calculations with OP and OPAL opacities", *Astrophys. J.* **420**, 880 - 883.

Kim, Y.-C., Demarque, P. & Guenther, D. B.: 1991, "The effect of the Mihalas, Hummer, and Däppen equation of state and the molecular opacity on the standard solar model", *Astrophys. J.* **378**, 407 - 412.

Kiriakidis, M., El Eid, M. F. & Glatzel, W.: 1992, "Heavy element opacities and the pulsations of β Cephei stars", *Mon. Not. R. astr. Soc.* **255**, 1P - 5P.

Kjeldsen, H., Bedding, T. R., Viskum, M. & Frandsen, S.: 1995, "Solarlike oscillations in η Boo", *Astron. J.* **109**, 1313 - 1319.

Kosovichev, A. G., Christensen-Dalsgaard, J., Däppen, W., Dziembowski, W. A., Gough, D. O. & Thompson, M. J.: 1992, "Sources of uncertainty in direct seismological measurements of the solar helium abundance", *Mon. Not. R. astr. Soc.* **259**, 536 - 558.

Kurtz, D. W.: 1995, "Seismology and cyclic frequency variability in roAp stars", in R. K. Ulrich, E. J. Rhodes Jr & W. Däppen (eds), *Proc. GONG'94: Helio- and Astero-seismology from Earth and Space*, ASP Conf. Ser. **76**, 606 - 617.

Kurucz, R. L.: 1991, "New opacity calculations", in L. Crivellari, I. Hubeny & D. G. Hummer (eds), *Stellar atmospheres: beyond classical models*, Kluwer, Dordrecht, 441 - 448.

Leibacher, J. and the GONG project team: 1995, "The Global Oscillation Network Group project", in R. K. Ulrich, E. J. Rhodes Jr & W. Däppen (eds), *Proc. GONG'94: Helio- and Astero-seismology from Earth and Space*, ASP Conf. Ser. **76**, 381 - 386.

Leibacher, J. & Stein, R. F.: 1971, "A new description of the solar five-minute oscillation", *Astrophys. Lett.* **7**, 191 - 192.

Libbrecht, K. G., Woodard, M. F. & Kaufman, J. M.: 1990, "Frequencies of solar oscillation", *Astrophys. J. Suppl.* **74**, 1129 - 1149.

Lubow, S. H., Rhodes Jr, E. J. & Ulrich, R. K.: 1980, "Five minute oscillations as a probe of the solar interior", in H. A. Hill & W. Dziembowski (eds), *Nonradial and nonlinear stellar pulsation, Lecture Notes in Physics* **125**, Springer-Verlag, Berlin, 300 - 306.

Lydon, T. J., Fox, P. A. & Sofia, S.: 1992, "A formulation of convection for stellar structure and evolution calculations without the mixing-length theory approximations. I. Application to the Sun", *Astrophys. J.* **397**, 701 - 716.

Lynden-Bell, D. & Ostriker, J. P.: 1967, "On the stability of differentially rotating bodies", *Mon. Not. R. astr. Soc.* **136**, 293 - 310.

Matthews, J. M.: 1993, "Observing the eigenmodes of δ Scuti and roAP stars", in T. M. Brown (ed.), *Proc. GONG 1992: Seismic investigation of the Sun and stars*, ASP Conf. Ser. **42**, 303 - 316.

Mihalas, D., Däppen, W. & Hummer, D. G.: 1988, "The equation of state for stellar envelopes. II. Algorithm and selected results", *Astrophys. J.* **331**, 815 - 825.

Mihalas, D., Hummer, D. G., Mihalas, B. W. & Däppen, W.: 1990, "The equation of state for stellar envelopes. IV. Thermodynamic quantities and selected ionization fractions for six elemental mixes", *Astrophys. J.* **350**, 300 - 308.

Monteiro, M. J. P. F. G., Christensen-Dalsgaard, J. & Thompson, M. J.: 1994, "Seismic study of overshoot at the base of the solar convective envelope", *Astron. Astrophys.* **283**, 247 - 262.

Monteiro, M. J. P. F. G., Christensen-Dalsgaard, J. & Thompson, M. J.: 1995a. "Helioseismic constraints on theories of convection", in R. K. Ulrich, E. J. Rhodes Jr & W. Däppen (eds), *Proc. GONG'94: Helio- and Asteroseismology from Earth and Space*, ASP Conf. Ser. **76**, 128 - 131.

Monteiro, M. J. P. F. G., Christensen-Dalsgaard, J. & Thompson, M. J.: 1995b. "Seismic properties of the Sun's superadiabatic layer. I. Theoretical modelling and parametrization of the uncertainties", *Astron. Astrophys.*, submitted.

Moskalik, P. & Dziembowski, W. A.: 1992, "New opacities and the origin of the β Cephei pulsation", *Astron. Astrophys.* **256**, L5 - L8.

Moskalik, P., Buchler, J. R. & Marom, A.: 1992, "Toward a resolution of the bump and beat Cepheid mass discrepancies", *Astrophys. J.* **385**, 685 - 693.

Noerdlinger, P. D.: 1977, "Diffusion of helium in the Sun", *Astron. Astrophys.* **57**, 407 - 415.

Pérez Hernández, F. & Christensen-Dalsgaard, J.: 1994a. "The phase function for stellar acoustic oscillations. II. Effects of filtering", *Mon. Not. R. astr. Soc.* **267**, 111 - 124.

Pérez Hernández, F. & Christensen-Dalsgaard, J.: 1994b. "The phase function for stellar acoustic oscillations – III. The solar case", *Mon. Not. R. astr. Soc.* **269**, 475 – 492.

Pamyatnykh, A. A., Vorontsov, S. V. & Däppen, W.: 1991, "A calibration of solar envelope models using the frequencies of intermediate-degree solar acoustic oscillations", *Astron. Astrophys.* **248**, 263 – 269.

Parker, P. D.: 1986, "Thermonuclear reactions in the solar interior", in P. A. Sturrock, T. Holzer, D. Mihalas & R. K. Ulrich (eds), *Physics of the Sun, vol. 1*, Reidel, Dordrecht, 15 – 32.

Paternò, L., Ventura, R., Canuto, V. M. & Mazzitelli, I.: 1993, "Helioseismological test of a new model for stellar convection", *Astrophys. J.* **402**, 733 – 740.

Petersen, J. O.: 1973, "Masses of double mode Cepheid variables determined by analysis of period ratios", *Astron. Astrophys.* **27**, 89 – 93.

Pidatella, R. M. & Stix, M.: 1986, "Convective overshoot at the base of the Sun's convection zone", *Astron. Astrophys.* **157**, 338 – 340.

Pinsonneault, M. H., Kawaler, S. D., Sofia, S. & Demarque, P.: 1989, "Evolutionary models of the rotating Sun", *Astrophys. J.* **338**, 424 – 452.

Pottasch, E. M., Butcher, H. R. & van Hoesel, F. H. J.: 1992, "Solar-like oscillations on α Centauri A", *Astron. Astrophys.* **264**, 138 – 146.

Proffitt, C. R.: 1994, "Effects of heavy-element settling on solar neutrino fluxes and interior structure", *Astrophys. J.* **425**, 849 – 855.

Proffitt, C. R. & Michaud, G.: 1991, "Gravitational settling in solar models", *Astrophys. J.* **380**, 238 – 250.

Rhodes Jr, E. J., Ulrich, R. K. & Simon, G. W.: 1977, "Observations of nonradial *p*-mode oscillations on the Sun", *Astrophys. J.* **218**, 901 – 919.

Rogers, F. J., Swenson, F. J. & Iglesias, C. A.: 1995, "OPAL Equation of State Tables for Astrophysical Applications", *Astrophys. J.*, in the press.

Rosenthal, C. S. & Christensen-Dalsgaard, J.: 1995, "The interfacial f mode in a spherical solar model", *Mon. Not. R. astr. Soc.*, in the press.

Rosenthal, C. S. & Gough, D. O.: 1994, "The solar *f*-mode as an interfacial mode at the chromosphere-corona transition", *Astrophys. J.* **423**, 488 – 495.

Rosenthal, C. S., Christensen-Dalsgaard, J. Nordlund, Å. & Trampedach, R.: 1995a. "Convective perturbations to solar oscillations: the f mode", in V. Domingo *et al.* (eds), *Proc. 4th SOHO Workshop: Helioseismology*, ESA SP-376, ESTEC, Noordwijk, in the press.

Rosenthal, C. S., Christensen-Dalsgaard, J., Houdek, G, Monteiro, M.J.P.F.G., Nordlund, Å. & Trampedach, R.: 1995b. "Seismology of the solar surface regions", in V. Domingo *et al.* (eds), *Proc. 4th SOHO Workshop: Helioseismology*, ESA SP-376, ESTEC, Noordwijk, in the press.

Rouse, C. A.: 1977, "On the first twenty modes of radial oscillation of the 1968 non-standard solar model", *Astron. Astrophys.* **55**, 477 – 480.

Roxburgh, I. W. & Vorontsov, S. V.: 1994, "Seismology of the solar envelope: the base of the convection zone as seen in the phase shift of acoustic waves",

Mon. Not. R. astr. Soc. **268**, 880 - 888.

Schatzman, E., Maeder, A., Angrand, F. & Glowinski, R.: 1981, "Stellar evolution with turbulent diffusion mixing. III. The solar model and the neutrino problem", *Astron. Astrophys.* **96**, 1 - 16.

Scherrer, P. H., Bogart, R. S., Bush, R. I., Hoeksema, J. T., Milford, P., Schou, J., Pope, T., Rosenberg, W., Springer, L., Tarbell, T., Title, A., Wolfson, J. & Zayer, I.: 1995, "Status of the Solar Oscillations Investigation – Michelson Doppler Imager", in R. K. Ulrich, E. J. Rhodes Jr & W. Däppen (eds), *Proc. GONG'94: Helio- and Astero-seismology from Earth and Space*, ASP Conf. Ser. **76**, 402 - 407.

Schiff, L. I.: 1949, *Quantum Mechanics*, McGraw Hill, New York.

Schmitt, J. H. M. M., Rosner, R. & Bohn, H. U.: 1984, "The overshoot region at the bottom of the solar convection zone", *Astrophys. J.* **282**, 316 - 329.

Scuflaire, R., Gabriel, M., Noels, A. & Boury, A.: 1975, "Oscillatory periods in the Sun and theoretical models with or without mixing", *Astron. Astrophys.* **45**, 15 - 18.

Seaton, M. J., Yan, Y., Mihalas, D. & Pradhan, A. K.: 1994, "Opacities for stellar envelopes", *Mon. Not. R. astr. Soc.* **266**, 805 - 828.

Severny, A. B., Kotov, V. A. & Tsap, T. T.: 1976, "Observations of solar pulsations", *Nature* **259**, 87 - 89.

Shaviv, G. & Salpeter, E. E.: 1973, "Convective overshooting in stellar interior models", *Astrophys. J.* **184**, 191 - 200.

Spergel, D. N. & Press, W. H.: 1985, "Effect of hypothetical, weakly interacting, massive particles on energy transport in the solar interior", *Astrophys. J.* **294**, 663 - 673.

Spiegel, E. A.: 1963, "A generalization of the mixing-length theory of turbulent convection", *Astrophys. J.* **138**, 216 - 225.

Steigman, G., Sarazin, C. L., Quintana, H. & Faulkner, J.: 1978, "Dynamical interactions and astrophysical effects of stable heavy neutrinos", *Astron. J.* **83**, 1050 - 1061.

Stein, R. F. & Nordlund, Å.: 1989, "Topology of convection beneath the solar surface", *Astrophys. J.* **342**, L95 - L98.

Stix, M. & Skaley, D.: 1990, "The equation of state and the frequencies of solar p modes", *Astron. Astrophys.* **232**, 234 - 238.

Swenson, F. J., Faulkner, J., Iglesias, C. A., Rogers, F. J. & Alexander, D. R.: 1994, "The classical Hyades lithium problem resolved?", *Astrophys. J.* **422**, L79 - L82.

Tassoul, M.: 1980, "Asymptotic approximations for stellar nonradial pulsations", *Astrophys. J. Suppl.* **43**, 469 - 490.

Thompson, M. J.: 1988, "Evidence for a thin perturbative layer near the base of the solar convection zone", in V. Domingo & E. J. Rolfe (eds), *Seismology of the Sun & Sun-like Stars*, ESA SP-286, 321 - 324.

Tomczyk, S., Schou, J. & Thompson, M. J.: 1995, "Measurement of the rotation rate in the deep solar interior", *Astrophys. J. Lett.*, in the print.

Turck-Chièze, S. & Lopes, I.: 1993, "Toward a unified classical model of the Sun: On the sensitivity of neutrinos and helioseismology to the microscopic physics", *Astrophys. J.* **408**, 347 - 367.

Ulrich, R. K.: 1970, "The five-minute oscillations on the solar surface", *Astrophys. J.* **162**, 993 - 1001.

Ulrich, R. K.: 1986, "Determination of stellar ages from asteroseismology", *Astrophys. J.* **306**, L37 - L40.

Ulrich, R. K. & Rhodes Jr, E. J.: 1977, "The sensitivity of nonradial p mode eigenfrequencies to solar envelope structure", *Astrophys. J.* **218**, 521 - 529.

Unno, W., Osaki, Y., Ando, H., Saio, H. & Shibahashi, H.: 1989, *Nonradial Oscillations of Stars, 2nd Edition* (University of Tokyo Press).

Vorontsov, S. V.: 1988, "A search of the effects of magnetic field in the solar five-minute oscillations", in J. Christensen-Dalsgaard & S. Frandsen (eds), *Proc. IAU Symposium No 123, Advances in helio- and asteroseismology*, Reidel, Dordrecht, 151 - 154.

Vorontsov, S. V.: 1991, "Asymptotic theory of acoustic oscillations of the sun and stars", *Astron. Zh.* **68**, 808 - 824 (English translation: *Sov. Astron.* **35**, 400 - 408).

Vorontsov, S. V. & Shibahashi, H.: 1991, "Asymptotic inversion of the solar oscillation frequencies: sound speed in the solar interior", *Publ. Astron. Soc. Japan* **43**, 739 - 753.

Vorontsov, S. V., Baturin, V. A. & Pamyatnykh, A. A.: 1991, "Seismological measurement of solar helium abundance", *Nature* **349**, 49 - 51.

Vorontsov, S. V., Baturin, V. A. & Pamyatnykh, A. A.: 1992, "Seismology of the solar envelope: towards the calibration of the equation of state", *Mon. Not. R. astr. Soc.* **257**, 32 - 46.

Wambsganss, J.: 1988, "Hydrogen-helium-diffusion in solar models", *Astron. Astrophys.* **205**, 125 - 128.

Winget, D. E.: 1991, "Asteroseismology of white dwarf stars with the Whole Earth Telescope", in G. Vauclair & E. Sion (eds), *White dwarfs, Proc. 7th European workshop on white dwarfs, Toulouse, France*, Kluwer, Dordrecht, 129 - 141.

Xiong, D. R. & Chen, Q. L.: 1992, "A nonlocal convection model of the solar convection zone", *Astron. Astrophys.* **254**, 362 - 370.

Zahn, J.-P.: 1991, "Convective penetration in stellar interiors", *Astron. Astrophys.* **252**, 179 - 188.

Zahn, J.-P.: 1992, "Circulation and turbulence in rotating stars", *Astron. Astrophys.* **265**, 115 - 132.

Testing solar models: The inverse problem

Douglas Gough

Institute of Astronomy & Department of Applied Mathematics
and Theoretical Physics, University of Cambridge, UK;
JILA, University of Colorado & National Institute of
Science and Technology, USA

FOREWORD

Broadly speaking, the inverse problem is the inverse of the forward problem. In the case of contemporary helioseismology, the forward problem is usually posed as that of determining the eigenfrequencies of free oscillation of a theoretical model of the sun. That problem is discussed by Christensen-Dalsgaard in this volume. I call inverting that problem the 'main' inverse problem. It is the one that I shall be discussing almost exclusively in this chapter. But also included in the forward problem must be the theoretical modelling of the oscillations as they really occur in the sun, forced, we believe, predominantly by the turbulence in the convection zone, and modulated by their nonlinear interactions with other modes of oscillation and by the perturbations they induce to the very convection that drives them, through variations in the turbulent fluxes of heat and momentum. The inverse of that problem is to derive from the fluid motion of the visible layers in the atmosphere of the sun, which I presume to be 'observed', estimates of the frequencies that the modes would have had had they not been disturbed by the other forms of motion. The outcome of that prior inversion provides the data for the main inverse problem.

This chapter is entitled: Testing solar models ... . By 'solar models' is meant any theoretical description of the sun that we might have in mind. The solar model discussed by Christensen-Dalsgaard in the preceding chapter is archetypical at present, and the reader is advised to keep that in mind when reading this chapter. However, many of the principles that I discuss are not restricted to such models, and will apply equally well to the more realistic models with which we expect to work in the future.

This chapter does not provide a detailed prescription for carrying out inverse calculations. For that the reader is referred to the various reviews and textbooks on the subject (e.g. Gough, 1985; Craig and Brown, 1986; Gough and Thompson, 1991; Thompson, 1995), and the research papers to which I shall refer later. Instead, it emphasizes the general principles by which inverse problems are formulated and solved, the assumptions that are made in that process, which limit their validity, and the meaning of the results of inverse calculations. By so doing I hope to provide a guide to appreciating the results, particularly for non-practitioners, so that those results can be used to advantage without misinterpretation.

1 INTRODUCTION

I shall adopt a rather loose view of what I mean by the (main) inverse problem: to determine some aspect of the structure of a solar model from the values of its eigenfrequencies. One cannot solve an inverse problem unless one understands the forward problem. It therefore behoves me at least to recall that problem, and to discuss it a little, before proceeding to argue backwards, otherwise there is a risk that my arguments will be flawed, or, if not, misunderstood. I therefore begin by writing down the wave equation

$$\Box\Psi = \mathbb{D}\Psi + F(\mathbf{r}, t)\,, \tag{1.1}$$

which is the basis of the study. Here Ψ is some wave function characterizing the oscillations, and $\Box$ is the adiabatic wave operator appropriate for a star, which can be derived from the discussions by, for example, Unno *et al.* (1989), Christensen-Dalsgaard and Berthomieu (1991) and Gough (1993). I shall discuss $\mathbb{D}$ and F soon. But let me point out straight away that neither $\Box$, $\mathbb{D}$ nor F depend explicitly on Ψ; the oscillatory motion under study is presumed to be of such low amplitude that linearization of the full equations of fluid dynamics, from which equation (1.1) was derived, is valid. Therefore I am already ignoring the interactions between different forms of oscillatory motion. That may not be a good approximation near the surface of the sun where the eigenfunctions of p modes are relatively large; although in the photospheric regions the interaction between two p modes is extremely small compared with most of the other phenomena we consider, the interaction of a p mode with the sum of all the other modes is not necessarily small. This is an issue that will need to be studied in some detail in the future. I might also point out, in passing, that there has been some attention devoted to seeking the sum of the interactions of a single mode (actually a g mode) with many p modes, for the purposes of detecting that g mode, but these attempts have been unsuccessful.

In writing the wave equation in the form (1.1) I have assumed that the dominant dynamics is represented by the adiabatic wave operator $\Box$. That operator depends on the structure of the solar model, which I shall refer to as the background (non-oscillating) state, and which I denote by $\mathbf{X}(\mathbf{r})$, the components of $\mathbf{X}$ being, say, the pressure p, density ρ, the first adiabatic exponent $\gamma_1 := (\partial \ln p/\partial \ln \rho)_{\rm ad}$, and other quantities that are usually considered to be small perturbations to the spherically symmetrical model of the preceding chapter, such as a steady background velocity field $\mathbf{v}(\mathbf{r})$, which is no doubt dominated by rotational flow $\mathbf{r}\times\boldsymbol{\Omega}$, and a possible magnetic field $\mathbf{B}(\mathbf{r})$. I have assumed that a (possibly rotating) frame of reference exists in which $\mathbf{X}$ is independent of time t. In that case there are genuinely separable solutions $\Psi(\mathbf{r}, t) = \Re[\Psi(\mathbf{r})e^{-i\omega t}]$ of the simple homogeneous adiabatic wave equation

$$\Box\Psi = 0\,, \tag{1.2}$$

having a purely sinusoidal dependence on t with frequency ω. The amplitude function $\Psi(\mathbf{r})$ satisfies

$$\mathcal{L}_\omega\Psi = 0\,, \tag{1.3}$$

where the three-dimension spatial differential operator $\mathcal{L}_\omega$ is obtained from the full wave operator $\Box$ by replacing $\frac{\partial}{\partial t}$ by $-i\omega$. Note that I have used the symbol Ψ to denote both the full wave function and its spatial part. I shall use ψ to represent the factor depending on r in the separated form $\Psi(\mathbf{r}) = \psi(r)Y_l^m(\theta,\phi)$ with respect to spherical polar coordinates (r,θ,ϕ) when the background state is spherically symmetrical, $\mathbf{X}(\mathbf{r}) = \mathbf{X}(r)$, where $Y_l^m(\theta,\phi) = P_l^m(\cos\theta)e^{im\phi}$ is a spherical harmonic of degree l and azimuthal order m, $P_l^m(\mu)$ being the associated Legendre function of the first kind. In that case $\mathcal{L}_\omega$ will represent the radial part of the corresponding three-dimensional operator with the same name. Context should dictate how the notation should be interpreted. In the spherically symmetric case, the solutions of equation (1.3), together with appropriate boundary conditions, admit discrete eigenfrequencies ω_{nl}, degenerate with respect to m. Christensen-Dalsgaard discusses in some detail in the preceding chapter how the values of ω_{nl} depend on various aspects the structure of the theoretical solar model. If the model were mildly aspherical, that discussion would refer to the spherically averaged structure $\bar{\mathbf{X}}(r) = (4\pi)^{-1}\int\int \mathbf{X}(r,\theta,\phi)\sin\theta\, \mathrm{d}\theta\mathrm{d}\phi$. Deviations of $\mathbf{X}$ from $\bar{\mathbf{X}}$, brought about by rotation, a magnetic field or large-scale (steady) circulation, split the degeneracy of the eigenfrequencies and distort the eigenfunctions. The modifications $(\delta\omega_{nlm}, \delta\Psi_{nlm})$ to the eigenmode (ω_{nl}, Ψ_{nl}) of the spherical model can be found by degenerate perturbation theory, associated with which is, of course, an inverse problem.

If one takes the Fourier transform of the full wave equation, one obtains

$$\mathcal{L}_\omega\Psi = \mathcal{D}_\omega\Psi + \mathcal{F}_\omega\,, \tag{1.4}$$

in an obvious notation. The quantities $\mathcal{D}_\omega$ and $\mathcal{F}_\omega$ represent all the physics that is omitted in equation (1.3). $\mathcal{D}_\omega$ contains the nonadiabatic terms, which are important only in and immediately beneath the atmosphere, except possibly for unobserved long-lived unstable g modes that may exist in the radiative interior of the sun. More important for my discussion are the effects of the turbulence in the convection zone. As is common in turbulence studies, I consider each of the turbulent terms in the equation of motion to be divided into two, an ensemble average, which I shall not define precisely, and a fluctuation from that average. Both terms are modulated by the oscillations. Note that this turbulence contains a magnetic field, which has both thin turbulent fibril components, larger-scale components in active regions, and also a global component. The ensemble averages provide a mean momentum flux (Reynolds and Maxwell stresses) and a mean heat (enthalpy) flux: the steady parts of those fluxes are presumed to be present in the equations for the solar model, the oscillatory perturbations are linearized and appear in $\mathbb{D}$ and its Fourier transform $\mathcal{D}_\omega$. The leading term of the turbulent fluctuation from the ensemble average is the inhomogeneous forcing term F in equation (1.1). The oscillatory perturbation to that term is always ignored, and I shall do so here too. Some day its role will be assessed. It would perhaps be premature to do so now, because we do not have reliable prescriptions for the larger terms $\mathbb{D}\Psi$ and F. Indeed, I should even point out that what prescriptions we do have for the turbulent fluxes have not been written explicitly in the simple-looking form $\mathbb{D}\Psi$, although I presume they could be. The outcome would be a rather complicated integro-differential operator. There are mixing-length prescriptions for the fluxes appearing in $\mathbb{D}$, and there are discussions of F which are mixing-length-like in spirit: in almost all studies the existence of the magnetic field is ignored. I should point out that most modern analyses of the homogeneous eigenvalue problem $(\mathcal{L}_\omega - \mathcal{D}_\omega)\Psi = 0$ find most eigenmodes (all the p and f modes and most of the g modes) to be stable. It is only those stable modes that I shall discuss. If there are any intrinsically unstable modes, we would need to understand how energy is extracted from them to limit their amplitudes to values comparable with those of the stable modes.

The influence of the inhomogeneous term $\mathcal{F}_\omega$ on equation (1.4) is to permit nontrivial solutions for all ω. Because the damping in $\mathcal{D}_\omega$ is small, the amplitude of Ψ is substantial only for values of ω that are close to the real parts of the eigenfrequencies of the homogeneous problem. Therefore a power spectrum of the solutions for given l and m exhibits a sequence of sharp peaks, each corresponding to a mode of order n. The amplitudes and shapes of those peaks, and their positions relative to the real parts ω_{nlm} of the eigenfrequencies, depend on $\mathcal{D}_\omega$ and $\mathcal{F}_\omega$. There is an interesting and very important inverse problem here: both to determine the values of ω_{nlm}, which are needed for the main inverse problem, and to learn about the acoustically relevant parts of the turbulent motion contained in $\mathcal{D}_\omega$ and $\mathcal{F}_\omega$. I shall not discuss that problem explicitly, though much of what I say about the main inverse problem pertains to it too. Instead, because of the serious uncertainties inherent in the turbulent interactions, an extremely important task in the main inverse problem is to devise means by which to make inferences that do not depend on those uncertainties.

It is fortunate that the principle uncertainties arising from the turbulence are very close to the surface of the sun. As discussed by Christensen-Dalsgaard, this property imposes a special functional form on the uncertain contributions to the eigenfrequencies which can be exploited when trying to make inferences about the deep interior. I shall return to that issue later. But in the meanwhile, let me point out that the thin boundary layer of intense turbulence at the top of the convection zone masks the atmosphere from the interior of the sun in which the waves that constitute the oscillation modes propagate. It is for that reason that I have paid little attention to the boundary conditions. Indeed, the uncertain influence of the turbulent boundary layer on the oscillation dynamics deep inside the star can be mimicked by a modification to the boundary conditions.

There are different ways in which the wave equation (1.1) can be formulated in detail – even the simple homogeneous wave equation (1.2) which will be the main object of study. Each can be transformed into another. In carrying out the mathematical manipulations for the transformations it is often necessary to make substitutions using some of the 'equilibrium' equations of the nonoscillating background state. Therefore, it goes without saying that the background state really must satisfy those equations. Those equations are basically the equation of 'hydrostatic' support, which must include the turbulent and magnetic stresses if the perturbations to those stresses are included in the wave equation (1.1), and the Poisson equation determining the gravitational potential Φ in terms of the density ρ. For studying adiabatic oscillations, the thermal equations need not be satisfied by the background model, because these control evolution on a timescale very much greater than the oscillation periods. Of course, the fact that the background state is not strictly static implies that strictly separable solutions with sinusoidal temporal variation probably do not exist; however, they are extremely good approximations which are certainly quite adequate for our purposes. Note, however, that if the equations for Φ and of hydrostatic support are not precisely satisfied, the derivations of the wave equation are invalid and one cannot assign to the stellar 'model' (if one is even then prepared to endow it with that name) a well defined set of eigenfrequencies.

In the case of a spherically symmetrical background state, the homogeneous differential equation $\mathcal{L}_\omega \psi = 0$ representing adiabatic oscillations is fourth-order, and is to be solved subject to two regularity conditions at the coordinate singularity $r = 0$ and two boundary conditions at the surface $r = R$. When solved numerically, the radial coordinate r is most commonly retained as the independent variable. When solved asymptotically for high n, the most natural coordinate to use for p modes is the acoustical radius $\tau = \int c^{-1}\,\mathrm{d}r$, where $c = (\gamma_1 p/\rho)^{1/2}$ is the sound speed.

1.1 About what can the eigenfrequencies tell us?

Since the wave equation depends on the background state only through the seismically relevant quantity $\mathbf{X} = (p, \rho, \gamma_1; \mathbf{v}, \mathbf{B})$, it can be only that quantity about which we can make inferences from the properties of the seismic oscillations. Christensen-Dalsgaard has discussed how the eigenfrequencies of a spherically symmetrical solar model depend on p, ρ and γ_1, and how certain classes of 'physical' perturbations influence p, ρ and γ_1, and consequently the frequencies. The variables p and ρ are related through the equation of hydrostatic support, so one can use just one of them to specify the background state, or some function of all three quantities, such as c^2. Note that in order to infer from the seismic model secondary variables, such as temperature, one needs to use the equation of state, which cannot be accomplished explicitly without knowing, or assuming, the chemical composition.

The existence of dynamical motion $\mathbf{v}$ and a magnetic field $\mathbf{B}$ necessarily implies that the background state is not spherically symmetrical. However, the influences of these phenomena on the background state and on the oscillations are small, and the effects can be determined by degenerate perturbation theory (e.g. Dziembowski and Goode, 1989, 1992; Gough and Thompson, 1989). That theory determines not simply the perturbed eigenfrequencies, but it also raises the degeneracy of the modes, determining which combinations of the $2l+1$ degenerate components of the separable solutions

$$\Psi_{nlm}(\mathbf{r}, t) = \psi_{nl}(r) P_l^m(\cos\theta) e^{im\phi} , \tag{1.5}$$

for given n and l and for $-l \leq m \leq l$, constitute actual 'modes' (i.e. motions with a given frequency ω_{nlm}) of even the unperturbed problem (strictly speaking, in the limit as the magnitude of the influence of $\mathbf{v}$ and $\mathbf{B}$ tends to zero). The largest, and the most commonly considered, perturbation to the eigenfrequencies comes from rotation (at least when measured with respect to an inertial frame of reference). Assuming the angular velocity $\mathbf{\Omega}$ throughout the star to be about a unique axis, then the separated forms (1.5) constitute the eigenfunctions when the axis of the coordinates is chosen to coincide with the axis of rotation. Then the integer quantities (n, l, m) continue to label the mode: they are good quantum numbers. Whatever the functional form of $\Omega(r, \theta) = |\mathbf{\Omega}|$, the perturbed eigenfrequency, $\delta\omega_{nlm} := \omega_{nlm} - \omega_{nl0}$, satisfies $\delta\omega_{nl(-m)} = -\delta\omega_{nl(+m)}$. This is obvious because reflection about a plane through the axis can be regarded either as $m \to -m$ or $\Omega \to -\Omega$. Moreover, rotation is the only perturbation with such symmetry, and therefore Ω can be investigated seismically in realistic circumstances (in which Ω is not the only symmetry-breaking agent) by using the component of $\delta\omega_{nlm}$ that is odd with respect to m. Once Ω is determined (assuming that to be possible) one can then calculate the effect of the terms quadratic in Ω (the centrifugal force both acting on the background model and contributing to the oscillation dynamics, and what might loosely be considered as the perturbation to the Coriolis force caused by the first-order rotational distortion to the eigenfunction), which cannot distinguish east from west and which therefore make a contribution to the frequency splitting by an amount which is an even function of m. What remains is, of course, also an even function of m, and is caused by the combination of the nonrotational component of $\mathbf{v}$ (associated with which there is no angular momentum), the field $\mathbf{B}$ and the (probably) associated aspherical contribution to c^2 and γ_1.

At this point I must point out that it can be shown that any even function of m representing $\delta\omega_{nlm}$ can be reproduced by an appropriate aspherical perturbation δc^2 to c^2 alone (hydrostatic equilibrium can always be satisfied formally with an appropriate distribution of chemical composition, which influences the relation between pressure and density, via the mean molecular mass in the case of a perfect gas). Therefore one cannot separate the influence of $\mathbf{v}$, $\mathbf{B}$ and δc^2 by observing normal-mode eigenfrequencies alone. One needs also to observe the deviation of the shapes of the eigenfunctions from spherical harmonics to obtain extra information, though whether that would be sufficient to unravel the perturbing agents, even in theory, is not known. The most fruitful approach in practice is likely to be one or all of the local procedures that are currently being developed and which I discuss briefly in section 8. But, in addition, non-seismic criteria are likely to be brought to bear on the issue, to offer a preference amongst seismically indistinguishable perturbations.

1.2 What cannot the eigenfrequencies tell us?

Part of what I said in the previous section concerned what one cannot learn by analysis of seismic frequencies alone. I introduced the concept of seismic indistinguishability, without actually defining it. At some level it should have been intuitively obvious, but permit me to expand a little. One might first introduce the idea of seismic equivalence: two solar models are seismically equivalent (with respect to frequency) if they support identical eigenfrequencies $\{\omega_{nlm}\}$. This property depends on the set $\{n, l, m\}$ of modes in the equivalence class; in practice it can contain only the modes whose frequencies have been measured, though one might extend it to those that might be measured in the future. Since there are only a finite number of p modes (the frequencies of pure, perfectly contained modes are bounded above because the atmosphere becomes leaky once the acoustic cutoff frequency is exceeded, and for any given frequency there is an upper bound to the degree l, given by the value corresponding to the eigenfrequency of the mode with $n = 1$), one cannot possibly invert the wave equation to determine $\mathbf{X}(\mathbf{r})$, since a function contains an infinite amount of information. Indeed, if there exists any function $\mathbf{X}$ that satisfies the data, there are also infinitely many more. Therefore any curve that is drawn to represent an inversion must be interpreted with considerable care.

It is worth mentioning that even were the solar atmosphere to be perfectly reflecting to all p modes, so that the sun could support an infinite set of frequencies $\{\omega_{nlm}\}$ which we might imagine to be known precisely, there could still be doubt that the structure could be determined seismically. I have in mind the simple question asked nearly thirty years by Marc Kac: Can one hear the shape of a drum? The question was asked of the simple classical wave equation with uniform sound speed, and the answer anticipated was negative. Indeed, it was shown recently that Kac's conjecture is correct, by Gordon, Webb and Wolpert (1992) who demonstrated the existence of isospectral drums of different shapes. The different 'drums' have quite different shapes. If, likewise, it were to be the case that other completely seismically equivalent solar models would have structures quite different from the models we consider, then we could probably rule them out on physical grounds.

To get some idea of what one cannot deduce about the structure of the sun, it is useful to consider the structure of the eigenfunctions. We know that p modes are essentially confined to an outer shell, between spheres of radii r_1 and r_2, the turning points of the modes. The upper turning point r_2 is near the surface, the lower turning point is where $r_1^{-1}c(r_1) \simeq w \equiv \omega/L$, where $L = l+\frac{1}{2}$. Basically, one cannot learn much outside that shell from seismic frequencies — the constraints imposed by hydrostatic support and the knowledge of $M_\odot$ and $R_\odot$ do provide some extension of the domain of diagnosis, however. Within the shell, one can learn about the structure by analysing data from groups of modes with varying w, which span different subshells, or by groups with the same value of w and varying n, whose eigenfunctions have more and more nodes and which sample in a manner somewhat analogous to the terms in a Fourier series. If one has measured all the modes of lowest order, one would expect, by analogy to Fourier representation, to capture the most slowly varying components of the structure. The converse is not necessarily the case, however, because in the absence of grave modes one might still be able to extract all the slowly varying information by using modes with varying w. However, it is evident that on the whole the smallest scale that might be resolved is likely to be given approximately by some combination of the separation Δr_1 between 'neighbouring' modes whose frequencies are available (i.e. adjacent modes, when they ordered monotonically in w, and hence in the associated value of r_1) and the smallest characteristic lengthscale of variation of the eigenfunctions. But that is only a guide. In principle, by judicious combination of many accurately measured modes one can resolve much more finely: one can see that that is the case by noticing that one can formally invert a Laplace transform, whose weight functions are real exponentials with scales of variation that are either of the order of the distance to the origin, or greater. However, there is much cancellation of data in the process, and the outcome is highly susceptible to data errors. Indeed, it is such error susceptibility that nowadays limits the resolution in practice. I shall discuss later how one can assess the local resolving power of data; a systematic global study has never been carried out.

It is the recognition of the existence of errors in the data that leads to the extension of the concept of seismic equivalence to seismic indistinguishability. If, for a given set of modes, some property of the structure could have been determined had the frequencies been known to arbitrary accuracy, but that to do so would involve constructing data combinations in which in practice all the significant information cancels, leaving only errors, then that property of the structure cannot be determined. Two representations of the sun that differ only by that property are seismically indistinguishable — they might also be called seismically equivalent, in a loose sense. Evidently, increasing either the extent or accuracy of the data can remove the indistinguishability.

2 THE IDEA OF INVERSION

Given that the forward problem, say that posed by the eigenvalue problem associated with equation (1.3), cannot be inverted to determine completely the structure $\mathbf{X}(\mathbf{r})$ of the sun in terms of the eigenfrequencies $\{\omega_{nlm}\}$, what is it that we understand by inversion? The answer to that question is a matter of opinion. I shall try to give some idea of the more commonly held views, some of which are held operationally, in the sense that they simply relate to specific inversion procedures.

Perhaps the most common procedure that is adopted under the name of inversion is to represent $\mathbf{X}$ by an example $\hat{\mathbf{X}}$ that is chosen explicitly to fit the data adequately, given their uncertainties. Since, if such an example exists, there is an infinity of alternative possibilities, one must adopt a criterion for selection. That is usually based on smoothness. Thus, for example, one might seek amongst those functions $\hat{\mathbf{X}}$ that adequately satisfy the data the one that minimizes some global measure P of curvature. P is sometimes called a penalty function. For spherically symmetrical configurations, radius r is a possible independent variable, and therefore one might choose $P = \int \hat{\mathbf{X}}'' \cdot \hat{\mathbf{X}}'' \mathrm{d}r$, where the prime denotes differentiation with respect to r, and the integral is from the centre to the surface of the solar model. Of course, one can generalize this to

$$P = \int_0^{s(R)} W(s) \left| \frac{\mathrm{d}^2 \hat{\mathbf{X}}}{\mathrm{d}s^2} \right|^2 \mathrm{d}s \,, \tag{2.1}$$

where $W \geq 0$ is an (arbitrary) weight function and $s(r)$ is a new coordinate. Part of the art of inversion is in choosing suitable functions W and s.

For an inversion in more than one dimension there is even greater flexibility. One might choose, for a two-dimensional inversion aimed at finding an acceptable example of the axisymmetric component $\hat{\mathbf{X}}$ of the structure,

$$P = \int \left[W_r(r,\mu) \left| \frac{\partial^2 \hat{\mathbf{X}}}{\partial r^2} \right|^2 + W_\theta(r,\mu) \left| \frac{\partial^2 \hat{\mathbf{X}}}{\partial \mu^2} \right| \right] r^2 \mathrm{d}r \mathrm{d}\mu \,, \tag{2.2}$$

where $\mu = \cos\theta$. Other penalty functions have been used, presumably for convenience, such as the 2-norm $\int \hat{\mathbf{X}} \cdot \hat{\mathbf{X}}$, which favours functions small in magnitude, or a function based on a measure of the overall square of the gradient of $\hat{\mathbf{X}}$, which favours functions that are flat.

Evidently, there are other means by which one might choose amongst possible functions $\hat{\mathbf{X}}$. One might incorporate into a procedure such as that described above particular features that one might expect on theoretical grounds to be present, such as an appropriate derivative discontinuity at the base of the convection zone. Alternatively, one might dispense with an explicit penalty function entirely, and replace it by the constraint that the sun is represented by a model that satisfies the equations, so-called standard or otherwise, of stellar evolution (including, if necessary, some simplified prescription for the internal dynamics to model rotation and large-scale asphericity). In the case of assuming spherical symmetry, this approach, in some form or another, has at least until recently been the most common. The procedure, in principle, is to characterize the theoretical models by a set $\{\alpha_i\}$ of uncertain parameters that arise in the theory, such as the mixing-length parameter, the solar age, the initial chemical composition and the host of uncertain parameters arising in the microphysics, which are discussed in the chapter by Bahcall, and then to adjust them to yield the best fit to the data. Just how one defines 'best fit', and also even how one defines 'adequate fit', in procedures such as this, or even procedures that simply incorporate penalty functions, is also, to some extent, part of the art of the subject. I should hasten to add, however, that this (in some sense extreme) model-calibrating approach is not normally regarded as falling into the category of inversion, at least amongst helioseismologists. The reason is probably partly that the theory of stellar evolution restricts the class of admissible functions $\hat{\mathbf{X}}$ too severely, thereby possibly eliminating any acceptable representation of the real sun if that model is flawed in some material way. Our failure to solve the neutrino problem is a stark warning that it is dangerous to accept such models uncritically.

It seems to me that a factor distinguishing inversion from model-calibration is attitude. In the case of calibrating models, one builds the best (one would hope) sequence of models one can, based on physical principles, and then selects from amongst them that model, or perhaps the group of seismically indistinguishable models, that reproduce the data most closely. This procedure has the obvious advantage of putting physics to the fore. On the other hand, one is usually not capable of incorporating all the relevant physics, and it is often difficult to divorce oneself from constraints imposed by the assumptions one has adopted. An obvious example is the common practice in discussions of the neutrino problem of restricting attention to so-called standard models of the sun, which are spherically symmetrical and essentially static, taking no cognizance of the internal dynamics. I presume that the hope is that the dynamics that actually operates in the sun is not important for the problem in hand. Indeed, from time to time arguments are advanced to support that view, but more often than not the modellers lose sight of them. Another example is a debate I encountered some 25 years ago at a scientific meeting in New York, concerning Dicke's oblateness measurements and their implications. Somebody had calculated the centrifugally induced oblateness theoretically, claiming that the greatest uncertainty in the result, about 30 percent if I recall correctly, arose from the uncertainty in the density distribution within the sun. There followed a discussion of what research should be carried out to improve the reliability of the density stratification. The stellar modellers, accustomed to building spherically symmetrical stars with no internal dynamics, had lost sight of the fact that at that time they did not know the angular velocity inside the sun to within a factor 10 or so, which made a contribution to the uncertainty in their calculations some 300 times greater than the contribution that they were so preoccupied in reducing. I tell this story not in ridicule of model-builders – we have probably all made mistakes of that kind – but to stress that we are all susceptible to accepting the dogma of a discipline without remembering to challenge it from time to time.

The attitude of inverters is to question everything. Of course, they don't succeed, for in practice one has to start from somewhere that is not too far from one's goal. However, some of the assumptions made by the modellers are relaxed, and that is healthy. Those assumptions need to be replaced by other assumptions, at least when searching for explicit representations $\hat{\mathbf{X}}$ of the sun. They are typically one of the smoothness criteria I discussed earlier, which are not only computationally convenient but are also justified on the ground that small-scale structure in $\hat{\mathbf{X}}$ cannot be resolved by the data. (I shall address later just how small that scale is.) The function $\hat{\mathbf{X}}$, it is sometimes argued, is therefore a smoothed representation of the true structure.

There are two remarks I wish to make about what I have just said. The first is that I regarded a smoothness criterion as an assumption, whereas, when I introduced it, it was simply a criterion for selecting one, or at most a finite number of representations $\hat{\mathbf{X}}$ from the infinite number potentially available. My change of view was deliberate. If you have read over it without having being brought abruptly to a halt, then you may be susceptible to falling into the trap of believing that $\hat{\mathbf{X}}$ actually resembles the sun. Why should one particular function selected from the infinity of possible choices be representative? The mitigating case is my second remark: that if smoothness is the selection criterion, then the chosen $\hat{\mathbf{X}}$ is indeed likely to be a smoothed-out version of reality. That is something we believe we can comprehend. But how is the smoothing carried out? That is a question I shall address below. The answer can be made quite explicit in the case of linearized inversions (and, since I cannot think nonlinearly except under rather simple circumstances, that is adequate at least for me). One can represent $\hat{\mathbf{X}}$ (or the deviation, presumed small, of $\hat{\mathbf{X}}$ from some known reference function $\mathbf{X}_0$) in terms of averages over essentially localized regions of space, weighted by functions $\mathcal{K}(r, r_0)$, sometimes called averaging kernels, which can be displayed.

These remarks have raised two more issues. The first is linearization. In many branches of mathematics it is difficult, even impossible, to be truly nonlinear, and the mathematics of inverse theory is no exception. Therefore, it is common to 'solve' problems as a sequence of linear iterations. I shall describe below how an iteration is carried out. The second issue concerns the representation of $\hat{\mathbf{X}}$ in terms of weighted averages. Typically, the quantities that the data represent are themselves weighted integrals of the structure $\mathbf{X}$. So what has 'inverting' them accomplished? It has simply combined those integrals into a new set, each of which is more easily digestible. Typically, though not always, the new weight functions $\mathcal{K}$ are localized in space, rather than being distributed over a substantial region as is characteristic of the weights determining the original data. The question immediately arises, therefore, of altering one's attitude further, and seeking data combinations that yield averaging kernels $\mathcal{K}$ that are most suited to answering whatever question one might have in mind. Different questions might require different averaging kernels. That step, which was first made by G. Backus and F. Gilbert, led to a transformation of our view of what we wish to accomplish: instead of trying to find one or a few functions $\hat{\mathbf{X}}$ that satisfy the data, one instead seeks easily comprehendable averages $\int \mathcal{K}\hat{\mathbf{X}}$ that are satisfied by all the functions $\hat{\mathbf{X}}$. If the problem has been posed correctly (what I really mean is that the assumptions that must necessarily have been made in representing the structure of the sun in mathematical terms do not lead to serious errors), then the actual structure $\mathbf{X}$ of the sun must have the same averages. Then one can almost dispense with the functions $\hat{\mathbf{X}}$, and regard the outcome of the inversion as a set of functionals of $\mathbf{X}$ itself.

One of the important ingredients of inversion is to arrive at some idea of the uncertainty in the outcome. This is often difficult to achieve: notwithstanding the inherent difficulties in estimating the uncertainties in the observational data, the manner in which those errors fold into the inversions depends on the assumptions of the underlying physical model. By how much those assumptions influence a particular inference one might wish to make from the inversion one has carried out must usually be judged by relaxing the assumptions. How well that can be accomplished depends largely on the extent of one's imagination: it depends on how far into the space of possibilities one's mind is able to penetrate. Fortunately, in the simple linear inverse problems which I have mentioned already, one can represent the uncertainties in a well defined mathematical form in terms of the uncertainties in the original data. I shall present some of those later in the chapter.

3 ON MODEL-CALIBRATION

The most common, and often the simplest method of using any form of data to make inferences about the physical state of the object under study is to fit a sequence of theoretical models to those data. The models $\mathcal{M}$ are defined by a set of parameters $\{\alpha_k\}$, and, for each set of values, the eigenfrequencies of oscillation $\omega_{\mathcal{M}\mathbf{i}}(\alpha_k)$ (in the case of seismic fitting) of modes $\mathbf{i}$ are calculated. The parameters are then adjusted to produce the best fit to the data.

The art of this procedure is to define what one means by 'best fit'. Evidently, the definition must depend on what physical question one has in mind. Even if the models were constructed with perfect physics, so that amongst the set is a model that faithfully represents the sun, inevitable errors in the data will lead to an error in the fitting: recognition that those errors are likely to be present leads to an uncertainty in the calibration. Let us suppose that we have reliable estimates of the variances of the errors. Then the calibration determines a region in parameter space within which the true sun is likely to lie. However, models are not perfect. Once one accepts that there is also uncertainty in the physics (i.e. in the necessarily approximate equations used to represent the physics), then one can no longer even calculate the extent of that region. What does one do then?

The most naive approach is to do nothing. Thus one might use all the data indiscriminately, and choose $\{\alpha_k\}$ such as to minimize

$$E(\alpha_k) = I^{-1} \sum_{\mathrm{i}} \sigma_{\mathrm{i}}^{-2} \left[\omega_{\mathrm{i}} - \omega_{\mathcal{M}\mathrm{i}}(\alpha_k)\right]^2 , \tag{3.1}$$

where σ_{i}^2 is the variance of the possible measurement error in the datum ω_{i}, and the summation is over all of the I available data. Such a calibration was first carried out explicitly for the sun by Christensen-Dalsgaard and Gough (1981), using only low-degree data, with a sequence of models for which the initial helium abundance Y was the only adjustable parameter, yielding Y somewhat greater than 0.25. But the minimum value of E was substantially greater than unity, which indicated that there was something wrong either with the models or with the observers' estimates of σ_{i}. (The authors supposed that it was with the models that something was the more seriously wrong.)

An obvious limitation of this procedure is that it provides no indication of the manner in which the models are deficient. For that, one needs to carry out further investigations. A variety of numerical studies of the response of eigenfrequencies to specific changes in the model are described in the extensive discussion by Christensen-Dalsgaard in the preceding chapter. These studies provide one with valuable experience. An alternative procedure is to inspect kernels associated with a variety of uncertain quantities that one may wish to vary; the derivations of such kernels are outlined in section 5. Approximate analytical methods, based on asymptotics or on simple models, are also of very great value. They can be used both to help interpret frequency mismatches and to design criteria by which to calibrate models. A second possible limitation might arise from putting too much emphasis on what has been measured, rather than on what one wishes to infer.

Permit me to digress a little on approximate analytical methods. It has been said that because they are not as precise as numerical calculations (I shall not digress further with a discussion of why that assumption is not always correct; it ought to be correct), they have no place in modern model calibrations. That statement is manifestly incorrect. If a numerical model is an apparently good representation of the sun, then analytical techniques can be used to estimate just the deviation of that model from the structure of the sun. The errors arising from the inadequacies of the analytical formulae contaminate the inferences of only the differences, and if those differences are small the error in the final model could be negligible. Indeed, the first helioseismic calibration was carried out in this manner, using the analytically determined eigenfrequencies of a deep plane-parallel envelope to scale numerically computed high-degree eigenfrequencies of a more realistic model envelope that do not fit the data in order to calibrate the adiabat deep in the convection zone. Further scaling, based this time on a sequence of numerical models, was then used to estimate the depth of the convection zone. An advantage of this method is that the analytical analysis often permits one to appreciate which aspects of the model are important in determining the eigenfrequencies and which are not. Consequently it provides insight into what the eigenfrequencies do and do not tell us. In the case of the calibration I have just mentioned, it was possible to be more confident in the reliability of what was being determined than it was from a numerical calibration with similar aims that was reported soon afterwards.

Another example of the use of the polytropic envelope was to exhibit that the influence of the outer layers of the sun influence mostly the higher-frequency modes, and to estimate the dependence on frequency of that influence. This was part of an early investigation of a helium-abundance calibration (Christensen-Dalsgaard and Gough, 1980) and was carried out to demonstrate how the absolute values of each of the low-degree cyclic frequencies $\nu_{n,l}$ depend sensitively on the uncertain upper boundary layer of the convection zone, whereas in the difference $\Delta_{nl}\nu = \nu_{n,l} - \nu_{n-1,l}$ the uncertainty largely cancels. Therefore $\Delta_{nl}\nu$ is a more robust diagnostic of the deep interior than are the absolute values of the frequencies themselves. It was that work that led to the scaling of frequency differences between sun and model by a scaled measure Q_{nl} of the inertia $\mathcal{I}$ of the mode which is now the norm when studying near-surface model perturbations (see the chapter by Christensen-Dalsgaard), for $\mathcal{I}$ is simply a measure of how the region of propagation, in which the energy resides, is connected acoustically to the surface, where the eigenfunctions are normalized. Of course, such properties could have been discovered from the full equations, together with their numerical 'solutions', but the fact is that that is not what happened. Moreover, that is not an isolated case. It illustrates how approximate analytical analyses facilitate understanding and at least accelerate, if not enable, the progress of the subject.

Let us now return to the model-calibration by high-degree modes. Figure 1 shows Deubner's early $\nu - l$ power spectrum, where $\nu = \omega/2\pi$ is cyclic frequency, on which is drawn the eigenfrequencies of two solar models with different values of the initial helium abundance Y. The modes responsible for the discernible ridges reside mainly in the adiabatically stratified region of the convection zone, where the stratification is polytropic. Therefore one might use the frequencies of a complete plane-parallel polytrope to try to understand the difference between the two theoretical curves. They are given by

$$\gamma_1^{-1}(\mu+1)^2 = \mu + [\mu - \gamma_1^{-1}(\mu+1)]\sigma^{-2} = 2n \tag{3.2}$$

(e.g. Gough, 1993), where $\sigma^2 = R\omega^2/gl$, R and g being the solar radius and the gravitational acceleration and μ being the polytropic index. Notice that Y, or its direct influence, is not evident. That tells us immediately that the calibration hinted at in Figure 1 must depend on a global property of the entire solar model.

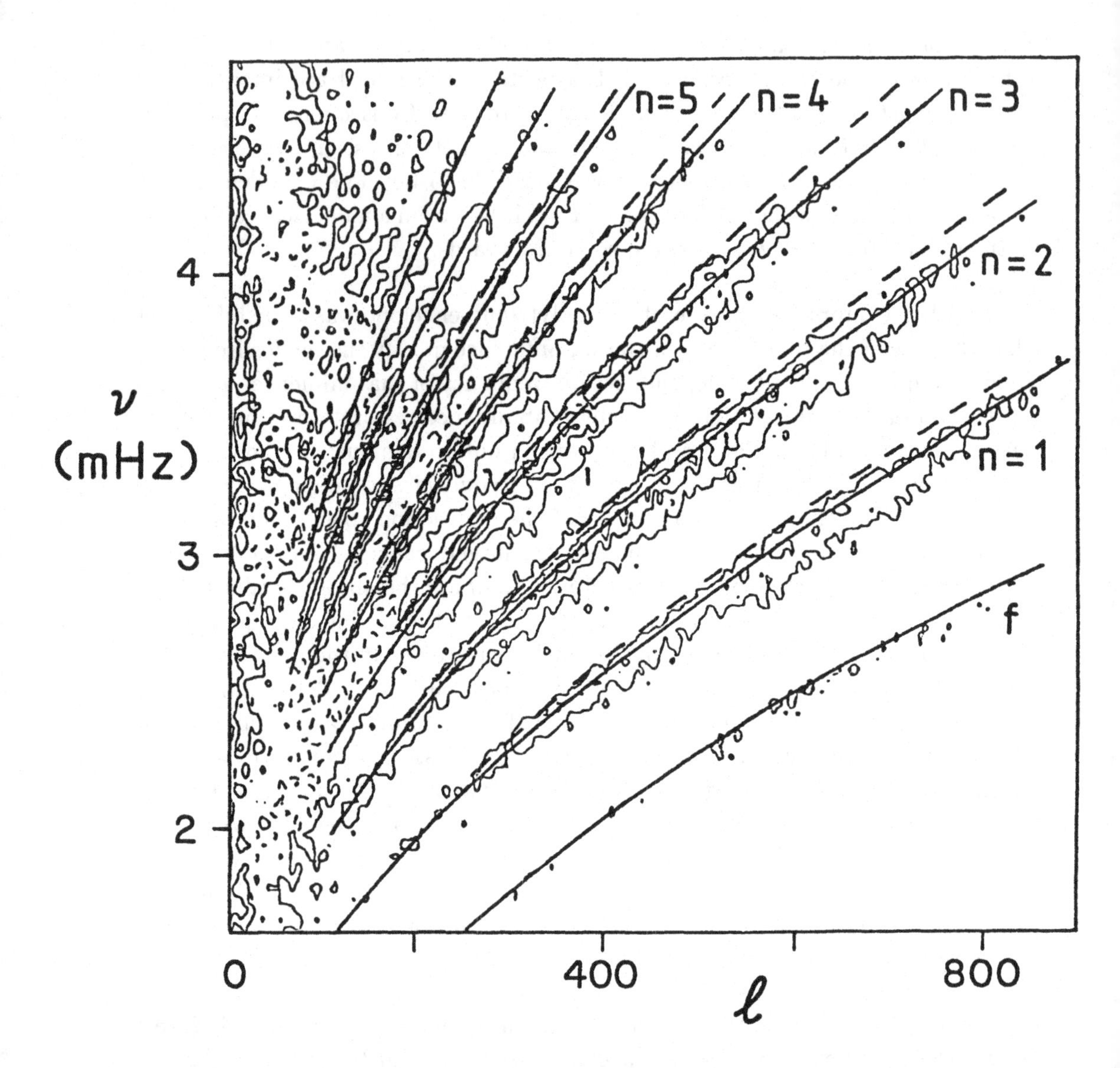

Figure 1: *Two-dimensional power spectrum of high-degree five-minute solar oscillations obtained by Deubner, Ulrich and Rhodes (1979); the closed curves are contours of constant power. The smooth continuous curves represent the cyclic frequencies* ν *(in mHz) of f and p modes of a model solar envelope with helium abundance* $Y = 0.25$, *computed in the manner of Berthomieu et al. (1980) using an equation of state derived from a free energy in which account is taken of the influence of electrostatic interactions on bound-state contributions to the partition function, in the Debye-Hückel approximation; the dashed curves correspond to* $Y = 0.19$. *The depths of the convection zones in the models are 220 Mm and 150 Mm, respectively (after Gough, 1982).*

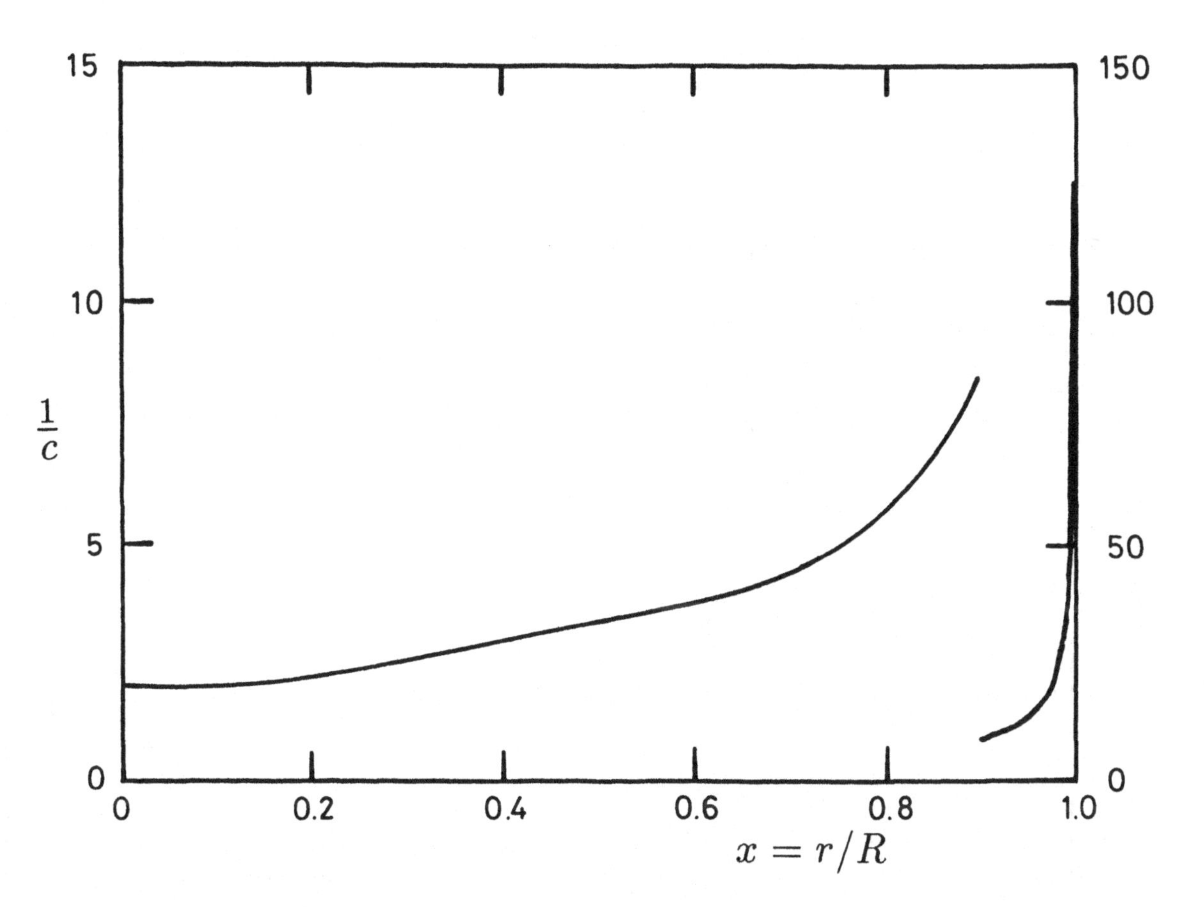

Figure 2: *Integrand for the quantity Δ appearing in equation (3.3), computed from a standard solar model; it is simply the inverse c^{-1} of the sound speed.*

Indeed, I have already pointed out that it was the adiabat and not Y that was first calibrated directly with the high-degree frequencies. The calibration was carried out using equation (3.2) by representing the convection zone, including its superadiabatic boundary layer, by a polytrope of index μ that is intermediate between the value 3/2 which is prevalent beneath the HeII ionization zone and the much greater values nearer the surface. Thus one recognizes that what is being measured is not actually Y, and that therefore to assess the reliability of any calibration requires an investigation of how solar evolution theory connects Y to the stratification of the convection zone, which depends essentially only on the equation of state and can be characterized by just a few integral properties of the upper boundary layer.

For the calibration using low-degree modes it is useful to inspect the asymptotic formula (Tassoul, 1980):

$$\nu_{nl} \sim (n + \tfrac{1}{2}l + \tfrac{1}{4} + \alpha)\Delta - (AL^2 - \delta)\Delta^2/\nu_{nl} + \dots . \quad (3.3)$$

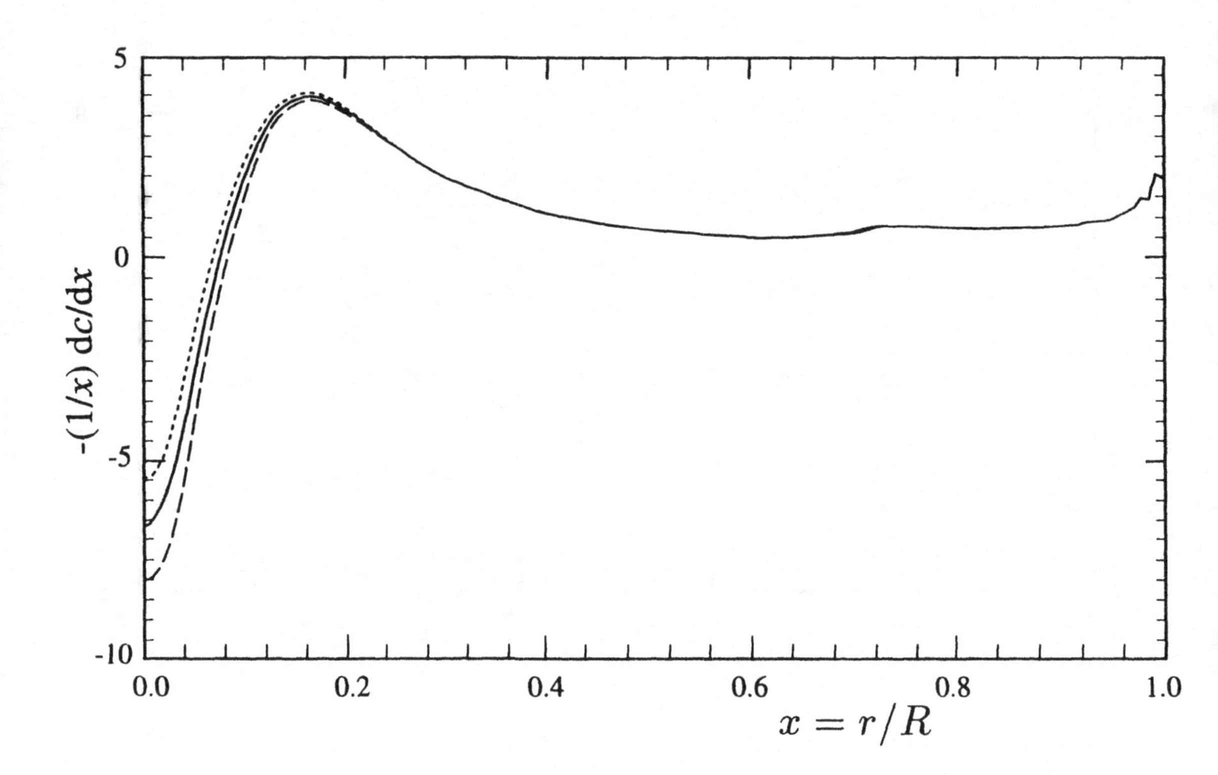

Figure 3: *Integrand, $-x^{-1}dc/dx$ where $x = r/R$, for the quantity A appearing in equation (3.3). It is plotted for three different ages of a standard solar model, 4.15 Gy (dotted), 4.60 Gy (continuous) and 5.10 Gy (dashed), to illustrate how it is influenced by the nucleosynthesis of helium in the core.*

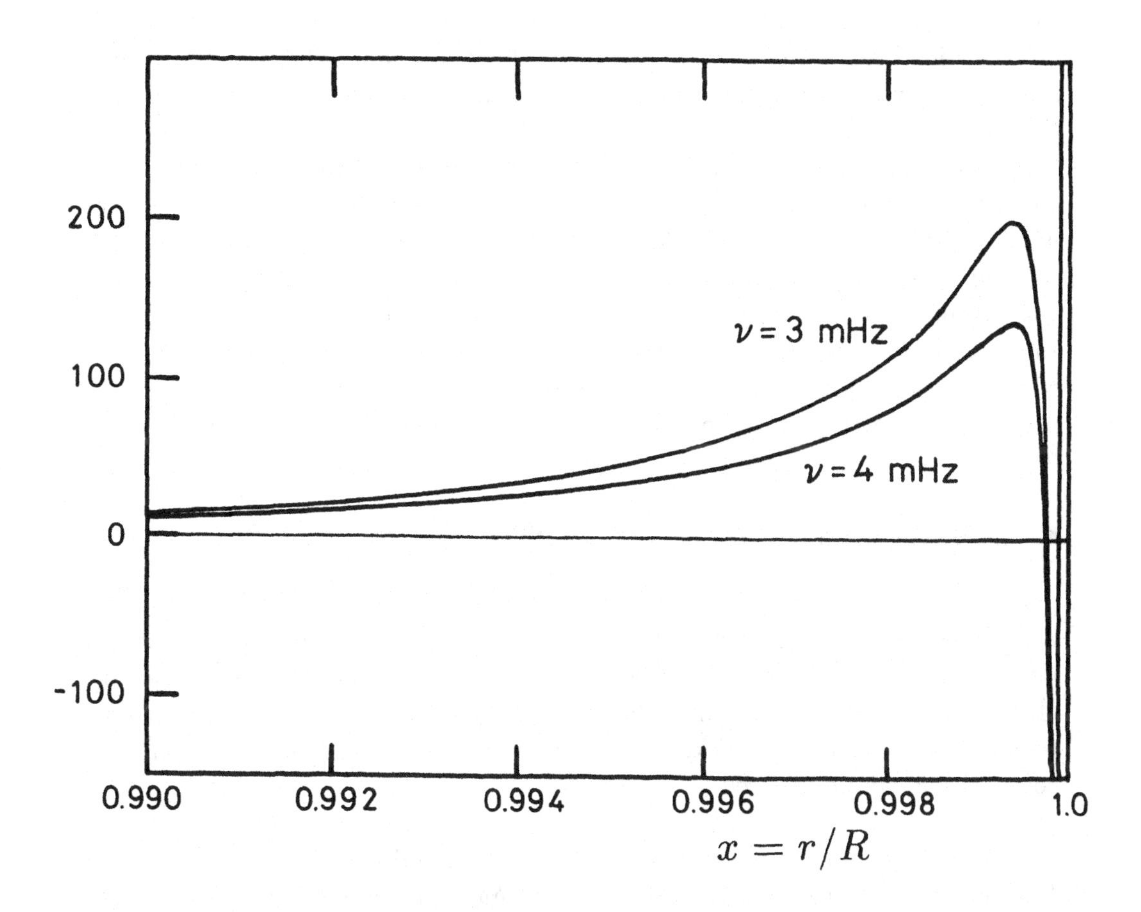

Figure 4: *Approximate integrands for the quantity $\alpha + \Delta\delta/\nu$ appearing in equation (3.3), for p modes of frequencies 3 and 4 mHz (from Gough, 1986).*

The constants α, Δ, and δ are integrals of the structure: expressions for Δ and A are given in the chapter by Christensen-Dalsgaard and depend only on the structure of the sun (and not on parameters associated with the mode of oscillation). The integrands are plotted in Figures 2 and 3. The quantity $\alpha + \Delta\delta/\nu_{nl}$ is more complicated: a simple expression is given by Gough (1986), and a more thorough investigation by Christensen-Dalsgaard and Pérez-Hernández (1992). Integrands at two frequencies are plotted in Figure 4. The important features to notice from the three figures are that $\alpha + \Delta\delta/\nu_{nl}$ is a property of the near-surface regions, that Δ is a global property of the sun, though weighted mainly near the surface where the sound speed is low (if the more appropriate acoustic radius were used as the independent variable, the integrand would be constant), and that A depends mainly on conditions in and immediately surrounding the energy-generating core. Thus we see how the so-called large separation $\Delta_{nl}\nu \simeq \Delta$, the small separation $\delta_{nl} \simeq (4l+6)A\Delta$ and the absolute values of the frequencies, which depend not only on Δ and A but also on the (uncertain) surface terms α and δ, are sensitive predominantly to different regions within the sun.

4 DIRECT ASYMPTOTIC INVERSIONS

An advantage of using asymptotic expressions is the relative ease with which inferences can be drawn. As will become clear soon, it also suggests a means of inspecting data for assessing what they imply. I restrict my discussion here to spherically symmetric background states, so that the spherical harmonic decomposition can be effected.

The starting point is the equations for the amplitude $\psi(r)$ of the scaled Lagrangian pressure perturbation $\Psi = u^{-1}\delta p = \psi(r)P_l^m(\cos\theta)e^{im\phi}$ and the corresponding amplitude $\Phi'(r)$ of the Eulerian perturbation to the gravitational potential. If Φ' is ignored, yielding the Cowling approximation, there is a single second-order differential equation for ψ (Gough, 1993):

$$\psi'' + K^2\psi = 0, \tag{4.1}$$

the prime here representing differentiation with respect to r, where

$$K^2 = \frac{\omega^2 - \omega_c^2}{c^2} - \frac{L^2}{r^2}\left(1 - \frac{\mathcal{N}^2}{\omega^2}\right), \tag{4.2}$$

in which

$$\omega_c^2 = \frac{c^2}{4\mathcal{H}^2}(1 - 2\mathcal{H}') - \frac{g}{h} \tag{4.3}$$

is the square of the acoustic cutoff frequency and

$$\mathcal{N}^2 = g\left(\frac{1}{\mathcal{H}} - \frac{g}{c^2} - \frac{2}{h}\right), \tag{4.4}$$

is a generalization, taking spherical geometry into account, of the square of the buoyancy frequency. In these expressions, the scaling function u is given by

$$u = \left(\frac{g\rho f}{r^3}\right)^{1/2}, \tag{4.5}$$

in which the discriminant

$$f = \frac{\omega^2 r}{g} + 2 + \frac{2}{H_g} - \frac{L^2 g}{\omega^2 r} \tag{4.6}$$

is positive where the mode behaves like a p mode and negative where it behaves like a g mode; also $H_g = -g(\mathrm{d}g/\mathrm{d}r)^{-1}$ is the scale height of the gravitational acceleration g, and h and $\mathcal{H}$ are scale heights of g/r^2 and u^2 respectively, defined with the minus sign as is H_g. Note that in the limit when spherical symmetry is unimportant, $\mathcal{H}$ reduces to the density scale height $H = (-\mathrm{d}\ln\rho/\mathrm{d}r)^{-1}$. This is a good approximation in the very surface layers. Contributions from the perturbed gravitational potential Φ' can be incorporated asymptotically into the expressions for ω_c^2, $\mathcal{N}$ and $\mathcal{H}$ without changing the basic mathematical structure of the problem, but I shall not discuss the details here.

The asymptotic eigenfrequency equation can be obtained from the JWKB approximation to the solution of equation (4.1) taking appropriate precautions in the neighbourhoods of the turning points. For high-frequency p modes there are essentially only two such turning points, r_1 and r_2, (the acoustic potential is complicated in the superadiabatic boundary layer at the top of the convection zone and may present several turning points to particularly the lower-frequency modes, but I shall absorb that complexity into a phase function $\tilde{\alpha}$), and the eigenfrequency equation can be written:

$$\int_{r_1}^{r_2} K \,\mathrm{d}r = \pi(n + \tilde{\alpha}), \tag{4.7}$$

where $\tilde{\alpha}$ is a function of ω which does not depend explicitly on l. Physically, this equation can be regarded as a resonance condition: that the augmentation of phase over the return path in the radial direction from r_1 to r_2 and back again, which is twice the integral on the left-hand side, is the sum of two phase jumps, $2\pi\tilde{\alpha}$, experienced on 'reflection' at the turning points plus an integral multiple of 2π. The equation can also be derived directly by ray theory, in which twice the left-hand side now arises as an integral of the phase gradient along a ray path from a point on a surface $\mathcal{S}$ of constant phase to another point on a surface $\mathcal{S}'$ on which the phase differs from that on $\mathcal{S}$ by an integral multiple of 2π; for a resonant mode, $\mathcal{S}'$ must coincide with $\mathcal{S}$ (e.g. Gough 1993). In evaluating that integral, phase jumps totalling $-2\pi\tilde{\alpha}$ modulo 2π must be added to account for the interference between neighbouring rays at the caustic surfaces

$r = r_1$ and $r = r_2$ (Keller, 1958), which if there were a single upper surface at which $K = 0$ would be approximately $2\pi - 2\tilde{\beta}$, where (in an approximation in which stratification is ignored in the local dispersion relation) $\tilde{\beta} = \frac{1}{2}\pi$ is the jump per caustic surface. An alternative, only slightly different view, useful for interpreting time-distance helioseismology, which I shall mention briefly again in my last lecture, yields (twice) equation (4.7) divided by ω: to carry the phase (at the phase speed) from $\mathcal{S}$ back to itself (i.e. $\mathcal{S}'$) must take an integral number of periods less an adjustment to take account of the phase augmentation at the caustics. That condition is equivalent to requiring that at resonance the time taken for a ray to propagate the phase, from a point A on the upper caustic surface, $r = r_2$ (which is essentially the phase at the observable surface of the sun) to a second point B at which the ray returns to that surface, be the same, modulo an integral number of periods (and adjusted for the caustic phase jumps) as the time taken for the interference pattern observable at the surface to propagate from A to B at its phase speed ω/k_h, where $k_h = L/r_2 \simeq L/R$ is the horizontal component of the wavenumber at the upper caustic surface (which is also essentially the wavenumber of the interference pattern on that surface). The resonance condition (4.7) is most easily derived in the approximate form obtained from equation (4.8) below by retaining only the first two terms on the right-hand side, for in this form stratification is essentially ignored and phase is assumed to travel along the ray path at the sound speed c. However, the danger of working in this approximation is that dispersion amongst the acoustic waves is ignored, and one is consequently more likely to loose sight of the distinction between phase speed and group velocity.

Equation (4.7) can be expanded for p modes about its asymptotic limit in which $k^2 \approx K^2 + L^2/r^2 \sim \omega^2/c^2$ is the simple local acoustic dispersion relation. The outcome is

$$n\pi\omega^{-1} \sim -\pi\omega^{-1}\alpha(\omega) + \mathcal{F}(w) + \omega^{-2}\mathcal{G}(w) + \ldots + L^2\mathcal{J}(\omega) + \ldots, \tag{4.8}$$

where $w = \omega/L$ and

$$\mathcal{F}(w) = \int_{r_1}^{R} (1 - a^2/w^2)^{1/2} \frac{\mathrm{d}r}{c} \tag{4.9}$$

in which $a = c/r$, and the explicit functional forms of $\mathcal{G}(w)$, $\mathcal{J}(\omega)$, etc., which make only small contributions to $n\pi\omega^{-1}$, are not of concern here. The function $\alpha(\omega)$ is related to $\tilde{\alpha}(\omega)$, and depends also on the stratification of the outer layers of the star near $r = r_2$ as represented by $\omega_c(r)$. By fitting frequency data $n\pi\omega^{-1}$, where ω is considered to be a function of n and w, to the functional form of the right-hand side of equation (4.8), the functions $\alpha(\omega)$, $\mathcal{F}(w)$, $\mathcal{G}(w)$, etc. can be obtained to within some uncertain constants (e.g. Christensen-Dalsgaard *et al.* 1989; Gough and Vorontsov, 1995). It is immediately apparent, for example, that α is certainly uncertain by an additive linear function of ω, say $\beta\omega$ where

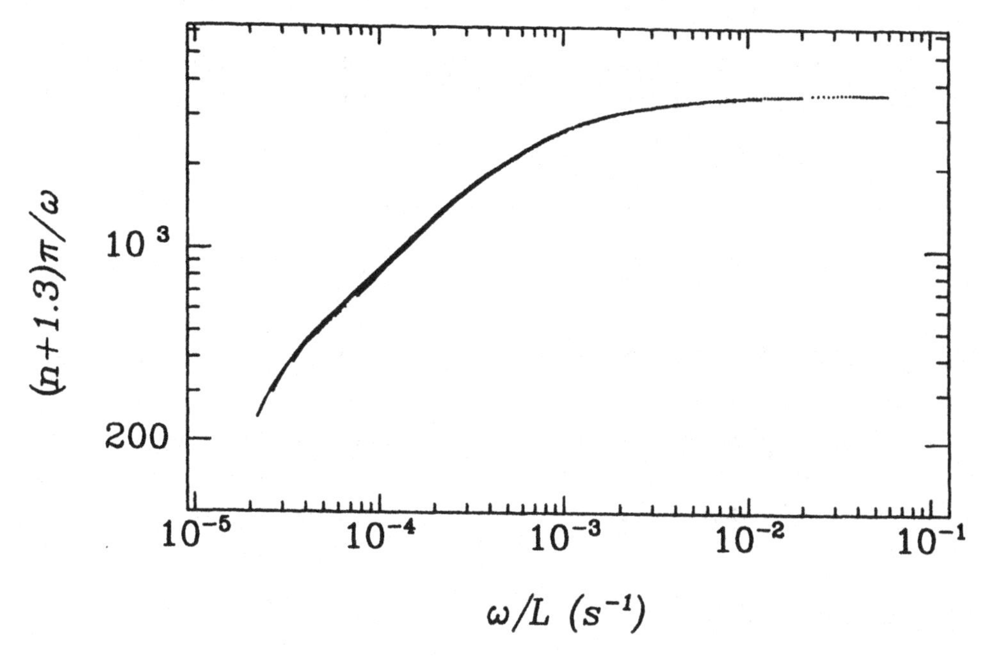

Figure 5: $(n + \alpha)\pi/\omega$, *plotted against* $w = \omega/L$ *for p modes whose frequencies are tabulated by Libbrecht et al. (1990). The quantity* α *has been taken to be 1.3; had a best-fitting* ω*-dependent* α *been adopted, it would have been difficult to perceive on this scale the deviations of the points from a single curve. The units of* ω *are* s^{-1}.

β is constant, the effect of which is to add an unknown constant $\pi\beta$ to $\mathcal{F}$. The quality of the fit can then be assessed by plotting, for example, $\pi\omega^{-1}(n + \alpha) - \omega^{-2}\mathcal{G}(w) - L^2\mathcal{J}(\omega)$, which should represent $\mathcal{F}(w)$, against w. An early example is illustrated in Figure 4.1, in which α was taken to be constant and the small terms $\mathcal{G}, \mathcal{J}$ etc. were ignored. The extent to which the outcome deviates from a single curve provides an estimate of the combined influence of random and systematic errors in the data and of inaccuracies in the asymptotic representation. Random data errors produce random scatter about the curve. Asymptotic inaccuracies and systematic data errors both yield systematic deviations, usually noticeable as an apparent multiplicity of curves. Determining which is which is a difficult task; progress can be made by extending the asymptotic sequence (4.8), but some care must be taken to ensure that that really does improve the representation. The systematic deviations barely discernible in Figure 5 are almost certainly due mainly to $\omega^{-2}\mathcal{G}(w)$ and $L^2\mathcal{I}(\omega)$.

Having determined $\mathcal{F}(w)$, one can then obtain information about the variation of sound speed through the star by inverting equation (4.9). Let us be aware, before proceeding with that inversion, that in order to carry it out we need to represent $\mathcal{F}$ as a function of a continuous variable w, and to do so requires representing the finite number of points plotted in Figure 5 by a continuous curve. This is the point at which an additional assumption must be introduced, namely that smooth interpolation is correct.

The inversion is accomplished by first differentiating equation (4.9):

$$\frac{\mathrm{d}\mathcal{F}}{\mathrm{d}w} = -w^{-3}\int_{a_s}^{w}(a^{-2}-w^{-2})^{-1/2}\frac{\mathrm{d}\ln r}{\mathrm{d}a}\mathrm{d}a\,, \tag{4.10}$$

where $a_s = a(R)$. The transformation of the independent variable from r to a is straightforward provided $a(r)$ is monotonic, which is so for stars with unmixed cores, such as is believed to be the case for the sun. If a were not monotonic, appropriate precautions must be taken, which I need not discuss here. Equation (4.10) is essentially Abel's integral equation, which is directly invertible, yielding

$$\frac{r}{R} = \exp\left[-\frac{2}{\pi}\int_{w_0}^{a}(w^{-2}-a^{-2})^{-1/2}\frac{\mathrm{d}\mathcal{F}}{\mathrm{d}w}\mathrm{d}w - \Lambda(w_0;a)\right]\,, \tag{4.11}$$

where w_0 is the lowest value of w for which there is a datum. This equation determines how r varies with $a = c/r$, which, particularly because the integral is determined numerically, could be inverted trivially to yield $c(r)$ if Λ were known. However, Λ is not known. It depends on the undetermined stratification of the outer layers above the lower turning point r_{10}, given by $a(r_{10}) = w_0$, of the shallowest mode. From its functional form, which I do not present here, it can be shown that, except near $a = w_0$, Λ is approximately constant; and from extrapolating c to the surface $r = R$ of the sun, Λ can be shown to be approximately $0.28(n_0+\alpha)/L_0$, where the values of n_0 and L_0 correspond to the shallowest mode (Gough, 1993). It is evident from the solution (4.11) that the effect of Λ is to multiply r/R by an uncertain scaling factor which, when L_0 is large, is close to unity.

The form of the expression (4.11) shows explicitly how the data determine the solution. As is evident from equation (4.9), each datum, in the form $\mathcal{F}(w)$, is an integral between the lower turning point $r = r_1$ and the surface of the sun. Consequently, conditions at any level depend on the difference between two such integrals at neighbouring values of r, which is why $\mathcal{F}$ is differentiated in equation (4.11). However, because the two neighbouring integrals (4.9), corresponding to different values of w, have different integrands, their difference depends not only on conditions near the lower turning points but also on the stratification above those turning points. Hence the weighted integral of $d\mathcal{F}/dw$ in equation (4.11) is formally over all modes that do not penetrate as deeply as the level of interest. However, in practice that is not strictly so, as is illustrated in Figure 6.5, because the process of determining $\mathcal{F}(w)$ as a function of a continuous variable is nonlocal.

A representation of $c^2(r)$ obtained from equation (4.11), using data of Duvall (1982) and Harvey and Duvall (1984), is illustrated in Figure 6 What is perhaps most obvious is the undulation at small r. This arises in part from systematic errors in $\mathcal{F}$ that result from not taking $\mathcal{G}$ and $\mathcal{J}$ into account when fitting $n\pi\omega^{-1}$ to the right-hand side of equation (4.8), and it is magnified by the differentiation required by equation (4.11). To be sure, that magnification is lessened by the subsequent integration, but the weighting function $(w^{-2} - a^{-2})^{-1/2}$ diverges at the lower turning point and therefore weakens the smoothing of the integration. Indeed, Abel integration can be regarded as being the square root of an integral (e.g. Gough, 1990), and therefore the principal term in the exponent in equation (4.11) can be considered to resemble the result of operating on $\mathcal{F}$ with the square root of differentiation.

Another point worth emphasizing is that the natural independent variable for acoustic diagnostics is acoustical radius τ, where $\mathrm{d}\tau = \mathrm{d}r/c$, as is evident from the integral in equation (4.9). With respect to this independent variable, the kernel in Figure 3.2 would be unity. Since sound speed is relatively high at low r, the region in which the inversion illustrated in Figure 4.2 is poor is acoustically quite small: for example, $r = 0.3R$ corresponds to an acoustic radius of only 0.125 T, where T is the acoustic radius of the photosphere. However, our prime interest is in the outcome of the inversion rather than how it was accomplished, so r is physically the more natural variable against which to plot. But we must appreciate the difficulty in penetrating with p modes into the very deep interior of the sun, and particularly into the energy-generating core.

There are two further points I wish to make about the inversion depicted in Figure 6. The first is that when the inversion was carried out it was noticed that in the outer regions of the radiative interior immediately beneath the convection zone, $0.5R \leq r \leq 0.7R$, the sound speed in the sun was greater than that in standard solar models of the time, by about 1 per cent. Unless the sun were in some transient thermally unbalanced state, the only plausible explanation was that the opacity must be substantially greater than had been provided by the astrophysical tables of the time (Christensen-Dalsgaard *et al.*, 1985). Indeed, this eventually led to the discovery of an error in the opacity calculations, the correction of which has not only brought solar models into closer agreement with helioseismic inferences but has also had substantial ramifications in other areas of the study of stellar pulsation.

The second point to which I wish to draw attention is the indication of the abrupt change in the second derivative of $c^2(r)$, magnified in the inset. This marks the transition from radiative to adiabatic stratification at the base of the convection zone, and has been analysed to determine the depth $R - r_{\rm c}$ of the con-

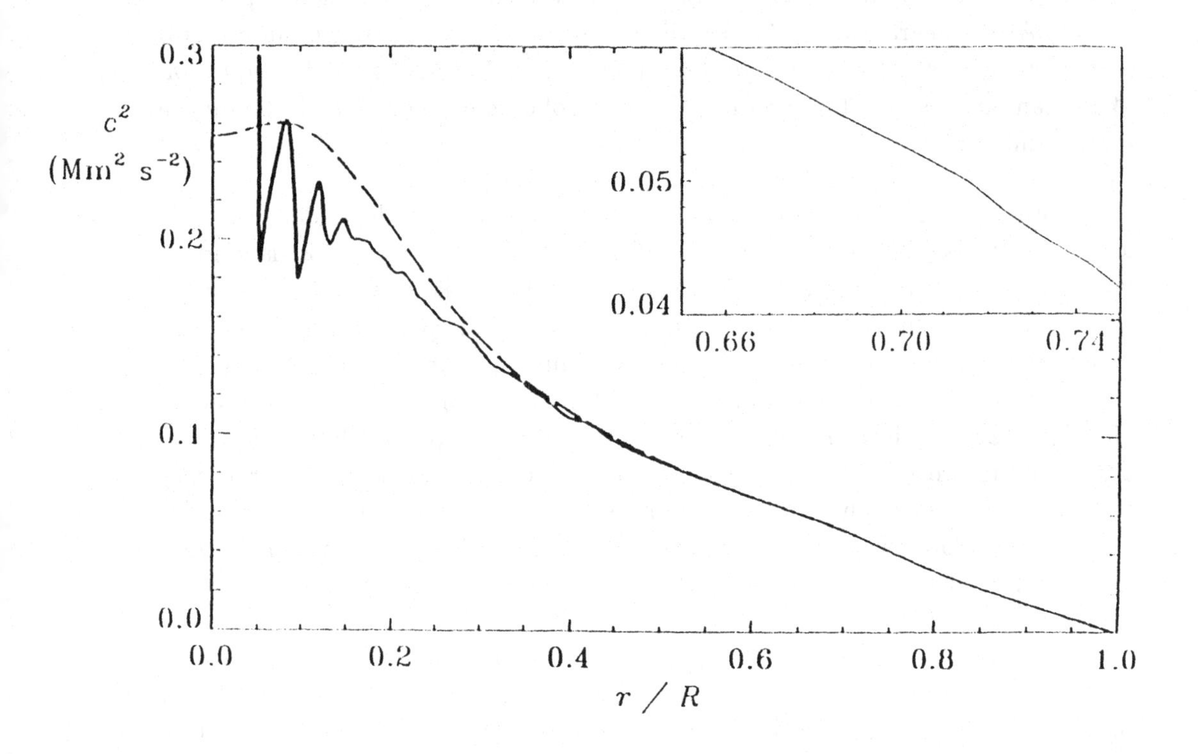

Figure 6: *Asymptotic inversion of solar p-mode frequencies by Christensen-Dalsgaard et al. (1986). The continuous curve is the square of the inferred solar sound speed, the dashed curve is c^2 in a standard solar model. The inset is an enlargement of just the solar inversion in the vicinity of the base of the convection zone.*

vection zone to within about 0.3 per cent of the solar radius: $r_c = 0.713 \pm 0.003R$ (Christensen-Dalsgaard, Gough and Thompson, 1991). Other asymptotic methods were also used in that determination, using linearized perturbations from a reference theoretical model of the sun, to which I turn in the next section.

Before leaving this subject I must point out that discontinuities, such as that encountered at the base of the convection zone, can be handled asymptotically, provided, of course, that the background state is sufficiently smooth either side of the discontinuity. Indeed, asymptotic methods have been used to investigate the nature of the transition between the radiative and the adiabatically stratified zones. Were standard local mixing-length theory to be more-or-less correct, the transition would occur abruptly at an interface. There, p, ρ, T and their first derivatives would be continuous, but there would be a discontinuity in the second derivative of T, and consequently in the second derivatives of ρ and c^2. This implies a discontinuity in the acoustic cutoff frequency ω_c, which would cause acoustic waves to be partially reflected at the interface. An alternative possibility is that the convective motion penetrates the boundary at which the local stability criterion changes, overshooting into the convectively stable region beneath. It is thought that the most likely form of overshoot would be such as to extend the adiabatically stratified region, with an abrupt transition into an almost unmodified radiative layer. That would lead to an interface with essentially discontinuous temperature and density gradients, and a consequent negative delta function in ω_c. This lower-order transition has the potential for stronger reflection, though if the discontinuity is small the reflection coefficient may be subject to cancellation between the effects of different processes. However, the phase jumps would be different. The effect of the reflection is to make the regions above and below the interface act as two strongly coupled cavities, which modulates the frequency spectrum as ω varies, moving the characteristic frequencies of the two cavities in and out of resonance with each other. This process has been studied asymptotically by Gough and Sekii (1993), Monteiro *et al.* (1993) and Roxburgh and Vorontsov (1993), and fitted to the observations to estimate the degree of overshoot. Thus, although the forward analysis was asymptotic, the inverse analysis was basically a parameter-fitting operation, which is different in style from the Abel inversion discussed earlier in this section. A parallel study of this problem based on a direct comparison of properties of numerically computed frequency spectra that are believed to be signatures of overshoot has also been carried out (Berthomieu *et al.*, 1993). Although there is some divergence amongst the authors' opinions about what the results of all these studies imply, I think it is fair to say that convincing evidence for overshoot has not been found. It should be remarked that although nobody believes that in reality the transition between the radiative and convective zones occurs discontinuously at a smooth interface, provided the thickness of the transition region is very much less than the inverse wavenumber of the modes, the influence on the eigenfrequencies would be essentially the same as it would have been had the transition been genuinely discontinuous.

It is perhaps worth pointing out that overshooting is unlikely to lead to the essentially spherical interface that was in mind when the seismic tests that I have just mentioned were carried out. It is more likely that the penetration of the convective flow is intermittent and spatially inhomogeneous. I suspect that local plumes occasionally penetrate into the stable regions, leading to a transition in stratification which, although it may be quite sharp, is horizontally and temporally variable, as is the case for convection in, for example, the earth's atmosphere. The influence of such a process on the oscillation modes has not been studied in detail, but it is not unlikely that the mean stratification experienced by the modes is actually smoother, not sharper, than it would have been in the absence of overshoot.

A potentially more sensitive test of the structure of the transition region could be made by g modes, if an adequate number could be detected, because the transition marks the boundary between the propagating and evanescent regions, which has a more direct influence on the eigenfrequencies. Some appropriate asymptotic analyses have been carried out (Zahn, 1970; Tassoul, 1980; Provost and Berthomieu, 1986; Ellis, 1988a,b), but without g-mode observations they cannot yet be put to use.

I conclude here by remarking that the direct asymptotic inversion discussed in this section does not depend on any specific reference model, and is therefore not subject to the biases which are known to arise from the use of such models. Therefore, although the appearance of inversions carried out by perturbation methods is usually cosmetically more pleasing to the eye, because it deviates by only a small amount from what we have been led to regard as a 'good-looking' theoretical model and is therefore itself good-looking in the same sense, it is not necessarily more reliable.

Notwithstanding that warning, other more precise methods certainly have their place, and indeed dominate the inversion industry. However, aside from model calibrations, they rely on linearization about some reference state. It is therefore appropriate now to describe how the linearization is accomplished, as a prelude to a discussion to how the constraints imposed by the seismic data can be processed for making reliable useful inferences.

5 FORMULATING LINEAR HELIOSEISMIC INVERSE PROBLEMS

My starting point is the variational principle for the frequency $\omega_{\mathbf{i}}$ of mode $\mathbf{i}$ of adiabatic oscillation of a star. Here, the label $\mathbf{i}$ would be (n, l, m) if the background state of the star were spherically symmetrical, or at least nearly so. Referred to a coordinate frame in which the structure of the star, including the motion within it, is steady, it may formally be written

$$\mathcal{I}\omega_{\mathbf{i}}^2 - 2\mathcal{R}\omega_{\mathbf{i}} - \mathcal{C} = 0\,, \tag{5.1}$$

where $\mathcal{I}, \mathcal{R}$ and $\mathcal{C}$ are mode-dependent integrals over the volume $\mathcal{V}$ occupied by the star which depend on the structure $\mathbf{X}(\mathbf{r})$ and appropriate eigenfunctions of oscillation. Typically, those eigenfunctions are the displacements $\boldsymbol{\xi}_{\mathbf{i}}$. Sometimes the integrals also contain contributions from the surface, depending on where the surface is considered to be. In that case it is usually the case that the integrals depend also on the eigenfunctions of the adjoint forward problem. I shall not refer to the adjoint functions explicitly in subsequent discussions, although the reader should realize that they may be present. Explicit forms for $\mathcal{I}, \mathcal{R}$ and $\mathcal{C}$ can be obtained from Lynden-Bell and Ostriker (1967). In the case of a nonrotating nonmagnetic spherically symmetrical star, with no internal rotation, $\mathcal{R} = 0$ and the result reduces to the comparatively simple form given by Ledoux and Walraven (1958).

Starting from a variational principle greatly simplifies the subsequent analysis. Regard $\mathbf{X}$ to be the structure of the sun, with associated frequencies $\omega_{\mathbf{i}}$ of oscillation, and let $\mathbf{X}_0$ be a reference theoretical model, with corresponding eigenfrequencies $\omega_{0\mathbf{i}}$. Then, provided the reference model is a close representation of the sun, the differences $\delta\omega_{\mathbf{i}} := \omega_{\mathbf{i}} - \omega_{0\mathbf{i}}$ of the solar frequencies from those of the reference are given approximately by the linearized equation

$$\frac{\delta\omega_{\mathbf{i}}}{\omega_{0\mathbf{i}}} \simeq \frac{\delta\mathcal{C} + 2\omega_{0\mathbf{i}}\delta\mathcal{R} - \omega_{0\mathbf{i}}^2\delta\mathcal{I}}{2\omega_{0\mathbf{i}}(\mathcal{I}\omega_{0\mathbf{i}} - \mathcal{R})}\,, \tag{5.2}$$

where $\delta\mathcal{I} := \mathcal{I}(\mathbf{X}, \boldsymbol{\xi}_{0\mathbf{i}}) - \mathcal{I}(\mathbf{X}_0, \boldsymbol{\xi}_{0\mathbf{i}})$, with similar definitions for $\delta\mathcal{R}$ and $\delta\mathcal{C}$, and the integrals in the denominator are evaluated for the reference model. Note that the variational integrals $\delta\mathcal{I}, \delta\mathcal{R}$ and $\delta\mathcal{C}$ are evaluated using only the eigenfunctions $\boldsymbol{\xi}_{0\mathbf{i}}$ of the reference model: in view of the variational principle, terms in the linearized perturbation of equation (5.1) that contain eigenfunction differences $\delta\boldsymbol{\xi}_{0\mathbf{i}}$ cancel. Not only does that lead to there being fewer terms in the expression (5.2) for $\delta\omega_{\mathbf{i}}$, but it also obviates the necessity to evaluate the eigenfunctions of the sun, albeit as linear perturbations from $\boldsymbol{\xi}_{0\mathbf{i}}$. The integrals $\delta\mathcal{I}, \delta\mathcal{R}$ and $\delta\mathcal{C}$ can be evaluated in terms of $\mathbf{X}_0$ and $\delta\mathbf{X} := \mathbf{X} - \mathbf{X}_0$. After linearization, equation

(5.2) can then be written in the form:

$$\delta \ln \omega_{\mathbf{i}} = \int_{\mathcal{V}} \mathbf{K}^{\mathbf{i}} \cdot \delta \mathbf{X} \mathrm{d}V \,. \tag{5.3}$$

Equation (5.3) expresses the difference between a frequency of the sun and that of the corresponding mode of the reference model as an integral of the structural difference $\delta\mathbf{X}$, weighted by a calculable kernel $\mathbf{K}^{\mathbf{i}}$ which depends only on the structure of the reference model and its eigenmode. To a set of such constraints can be added the equation requiring that the total mass of the model (which I presume to be $M_\odot$) is unchanged by the variation; that constraint is of the same form as equation (5.3) but with zero on the left-hand side, and with a kernel satisfying $\mathbf{K} \cdot \delta\mathbf{X} = 4\pi r^2 \delta\rho$. Typically the seismic kernels $\mathbf{K}^{\mathbf{i}}$ are oscillatory, extending at least over the region of the star within which the waves that constitute the modes can propagate. Examples are illustrated in Figure 5.1. Thus, given a measurement of $\omega_{\mathbf{i}}$, $\delta\omega_{\mathbf{i}}$ is known. Consequently, a set of equations (5.3) corresponding to the observed modes $\{\mathbf{i}\}$ constrains the function $\delta\mathbf{X}$. As I explained in the previous section, it is the aim of inverse theory to re-express those constraints in a form that is more easily comprehendable. I shall postpone to subsequent sections how one goes about achieving that aim. It goes without saying that in observing $\omega_{\mathbf{i}}$ one must have identified the mode correctly. Otherwise the kernel $\mathbf{K}^{\mathbf{i}}$ would be inappropriate, and the inferences would be invalid.

By way of illustration, I sketch how equation (5.3) can be derived for a spherically symmetrical star of radius R under the idealized condition of it having a perfectly reflecting surface such that all the surface integrals vanish identically. Then $\mathcal{R} = 0$,

$$\mathcal{I} = \int_0^R \rho \boldsymbol{\xi} \cdot \boldsymbol{\xi} \, \mathrm{d}V \tag{5.4}$$

and

$$\begin{aligned} \mathcal{C} = \int_0^R & \left[\rho c^2 (\mathrm{div}\, \boldsymbol{\xi})^2 + 2\rho \boldsymbol{\xi} \cdot \mathbf{g} \; \mathrm{div}\, \boldsymbol{\xi} + \boldsymbol{\xi} \cdot \nabla \rho \; \boldsymbol{\xi} \cdot \mathbf{g} \right] \mathrm{d}V \\ & -4\pi G \int_0^R \int_0^R |\mathbf{r} - \mathbf{r}'|^{-1} \, \mathrm{div}\, \rho \boldsymbol{\xi} \; \mathrm{div}\, \rho' \boldsymbol{\xi}' \mathrm{d}V \mathrm{d}V' \end{aligned} \tag{5.5}$$

where ρ, c and $\mathbf{g}$ are the density, adiabatic sound speed and gravitational acceleration in the reference model, and here $\rho', \boldsymbol{\xi}'$ refer to the dummy variable $\mathbf{r}'$ (rather than to differentiation). For the spherically symmetrical model one can write

$$\boldsymbol{\xi}(\mathbf{r}) = \Re \left[\left(\xi(r) P_l^m(\cos\theta), \; L^{-1}\eta(r) \frac{dP_l^m}{d\theta}, \; imL^{-1}\eta(r) \mathrm{cosec}\theta \; P_l^m \right) e^{im\phi} \right], \tag{5.6}$$

where $L^2 = l(l+1)$ and the vertical and horizontal displacement amplitudes $\xi(r)$ and $\eta(r)$ are real and depend only on n and l; the eigenfrequencies are

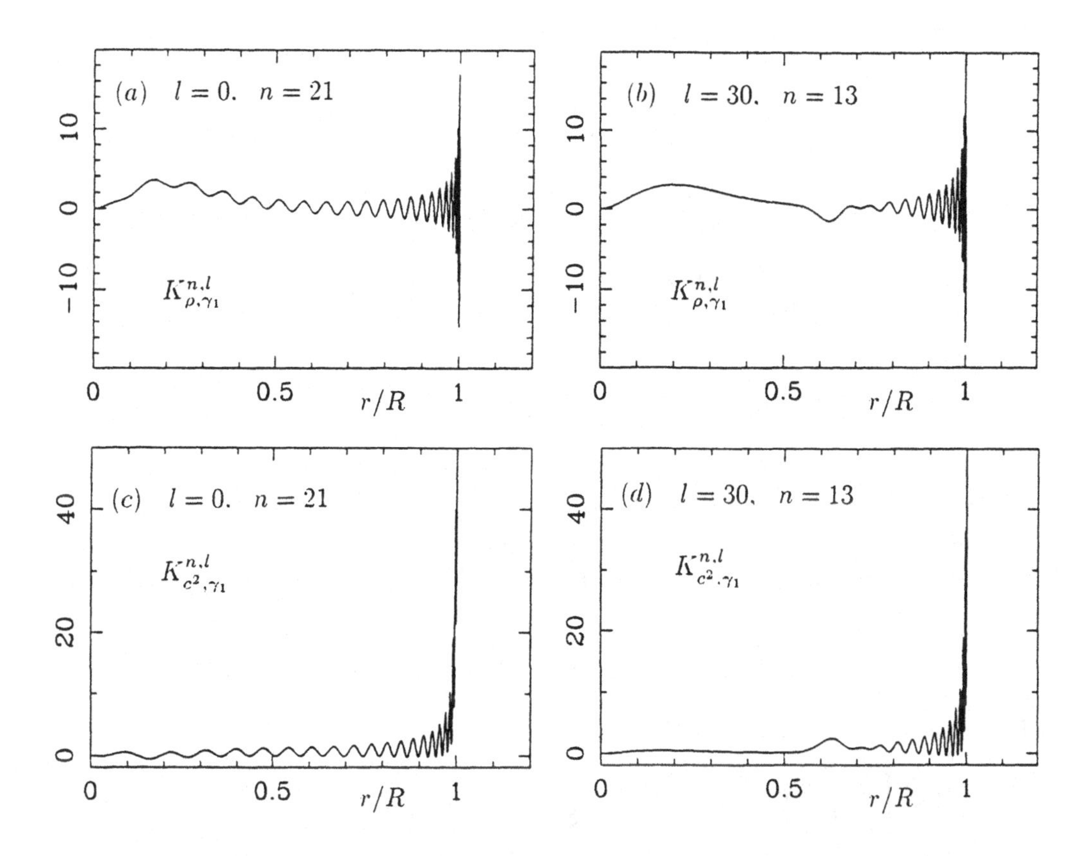

Figure 7: *Components of kernels* $\mathbf{K^i}$ *for eigenfrequency perturbations of two solar p modes with frequencies near 3 mHz. Panels (a) and (b) are the kernels* $K^{n,l}_{\rho,\gamma_1}(r)$ *for relative density perturbations at constant adiabatic exponent* γ_1*; (c) and (d) are relative sound-speed kernels* $K^{n,l}_{c^2,\gamma_1}(r)$*, also at constant* γ_1 *(from Gough and Kosovichev, 1988). Since p-mode frequencies depend predominantly on sound speed, the kernels* $K^{n,l}_{c^2,\gamma_1}$ *are concentrated in the propagating region above the lower turning point; this is essentially the entire star for monopole* ($l = 0$) *modes, and is in* $r > r_1 \simeq 0.6R$ *for* $p_{13}(l = 30)$*. The density kernel* $K^{13,30}_{\rho,\gamma_1}$ *is noticeably different from zero below* $r = r_1$ *because the perturbed hydrostatic equation (5.11), together with the equation of conservation of mass, was used in its derivation; it can be reduced almost to zero in that region by subtracting an appropriate multiple of the relative density kernel,* $4\pi r^2\rho$*, associated with mass conservation.*

degenerate with respect to m. Substituting into equation (5.5) and formally perturbing the structure of the background state then yields

$$\frac{1}{4\pi}\delta C = \int_0^R \left[\rho c^2\chi^2 r^2\left(2\frac{\delta c}{c}+\frac{\delta\rho}{\rho}\right) - \frac{\mathrm{d}}{\mathrm{d}r}\left(\xi^2 r^2\frac{\mathrm{d}p}{\mathrm{d}r}\right)\frac{\delta\rho}{\rho} + \left(2\xi\chi+\rho^{-1}\xi^2\frac{\mathrm{d}\rho}{\mathrm{d}r}\right)r^2\frac{\mathrm{d}\delta p}{\mathrm{d}r}\right]\mathrm{d}r$$
$$+\int_0^R \left[\left(\rho\chi+\xi\frac{\mathrm{d}\rho}{\mathrm{d}r}\right)r^2\delta\Phi' + \Phi' r^2\left(\chi\delta\rho+\xi\frac{\mathrm{d}\delta\rho}{\mathrm{d}r}\right)\right]\mathrm{d}r\,, \tag{5.7}$$

in which

$$\chi = \frac{1}{r^2}\frac{\mathrm{d}}{\mathrm{d}r}\left(r^2\xi\right) - \frac{L^2}{r}\eta \tag{5.8}$$

is the amplitude of div$\boldsymbol{\xi}$ and $\Phi'(r)$ is the radial part of the perturbation to the gravitational potential associated with the oscillation. The latter is related to the density perturbation by the Poisson equation, which may be rewritten in terms of the displacement eigenfunction using the equation of conservation of mass:

$$\frac{1}{r^2}\frac{\mathrm{d}}{\mathrm{d}r}\left(r^2\frac{\mathrm{d}\Phi'}{\mathrm{d}r}\right) - \frac{L^2}{r^2}\Phi' = 4\pi G\left(\rho\chi+\frac{\mathrm{d}\rho}{\mathrm{d}r}\xi\right)\,, \tag{5.9}$$

which can be solved for Φ' in the form

$$\Phi' = \frac{-4\pi G}{2l+1}\left[\int_0^r\left(\frac{r'}{r}\right)^{l+1} r'\left(\rho'\chi'+\frac{\mathrm{d}\rho}{\mathrm{d}r}\xi'\right)\mathrm{d}r' + \int_r^R\left(\frac{r}{r'}\right)^{l} r'\left(\rho'\chi'+\frac{\mathrm{d}\rho}{\mathrm{d}r}\xi'\right)\mathrm{d}r'\right]. \tag{5.10}$$

The perturbed Φ', namely $\delta\Phi'$, arising from the perturbation $\delta\mathbf{X}$ to the background state, can then be obtained as weighted integrals of $\delta\rho$ and $\frac{\mathrm{d}\delta\rho}{\mathrm{d}r}$ by perturbing equation (5.10), and it can then be substituted into equation (5.7). The derivative $\frac{\mathrm{d}\delta\rho}{\mathrm{d}r}$ can be removed by integration by parts, and $\frac{\mathrm{d}\delta p}{\mathrm{d}r}$ can be eliminated using the perturbed equation of hydrostatic support:

$$\frac{\mathrm{d}\delta p}{\mathrm{d}r} = -\frac{G}{r^2}\left(\int_0^r 4\pi r'^2\rho'\mathrm{d}r'\right)\delta\rho - \frac{G\rho}{r^2}\int_0^r 4\pi r'^2\delta\rho'\mathrm{d}r'\,. \tag{5.11}$$

After some rearrangement, one obtains equation (5.3) in the desired form:

$$\delta\ln\omega_{\mathrm{i}} = \int_0^R (K_{c,\rho}^{n,l}\delta\ln c + K_{\rho,c}^{n,l}\delta\ln\rho)\mathrm{d}r\,. \tag{5.12}$$

Expressions for the kernels $K_{c,\rho}^{n,l}$ and $K_{\rho,c}^{n,l}$ are given by Gough and Thompson (1991), and an example is plotted by Christensen-Dalsgaard in Figure 3 of the preceding chapter.

It is sometimes useful to express $\delta\omega_{\mathrm{i}}$ in terms of the perturbation of a structure variable, such as ρ, and the perturbation of a thermodynamic coefficient such as γ_1 which does not appear explicitly in the hydrostatic equations. Thus one can substitute

$$\delta \ln c = \tfrac{1}{2}(\delta \ln \gamma_1 + \delta \ln p + \delta \ln \rho) \tag{5.13}$$

into equation (5.12) and express $\delta \ln p$ in terms of $\delta \ln \rho$ using the integral of equation (5.11). Two such kernels are illustrated in Figure 7. To obtain $\delta\omega_{\mathrm{i}}$ in terms of δc and $\delta\gamma_1$ is more complicated, for then, instead of transforming all of $\delta \ln p$ one must transform just that amount to arrange that $\delta \ln p$ and $\delta \ln \rho$ appear together as a simple sum, so that they can be eliminated in favour of $\delta \ln c^2$ and $\delta \ln \gamma_1$. The calculation of the appropriate combination requires the solution of a second-order differential equation (Gough, 1993). The outcome is illustrated in Figure 7 for the same two modes as the density kernels.

A more flexible procedure is to relate any pair of variables, say $\mathbf{z}_1^{\mathrm{T}} = (\delta \ln c, \delta \ln \rho)$, for which one already knows the data kernels $\mathbf{K}_1^{\mathbf{i}}$, to a new pair $\mathbf{z}_2^{\mathrm{T}}$ via the variables $\mathbf{y}$ that appear in any of the constraints that apply to the equilibrium model, and to project those constraints onto an auxiliary variable $\mathbf{w} = (w_1, w_2, w_3)$ to derive the new kernels $\mathbf{K}_2^{\mathbf{i}}$; the seismic constraints can be cast into an appropriate form if the function $\mathbf{w}$ satisfies an inhomogeneous differential equation whose homogeneous operator is the adjoint of the operator defining those constraints and whose inhomogeneous term contains $\mathbf{K}_1^{\mathbf{i}}$ (e.g. Marchuk, 1977; cf. Masters, 1979). Thus one could use just the hydrostatic constraints to obtain $K^{n,l}_{c^2,\gamma_1}$ and $K^{n,l}_{\gamma_1,c^2}$, or one can accept also the equations of steady energy production and energy transfer (if one accepts the formulae for energy generation and for opacity, and if one is prepared to believe that the sun is truly in thermal balance with no transport by macroscopic motion beneath the convection zone) to derive kernels for, say, the perturbations δY and δZ of helium and heavy elements (Gough and Kosovichev, 1990). By way of illustration, I sketch the procedure for deriving kernels $\mathbf{K}_2^{\mathbf{i}}$ for $(\delta v, \delta \ln \gamma_1)$, where $v = rg^{-1}N^2 = \gamma_1^{-1}\mathrm{d}\ln p/\mathrm{d}\ln r - \mathrm{d}\ln\rho/\mathrm{d}\ln r =: H^{-1} - \gamma_1^{-1}V$ is a dimensionless parameter of convective stability, from the kernels $\mathbf{K}_1^{\mathbf{i}} = (\mathbf{K}^{\mathbf{i}}_{c,\rho}, \mathbf{K}^{\mathbf{i}}_{\rho,c})$ for $\mathbf{z}_1^{\mathrm{T}} = (\delta \ln c,\ \delta \ln \rho)$. Perturbing the definition of v yields

$$\delta v = -\gamma_1^{-1}V(2\delta \ln \rho + \delta \ln \widetilde{m} + 2\delta \ln c) - \mathrm{d}\,\delta \ln \rho/\mathrm{d}\ln r, \tag{5.14}$$

where $\widetilde{m}$ is the mass enclosed in the sphere of radius r of the equilibrium model, whose perturbation satisfies

$$\frac{\mathrm{d}\ln \delta\widetilde{m}}{\mathrm{d}\ln r} = U(\delta \ln \rho - \delta \ln \widetilde{m}); \tag{5.15}$$

the functions U and V are given by

$$U = \frac{4\pi r^3 \rho}{\widetilde{m}}, \qquad V = \frac{G\widetilde{m}\rho}{rp}. \tag{5.16}$$

Together with the perturbed hydrostatic equation in the form

$$\frac{\mathrm{d}\delta \ln p}{\mathrm{d}\ln r} = V(\delta \ln p - \delta \ln \rho - \delta \ln \widetilde{m}), \tag{5.17}$$

equations (5.14) and (5.15) can be written in the vector form

$$\frac{\mathrm{d}\mathbf{y}}{\mathrm{d}\ln r} - \mathbf{A}\mathbf{y} = \mathbf{B}\mathbf{z}_2 \,, \tag{5.18}$$

where

$$\mathbf{y} = \begin{pmatrix} \delta \ln p \\ \delta \ln \rho \\ \delta \ln \widetilde{m} \end{pmatrix}, \quad \mathbf{z}_1 = \begin{pmatrix} \delta \ln c \\ \delta \ln \rho \end{pmatrix}, \quad \mathbf{z}_2 = \begin{pmatrix} \delta v \\ \delta \ln \gamma_1 \end{pmatrix}, \tag{5.19}$$

amongst which I have included $\mathbf{z}_1$ for completeness, and

$$\mathbf{A} = \begin{pmatrix} V & -V & -V \\ 0 & -2\gamma_1^{-1}V & -\gamma_1^{-1}V \\ 0 & U & -U \end{pmatrix}, \quad \mathbf{B} = \begin{pmatrix} 0 & 0 \\ -1 & 2\gamma_1^{-1}V \\ 0 & 0 \end{pmatrix}.$$

Let the original perturbation $\mathbf{z}_1$ be related to $\mathbf{y}$ and $\mathbf{z}_2$ by

$$\mathbf{z}_1 = \mathbf{C}\mathbf{y} + \mathbf{D}\mathbf{z}_2 \,. \tag{5.20}$$

In this example, $\mathbf{C}$ and $\mathbf{D}$ are matrices which can be obtained by using equation (5.13) to express $\delta \ln c$ in terms of the other variables:

$$\mathbf{C} = \begin{pmatrix} \frac{1}{2} & \frac{1}{2} & 0 \\ 0 & 1 & 0 \end{pmatrix}, \quad \mathbf{D} = \begin{pmatrix} 0 & \frac{1}{2} \\ 0 & 0 \end{pmatrix}. \tag{5.21}$$

It is always possible to find an equation of the form (5.20) if it is possible to relate the perturbation $\mathbf{z}_2$ to $\mathbf{z}_1$. However, in general $\mathbf{C}$ and $\mathbf{D}$ can be matrix differential operators; moreover, it is evident from the following analysis that equation (5.20) need not be satisfied in the strong sense, but only in projection with all the kernels:

$$\delta \ln \omega_{\mathrm{i}} = \int_0^R \mathbf{K}_1^{\mathrm{i}} \cdot \mathbf{z}_1 \mathrm{d}r = \int_0^R \mathbf{K}_1^{\mathrm{i}} \cdot (\mathbf{C}\mathbf{y} + \mathbf{D}\mathbf{z}_2) \mathrm{d}r \,. \tag{5.22}$$

(It is also evident how formally to generalize the argument to aspherical reference configurations, by replacing integration with respect to r by integration over the volume $\mathcal{V}$ of the star.) The kernels $\mathbf{K}_2^{\mathrm{i}}$ are now obtained by seeking a function $\mathbf{w}$ that satisfies

$$\int_0^R \mathbf{K}_1^{\mathrm{i}} \cdot \mathbf{C}\mathbf{y} \mathrm{d}r = \int_0^R \mathbf{w} \cdot \mathbf{B}\mathbf{z}_2 \mathrm{d}r = \int_0^R (\mathbf{B}^{\dagger}\mathbf{w}) \cdot \mathbf{z}_2 \mathrm{d}r \,, \tag{5.23}$$

where the superscript † denotes adjoint (in this example $\mathbf{B}^\dagger$ is simply the transpose $\mathbf{B}^{\mathrm{T}}$ of $\mathbf{B}$). The constraints (5.22) can then be written in the requisite form

$$\delta \ln \omega_{\mathrm{i}} = \int_0^R \mathbf{K}_2^{\mathrm{i}} \cdot \mathbf{z}_2 \mathrm{d}r \tag{2.24}$$

after integrating the second term in parentheses by parts and presuming the integrated terms to vanish, where

$$\mathbf{K}_2^{\mathrm{i}} = \mathbf{B}^\dagger \mathbf{w} + \mathbf{D}^\dagger \mathbf{K}_1^{\mathrm{i}} \tag{5.25}$$

is the kernel which was sought. It remains only to find $\mathbf{w}$. That is accomplished by projecting it onto the constraints (5.18):

$$\int_0^R \mathbf{w} \cdot \frac{\mathrm{d}\mathbf{y}}{\mathrm{d}\ln r} \mathrm{d}r = \int_0^R \mathbf{w} \cdot (\mathbf{A}\mathbf{y} + \mathbf{B}\mathbf{z}_2) \mathrm{d}r\,, \tag{5.26}$$

eliminating $\mathbf{z}_2$ with equation (5.23), and integrating by parts to give

$$-\int_0^R \mathbf{y} \cdot \frac{\mathrm{d}\mathbf{w}}{\mathrm{d}\ln r} \mathrm{d}r = \int_0^R (\mathbf{A}^\dagger \mathbf{w} + \mathbf{C}^\dagger \mathbf{K}_1^{\mathrm{i}}) \cdot \mathbf{y} \mathrm{d}r\,. \tag{5.27}$$

This equation is satisfied if

$$\frac{\mathrm{d}\mathbf{w}}{\mathrm{d}\ln r} + \mathbf{A}^\dagger \mathbf{w} = -\mathbf{C}^\dagger \mathbf{K}_1^{\mathrm{i}}\,, \tag{5.28}$$

the operator on the left-hand side being evidently adjoint to the corresponding operator in equation (5.18). Equation (5.27) is satisfied if the integrated parts $r\mathbf{y} \cdot \mathbf{w}$ vanish at $r = 0$ and $r = R$. In general, that condition provides boundary conditions to be applied to the solution of equation (5.28). In the case considered here, it requires that w_2 and w_3 be regular at the coordinate singularity $r = 0$, and that $w_1(R)$ be chosen to render $\mathbf{y} \cdot \mathbf{w} = 0$ at $r = R$.

The quantity γ_1 can be expressed in terms of p, ρ and the chemical composition X_α. (I use the same symbol as I do for the generic components X_i of the structure function $\mathbf{X}$, but I use the Greek index, which is to be regarded as taking values in a subset of the range of i.) Then one can write $\delta \ln \gamma_1 = \gamma_p \delta \ln p + \gamma_\rho \delta \ln \rho + \gamma_\alpha \delta \ln X_\alpha$, summation over repeated Greek indices being understood, where $\gamma_p := (\partial \ln \gamma_1 / \partial \ln p)_{\rho, X_\alpha}$, $\gamma_\rho := (\partial \ln \gamma_1 / \partial \ln \rho)_{p, X_\alpha}$ and $\gamma_\alpha := (\partial \ln \gamma_1 / \partial \ln X_\alpha)_{p,\rho}$. Then, as before, one can eliminate either one or both of p and ρ in favour either of the other, or of another structure variable such as c^2. Since the relation between p, ρ and γ_1 is only weakly dependent on the abundances of the heavy elements, as a first approximation one can ignore that dependence and retain only the dependence on the abundance Y of ^{4}He. Then the frequency constraint takes a form such as

$$\delta \ln \omega_{\mathrm{i}} \simeq \int_0^R (K_{c^2,Y}^{\mathrm{i}} \delta \ln c^2 + K_{Y,c^2}^{\mathrm{i}} \delta Y) \mathrm{d}r. \tag{5.29}$$

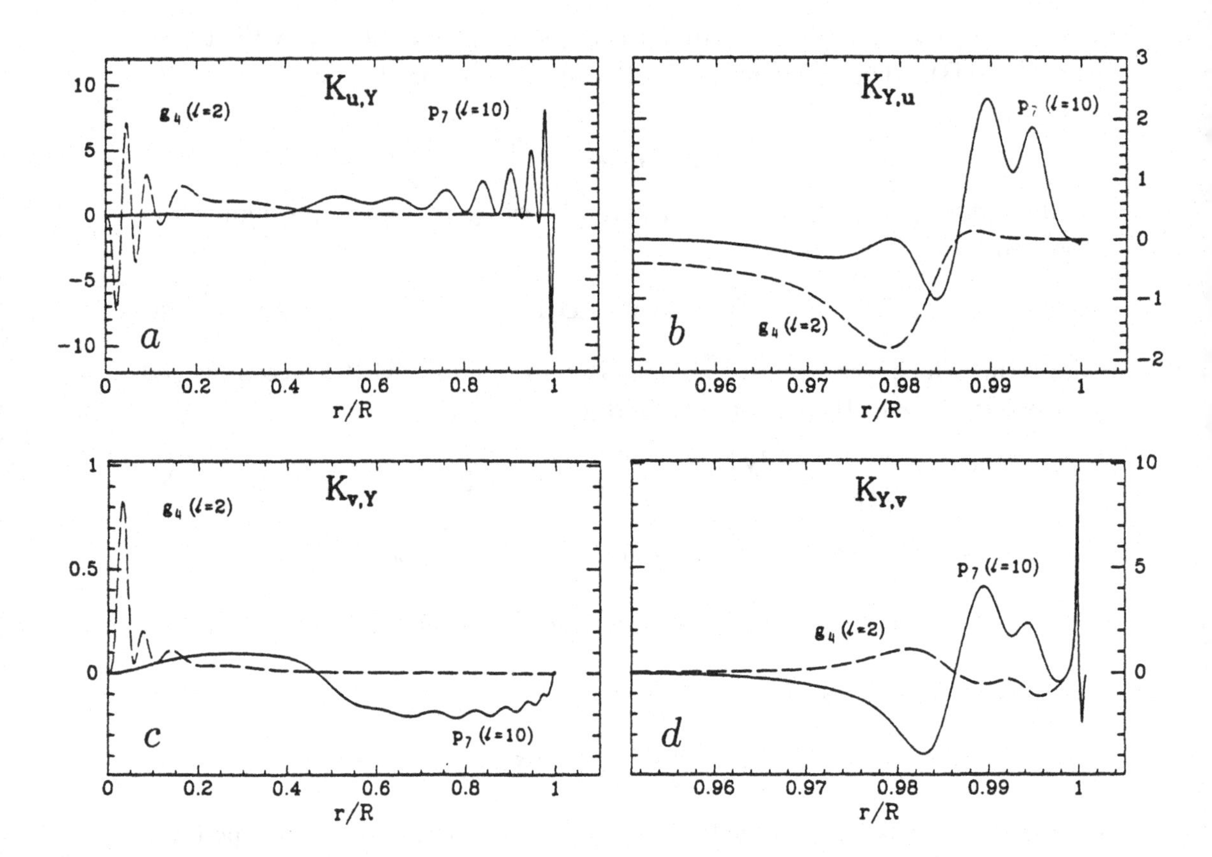

Figure 8: *Kernels* $K_{u,Y}^{n,l}$, $K_{Y,u}^{n,l}$ *and* $K_{v,Y}^{n,l}$ *and* $K_{Y,v}^{n,l}$ *for* $p_7(l = 10)$ *and* $g_4(l = 2)$, *where* $u = p/\rho = c^2/\gamma_1$ *and* $v = rg^{-1}N^2$ *is a measure of convective stability.*

Such an expression is particularly useful in the convection zone, where the chemical composition is well mixed and, although its value is not known *a priori*, δY is known to be constant and can be taken outside the integral if need be. That is so also of the abundance perturbations of other chemical species that appear in the full equation. Examples of $K_{c^2,Y}$ and K_{Y,c^2} are illustrated in Figure 8.

It should be noticed that the status of equation (5.29) is not the same as that of equation (5.12) and the other similar equations that are discussed immediately following it: whereas equation (5.12) and its relatives are truly primary in the sense that they depend only on dynamical variables, and their derivations require the use of only the hydrostatic constraint, both explicitly in the manipulations of the integrals and implicitly in the acceptance of the wave equation (1.3), equation (5.29) is in some sense secondary, for it involves a nondynamical variable Y. To derive it requires the use of the equation of state, which was not used previously. Thus we have now the beginnings of a procedure for testing equations of state. The original intention for equation (5.29) was to use it to determine the helium

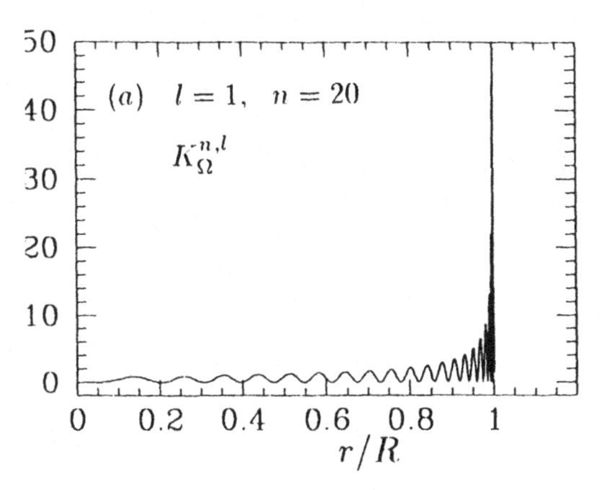

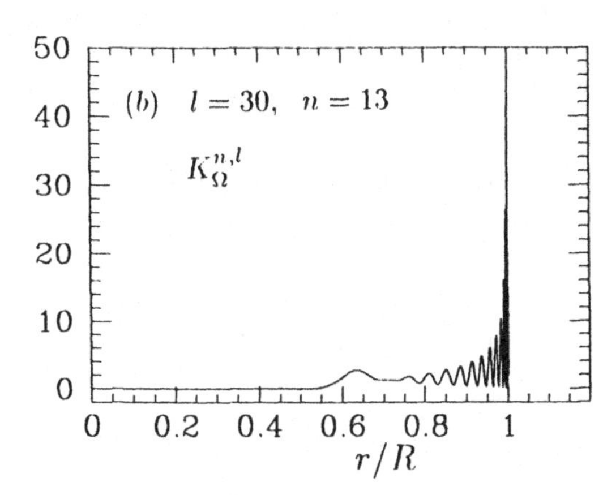

Figure 9: *Rotational splitting kernels $K_\Omega^{20,1}$ and $K_\Omega^{13,30}$ for the spherically symmetrical angular velocity $\Omega(r)$, defined such that $m^{-1}\delta\omega_{n,l,m} = \int K_\Omega^{n,l}\Omega\mathrm{d}r$, the range of integration being from the centre to the surface of the star. Notice the very close similarity between $K_\Omega^{13,30}$ and the sound-speed kernel $K_{c^2,\gamma_1}^{13,30}$ illustrated in Figure 7d; indeed, the corresponding asymptotic envelopes, to the order discussed in sections 4 and 6.3, are identical.*

abundance in the convection zone. However, Dziembowski *et al.* (1992) and Kosovichev *et al.* (1992) discovered some inconsistencies in the inversions for Y, which led them to infer the existence of an error in the equation of state.

Perturbations that maintain spherical symmetry also retain the degeneracy of the eigenfrequencies with respect to azimuthal order m. It takes asphericity to remove that degeneracy. The greatest contribution to the frequency splitting is produced by rotation. Provided that the splitting is small compared with the absolute value of the eigenfrequency, then the solution to equation (5.1) can be expanded about that with $\mathcal{R} = 0$, yielding for the frequency splitting $\delta\omega_\mathrm{i} = \omega_\mathrm{i} - \omega_{0\mathrm{i}}$,

$$\delta\omega_\mathrm{i} \simeq \frac{\mathcal{R}}{\mathcal{I}}, \tag{5.30}$$

where $\omega_{0\mathrm{i}}^2 = \mathcal{C}/\mathcal{I}$. The integral $\mathcal{R}$ is a linear functional of the angular velocity $\mathbf{\Omega}(\mathbf{r})$. Because the rotational distortion to the structure of the background state comes from the centrifugal force, which is quadratic in Ω, to first order in Ω it is adequate to use the structure of the corresponding nonrotating solar model in the evaluation of the integrals $\mathcal{I}$ and $\mathcal{R}$. If one assumes that the axis of rotation is identical at all radii r, then it is expedient to take that as the coordinate axis, so that the only velocity in the background state is steady rotation with angular velocity $\Omega(r,\theta)$. In that case, functions (5.6) constitute the zero-order eigenfunctions, and azimuthal order m is a well defined quantum number. The form for $\mathcal{R}$ can be calculated from the analysis of Lynden-Bell and Ostriker (1967):

$$\mathcal{R} = \int_{\mathcal{V}} [m\boldsymbol{\xi} \cdot \boldsymbol{\xi} + \mathbf{k} \cdot (\boldsymbol{\xi} \times \boldsymbol{\xi}^*)] \, \rho \Omega \, \mathrm{d}V \,, \tag{5.31}$$

where $\boldsymbol{\xi}^*$ is the azimuthal conjugate of $\boldsymbol{\xi}$, obtained by replacing ϕ by $\phi - \pi/2$ in the representation (5.6), and $\mathbf{k}$ is a unit vector in the direction of the rotation axis.

Thus equation (5.31) is in the same linear form,

$$\delta\omega_{\mathrm{i}} = \int_{\mathcal{V}} K^{\mathrm{i}}_{\Omega} \Omega \mathrm{d}V \,, \tag{5.32}$$

as the constraints (5.3), except that the kernel is a scalar. Two examples are illustrated in Figure 9. As is the case with the sound-speed kernel, its magnitude is substantial only in the propagation region $r > r_1$. It is important to recognize that, after substituting the eigenfunctions (5.6) into equation (5.31), the quantity $\mathcal{R}$ is an odd function of m. All axisymmetric symmetry-breaking agents other than rotation cannot distinguish between east and west, and therefore contribute an even component to the degeneracy splitting. They too can be written in the form (5.3). However, they cannot be distinguished from one another from the oscillation eigenfrequency splitting alone. Any perturbation, such as might arise from large-scale convective flow or from a magnetic field, can be mimicked by an appropriate nonuniformity in sound speed. Therefore to distinguish between them requires the use of additional information. That information might come from the distortion to the eigenfunction, which is likely to be used in local analyses. Alternatively, nonseismic information might be brought to bear. However, it is unlikely that an unambiguous interpretation of the even component of the splitting will be forthcoming in the immediate future, partly because the splitting is small and therefore difficult to measure accurately; consequently even the seismically equivalent sound-speed variation that might produce it will not be well determined. Only the angular velocity Ω can be inferred with some confidence, because nothing else contributes to the odd component of the splitting. Once Ω is determined, one can then calculate its contribution to the even component of the splitting, which comes from terms quadratic in Ω, both in the distortion of the background state and in the advection terms in the oscillation momentum equation. This can be subtracted from the observed splitting, leaving the remainder to be interpreted in terms of other sources of asphericity.

Finally I should raise the issue of nonaxisymmetric asphericity. In that case, degenerate perturbation theory must be used to calculate the eigenmodes (e.g. Gough and Thompson, 1990; Lavely and Ritzwoller, 1992; Gough, 1993). The zero-order eigenfunctions are no longer given by equation (5.6), but instead by linear combinations of the $2l+1$ such functions with different m but like n and l. There are $2l+1$ distinct combinations, though the asphericity may not have enough asymmetry to demand their forms uniquely. The azimuthal order m is no longer a quantum number, and the labelling must be modified. I shall not discuss the analysis of the frequency splitting, because the study is in its infancy, and extracting useful information about the background state is difficult (Lavely and Ritzwoller, 1993). It is more likely that information will, at least in the first instance, come from local analyses of the oscillations.

6 ON 'SOLVING' LINEARIZED INVERSE PROBLEMS IN ONE DIMENSION

As I have already pointed out in section 2, the goal in 'solving' linear inverse problems is to find a rearrangement of the linearized constraints

$$\delta \ln \omega_{\mathrm{i}} = \int_0^R \mathbf{K}^{\mathrm{i}} \cdot \delta \mathbf{X} \, \mathrm{d}r \tag{6.1}$$

that is potentially more easily comprehendable than the raw constraints themselves. Typically this is accomplished in one or both of two quite different ways: either (i) by seeking an example of the vector function $\delta\mathbf{X}$, having components δX_i (the scalar subscript i should not be confused with the mode identifier $\mathbf{i}$), that has some predetermined properties and that satisfies the constraints (6.1) to an acceptable degree, or (ii) by seeking sets of coefficients $\{c_{\mathrm{i}}\}$ such that the kernel combinations $\sum_{\mathrm{i}} c_{\mathrm{i}} \mathbf{K}^{\mathrm{i}}$ have some desired properties. I shall describe the alternatives separately. I restrict myself in this section to one-dimensional problems, in which $\delta\mathbf{X}(r)$ is the deviation of the spherically averaged structure of the sun from the structure of a spherically symmetrical reference model. That deviation might be in the stratification itself, or it might be the (scalar) angular velocity $\Omega(r)$. The motivation for the restriction is primarily simplicity, and secondarily because helioseismologists have much more experience with the one-dimensional problem than they do with its two- and three-dimensional counterparts.

In any process of inference it is essential to estimate the uncertainty in the outcome. Errors can arise from inaccuracies in the linearized equation (6.1) and in the computation of the eigenfrequencies of the reference model and their associated kernels $\mathbf{K}^{\mathrm{i}}$, from errors ϵ_{i} in the observations d_{i} of $\delta \ln \omega_{\mathrm{i}}$, and from errors in the interpretation of the inversion. I shall assume that the first of these is not at issue. I shall presume that the principles by which the reference model was constructed are correct. It might eventually come to pass that the final inferences are inconsistent with this presumption, but that can be known for sure only if the uncertainties in all the other steps in the inference process have been dealt with adequately: in that eventuality one must materially revise the reference model, which itself results in, or from, a genuine increase in our understanding of the sun, which is the true purpose of the endeavour. I shall certainly presume that the computation of $\omega_{\rm i}$ and $\mathbf{K}^{\rm i}$ is correct; the procedure is now standard and straightforward, requiring little more than (a fair amount of) care, so there should be no good reason for erring at that stage. Moreover, if nonlinearities are great enough for the linearized equation to be inadequate, I shall for the moment assume that a sequence of progressively improved reference models, defined by $\mathbf{X} + \delta\mathbf{X}$, could be constructed by iteration.

It remains, therefore, firstly to take uncertain errors $\epsilon_{\rm i}$ in the data $d_{\rm i}$ explicitly into account, and secondly to interpret correctly the results of the inversion procedure. To accommodate the former, I write the constraints that are actually available as

$$d_{\rm i} = \int_0^R \mathbf{K}^{\rm i} \cdot \delta\mathbf{X} {\rm d}r + \epsilon_{\rm i} \, . \tag{6.2}$$

I shall need to know something about the properties of the errors $\epsilon_{\rm i}$. I shall assume that they are randomly distributed, with zero mean and covariance matrix $E_{\rm ij}$.

I must point out right away that $E_{\rm ij}$ is not at all easy to determine. Errors are often estimated from the scatter of the data about some presupposed function of (n, l, m) which might simply be a smooth interpolant. Of course smoothness need not necessarily be the correct paradigm, particularly when one realizes that there may be discontinuities or near-discontinuities in some aspect the stratification, such as there could be at the base of the convection zone. As I mentioned in section 4, that modulates the frequency spectrum, yielding a deviation from a smooth function which might risk being mistaken for noise. It is also usual to suppose that the errors are uncorrelated, so that $E_{\rm ij} = {\rm diag}(\sigma_{\rm i}^2)$, $\sigma_{\rm i}^2$ being the variance of $\epsilon_{\rm i}$. There is evidence in some data that that assumption is not correct; I shall illustrate in section 8 the pitfalls of making such an error in the error estimate. In addition, some data-analysis procedures introduce bias, which is rarely discussed. And one must be aware that there may also be contamination by systematic measurement errors, which can be extremely difficult to detect. Given all this uncertainty, it is essential to consider the results of a variety of inversions, in which the errors are treated somewhat differently, to help guard against too hasty a conclusion.

6.1 Regularized least-squares fitting (RLSF)

The shorter acronym RLS, which is etymologically inconsistent with the others, is often used to specify this procedure. I prefer the full acronyms, because they emphasize the fundamental differences between the approaches of the methods. The objective of this procedure is to select one of the functions δX_i that cause equation (6.2) to be satisfied. First one expresses δX_i as a linear combination of a chosen finite set of N basis functions $\phi_{(i)j}(r)$, $j = 1, ..., N$,

$$\delta X_i = \sum_{j=1}^{N} \alpha_{ij} \phi_{(i)j} , \tag{6.3}$$

where α_{ij} are constants to be determined. Those functions may or may not be the same for each component i of $\delta\mathbf{X}$. For example, one might choose piecewise-constant functions on a dissection $\{r_j\}$ of the interval $[0, R]$ of r:

$$\phi_{(i)j}(r) = \begin{pmatrix} 1 & \text{if} & r_{j-1} < r < r_j, \\ 0 & \text{otherwise.} & \end{pmatrix} \tag{6.4}$$

Other simple alternatives are a continuous set of piecewise linear functions or a set of splines.

If the data errors are presumed to be uncorrelated, one chooses the coefficients α_{ij} to minimize the L_2-mismatch of the integral in equations (6.1) and (6.2) from the data, weighted by the estimated inverse variances of the errors. Usually one offsets the mismatch with a penalty function $P(\delta\mathbf{X})$ of the kind I discussed in section 2. Thus, for I data d_{i} one might minimize

$$\mathcal{E} = \chi^2 + \lambda P , \tag{6.5}$$

where

$$\chi^2 = (I - N)^{-1} \sum_{\mathrm{i}} \sigma_{\mathrm{i}}^{-2} \left(d_{\mathrm{i}} - \sum_{ij} \alpha_{ij} \int_0^R K_i^{\mathrm{i}} \phi_{(i)j} \mathrm{d}r \right)^2 \tag{6.6}$$

and in which λ is a positive parameter to be varied at will and the expansion (6.3) has been used to evaluate P. (I omit the second subscript of $\mathbf{K}$ for convenience.)

If P is quadratic, as is the case of the specific examples mentioned in section 2, this leads to a set of linear equations for α_{ij}, which I write as

$$\sum_{(kl)} M_{(ij)(kl)} \alpha_{(kl)} = \sum_{\mathrm{i}} B_{(ij)\mathrm{i}} d_{\mathrm{i}} / \sigma_{\mathrm{i}} , \tag{6.7}$$

in which $\mathbf{M}$ is a linear function of λ. The solution is a linear combination of the data. Note that one can consider (i, j), (and (k, l)) and $\mathbf{i}$ to be ordered into one-dimensional sequences, so that $\mathbf{M}(\lambda)$ and $\mathbf{B}$ are ordinary matrices. I have introduced the parentheses simply to draw attention to that: $\alpha_{(kl)}$ is the same quantity as α_{kl}. Notice also that because the process of obtaining the data d_{i} contains many separate operations, the errors ϵ_{i} are likely to be approximately Gaussian distributed, so the L_2-minimum is likely to be close to the maximum likelihood. Whether or not one has achieved an acceptable fit to the data can be judged from the value of χ^2 that is obtained.

The introduction of the penalty function P is called regularization, often Tikhonov regularization (cf. Tikhonov and Arsenin, 1977). There are other forms of regularization that have been used, such as omitting P from the expression (6.5) for $\mathcal{E}$ and restricting $\alpha_{(ij)}$ to lie in the subspace spanned by the right vectors $w_{(ij)}$ associated with the largest singular values $\mu_{(ij)}$ of the matrix $\mathbf{M}(0)$. (They are called singular vectors.) The smallest singular value accepted is what restricts the subspace, and thus takes the place of λ. This procedure is actually not unlike that of Tikhonov regularization: rather than truncating abruptly an expansion of $\alpha_{(ij)}$ in terms of the singular vectors $w_{(ij)}$, Tikhonov regularization multiplies each term by a filter factor which is close to unity for large $\mu_{(ij)}$ and is small for small $\mu_{(ij)}$, with a smooth transition near a value which depends on the choice of λ, the precise nature of the transition depending on the choice of P. The reader is referred to the texts by Tikhonov and Arsenin (1977) and Craig and Brown (1986), and to the papers by Hansen (1992) and Christensen-Dalsgaard, Hansen and Thompson (1993) for technical details. The outcome relates $\alpha_{(ij)}$ to the right-hand side $\mathbf{B}\boldsymbol{\delta}$ of equation (6.7), where $\boldsymbol{\delta}$ is the vector of error-reduced data, having components $d_{\mathrm{i}}/\sigma_{\mathrm{i}}$, by the matrix equation

$$\alpha = \mathbf{M}_{\mathcal{L}}^{-1}\boldsymbol{\delta} \tag{6.8}$$

where $\mathbf{M}_{\mathcal{L}}^{-1}$ is sometimes called the Lanczos (1961) or Moore-Penrose inverse of $\mathbf{M}$ (cf. Rao and Mitra, 1971).

Another procedure that has been adopted is to use the resolution of expansion (6.3) as a control on regularization. For example, one can choose basis functions (6.4) on a nonuniform dissection of N subintervals of the interval $[0, R]$ of r, the number N now playing the role of the regularization parameter. A procedure to define how the dissection intervals vary with r based on the effective wavelengths of the kernels has been used by Gough and Nejad (see Gough and Toomre, 1988); alternatively, guided by asymptotics of the kind discussed in sections 4 and 6.3, one might use the density of lower turning points of the modes constituting the data set.

6.2 Spectral expansions and the annihilator

In the early days of the subject, a clear distinction was made between what can and what cannot be determined by the data. Thus, at the outset the space of possible functions $\delta\mathbf{X}$ was divided into two: a subspace $\mathcal{S}$ of functions which in principle might be mapped to or from the frequency differences $\delta\ln\omega_{\mathbf{i}}$ via equation (6.1), and its complement $\mathcal{A}$, called the annihilator, all functions $\mathbf{F}^+$ in which are orthogonal to the kernels: $\int \mathbf{K}^{\mathbf{i}}\cdot\mathbf{F}^+\mathrm{d}r = 0$. ($\mathcal{S}$ here is not to be confused with the asymptotic surface of constant phase of section 4.) Evidently, the separation into $\mathcal{S}$ and $\mathcal{A}$ depends upon which modes are represented in the data set. Because the constraints have been linearized, any function $\mathbf{F}^+$ in $\mathcal{A}$ can be added to a putative $\delta\mathbf{X}$ without changing $\delta\ln\omega_{\mathbf{i}}$. Evidently, the seismic constraints provide no information whatever about the component of the actual function $\delta\mathbf{X}$ in $\mathcal{A}$, and under the influence of the work of Backus and Gilbert (1967) consideration was given to restricting the inversion to only the component $\mathbf{F}$ of $\delta\mathbf{X}$ that lies in the subspace $\mathcal{S}$. That confines the inference to functions one can hope to measure. The subspace $\mathcal{S}$ is spanned by the kernels $\mathbf{K}^{\mathbf{i}}$. Thus, one chooses $\phi_{(i)j} = K_i^{\mathbf{i}}$ (with a 1-1 mapping between j and $\mathbf{i}$) in equation (6.3), and carries out the data fitting by least squares, as was described in the previous section.

The outcome of the RLSF procedure is a set of coefficients in the expansion (6.3) each of which is a linear combination of the data. Had no regularization condition been imposed, some of those combinations might have contained very large, almost cancelling terms, causing them to be dominated by the data errors. Evidently, such combinations cannot be determined by the data, and therefore a procedure must be adopted for dealing with them. The Backus–Gilbert philosophy is to remove them from consideration. Thus, the annihilator is expanded to include those unmeasurable components of $\delta\mathbf{X}$, and consideration is restricted to functions in a reduced space $\mathcal{S}^*$. That is precisely what is accomplished by truncating the singular-value decomposition of the matrix equation (6.7) by rejecting contributions from vectors associated with small singular values. Of course, one must decide in advance just where to truncate, which is equivalent to deciding on the boundary between $\mathcal{S}^*$ and the corresponding augmented annihilator $\mathcal{A}^*$. Tikhonov regularization instead applies a smooth filter to the expansion, essentially blurring the boundary between the two subspaces with a continuously varying weighting function. Hansen, Sekii and Shibahashi (1992) have discussed some effects of that transition.

Restricting attention to $\mathcal{S}^*$ has fallen out of fashion in helioseismology, in the main, I suspect, because abstract function-space restrictions are difficult to understand. The principal objective of inversion is to obtain information that is useful, rather than mathematically elegant. Therefore it is usual to adopt other criteria, such as simplicity, for choosing the basis $\phi_{(i)j}$. Such bases invariably have components in $\mathcal{A}^*$, and therefore there is a sense in which the penalty function P plays a greater role than does singular-value truncation. Indeed, Christensen-Dalsgaard, Schou and Thompson (1990) have declared that this leads to better inversions, on the ground that if the true $\delta\mathbf{X}$ is smooth, then including some smooth basis functions from $\mathcal{A}^*$ in the expansion (6.3), together with a choice of P that favours smoothness, leads to a representation of $\delta\mathbf{X}$ whose pointwise values can be closer to the true values.

One should not loose sight of the possibility that an arbitrarily chosen basis $\phi_{(i)j}$ may not span $\mathcal{S}^*$. That could lead to a failure to discover some property of $\delta\mathbf{X}$. However, one can guard against that by investigating the robustness of the inversions against augmention of the basis.

6.3 Linearized asymptotic inversions (LAI)

Asymptotic representations of the displacement eigenfunction (5.6), which can be obtained directly from the asymptotic solutions of equation (4.1), can be substituted into the kernels discussed in section 5 to yield relatively simple expressions for the integral constraints (6.1). Alternatively, those constraints can be constructed directly by asymptotic perturbation theory (e.g. Gough, 1993). The outcome is often, though not always, in the form of an Abel integral equation, or can easily be transformed into one. Consequently, after interpolating data as in section 4, inversion for a component of $\delta\mathbf{X}$ (typically sound speed or angular velocity) can be carried out. If the reference model structure $\mathbf{X}$ is smooth, the iterated model $\mathbf{X} + \delta\mathbf{X}$ tends to be smooth too, even if $\delta\mathbf{X}$ contains undulations of the kind evident in Figure 4.2, because $|\delta\mathbf{X}| \ll |\mathbf{X}|$. For some people the outcome is thereby cosmetically more acceptable. However, unlike the direct inversions of section 4, the outcome depends to some extent on the assumptions that have been adopted in constructing the reference model. Indeed, failure to appreciate the importance of that dependence has led to overestimation of the reliability of the inversion. Of course, it is also necessary that the conditions for the validity of the asymptotic expansions be satisfied: namely, that the background state varies slowly except possibly at an explicitly recognized finite set of points at which $\mathbf{X}$ or its derivatives can be discontinuous.

It is instructive to record the leading-order asymptotic formula. A spherically symmetrical perturbation δc to the sound speed leads to frequency changes $\delta\omega$

given approximately by

$$\frac{\delta\omega}{\omega} \simeq S^{-1}\int_{r_1}^{R} \frac{\omega}{r^2c^2K}\frac{\delta c}{c}r^2\mathrm{d}r = S^{-1}\int_{r_1}^{R} a^{-1}(a^{-2}-w^{-2})^{-1/2}\frac{\delta c}{c}\frac{\mathrm{d}r}{c}, \tag{6.9}$$

where $K = \left(\frac{\omega^2}{c^2}-\frac{L^2}{r^2}\right)^{1/2} =: \left(k^2-\frac{L^2}{r^2}\right)^{1/2}$ is the vertical component of the local wavenumber, and

$$S(w) = \int_{r_1}^{R}\frac{\omega\mathrm{d}r}{c^2K} = \int_{r_1}^{R} a^{-1}(a^{-2}-w^{-2})^{-1/2}\frac{\mathrm{d}r}{c}. \tag{6.10}$$

This implies that $\delta\ln\omega$ is a function of w alone: $\delta\ln\omega = S^{-1}f(w)$, say. Equation (6.9) can be obtained directly, simply by perturbing equation (4.8) with only $\mathcal{F}$ retained on the right-hand side. Moreover, it is evidently essentially in the form of an Abel integral. The frequency perturbation due to a spherically symmetrical angular velocity $\Omega(r)$ is obtained by replacing $\delta c/c$ by $m\Omega/\omega$. Corresponding formulae, involving also integrals over colatitude θ, can be obtained for latitudinally varying axisymmetrical perturbations. In terms of ray theory, these formulae can be interpreted as being (volume) averages of the perturbation, $\delta c/c$ or $m\Omega/\omega$ for example, weighted by the relative time the acoustical disturbance travelling at the group velocity c^2k/ω spends in any given volume element.

Equation (6.9) could be inverted only if $f(w)$ were known for all w down to $a_\mathrm{s} = a(R)$, which is never possible. However, if one chooses to extend f from the smallest value w_0 of w for which there is a mode down to $w = a_\mathrm{s}$, which is tantamount to assuming the structure of the sun to be known from the surface down to the lowest turning point of the shallowest mode, the inverse of equation (6.9) can be written

$$\frac{\delta c}{c} = \frac{2a^3}{\pi}\left(-\frac{\mathrm{d}\ln a}{\mathrm{d}\ln r}\right)\left[\frac{f(a_\mathrm{s})}{a_\mathrm{s}\sqrt{a^2-a_\mathrm{s}^2}} + \int_{a_\mathrm{s}}^{a}\frac{(\mathrm{d}f/\mathrm{d}w)\mathrm{d}w}{w(a^2-w^2)^{1/2}}\right]. \tag{6.11}$$

When $a \gg a_\mathrm{s}$, the first term in square brackets and the contribution to the integral from $w < w_0$ are both small compared with the rest of the integral, and equation (6.11) can be approximated by

$$\frac{\delta c}{c} = \frac{2a^3}{\pi}\left(-\frac{\mathrm{d}\ln a}{\mathrm{d}\ln r}\right)\int_{w_0}^{a}\frac{(\mathrm{d}f/\mathrm{d}w)\mathrm{d}w}{w(a^2-w^2)^{1/2}}. \tag{6.12}$$

6.4 Optimally localized averaging (OLA)

Both the RLSF and the LAI yield representations $\mathbf{F}$ of $\delta\mathbf{X}$ that at any point r are linear combinations of the data. Thus, component i of $\mathbf{F}$ can be written in the form

$$F_i(r) = \sum_{\rm i} c_{i{\rm i}} d_{\rm i} \,, \tag{6.13}$$

in which the coefficients $c_{i{\rm i}}(r)$ are independent of the data $d_{\rm i}$. What is such a representation really telling us? If one substitutes for $d_{\rm i}$, using equation (6.2), one obtains

$$F_i(r) = \sum_j \int_0^R \mathcal{K}^i_j \delta X_j \mathrm{d}r + \sum_{\rm i} c_{i{\rm i}} \epsilon_{\rm i} \,, \tag{6.14}$$

where

$$\mathcal{K}^i_j = \sum_{\rm i} c_{i{\rm i}} K^{\rm i}_j \,. \tag{6.15}$$

Thus, each value of $F_i(r)$ is actually the sum of weighted integrals of the components of $\delta\mathbf{X}$. Evidently, if the function F_i is to be a good representation of δX_i, in the sense that its value at any point $r = r_0$ is close to the value of $\delta X_i(r_0)$, whatever the function $\delta\mathbf{X}$, then, because the weighting functions (usually called averaging kernels) $\mathcal{K}^i_j$ are independent of $\delta\mathbf{X}$ (and of the data), $\mathcal{K}^i_i$ must be large near $r = r_0$ and small elsewhere, and $\mathcal{K}^i_j$ must be small everywhere when $j \neq i$. How good the representation is likely to be can therefore be assessed by inspecting the averaging kernels $\mathcal{K}^i_j$. Examples are illustrated in Figure 6.1 for inversions of constraints (5.32) for a spherically symmetrical angular velocity. Sure enough, the kernels tend to be localized near $r = r_0$; however, they also have sidelobes, some with substantial contributions near the surface $r = R$ of the sun where the sound speed is small and the original data kernels $\mathcal{K}^{\rm i}_\Omega$ are all large. Therefore, one is immediately confronted with a difficulty in interpreting the inversions. For many people, including me, nonlocal averages can be difficult to understand, particularly when the averaging kernel is not everywhere of the same sign. It is important to realize that because one can never analyse more than a finite number of seismic data, each of which is an integral of $\delta\mathbf{X}$ weighted with a continuous function $\mathbf{K}^{\rm i}$, one can never infer a point value of $\delta\mathbf{X}$ from the data: the inference is always an average over an extended range of r.

Granted that inversion is supposed to be useful, it seems more natural, therefore, to subordinate the desire to satisfy the constraints to the more important requirement that the averaging kernels $\mathcal{K}^i_j$ have properties that render the averages (6.14) easy to interpret. If one wishes to have a sense of the value of $\delta\mathbf{X}$ in the vicinity of $r = r_0$, one would try to construct a kernel that resembles the delta function $\delta(r - r_0)$. One might desire no negative sidelobes, and one would certainly wish $\mathcal{K}^i_j(r; r_0)$ to be small whenever r is far from r_0 or $j \neq i$. Backus and Gilbert (1968) pioneered the construction of such kernels. Initially their purpose was merely to determine how localized is the information contained in the data (the resolving power), rather than how to extract that information. But subsequently the averages were used explicitly as a means of representing $\delta\mathbf{X}$ (Backus and Gilbert, 1970).

The procedure proposed originally by Backus and Gilbert was for the case of a scalar function δX. In that case $\mathcal{K}$ has no indices. The idea was to seek directly an averaging kernel $\mathcal{K}(r; r_0)$ that is small far from $r = r_0$, yet whose integral is unity. In principle, this could be accomplished by minimizing the integral $\int_0^R D(r - r_0)\mathcal{K}^2(r; r_o)\mathrm{d}r$ subject to the normalization constraint

$$\int_0^1 \mathcal{K}(r; r_o)\mathrm{d}r = 1\,. \tag{6.16}$$

The function $D(r - r_0)$ vanishes at $r = r_0$ and is positive elsewhere; therefore $\mathcal{K}^2$ could be large near $r = r_0$, enabling the constraint (6.16) to be satisfied without contributing substantially to the integral to be minimized, thereby producing a function with the desired properties. Evidently the precise form of D should not be crucial to the outcome, and it is found, provided a sufficient variety of data kernels K^i are available, that that is so. Most commonly, $D(r - r_0) = 12(r - r_0)^2$ is adopted, because it is computationally convenient. The outcome of the minimization is again a series of the form (6.15).

The procedure as I have just described it is not satisfactory, because associated with the most delta-like function $\mathcal{K}$ are coefficients c_i with large magnitudes (with different signs, for the inevitable cancellation). This leads to the sum (6.13) being dominated by the data errors, because one cannot expect them to cancel in the way that the combination of the integrals must. Consequently a compromise between localization and error domination must be adopted. That is achieved by minimizing instead

$$12\int_0^R (r - r_0)^2\mathcal{K}^2(r; r_0)\mathrm{d}r + \lambda \sum_{\mathrm{ij}} E_{\mathrm{ij}} c_\mathrm{i} c_\mathrm{j}\,, \tag{6.17}$$

in which I have explicitly adopted the quadratic representation of D; as earlier, E_{ij} is the covariance matrix of the errors, and the so-called tradeoff parameter λ is to be chosen to produce the desired compromise. Examples of such optimally localized kernels, again for angular velocity $\Omega(r)$, are illustrated in Figure 10.

When the function δX_i has more than one component, the quantity to be minimized for obtaining localized information about component i is

$$12\int_0^R (r - r_0)^2\left[\mathcal{K}_i^i(r; r_0)\right]^2\mathrm{d}r + \sum_{k\neq i} \mu_{ik}\int_0^R (\mathcal{K}_k^i)^2\mathrm{d}r + \lambda_i \sum_{\mathrm{ij}} E_{\mathrm{ij}} c_{i\mathrm{i}} c_{i\mathrm{j}}\,, \tag{6.18}$$

in which $\mathcal{K}_k^i$ is given by equation (6.15). One attempts to localize $\mathcal{K}_i^i$ and have $\mathcal{K}_k^i$ $(k \neq i)$ small everywhere, without serious contamination with errors, by adjusting the tradeoff parameters μ_{ik} and λ_i.

Other functionals are used to localize kernels. Backus (1970) has considered a large class of them. For example, one can replace the first term in equation

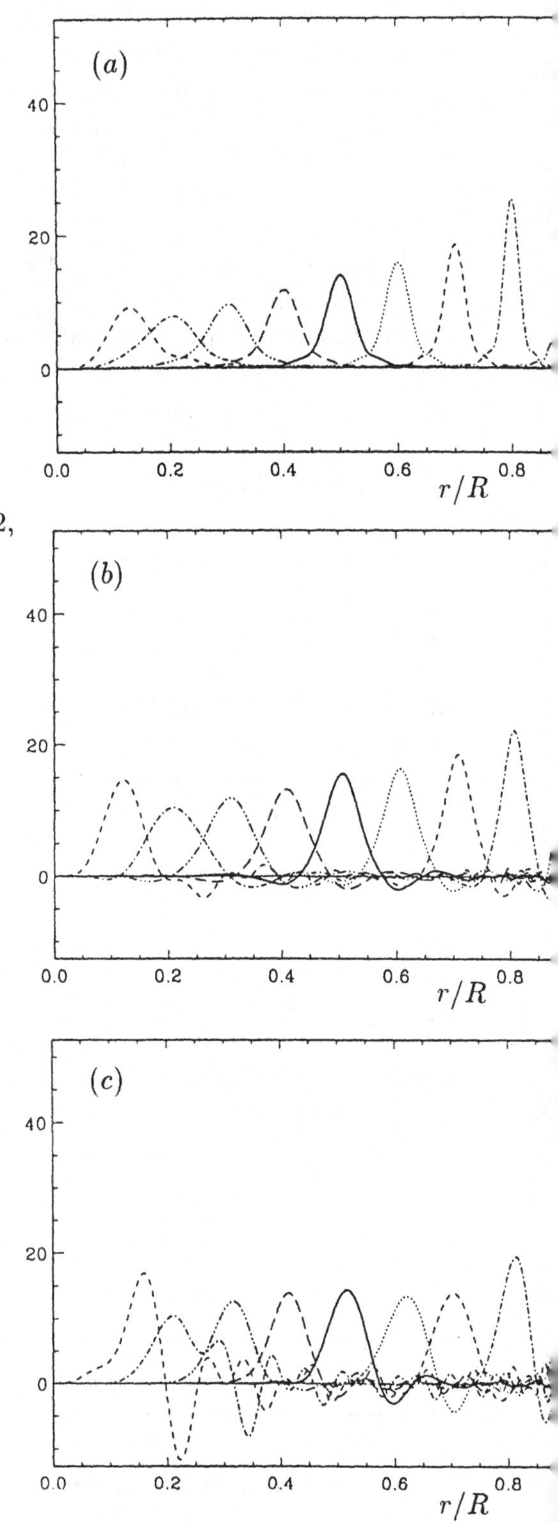

Figure 10:*Sequences of averaging kernels* $\mathcal{K}_\Omega(r; r_0)$ *about radii* $r_0/R = 0.1, 0.2, \ldots, 0.9$, *by (a) OLA, (b) RLSF, (c) LAI, computed from a set of 834 five-minute p modes of degrees ranging from 0 to 200 (from Christensen-Dalsgaard et al., 1990). In the case of OLA, the radii* r_0 *are the radii about which it was intended to centre the kernels, rather than the actual centres* $\bar{r}$ *(which in the examples illustrated here are close to* r_0*); in the cases of RLSF and LAI, they are the radii at which the function resulting from the inversion is equal to its corresponding average. In all cases the kernel centred at* $r_0/R = 0.5$ *is shown as a bolder curve. For the reason explained in the text, the kernels from LAI are not insignificant throughout the propagation region, though they are substantially greater in the vicinity of the lower turning point. The same is true, but to a lesser extent, of the kernels from RLSF.*

(6.17) or (6.18) by

$$\int_0^R [\mathcal{K}(r;r_0) - \mathcal{D}(r;r_0)]^2 \mathrm{d}r, \tag{6.19}$$

(ignoring indices on $\mathcal{K}$) in which $\mathcal{D}$ resembles the delta function $\delta(r - r_0)$. Oldenberg (1976) used this functional, with $\mathcal{D}(r;r_0) = \delta(r - r_0)$, for calculating Fourier transforms from noisy incomplete data, partly because it is computationally convenient. Pijpers and Thompson (1992) used it for helioseismological inversion, originally also for computational convenience, but with $\mathcal{D}$ chosen to be a Gaussian function which more closely resembles what one can expect to achieve for $\mathcal{K}$. They called this specific method: subtractive optimally localized averaging (SOLA). Subsequently it was found that the extra flexibility afforded by adjusting the target function $\mathcal{D}$ can permit one to obtain localized kernels $\mathcal{K}$ more closely resembling the desired paradigm, such as having no negative sidelobes, than would be possible with equation (6.17) or (6.18). Consequently this example of OLA is becoming more popular.

I should point out that under some circumstances one might wish kernels to satisfy criteria other than being the most localized. This is the case particularly when the data are too sparse or too inaccurate to obtain well localized sidelobe-free averaging kernels. Under such circumstances one might prefer, for example, to construct a kernel that is strictly positive in one region of the sun and strictly negative elsewhere, in order to be able to make a reliable statement about the sign of the mean variation, or of differences of certain extreme values, of a component of $\delta\mathbf{X}$. Alternatively, one might try to design a kernel that is sensitive to, say, a discontinuity in the gradient of $\delta\mathbf{X}$ (Pijpers and Thompson, 1994).

Finally, I wish to point out that what OLA is doing is to combine the data into a more palatable form. The outcome is to be regarded as no more than what it is: a set of weighted averages of all possible function $\delta\mathbf{X}$ that satisfy the data. Therefore there is no question as to whether, as a procedure, it is right or wrong; it is only in how one might subsequently choose to interpret those averages, perhaps in terms of specific realizations of $\delta\mathbf{X}$, as is often one's wont, that there is room for making mistakes.

6.5 Uncertainty and tradeoff

The task of choosing suitable tradeoff parameters is much of the art of inversion. Consequently there is diversity of opinion about how best it should be carried out. The philosophy of RLSF causes attention to be focussed mainly on how well the solution reproduces the data. Thus one might choose λ in equation (6.5) to be such that $\chi^2 \simeq 1$. This provides the greatest smoothing (if P penalizes a measure of curvature) consistent with the data being reproduced on average to within a standard deviation. More sophisticated criteria, also based on a measure of how well the data are reproduced by the solution, have been suggested. For example, the measure of misfit could be replaced by $\hat{\chi}^2$, which is defined as is χ^2 in equation (6.6) but with $I - N$ replaced by trace$(\mathbf{I} - (\mathbf{M}^{\mathrm{T}}\mathbf{M})^{-1}\mathbf{M}^{\mathrm{T}}\mathbf{B}\boldsymbol{\delta})$, the so-called equivalent (number of) degrees of freedom for error (Golub *et al.*, 1979; Craig and Brown, 1985), in which $\mathbf{M}^{\mathrm{T}}$ is the transpose of the matrix $\mathbf{M}$ appearing in equation (6.7), $\mathbf{I}$ is the identity matrix and $\boldsymbol{\delta}$ is the vector of error-normalized data, whose components are $d_{\mathrm{i}}/\sigma_{\mathrm{i}}$. However, such criteria take little cognizance of how well the solution might represent the actual function $\delta\mathbf{X}$, though Barrett (1993) has tried to explain how well two tradeoffs, one based on setting $\hat{\chi}^2 = 1$, the other on minimizing $\hat{\chi}^2$, reproduce the data in an artificial helioseismological RLSF example. Indeed, such average-error criteria could suffer the disadvantage of leading to poorer representations of some aspects of $\delta\mathbf{X}$ as the number I of data is increased. For example, if one were interested in $\delta\mathbf{X}$ in the vicinity of $r = 0.5R$, say, then by adding a large number of eigenfrequencies of modes of high degree, which are determined by only the structure of the very outer layers of the sun, the deviation of the frequencies of the more deeply penetrating modes that are required for determining $\delta\mathbf{X}$ in the region of interest could be permitted to increase at constant $\hat{\chi}^2$ if the additional less relevant data were reproduced very well by $\delta\mathbf{X}$. Indeed, since we do not understand the physics of the outer layers very well, reproducing the frequencies precisely may even be forcing the theoretical model merely to comply with incorrect constrains near the surface, at the expense of degrading the inferences in regions in which we understand the physics better.

It is more logical to choose tradeoff parameters on the basis of the properties of the 'solutions' they render. Thus, in the case of OLA, if one were to know the covariance matrix of the errors, one could compute the resulting standard deviations of the possible errors in the averages, and determine how they vary with the characteristic width of the averaging kernels centred at some desired point. Figure 11 illustrates how constraining the procedure to decrease the errors, by increasing λ in expression (6.17), leads to broader averages. Indeed, as the number N of independent data becomes very large, the error becomes very large as the minimum kernel-width is approached; and the so-called tradeoff curves in Figure 11b become nearly constant as the minimum error is approached. One can then choose any or several points on the curve according to one's requirements, though evidently it is imprudent to be very near to the most extreme localization or the lowest error. For general-purpose presentation, a point in the knee of the curve is usually selected.

Of course this procedure for selecting tradeoff parameters does not depend

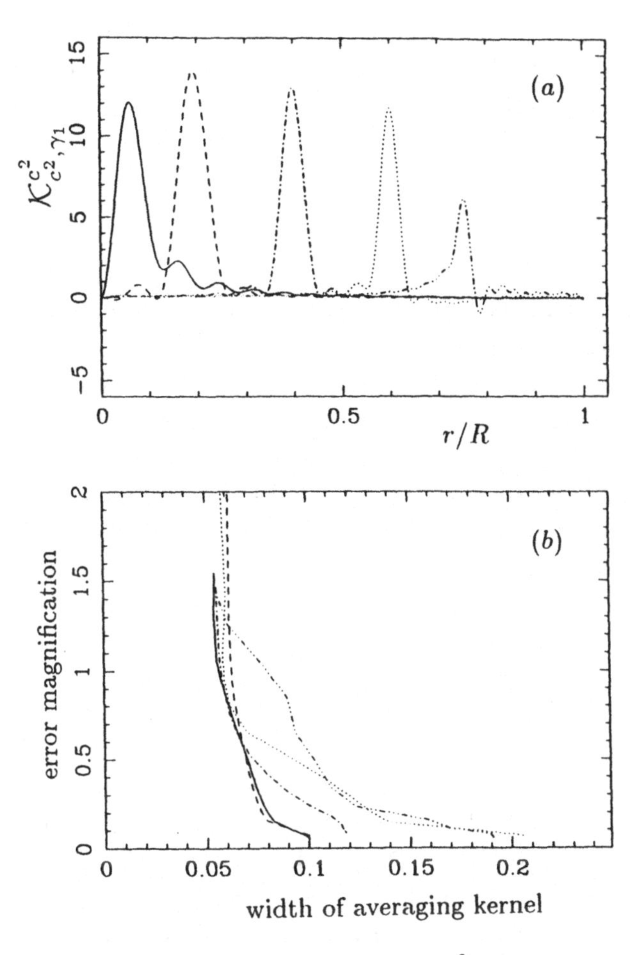

Figure 11: *(a) Optimally localized averaging kernels* $\mathcal{K}_{c^2,\gamma_1}^{c^2}$ *for the square of the sound speed, computed from the p modes in the compilation by Duvall et al. (1988) whose turning points lie beneath the convection zone, obtained for the inversions of the kind reported by Gough and Kosovichev (1988). Sidelobes are particularly noticeable in the kernels at the extremeties of the range of radii* r_0 *about which it is possible to centre them. Despite the lack of high-degree modes, it is possible to eliminate the large contributions to the individual mode kernels, evident in Figures 7 and 8, near the surface of the star. (b) Tradeoff curves for the kernels represented in (a): the error magnification factor* $\Sigma(c_i^2\sigma_i^2)^{1/2}/\Sigma\, c_i\sigma_i$ *is plotted against the characteristic kernel width, defined as the distance between the first and last quartile points (such that the first and the last quarters of the integral of* $\mathcal{K}_{c^2,\gamma_1}^{c^2}$ *lie outside the interval defining the width), for sequences of kernels centred at* $r_0/R = 0.05, 0.20, 0.40, 0.60$ *and 0.75, using the same linestyles as in (a). Notice that with this mode set, which lacks modes of high degree that are confined to the outer regions of the sun, error magnification tends to be greater for averages centred at larger radii.*

on the manner in which the averaging kernels were constructed. So it could equally well be used for RLSF, and for most purposes that would be more useful. However, in practice it is rare.

6.6 RLSF v OLA

How does the quality of the results of the two quite different approaches, RLSF and OLA, compare? Provided one accepts that what both procedures actually do is to end up with a new set of averaging kernels, the answer lies entirely in the form of those kernels. Thus, for example, one compares the kernels in Figures 10a and 10b, and also 10c if one wishes to bring asymptotic inversion into the comparison. What more one can say than that OLA yields kernels more localized than those of the other methods requires an excursion into function space, averaging classes of putative functions with the kernels, either by imagination or by direct computation. One is finally always reduced to considering specific examples.

Averages weighted with well localized kernels, such as those in Figure 10a, are relatively straightforward to interpret. However, I must emphasize that one must not be too naive. Broadly speaking, one could imagine a curve through those averages (it is easy to draw that in one's mind's eye when looking at Figure 12, for example) and then imagine possible solutions $\delta\mathbf{X}$ which might oscillate about that curve in such a way that the contribution from the fine structure to those averages is zero. Notice that the curve one has in one's mind is a sort of blurred (i.e. smoothed) version of all the possible solutions $\delta\mathbf{X}$, but it is not necessarily equal to any of them. The range of possibilities permitted by RLSF is, at least for me, harder to imagine, because a deviation from a putative $\delta\mathbf{X}$ in one location induces a change elsewhere. However, there are others who think otherwise. There are some people who consider it preferable to have at least one function that satisfies the data from which to launch one's imagination, as is provided by RLSF, than none. I am more worried by being biased by a single member of a class of functions that in many ways may not be typical.

As an example, I illustrate the results of some early inversions of artificial data for the density of a simple piecewise polytropic solar model in which γ_1 was considered to be a known constant (actually 5/3). In Figure 12 are plotted optimally localized averages (dots) and functions obtained by least-squares fitting of the data (dashed lines), for two different sets of only 20 error-free data. The continuous line is the actual answer. The localized averages are plotted against

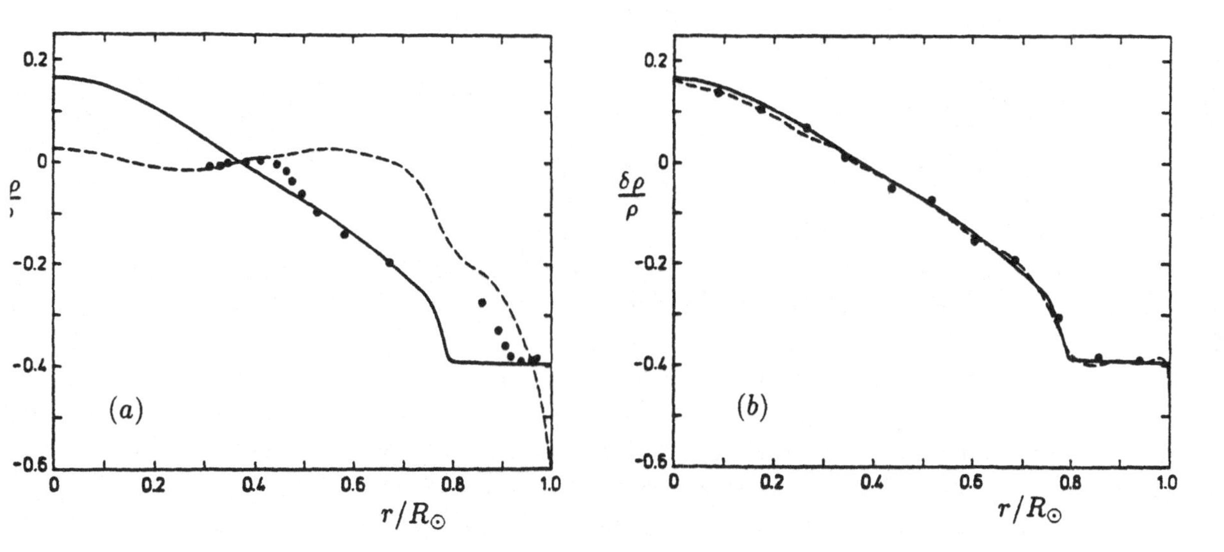

Figure 12: *Density inversions of artificial error-free frequencies of 20 low-degree modes of a simple piecewise polytropic model of the sun using a similar model as a reference, carried out by Cooper and Gough (cf. Cooper, 1981). Both models have polytropic interiors of index 3 matched continuously onto polytropic envelopes of index 3/2 at* $r = r_c$, *with* $r_c = 0.75R$ *in the proxy sun and* $r_c = 0.80R$ *in the reference. The dots in the figures denote optimally localized averages and are plotted against the actual centres* $\bar{r}$ *of the averaging kernels; the dashed curves were obtained by least-squares fitting of a piecewise constant function, using truncated Moore-Penrose inversion for regularization. The inversions used frequences (a) of p modes of* $1 \leq n \leq 10$ *with* $l = 1$ *and 2, and (b) of p modes and g modes of* $1 \leq n \leq 5$ *with* $l = 1$, *and also p modes with* $1 \leq n \leq 5$, *g modes with* $1 \leq n \leq 4$ *and the f mode, all with* $l = 2$. *In case (a) neither method yields very well localized kernels, and the kernels from the two methods are rather different; moreover, neither method yields kernels centred at small values of* r, *the kernels corresponding to all points at small radii of the least-squares fitted function having centres* $\bar{r}$ *at about the same place, which is why the function is almost constant in the inner half of the model. The presence of g modes in case (b) permits good localization, considering the low number of modes used for the inversions, and both methods yield approximtely the same result. It is interesting that the results in this case are so good, given the large values of* $\delta\rho/\rho$ *and the fact that the problem has been linearized; the reason is that for a sequence of piecewise polytropes in which only* r_c *varies, the eigenfrequencies vary nearly linearly with* r_c, *when* r_c *is within the range bounded by the values in the two models.*

the actual 'centres' $\bar{r}$ of the averaging kernels $\mathcal{K}$, defined such that

$$\int_0^{\bar{r}} \mathcal{K} dr = \int_{\bar{r}}^{R} \mathcal{K} dr = \tfrac{1}{2} , \tag{6.20}$$

rather than against the target value r_0 about which the attempt to centre them was made. There are no dots in Figure 12a for $\bar{r}/R < 0.3$ because it was not possible to construct such kernels, despite attempts with $r_0/R < 0.3$. Evidently, the data cannot yield localized information in that region. However, the dashed curve extends over the entire range of r.

Figure 12a shows inversions of low-degree p-mode frequencies: the kernels from neither method are well localized, so the values of the averages are not always very close to local value of the functions whose averages they represent. (Perhaps I should stress that the dots are averages of both the continuous and the dashed lines, since both satisfy the data.) Which of the representations is more useful is perhaps a matter of opinion. (My opinion is that neither is particularly useful without knowledge of the associated averaging kernels.)

A potential disadvantage of RLSF is that it always yields a solution. Thus, in Figure 6.3a the dashed line exists in the region $\bar{r}/R < 0.3$ in which the data provide no localized information. The unwary might be tempted to make local (invalid) interpretations of the form of $\delta\rho/\rho$, which was actually dictated almost entirely by the form of the regularization that was adopted, and not by the data.

Figure 12b illustrates inversions of a data set containing a combination of p-mode and g-mode frequencies. The results appear to be substantially better, and indeed inspection of the averaging kernels reveals that they are: the averaging kernels, both from OLA and RLSF, are more highly localized.

There are several lessons to be learnt from this exercise. The first is that g modes, which have kernels weighted preferentially in the innermost regions of the sun, quite naturally permit one to make inferences for small values of r. The realization of that is what has provoked the search for internal g modes of the sun. The second is that g modes, which have kernels preferentially weighted in the innermost regions of the sun and which are evanescent in the convective envelope, also improve the inferences in the outer layers of the sun. The reason is that the low-degree p-mode kernels are superficially similar, whereas the g-mode kernels are quite different from them. Combinations of p-mode kernels alone that cancel in the interior also tend to cancel elsewhere, and localization is difficult. The addition of g modes widens the class of combinations that cancel near the core, yielding a broader range of functions for localizing information elsewhere. Of course g modes are not required for determining the stratification of the outer layers of the sun, because in practice we have high-degree p-mode data available (which were not used in these inversions). However, for other stars, for which modes of only low degree are detectable, that worth of g modes would be valued.

The third lesson to be learnt is that when RLSF and OLA appear to yield similar results, in the sense that the averages are closer to the point values of the RLSF solution, then both tend to be closer to the correct function. This has since been demonstrated for a wide range of helioseismological inversions, and illustrates the need for using several different methods. Why should that be? It seems that if OLA demonstrates that information can be localized, then by RLSF it actually is. It is intuitively obvious that if the optimal kernels are well localized, then at least one of the smoothest functions that satisfies the data (e.g by RLSF) must pass almost through those averages. The following more formal argument of T. Sekii (personal communication) elucidates how that can be the case, though it doesn't prove that it must be.

Suppose we express δX_i as an expansion (6.3) of piecewise-constant functions $\phi_{(i)j}$ defined by equation (6.4), or of any other convenient localized basis, and let us ignore further regularization. The constraints (6.2) can then be written, ignoring errors, in the matrix form

$$\boldsymbol{\delta} = \mathbf{M}\boldsymbol{\alpha}\,, \tag{6.21}$$

where $\boldsymbol{\alpha}$ is the vector of expansion coefficients and $\boldsymbol{\delta}$ represents the error-weighted data. The objective of OLA is to find a linear combination of the data:

$$\mathbf{A}\boldsymbol{\delta} = \mathbf{A}\mathbf{M}\boldsymbol{\alpha} \tag{6.22}$$

which one hopes will represent $\boldsymbol{\alpha}$ in a localized way, whatever the data (and therefore whatever the solution $\boldsymbol{\alpha}$). Strictly, this requires $\mathbf{AM} = \mathbf{I}$, though since perfect localization is unlikely to be possible (notice that my expansion (6.3) has yielded a discrete rather than a continuous representation, so perfect localization, in the sense of finding combinations that determine each component of $\boldsymbol{\alpha}$ uniquely, is formally not impossible; it requires merely that the matrix $\mathbf{M}$ be nonsingular, or, in practical cases, well conditioned, though in practice this is usually not the case) we must be content with $\mathbf{AM}$ being diagonally dominated with nearly unit diagonal elements. Then $\mathbf{A} \simeq \mathbf{M}_{\mathrm{L}}^{-1}$, namely the left-inverse of $\mathbf{M}$.

Least-squares fitting requires one to minimize

$$\chi^2 \propto (\mathbf{M}\boldsymbol{\alpha} - \boldsymbol{\delta}) \cdot (\mathbf{M}\boldsymbol{\alpha} - \boldsymbol{\delta})\,, \tag{6.23}$$

which is achieved when

$$\boldsymbol{\alpha} = (\mathbf{M}^{\mathrm{T}}\mathbf{M})^{-1}\mathbf{M}^{\mathrm{T}}\boldsymbol{\delta} \equiv \hat{\mathbf{A}}\boldsymbol{\delta}\,, \tag{6.24}$$

which is also a linear combination of the data. If we are to succeed in making χ^2 small, $\boldsymbol{\delta}$ must be close to $\mathbf{M}\boldsymbol{\alpha}$, whatever its value, and hence $\boldsymbol{\alpha} \simeq \hat{\mathbf{A}}\mathbf{M}\boldsymbol{\alpha}$. This is achieved at least when $\hat{\mathbf{A}} = \mathbf{M}_{\mathrm{L}}^{-1}$, which is the OLA result.

Once one accepts that tailoring averaging kernels to meet some desired criteria is the most natural procedure, why does one not always adopt it? The reason is principally one of ease of execution. To carry out a RLSF requires inverting matrices of dimension $N \times N$, which can be kept under reasonable control; LAI requires a procedure for interpolation, which need not even involve very large matrices. OLA, on the other hand, when carried out properly, requires the inversion of matrices of dimension $I \times I$, which for very large data sets can be quite unmanageable. Progress in handling such large matrices is urgently needed. Christensen-Dalsgaard and Thompson (1993) have made the attractive suggestion of working with a reduced data set, each composed of combinations of the entire data set projected onto singular vectors of the matrix $\mathbf{M}$. The procedure ought to produce a reduced data set that contains all the information that is accessible from the original data. The preprocessed data, and their associated kernels, are used for inversion in the same manner as the original data. Because the data set is much smaller, the amount of computation is greatly reduced, requiring, in addition to matrix multiplications, the inversion of a matrix of dimension only $N \times N$. Such a procedure may nonetheless still prove expensive in two dimensions (see section 7). In that case it may be possible to iterate about a subset of nonzero coefficients obtained by a simpler method, such as LAI.

6.7 On the representation of localized averages

The presentation of the inversions in Figure 12 is incomplete, for it gives no information about the averaging kernels. It is often useful to include an indication of the extent of the interval over which the averaging is made, by means of a horizontal bar, as in Figure 13, which presents an OLA inversion of real solar data. In this figure, the characteristic widths of the kernels were taken to be the extents of the sum of the second and third quartiles. There are also vertical bars, which represent ±one standard error. (There are no such errors associated with Figure 12, if the inversions are regarded as averages, because error-free data were used.) They were computed as $(\sum_i c_{ii}^2\sigma_i^2)^{1/2}$, the errors ϵ_i having been assumed to be independent with standard deviations σ_i. Thus, the centre of each cross represents the value of the expectation of an average over a kernel whose width is indicated by the horizontal bar; moreover, there is a 68 per cent chance that the actual value of any selected average, taken in isolation, lies within the range indicated by the vertical bar (provided the errors of the averages are normally distributed, which they certainly would be if the data errors ϵ_i are normally distributed).

It should be appreciated that the errors in the inversions at each point are not independent. That is obviously so in general: if two points about which averages are obtained are imagined to approach one another, the coefficients c_{ii} also approach one another; not only do the values of the averages converge, but so also do the errors, which eventually become perfectly correlated as the two points meet. Therefore the corridor defined by the envelope of the ends of the error bars, which one can easily visualize when a graph of closely spaced inversions is presented, must be interpreted with care. It does not represent, for example, the region within which all the averages lie with 68-per-cent probability. That would require all the errors to be correlated perfectly. However, one could draw such a corridor, or simply extend the error bars by the appropriate factor, given the covariance matrix of the errors, and to do so might for some purposes give a more realistic impression of the overall properties of the function $\delta\mathbf{X}$. Gough, Sekii and Stark (1995) show how such extensions of the error bars can be estimated.

It goes without saying that LAI and inversions by RLSF could also be represented in this way. However, because the kernels are localized less well, often having substantial sidelobes far from their centres, the significance of the horizontal bars would need to be interpreted with rather more care.

6.8 The direct product of inversion algorithms

The inversion of linearized constraints leads to a set of averages $\overline{\delta\mathbf{X}}(\bar{r})$ of the structure function $\delta\mathbf{X}(r)$ each of which is a linear combination of the data d_i:

$$\overline{\delta\mathbf{X}}(\bar{r}) = \sum_i \mathbf{c}_i(\bar{r}) d_i\,. \tag{6.25}$$

The coefficients $\mathbf{c}_i$ do not depend on the data. Therefore, once they have been calculated, averages $\overline{\delta\mathbf{X}}$ from any set of data can easily be evaluated from equation (6.25). Even though it may have taken a great deal of computational effort to determine $\mathbf{c}_i$ in the first place, subsequent inversions of similar data sets can be carried out with no difficulty since these involve simply evaluating the scalar product on the right-hand side of equation (6.25).

The values of $\mathbf{c}_i$ do depend on which modes are included in the data set, and on the statistics of their likely errors. Indeed, the extent of the data set and the error statistics appear explicitly in the quantities defined by equations (6.5), (6.6) and (6.17) or (6.18) that are minimized in carrying out RLSF and

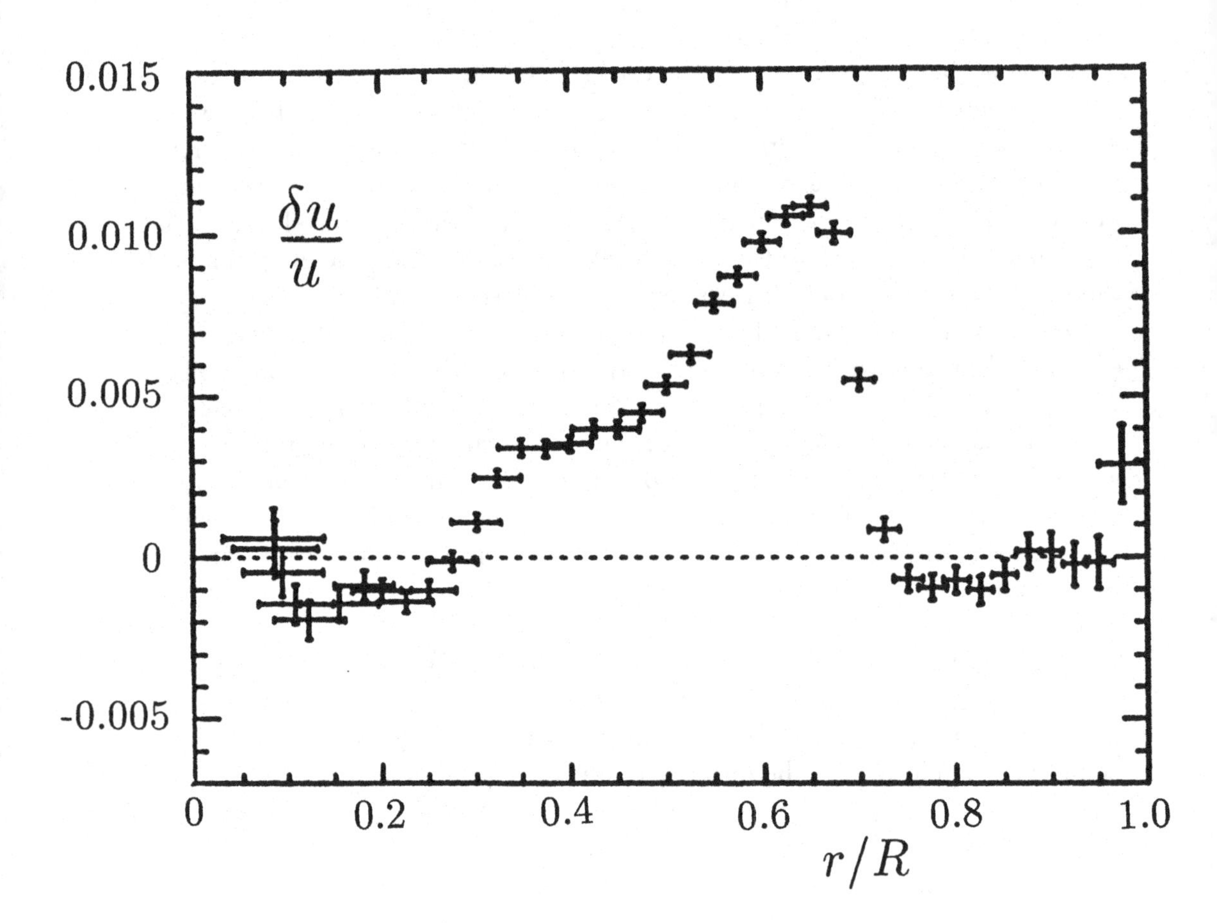

Figure 13: *Optimally localized averages of the difference δu from a reference model of the quantity $u(r) = p/\rho$, obtained from the frequencies of 16 p modes with $l \leq 2$ and with frequencies between 2.5 and 3 mHz obtained from IPHIR data (Toutain and Fröhlich, 1992) and 598 p modes observed by Libbrecht et al. (1990) with frequencies between 1.5 and 3 mHz and with $4 \leq l \leq 140$ (from Gough and Kosovichev, 1995). The horizontal bars represent the widths of the averaging kernels, the vertical bars each extend to ± 1 standard error from the value of the average.*

OLA; they are taken into account in asymptotic inversions at the outset when the function of $w = \omega/L$ representing the right-hand side of equation (6.9) – or the function $\mathcal{F}(w)$ defined by equation (4.9) in the case of full inversions – is fitted to the data.

It is of some interest to inspect the values of these coefficients. With asymptotic ideas in mind, they are plotted in Figure 14 against w, for a LAI for angular velocity Ω at $r = 0.5R$, and for inversions by RLSF and OLA. The LAI coefficients (represented by the continuous curves) are largest in magnitude near the lower turning point (corresponding to $\hat{w} := \nu/2\pi L \simeq 137\mu\text{Hz}$), as one expects from equation (6.11). The RLSF and OLA exhibit a similar tendency. But they both also rely quite heavily on modes that penetrate much more deeply. It is curious that RLSF hardly uses the more shallowly penetrating modes, even less than LAI. On the other hand, OLA utilizes them extensively to remove the sidelobes from the averaging kernel; Christensen-Dalsgaard, Schou and Thompson (1990) report that if those high-degree modes are excluded from the data set, the optimally localized kernels deteriorate significantly, developing negative sidelobes nearer the surface and taking on the appearance of RLSF kernels.

7 LINEARIZED TWO-DIMENSIONAL INVERSIONS

It is straightforward to formulate two-dimensional generalizations of the methods discussed in the previous section. Putting them into operation is quite another matter, at least in the case of RLSF and OLA. The reason is that the dimensions of the matrices that need to be inverted become very large. Discussions in the literature have been concerned mainly with the inversion of rotational splitting data to determine the sun's angular velocity $\Omega(r,\theta)$. I shall do likewise. It is the simplest 2-d helioseismological problem. However, from the point of view of inversion, it contains essentially all the salient new features of the problem of determining the axisymmetric component of the asphericity in the structure. The rotational splitting constraints (5.30) can be rewritten in terms of the independent variables r and $\mu = \cos\theta$ as

$$d_{\mathrm{i}} = \int_{-1}^{1} \mathrm{d}\mu \int_{0}^{R} \mathcal{K}^{\mathrm{i}}\Omega\, \mathrm{d}r \tag{7.1}$$

in which I have absorbed a geometrical factor into $\mathcal{K}^{\mathrm{i}}$ (and dropped its subscript).

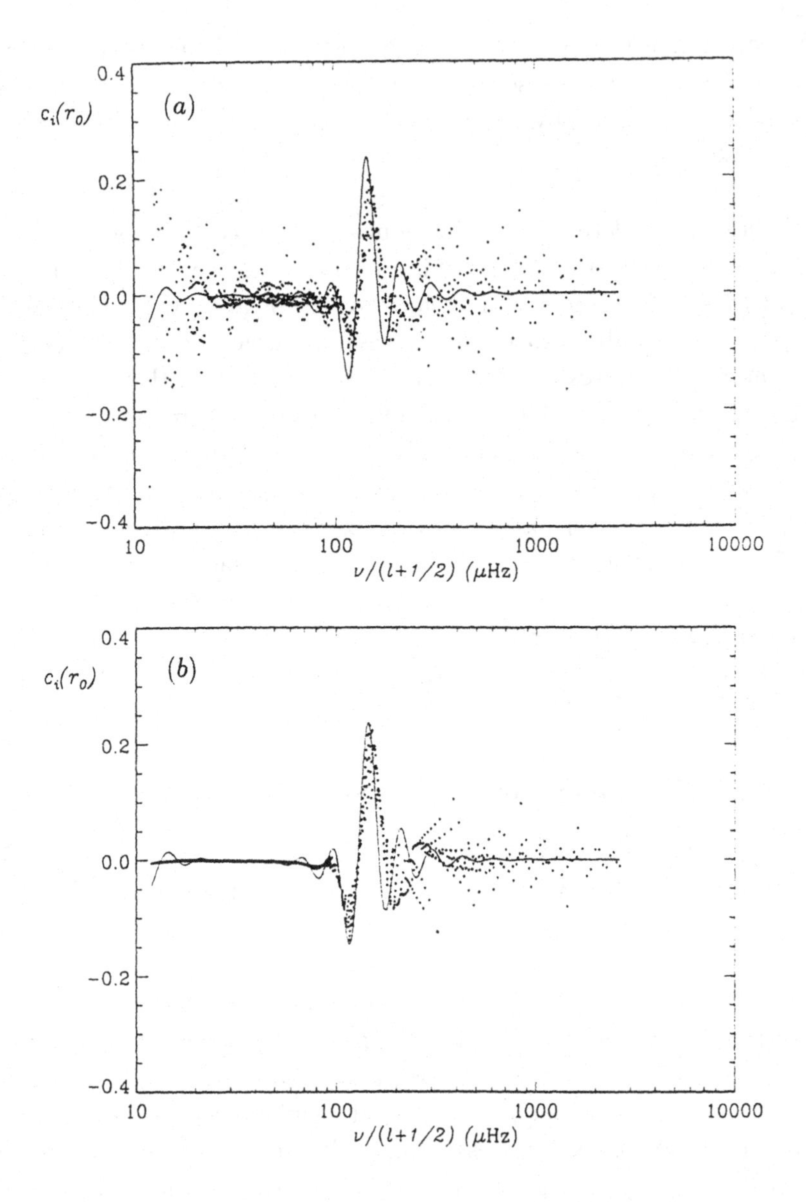

Figure 14: *Coefficients c_i used for the construction of the angular-velocity averaging kernels that are shown in Figure 6.1 for $r_0/R = 0.5$, at which $c/2\pi r \simeq 137\mu Hz$; (a) OLA, (b) RLSF, plotted against $\hat{w} = w/2\pi = \nu/L$, where ν is cyclic frequency. The continuous line represents the corresponding coefficients for LAI (from Christensen-Dalsgaard et al., 1990). Although the asymptotic formula for rotational splitting of a multiplet, namely equation (6.9) with $\delta c/c$ replaced by $m\Omega/\omega$, depends on the function $\Omega(r)$ only for $r > r_1$, where r_1 is the lower turning point of the mode, and its inverse (cf. equation 6.11) appears to depend only on modes with $r_1 > r$, in practice the inversion yields a value of $\Omega(r)$ that depends also on data with $r_1 < r$, and consequently with $w > c/r$. The reason must be that the fitting of the continuous curve to the data is necessarily nonlocal, the value at any value of w depending on data over a range of w.*

7.1 So-called fully two-dimensional methods

The application of RLSF to the constraints (7.1) is quite obvious. One simply expresses Ω as an expansion in some set of basis functions $\phi_j(r, \mu)$, which might, for example, be piecewise unit functions each of which takes a nonzero constant value in a single grid cell of a discretization of r and μ, and regularizes with a two-dimensional penalty function, typically of the kind discussed in section 2. The first helioseismic examples of the method used in this way were carried out by Sekii (1990, 1991) and Schou (1991). They were principally as feasibility studies, because the m dependence of the solar splitting was (and still is) known with only low resolution. A more detailed discussion of this, and some other methods, has been presented by Schou, Christensen-Dalsgaard and Thompson (1994). As with RLSF in one dimension, the resulting averaging kernels tend to suffer from being uncomfortably large near the surface of the sun. An example of an averaging kernel is illustrated in Figure 15.

Formally, OLA is equally straightforward to generalize: in principle one simply seeks to find a linear combination of data kernels that is optimally localized in two dimensions, by minimizing expressions similar to (6.17) or (6.18) in which the integrals are now over both r and μ.

The expense of the execution of RLSF can be controlled by restricting the number N of basis functions ϕ_j used to represent the solution. In practice, the number is very much less than the number I of data. Full OLA, on the other hand, requires the inversion of matrices of dimension $I \times I$, if all the data are to be used; so far as I am aware, it has never been executed directly on a usefully large data set. However, optimally localized 2-d kernels, using SOLA, have been constructed from a reduced data set, constructed by Christensen-Dalsgaard *et al.* (1995) with the singular-matrix preprocessing of Christensen-Dalsgaard and Thompson (1993) which I mentioned at the end of subsection 6.6.

7.2 Hybrid methods

To date, most frequency-splitting data have been presented in terms of an expansion of the kind

$$d_i = (\omega_{nlm} - \omega_{nl0})/2\pi = \sum_{j=1}^{J} a_j(n,l)\mathcal{P}_j(m), \tag{7.2}$$

where $\mathcal{P}_j$ is a polynomial of degree j whose coefficients depend on l; only the odd-degree terms are used in this discussion. Most commonly $\mathcal{P}_j(m)$ is $LP_j(m/l)$

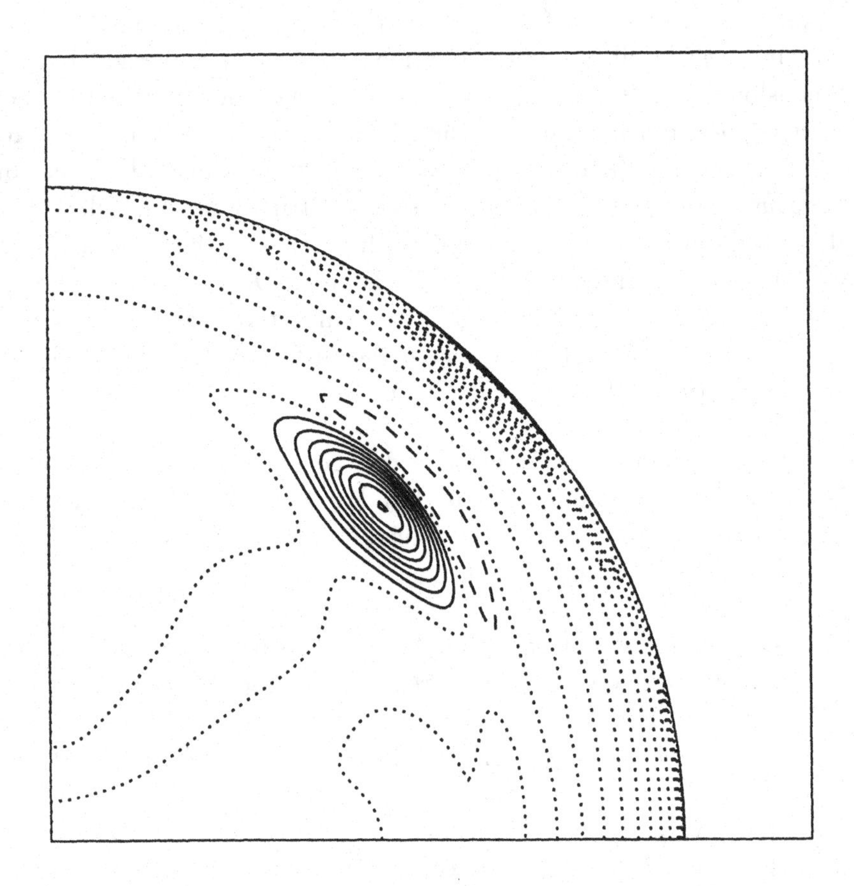

Figure 15: *Contours of an averaging kernel for angular velocity from 2-d RLSF of rotationally split frequencies of all modes with* $1 \leq l \leq 200$ *and* $1mHz \leq \nu \leq 4mHz$*, corresponding to the point* $r/R = 0.72$*,* $\mu = 0.707$*. The mode set contains a total of nearly* 2×10^5 *splitting frequencies from 2413 multiplets. The continuous contours are at positive values of the kernel, the dashed contours at negative values, and the dotted contours at zero (from Schou et al., 1994).*

or $LP_j(m/L)$, where P_j is a Legendre polynomial, and J is typically 6 or 12, yielding 3 or 6 terms for inversion for Ω. Such an expansion leads naturally to expanding the latitudinal variation of Ω in polynomials of μ^2, such as

$$\Omega(r,\theta) = \sum_{s=0}^{s_{\max}} \Omega_s(r) P_{2s}(\mu) . \tag{7.3}$$

The number $s_{\max} + 1$ of terms in this expansion is the same as the number of odd-degree terms in (7.2), namely $J/2$ if J is even. In that case $\Omega_{s_{\max}}(r)$ depends only on $a_J(n,l)$, and Ω_s depends only on linear combinations of a_{2j} with $s+1 \le j \le J/2$ (e.g. Brown *et al.*, 1989). Thus the inversion can be carried out as a sequence of one-dimensional inversions with respect to the radial coordinate r, starting with $\Omega_{J/2}$, and continuing sequentially with the coefficients Ω_s of the polynomials of lower and lower degree. Each inversion uses information from the previous one. Ritzwoller and Lavely (1991) have recommended using instead an expansion of $\Omega(r,\theta)$ based on vector spherical harmonics, for then there is an appropriate set of polynomials $\mathcal{P}_j$ in which to expand $d_{\rm i}$ that decouples the inversion. To my knowledge, no real advantage, other than mathematical elegance and perhaps a little reduction in work, has yet been demonstrated for this basis.

Unlike OLA and RLSF, there is a marked potential asymmetry in the manner in which the inversions are carried out with respect to the two coordinates μ and r. Inversions for the radial dependence of the coefficients Ω_s can be carried out by any of the one-dimensional procedures: OLA, (L)AI or RLSF. But the inversion with respect to μ is carried out by fitting a set of functions to the data, and at present that is typically accomplished as a data-processing operation by the observers. The task of the inverters (who are usually not the observers) is usually just to translate the data-expansion (7.2) into the corresponding expansion (7.3) of the angular velocity. The original Backus-Gilbert philosophy is almost always adopted for effecting that translation: to exclude functions in the annihilator. However, one can equally well introduce further terms into the expansion (7.3), and determine the coefficients by some regularization procedure.

Of course, the fitting of the expansion (7.2) might have been carried out as a RLSF to estimated eigenfrequencies, in which case, if RLSF is used also for the radial inversions, the sole asymmetry in the procedure is social. However, usually the fitting procedure is not so straightforward. The fitting is usually carried out as an integral part of the data processing, by fitting the expansion directly to a power spectrum, for example, rather than to estimated frequencies. The outcome is a product of the model of the oscillations that forms the basis, either explicitly or implicitly, of what in the Foreword I called the prior inversion. It is perhaps the transference of this step from the main to the prior inversion that induced Schou, Christensen-Dalsgaard and Thompson (1994) to dub what remains to be accomplished by the inverters a 1.5-d inversion. The final outcome, of course, is in principle no less two-dimensional than that of any of the other methods.

7.3 Asymptotic inversions

An asymptotic approximation to the rotational-splitting constraint (7.1) can be expressed in terms of a double Abel integral, one component with respect to a^2, as is equation (4.9) and the corresponding equation that results from the linearization of equation (4.8) about a reference model, the other with respect to μ^2. The formula can be obtained either as a direct asymptotic solution of the two-dimensional wave equation (Gough, 1993) or by substituting asymptotic forms for the spherical harmonics and the radial dependence of the oscillation eigenfunctions that appear in the exact linearized expression (7.1) and evaluating the integrals asymptotically (Kosovichev and Parchevski, 1988). It can be inverted simply by inverting the 1-d Abel integrals successively. The asymptotic splitting formula predicts that the splitting is a function of $w = \omega/L$ and m/L alone. Thus, once the data are fitted to a surface in $w - m/L$ space, the integrals in the inverse formula can be evaluated immediately. A prelimminary investigation of the method, using artificial data, was reported by Sekii and Gough (1993). The method works well throughout much of the sun, but it fails near the centre where the asymptotic representation is not so good. The results can be improved by adjusting the splitting data associated with the more deeply penetrating modes by an asymptotic estimate of the deviation of the Abel kernels from the full kernels.

7.4 $1 \otimes 1$ inversions

The asymptotic factorization of the splitting kernel into the product of a function of a^2 (and hence of r) and a function of μ motivated Sekii (1993a,b) to seek optimally localized averaging kernels in almost factorized form. The details of the procedure rely on the fact that the exact splitting kernels K^{i} can each be expressed as the sum of two terms, the first of which dominates almost everywhere and can be factorized into the product of a function of r and a function W^{i} of μ. Therefore one can attempt first to find, for every (n, l) multiplet, localized combinations associated with each m-component of the functions W^{i} that are essentially the same for all multiplets. Then, regarding those averages as functions of r, one can perform an inversion of them. In particular, one can carry out the second inversion also as an OLA, and thereby mimic a 2-d OLA as the product of two 1-d OLA. The outcome is, as usual, a linear combination of the original data kernels. Finally, having determined the coefficients of an optimal combination of the leading terms of the kernels, the corresponding exact averaging kernel can be computed as the same combination of the full kernels. In practice it is best to use SOLA for at least the localization in μ, because then one has more control over maintaining uniformity amongst multiplets (Sekii, 1995). The degree of localization of the target functions can be adjusted, depending on the depth of the intended localization in r.

Of course, one doesn't succeed in obtaining precisely separable kernels by this means, but that is beside the point. Separability of averaging kernels is not even a particularly desirable property, on the whole. It was invoked merely to motivate a procedure. The power of the method, as in the case of 1-d inversions, is to be judged by how well it produces readily interpretable results. That can be accomplished by inspecting the averaging kernels, one of which is illustrated in Figure 16. Unlike that in Figure 15 from RLSF, it is free from sidelobes, and is not contaminated by substantial peaks or troughs near the surface. I must emphasize that unfortunately Figures 15 and 16 are not directly comparable, because they were obtained from different mode sets. However, recent comparisons with identical mode sets confirm the impression given by these figures: that, by the usual criteria of localization and lack of negative sidelobes, $1 \otimes 1$ OLA is superior to 2-d RLSF.

The quality of the $1 \otimes 1$ averaging kernels is not so great nearer the centre of the sun, where the asymptotic separability is not well satisfied by the full kernels. However, there are not a great many deeply penetrating modes, so in the central regions one can carry out an essentially full two-dimensional OLA, using the individual deeply penetrating low-degree kernels coupled with combinations of high-degree kernels tailored to mimic the μ dependence of the kernels of low degree (Sekii, 1993b). Sekii is currently refining the details of the procedure.

7.5 A comment

As is the case with all inversions, 2-d inversions for angular velocity and corresponding inversions for aspects of structural asphericity should be carried out by many methods, because different inversion procedures can reveal different properties. The term 'full(y) 2-d' has been used in the literature recently, often in conjunction with RLSF. That unfortunate adumbration has misled readers into suspecting that it implies superiority. As is evident from my foregoing discussion, that is hardly the case. Indeed, by being fully two-dimensional, in the sense that the computer must disentangle the information in both dimensions without

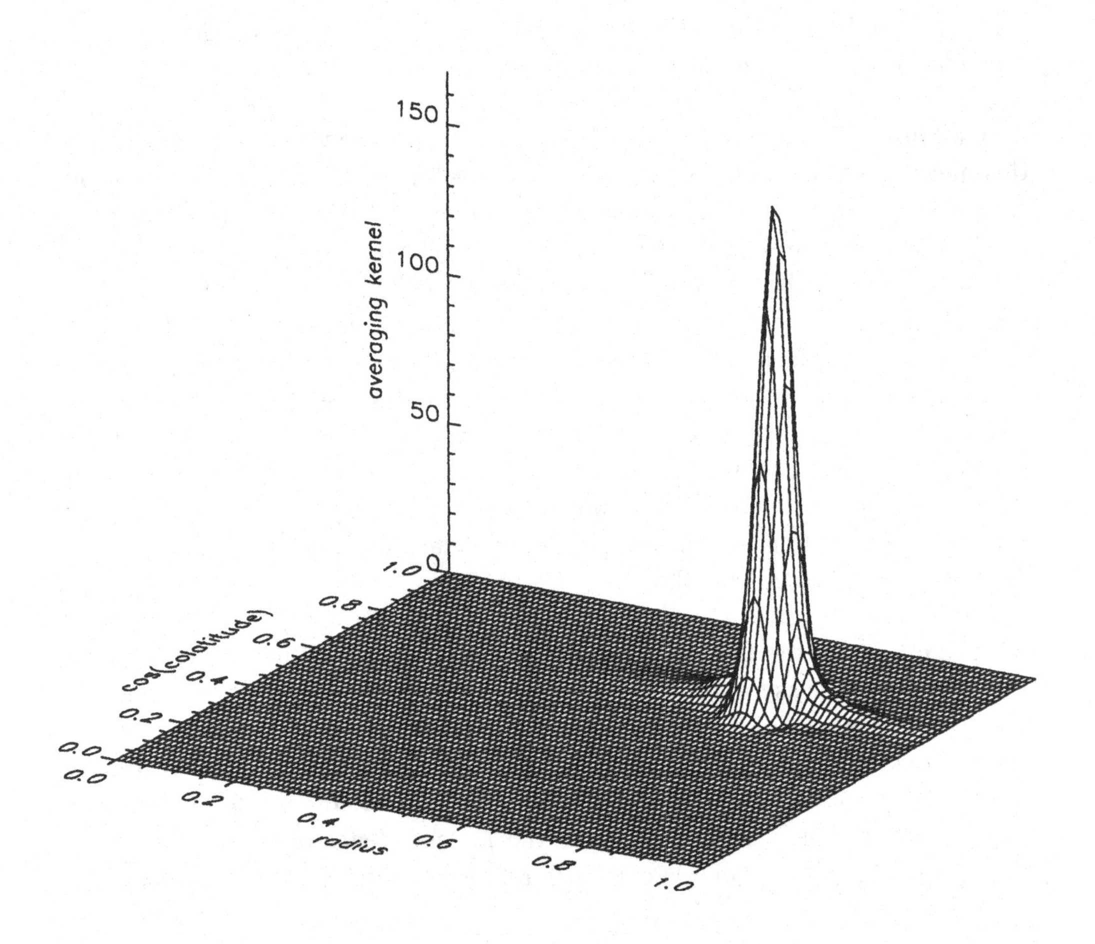

Figure 16: *Averaging kernel for a* $1 \otimes 1$ *OLA of angular velocity about the target point* $r/R = 0.72$, $\mu = 0.707$ *, the same point as that of Figure 7.1. The mode set is that of the second GONG hare-and-hounds exercise (Gough and Toomre, 1993); it contains* 7×10^4 *splitting frequencies of modes with degree up to 250. SOLA was used for the angular localization, and the Backus-Gilbert OLA in the radial direction (after Sekii, 1995). Note the absence of sidelobes.*

assistance from elsewhere, a severely inferior property of the RLSF procedure is immediately manifest: that it is computationally expensive. With sufficient terms in the expansions, hybrid inversions of the kind discussed in subsection 7.2 can be just as much two-dimensional, and in principle they might yield equally good results. The outcome of $1 \otimes 1$ OLA is also fully two-dimensional; it is merely the procedure that is not. As I have stressed before, the quality of an inversion is judged mainly by the quality of the inferences that can be drawn from it, and in many respects, of the methods discussed here, $1 \otimes 1$ OLA scores the highest, particularly when it is coupled with a 2-d OLA in the radiative interior. Moreover, it scores bonus points because it requires relatively modest computing resources.

It is perhaps worth remarking also that further hybridization is possible, by using $1 \otimes 1$ OLA or RLSF with the coefficients a_j in the expansions (7.2), or combinations of such coefficients and individual mode frequencies. All that is required is the kernels associated with those coefficients. Determining them precisely is not necessarily straightforward. At today's level of sophistication it requires relating the data processing that generated the a_j to the mode kernels. This is approximated by finding a combination of frequencies that fits the coefficients, perhaps by least squares, and constructing the same combination of mode kernels. But in the future a more realistic model of the acquisition and subsequent processing of the data may be called for, which could require dynamical studies beyond those of normal modes.

8 TOWARDS REALISM

I conclude these lectures as I began: by raising some of the complicated physical issues that have been ignored, or that have at least been withdrawn to the subconscious. Many of them make only small contributions to the eigenfrequencies. But measurements are becoming much more precise, and, we hope, more accurate; and the inferences of the properties of the sun that we are trying to make are becoming more and more subtle. Many of these inferences would be seriously invalidated if those contributions to the eigenfrequencies were not accounted for. Therefore, future research must face a plethora of complications that in the past have been regarded as being insignificant. The objective will be partly to eliminate them, and thereby improve the accuracy with which we can make simple deductions about the variation of the quantities we normally consider in theoretical models of the sun. However, in the process of eliminating them we shall gain information about them, and hence be in a position to use the sun as a laboratory for studying a range of physical problems under conditions that cannot be achieved on earth.

8.1 Uncertain influence of the surface layers

What interests most stellar physicists is the spherically averaged structure $\bar{\mathbf{X}}(r)$ of the sun, particularly those aspects of it that are described in the so-called standard solar models. Those are determined by the average multiplet frequencies, ω_{nl}.

I described in section 5 how knowledge of the eigenfrequencies imposes integral constraints on the possible functions $\bar{\mathbf{X}}$. I pointed out that realistic boundary conditions, rather than idealizations of them which are often used implicitly, if not explicitly, in the numerical computations, render the forward problem non-self-adjoint, and formally invalidate the kernels in the linearized constraints. However, I did not consider the implications of that explicitly, because, as I discussed in the introduction, there are uncertainties in the physics of the outer layers of the sun, particularly in the upper superadiabatic boundary layer of the convection zone, that disconnect knowledge of the eigenfunctions deep in the interior of the sun from conditions at the boundary. Viewed from the interior, the effect is as to render uncertain the boundary conditions that should be applied to the solutions of a differential equation whose structure deep down is known. Evidently, this renders uncertain the actual values of the eigenfrequencies, which is why naive direct comparisons of the values of theoretical eigenfrequencies with observation is of very limited use at best, and more typically in the hands of the unwary, is outright misleading. However, knowledge of the functional form of the dependence of the uncertainty on n and l can permit one to eliminate that uncertainty from the inversions. This is accomplsihed essentially by finding combinations of the eigenfrequencies that are independent of the boundary conditions, at least when the differences between the boundary conditions that are applied and those that should have been applied are small enough for linearization to be valid.

To see how this comes about, consider the linearized variational integral relation (5.2), which in the absence of rotation becomes

$$\delta \ln \omega_{\mathrm{i}} \simeq \frac{\delta \mathcal{C} - \omega_{0\mathrm{i}}^2 \delta \mathcal{I}}{2\omega_{0\mathrm{i}}^2 \mathcal{I}}, \tag{8.1}$$

where $\delta\mathcal{C}$, $\delta\mathcal{I}$ and $\mathcal{I}$ are integrals. I omit rotation solely for simplicity of exposition. Provided the horizontal scale of variation, $l^{-1}R$, of the eigenfunction of mode $\mathbf{i}$ is much greater than the acoustically relevant scale heights of the uncertain region, which is certainly the case if l is a few hundred or less (the precise value depends on the required level of precision, which is continually increasing), then the eigenfunctions, regarded as functions of ω and l, are independent of l in that region. Consequently, the uncertainties in the perturbation integrals $\delta\mathcal{C}$

and $\delta\mathcal{I}$ in the numerator of the right-hand side of equation (8.1) are functions of ω alone, and therefore the uncertainty in the relative frequency perturbation $\delta\ln\omega_{\rm i}$ is of the form $F(\omega)/\mathcal{I}$, where F depends on ω but not explicitly on l. Equation (8.1) can therefore be rewritten as

$$\delta\ln\omega_{\rm i} = \frac{\delta_0\mathcal{C} - \omega_{0\rm i}^2\delta_0\mathcal{I}}{2\omega_{0\rm i}^2\mathcal{I}_0} + \frac{F(\omega)}{\mathcal{I}_0}, \tag{8.2}$$

where $\mathcal{I}_0$ is the modal inertia of the reference model, and $\delta_0\mathcal{C}$ and $\delta_0\mathcal{I}$ are the differences between the integrals $\mathcal{C}$ and $\mathcal{I}$ for the sun and the reference, those integrals being explicit functionals of those aspects of the sun or the reference model that have been accounted for in the reference model. In other words, the explicit forms for the kernels $\mathbf{K}^{\rm i}$ are now deemed to be correct, and all the physics that has been either missed or misrepresented in the surface layers has been absorbed into the unknown function F. Inversions can now be carried out using the constraints (8.2) instead of (8.1). They are weaker than the constraints (8.1) because F is regarded as an arbitrary function of ω. Therefore the inferences that are drawn from them, though apparently less tight, are more reliable, because they are much more weakly dependent on the uncertain physics of the surface layers. Inversions using these constraints were first carried out by Dziembowski, Pamyatnykh and Sienkiewicz (1990) using RLSF and by Däppen *et al.* (1991) using OLA.

The recognition that there is a contribution $F/\mathcal{I}$ to the constraints (8.2) that at present is intrinsically irreducible makes us realize that even after iteration the numerical values of the eigenfrequencies of the reference model (whose interior closely resembles that of the sun) cannot be made to be arbitrarily close to the frequencies of oscillation of the sun, whatever the function $F(\omega)$. Therefore, with increasing precision, the linearized first term on the right-hand side of equation (8.2) may not be adequate. However, the presumed functional dependence of $\delta_0\mathcal{C}$ and $\delta_0\mathcal{I}$ on the eigenfunction $\boldsymbol{\xi}_{0\rm i}$, and the equations satisfied by $\boldsymbol{\xi}_{0\rm i}$ interior to the uncertain outer layers, are probably reliable. Consequently, the values of the expressions for $\delta_0\mathcal{C}$ and $\delta_0\mathcal{I}$ can be improved by using eigenfunctions $\boldsymbol{\xi}_{\rm i}$ computed against the background state of the reference model but with the observed frequencies $\omega_{\rm i}$; this will, of course, require the relaxing of the theoretical boundary conditions. By so doing the seismically accessible aspects of the reference model could in principle be made to converge to those of the sun by iteration, provided, of course, there were no other errors in the physics of the reference model.

In theory, the kernels $\mathbf{K}^{\rm i}$ obtained by this procedure would depend on the observations, which would require them to be recomputed for every set of data. To avoid that, one can expand to as many orders as desired the eigenfunctions $\boldsymbol{\xi}_{\rm i}$ about those, $\boldsymbol{\xi}_{0\rm i}$, of the reference model (with the reference eigenfrequencies):

$\xi_i = \xi_{0i} + \xi_{\omega i}\delta \ln \omega_i + ...$, where $\xi_{\omega i} \equiv (\partial \xi_i / \partial \ln \omega)_l$, ... ; substitute into $\mathbf{K}^i$ and regroup the terms to make $\delta \ln \omega_i$ the subject of the modified equations (8.2). By this device a revised set of kernels that are independent of the data is obtained, which can be entered into inversion procedures in the usual way.

It goes without saying that these considerations concerning the surface layers and the elimination of their influence apply equally forcefully to aspherical inversions.

8.2 Correlated errors

All inversions depend not only on the data d_i, but also on the statistics of the errors. The latter are often very difficult to ascertain. Indeed, the recognition of that fact must render suspect any declared preference for the outcome of a tradeoff criterion that depends heavily on those statistics, and is the reason why helioseismologists have sometimes resorted to other criteria, such as (yet again) smoothness of the final represeentation (e.g. Gough, 1985).

The analysis of the data from observations usually produces estimates of the values of the variances σ_i^2 of the probable errors ϵ_i in those data. As is evident, I trust, from my foregoing discussion, a tradeoff with insufficient error suppression usually results in erratic inversions, which can usually be recognized. What is much more dangerous, however, because it is more difficult to recognize, is error correlation. Estimation of the covariance matrix E_{ij} requires a very careful analysis of the data-acquisition and data-analysis procedures, which is complicated and is itself fraught with uncertainty. Error correlation is difficult to assess, and is usually ignored. It is rare to do more than merely caution against the recommendation to assume that the errors are independent. Yet we are aware that extracting frequencies from power spectra is susceptible not only to bias but also to correlation when power from different unresolved modes and their aliases overlap. Schou (1993) and Bachmann *et al.* (1993), for example, have discussed evidence for bias in their rotational splitting data, and have investigated possible causes. If a cause for bias can be identified and analysed, then one has a hope of estimating not only its influence on the expectations of the data values, but also the correlations it imposes on the random errors.

To illustrate the danger that can befall a naive inverter, and the extent to which it can be averted with a good estimate of the covariance matrix, I report the result of a blind test with artificial data carried out recently in collaboration with T. Sekii. I take the opportunity at this point to digress a little to remark on the importance of tests being blind. It is very often the case that, when one knows the function **X** from which artificial data have been created, one can just perceive amid the noise in the inversions subtle features of that function, yet correctly recognize otherwise similar features in the inversion which one knows to be no more than a product of the noise. However, when one does not know **X**, one needs to study the robustness of any inference to variations in the inversion procedure, including the assumptions that are adopted in the formulation of the integral constraints, and to judge very carefully the outcome. Of course, this is true too of inversions of real data. But in that case the outcome is to some extend a matter of opinion. When the 'true' answer is known by someone else, the accuracy of the inverter's interpretation can be tested. As is evident from the 'hare-and-hounds' exercises that were carried out in preparation for inverting GONG data (Gough and Toomre, 1988, 1993), one can both overinterpret the data by insisting on the reality of a feature which is actually a product of noise, and underinterpret it by rejecting a feature which is judged to be insignificant (even when that feature is explicitly pointed out to the hound by the hare). This emphasizes how inversion is not merely a mathematical procedure, which any competent helioseismologist adequately equipped with computing facilities can undertake, but entails also the interpretation of the results of the mathematics, which is much more difficult.

I return now to the particular blind test concerning correlated errors. To keep matters simple, the artificial data were from a scalar function in one dimension: specifically, rotational splitting produced by a spherically symmetrical angular velocity $\Omega(r)$. Indeed, to make matters really simple, Ω was taken to be constant, $\Omega = \Omega_0$, though of course the inverter was not told that. He was, however, given the value of Ω_0 and asked to express $\Omega(r)$ in units of it, in order that his inferences could readily be assessed. The inverter was aware that 'random' errors had been added to the data, but he was not told their statistical properties. Instead, he was asked to estimate the uncertainty in his inferences under the assumption that there was no confusing sharp feature in the function Ω. In this pilot study, to ensure (in the hare's mind) a spurious outcome, a highly coherent covariance matrix was adopted: $E_{ij} = \sigma^2 A_{ij}$, where

$$A_{ij} = \cos[k^2(\hat{w}_i^2 - \hat{w}_j^2] \,, \tag{8.3}$$

where $\hat{w}_i = \nu_i/L$, the cyclic frequency ν_i being expressed in mHz. The value of the constant k, which is unrelated to the asymptotic wavenumber k in sections 4 and 6.3, was taken to be 7.5. With this choice, according to asymptotic theory, an undulating correlation between data errors associated with modes of varying penetration depth should arise, with approximately one wavelength spanning the extent of the convection zone.

As anticipated, the inverter assumed the errors to be uncorrelated – he could hardly proceed otherwise without more information. He inverted by OLA and

RLSF, using a wide range of tradeoff parameters. An example of a function fitted by least squares, regularized with the penalty function $P = \int(\Omega'')^2 dr$, is illustrated in Figure 17a; OLA is similar. Included, as dashed lines, are 1-σ estimates of the pointwise uncertainty. There was also an estimate of the variance of the errors in the frequency splitting, but that is difficult to interpret. Less severely regularized fits yield a similar representation, but with a superposed rapidly oscillating component and a correspondingly augmented error estimate. The inverter, having been assured of there being no complicated feature in Ω, concluded that Figure 8.1a probably represented Ω more-or-less within the limits indicated in the figure.

The inverter was then told that the data errors were correlated, and was given the functional form (8.3) of the covariance matrix. He was invited to repeat the inversion, estimating σ in the process. An example of RLSF is shown in Figure 17b, with the same tradeoff as for Figure 17a. (I should point out that in carrying out RLSF the definition (6.6) of the quadratic form χ^2 must be generalized by replacing the implied diagonal matrix whose elements are σ_i^{-2} by the inverse of the covariance matrix). It agrees everywhere with the correct function $\Omega/\Omega_0 = 1$, within about two standard deviations of the estimate 1-σ error. Moreover, the estimated value of the factor σ^2 relating E_{ij} to A_{ij} was within 10 per cent of the actual value.

It is unlikely that such a covariance matrix so highly coherent over the entire data set would arise in practice. Nevertheless, this exercise does demonstrate, admittedly in an exaggerated way, how error correlation can deceive by producing a plausible yet spurious result. It is interesting to note that in this example the false assumption of no correlation amongst the errors led not only to a spurious inference, but also suppressed the estimate of the errors in the representation of $\Omega(r)$, as indicated by the distance between the dashed curves. On a positive note, the exercise also demonstrates how well the inversion procedures can work when the error statistics are well estimated. It emphasizes how important it is not only for the magnitudes of potential data errors to be estimated, but also the correlations between them. Although the carrying out of those estimates can be a formidable task, perhaps even more formidable that obtaining the values of the data, it is evident that it is absolutely essential for it to be accomplished if reliable inferences are to be drawn.

8.3 On the stratification of the convection zone

There has been considerable discussion of the stratification of the convection zone. The principal motivation has resulted from the fact that beneath the up-

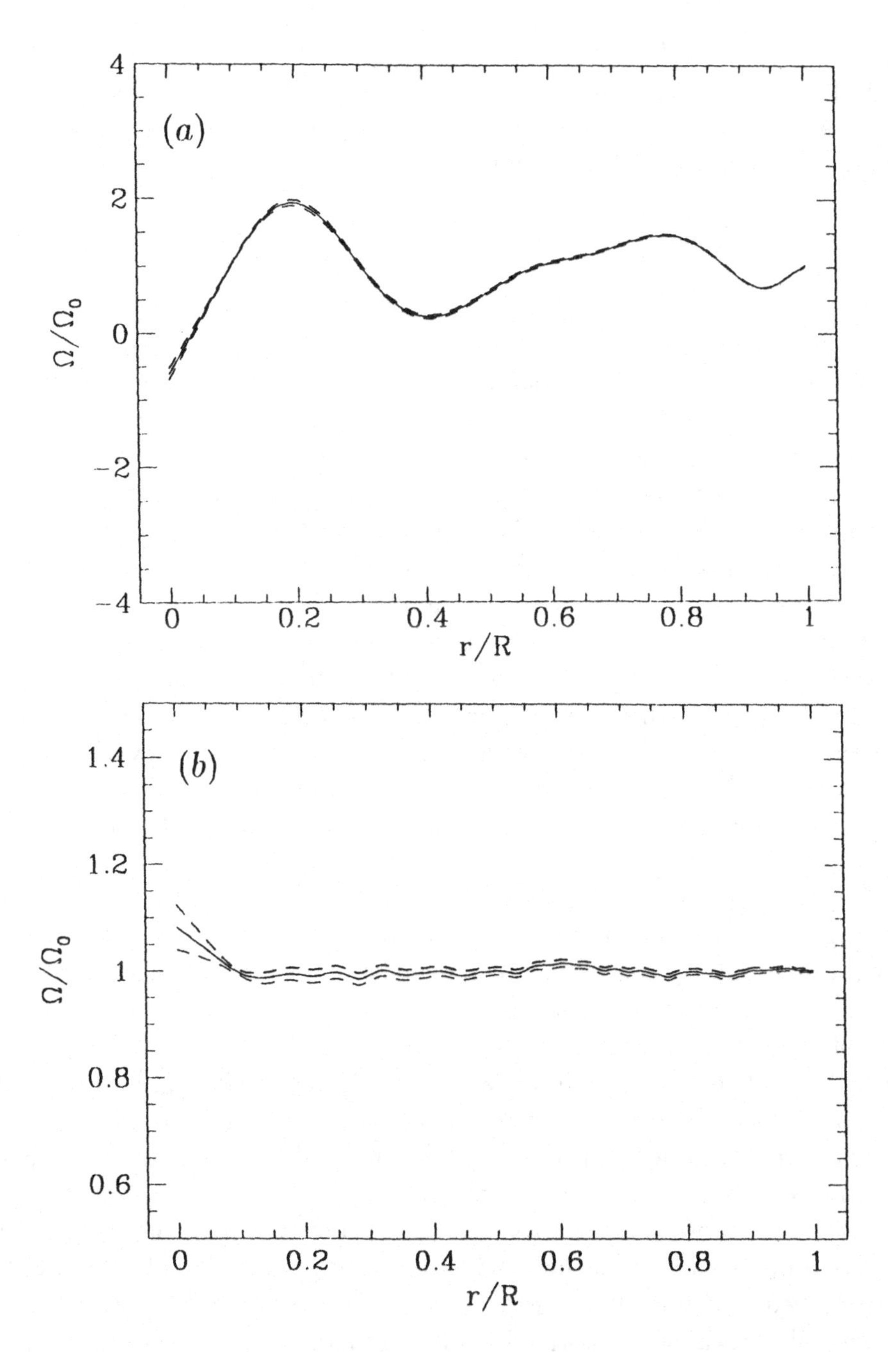

Figure 17: *Inversions by RLSF of artificial rotational splitting frequencies computed from the constant angular velocity $\Omega(r) = \Omega_0$. Correlated random errors with covariance matrix $E_{ij} = \sigma^2 A_{ij}$ were added to the data, where A_{ij} is given by equation (8.3) and $\sigma = 0.05\Omega_0$. (a) inversion in which the errors were assumed to be independent, (b) inversion in which the form A_{ij} of the correlation matrix was correctly taken into account.*

per superadiabatic boundary layer the zone is very close to being adiabatically stratified. This constraint permits one to relate the seismically accessible stratification directly to the thermodynamics of the gas, and thereby to enable one to measure certain aspects of the equation of state.

The earliest approach was by asymptotic inversion. Specifically, it was shown that

$$\widehat{W}(r) := \frac{r^2}{G\widetilde{m}} \frac{dc^2}{dr} = \Theta := \frac{1 - \gamma_\rho - \gamma}{1 - \gamma_{c^2}} \tag{8.4}$$

in the region in which the stratification is adiabatic, where $\gamma_\rho := (\partial \ln \gamma_1 / \partial \ln \rho)_{c^2}$ and $\gamma_{c^2} =: (\partial \ln \gamma_1 / \partial \ln c^2)_\rho$, both derivatives being taken at constant chemical composition, where $\widetilde{m}(r)$ is the mass contained within the sphere of radius r, and G is the gravitational constant. Thus a seismic inference of c^2 and $\widetilde{m}$ is equivalent to the inference of the variation of the value of the purely thermodynamic function Θ through that region. The sound speed c^2 is obtainable directly from a direct asymptotic inversion of the kind described in section 4, and $\widetilde{m}(r)$ can be obtained from it by integrating the equation of hydrostatic support. Because the mass of the layers overlying any point in the convection zone is small, particularly if that point lies in the upper parts of the zone – the total mass of the convection zone is only about 0.02 $M_\odot$ –, the estimate of the deviation of $\widetilde{m}$ from $M_\odot$ need not be sophisticated; indeed, it is quite common practice to use for simplicity the diagnostic $\widetilde{W}(r)$, in which $\widetilde{m}$ is replaced by $M_\odot$ in equation (8.4).

The original intent was to measure the helium abundance Y in the convection zone, essentially by comparing $\widetilde{W}$ inferred from solar data with a grid of corresponding functions inferred using the same combination of eigenfrequencies of a grid of theoretical model envelopes in which Y and the mixing-length parameter α vary. This is a straightforward example of a model calibration that relies on asymptotic analysis only to design a combination of data that approximates the thermodynamic quantity Θ, and which is only very weakly dependent on uncertainties in the models that do not relate directly to that quantity. The errors in the asymptotic representation of $\widetilde{W}$ do not influence the calibration directly, but only indirectly to the extent that they permit a small contamination by extraneous factors that might not have been accurately represented by the theoretical models. I stress this because the method has been erroneously judged as relying directly on the ability of the asymptotic analysis to reproduce the function $\widetilde{W}$, thereby limiting the accuracy with which the calibration can be performed. That judgement is false: the calibration itself uses only theoretical eigenfrequencies that are computed numerically; the asymptotic analysis is used to interpret the result of the calibration.

Determining $\widetilde{W}$ requires an accurate knowledge of the oscillation frequencies because it contains a derivative of the sound speed, and hence requires yet more subtle combinations of data, with greater cancellation and consequent greater susceptibility to error. This is evident from the asymptotic formula, which requires yet another differentiation of the function $\mathcal{F}(w)$ defined by equation (4.9):

$$\widetilde{W} = \frac{2r^3a^2}{GM_\odot}\left[1 + \left(\frac{\mathrm{d}\ln r}{\mathrm{d}\ln a}\right)^{-1}\right], \tag{8.5}$$

where $r(a)$ is given by equation (4.11) and

$$\frac{\mathrm{d}\ln r}{\mathrm{d}\ln a} \sim -\frac{2}{\pi}\int_{a_s}^{a}(w^2-a^2)^{-1/2}\left[\frac{a}{w}\frac{\mathrm{d}}{\mathrm{d}w}\left(w\frac{\mathrm{d}\mathcal{F}}{\mathrm{d}w}\right)+\frac{w}{a}\frac{\mathrm{d}\mathcal{F}}{\mathrm{d}w}\right]\mathrm{d}w$$
$$-\frac{2}{\pi}\frac{\mathrm{d}\mathcal{F}}{\mathrm{d}w}\bigg|_{a=a_s}\frac{a}{\sqrt{a^2-a_s^2}}. \tag{8.6}$$

As is evident from the discussion in section 6 of the inversion coefficients c_i, any other (nonasymptotic) inversion method will require similar data combinations.

I should stress that the original application of this method to solar data (Däppen, Gough and Thompson, 1988) used only the leading two terms in the expansion (4.8) to establish $\mathcal{F}(w)$, and was therefore more susceptible to erroneous modelling of the surface layers than is nowadays necessary. A more recent application (Baturin *et al.*, 1994), using all the terms quoted explicitly in equation (4.8), yields a much more reliable representation of $\widetilde{W}$. However, at present it cannot straightforwardly be converted into a helium abundance, because the model calibration cannot reproduce the solar $\widetilde{W}$. (I speak loosely here; by $\widetilde{W}$ I now mean that combination of frequencies that would reproduce the value of $\widetilde{W}$ were the asymptotics to be exact.) This highlights a fundamental error in the theoretical modelling: either the solar convection zone is not both adiabatically stratified and supported hydrostatically almost totally by the gradient of gas pressure, which is most unlikely, or the equation of state is incorrect. Granted either, the calibration for Y must be in doubt.

The disparity is illustrated in Figure 18, in which the solar $\widetilde{W}$ is compared with the corresponding values (again computed as corresponding combinations of p-mode frequencies) for best-fitting model envelopes. In Figure 18a the theoretical model was computed using an equation of state based on the so-called chemical picture (e.g. Christensen-Dalsgaard and Däppen, 1992) in which straightforward Debye-Hückel screening was adopted in the partition function for bound states but no additional truncation was used; in Figure 8.2b the internal partition functions for H, He and He^+ were truncated by a simple formula which depends on the ratio of the mean interparticle distance to the corresponding Bohr radii r_{Bi} of species i (H, He or He^+), and which prevents bound species

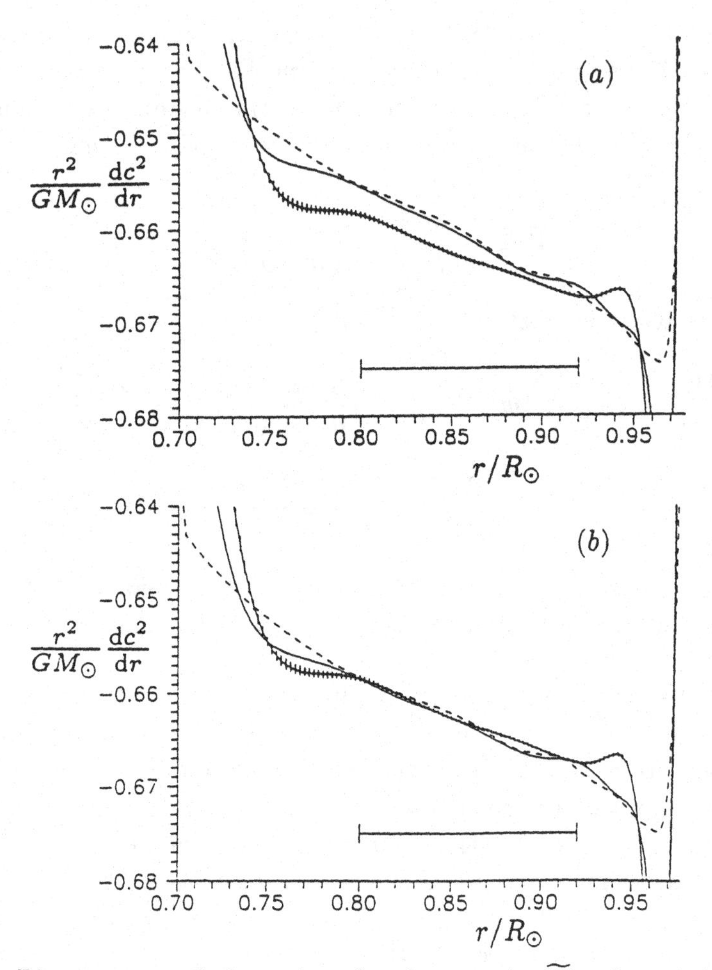

Figure 18: *Direct asymptotic inversions for the quantity $\widetilde{W}$ using 1415 frequencies between 1.1 and 4.0 mHz of high-degree modes of Libbrecht et al. (1990) with lower turning points r_1 satisfying $r_1/R < 0.997$. The sound speed was obtained by inverting the function $\mathcal{F}(w)$ obtained by fitting to the data all the terms explicitly included in the asymptotic expression (4.8). The curve with error bars was obtained from solar data, the other continuous curve from the eigenfrequencies (of the same mode set) of a model solar envelope, for which the exact function $\widetilde{W}$ is given by the dashed line. The inversions fail near the base of the convection zone because the number and distribution of data constraining the fit of $\mathcal{F}$ (and, more pertinently, its first and second derivatives) is inadequate. Additionally, there are errors near the upper extremity arising from the lack of knowledge of the structure of the surface layers, whose influence diminishes with depth. The solar envelope models were computed with an equation of state with Debye-Hückel electron screening but (a) no explicit 'pressure ionization', (b) 'pressure ionization' produced by a reduction to bound-state occupation numbers by the factor exp $(-\bar{r}^2/\alpha^2 r_{\mathrm{B}i}^2)$, where $\bar{r}$ is the mean interparticle distance, $r_{\mathrm{B}i}$ is the appropriate Bohr radius and $\alpha = 1.67$. Neutral helium was treated in the hydrogenic approximation (from Baturin et al., 1994).*

from overlapping severely by inducing ionization – so-called pressure ionization – when the density is high. Attention should be directed in the figures to the region $0.80 < r/R_\odot < 0.92$ indicated by the horizontal bars, beneath the HeII ionization zone, where the comparison is thought to be most reliable (i.e. where the asymptotic inversion for $\widetilde{W}$ most closely reproduces the actual value when theoretical eigenfrequencies are used, and therefore, by inference, where the model calibrations are the least contaminated by extraneous factors). Agreement is substantially better, though not perfect. Several calibrations were carried out, and the best fit was obtained for effective radii of the bound species of about $1.7r_{\mathrm{B}i}$. It would be premature at this stage to conclude that the effective radii of bound species in a dense (by terrestrial standards) plasma have been measured, for there remain other uncertainties in the equation of state that need to be investigated. What is really exciting about this result is that it shows that the influence of small corrections to γ_1 by phenomena such as the quantum exclusion in what, by normal standards of quantum degeneracy, is a dilute plasma, can be detected in the solar oscillation spectrum. This opens the door for further utilization of the sun as a physics laboratory.

I must point out that the investigation I have just described was not the first to produce seismic evidence for the inadequacy of modern equations of state. A sequence of direct attempts to obtain Y by OLA and RLSF of constraints of the form (5.29), but for technical reasons with c^2 replaced by $u = p/\rho$, modified by the addition of $F(\omega)/\mathcal{I}_0$ as in equation (8.2) to remove most of the contamination from the uncertain surface layers, revealed inconsistencies that almost demanded an inadequacy of all the equations of state that were used (Kosovichev *et al.*, 1992), a conclusion that was upheld by Dziembowski *et al.* (1992). More recently Elliott (1995) has carried out OLA inversions for γ_1; he used the convective stability parameter $A^* = \gamma_1^{-1} d\ln p/d\ln r - d\ln\rho/d\ln r \quad (= v)$ as a subsidiary variable, because A^* vanishes throughout most of the convection zone where the stratification is nearly adiabatic, and the optimal averages of γ_1 are quite weakly contaminated by errors in the stratification. He compared inversions of solar data with inversions of corresponding frequencies of models based on the MHD prescription (Mihalas *et al.*, 1988) of the equation of state and the Livermore prescription (Rogers, 1986; Iglesias, Rogers and Wilson, 1987, 1992). The MHD prescription was found to be inadequate, as Dziembowski *et al.* (1992) and Kosovichev *et al.* (1992) had found previously; however, provided the total heavy-element abundance Z is permitted to be a (constant, yet) adjustable unknown quantity in the convection zone, the Livermore prescription appears to agree with the data to within the current uncertainties. Of course, these inversions also yield a value of Y in the convection zone. Like the model calibrations based on asymptotically designed combinations of frequencies, they are relatively free from contamination by the influence of outer layers of the sun on the frequencies. But, like all attempts to determine Y seismologically, they do depend on the equation of state.

An additional calibration for Y has been carried out using the asymptotic phase $\tilde{\alpha}(\omega)$ of equation (4.7), which is strongly influenced by the uncertain outer layers of the sun (Vorontsov *et al.*, 1991; Christensen-Dalsgaard and Pérez Hernández, 1991, 1992, 1994a,b; Basu and Antia, 1995). It is discussed in the previous chapter in this book by Christensend-Dalsgaard. As Christensen-Dalsgaard points out, its reliability rests on having successfully eliminated those aspects of the phase that depend on the uncertainties in the outer layers, and although Christensen-Dalsgaard offers cogent arguments to justify why that might be so if the physics of the theoretical models whose properties he studies have not been misrepresented too seriously, there must remain some doubt. Therefore, this calibration for Y is less reliable. That is not to belittle the analysis: its purpose is primarily for a study of the outer layers, both for its own sake and to improve our ability to study the interior. That the calibration gives a result which is very similar to those of the other methods is therefore more a confirmation of the analysis of the influence of the outer layers than it is of the reliability of the value of Y that it yields.

Whilst I am on the subject of the surface layers, it is worth pointing out that some aspects of the lateral inhomegeniety produced by the convection zone should be learnt from a study of the f modes. Because those modes, at least in their pure form, would have frequencies ω that at high degree become essentially independent of the stratification of the sun were the sun to be laterally homogeneous, namely $\omega^2 = \omega_{\mathrm{f}}^2 := gL/R$, one might hope that inhomogeneity could be measured from the deviations of ω from ω_{f}. Then, what has been learnt might then be applied to inferring the influence on p-mode frequencies. Murawski and Roberts (1993a,b) and Rosenthal *et al.* (1995) have made some progress in studying this issue. However, it may be complicated by a tendency for the modes to concentrate about regions of rapid vertical density variations (Rosenthal and Gough, 1994).

8.4 Mode distortion

There are many other complicating processes, present in the real sun but often excluded from our current models of it, which influence the eigenfrequency spectrum. They relate not only to the microphysics but also to the macroscopic dynamics: large-scale flow, magnetic fields, temporal variation of the background state, to mention but a few. I shall not discuss them here, but I mention them to emphasize that we are very far from having even developed the tools necessary to infer conditions in the solar interior to the level of precision and detail required for addressing some of the important issues concerning the structure and dynamics of the sun – the neutrino problem and the dynamics of the solar cycle, for example – not to mention that we still await adequately extensive and accurate data. Helioseismology is only just beginning. However, I do not wish the reader to be left with the impression that frequencies are everything. If they were, then, as I discussed at the beginning of these lectures, there would be many aspects of the structure that would never be unravelled even in principle by seismic means alone. But there is other information in the oscillations, namely the alignment of the eigenfunctions and the deviation of the eigenfunctions from the ideal spherically harmonic structure that they would have had had the sun been spherically symmetrical, and the deviation of the temporal behaviour from being purely sinusoidal oscillations. The former are produced by asphericity, the latter by temporal variation of the background state; both are produced also by nonlinear interactions between the modes. Without information of such deviations, we would never be able to investigate some of the most basic issues, such as distinguishing between a magnetic field, a flow field and a sound-speed inhomogeneity, determining where in the sun the modes are excited, or even detecting a north-south asymmetry in the angular velocity.

There are several techniques that have been tried or are under development for detecting such properties of the eigenfunctions. I have already mentioned the (unsuccessful) attempt to detect a signature of a nonlinear periodic g mode in the core as a frequency modulation of p modes. I should mention also the fact that the quite common mistake that observers make in confusing east and west has caused them to notice, once the mistake has been rectified, that after projecting the solar data onto spherical harmonics, temporal power spectra have cleaner peaks when the axis of the spherical harmonics is aligned correctly with the axis of the photospheric rotation. That implies that the alignment of the eigenfunctions by the dominant genuine dynamical symmetry-breaking agent, which is a combination of internal rotational shear and asphericity of the outer layers of the sun, is aligned with the rotation of the photosphere at least to within twice the angle of inclination of that rotation axis from the normal to the plane of the ecliptic. It would be interesting to determine whether the internal angular velocity itself is also so aligned, and if not, to determine the direction of the total angular-momentum vector.

There are several methods under development for determining the distortion of the eigenfunctions. One obvious naive way might be to project spatially resolved seismic data onto functions that deviate from spherical harmonics – the putative eigenfunction – and to seek the deviation that yields peaks in the temporal power spectrum with the smallest sidelobes. That is likely to be difficult, partly because one is always doomed to have leakage into the projection from other modes (so long as observations are from a single vantage point), and partly because the background state is not invariant in time, both due to convection and other forms of motion that are distinct from the oscillations, and due to the ensemble of all the modes of oscillation other than the one onto which one is trying to project the data.

The remaining methods are all local. Either observations over only a limited patch of the visible solar disk are taken into account for analysis, or observations over the entire disk are processed in such a way as to eliminate certain modes in a patch, and the remaining signal is analysed locally. Conceptually, the most straightforward procedure, at first sight, is so-called ring analysis (Hill, 1988). High-degree p modes are regarded as an ensemble of horizontally propagating waves trapped in the waveguide formed by the spherical shell between the turning points. A power spectrum of the motion with respect to the two horizontal spatial dimensions and time has maximum power on surfaces whose shape is that of a set of distorted nested trumpets about the time axis, which intersect surfaces of constant frequency ω in rings. Were the sun to be spherically symmetrical, those rings would be circular, and centred at the origin. But advection displaces them and elongates them; and further distortion is produced by horizontal inhomogeneity of the background state, including shear, and by a horizontal component of a magnetic field. To date, only the displacement has been measured. This can be related to a mode-weighted mean horizontal flow. By combining such means corresponding to different modes, an inversion for the depth variation of that flow can be carried out by one of the standard procedures. Finally, by considering different patches on the surface of the sun, Patrón *et al.* (1995) have obtained a three-dimensional representation of the subsurface flow.

An obvious disadvantage of this method is that it is highly susceptible to the effects of beating between waves of slightly different wavenumber or frequency, which can bias the outcome. How serious that bias might be has not been properly investigated. An attempt to overcome it, and at the same time to measure in greater detail the spatial variation of the wavenumber at any given frequency, has been made by Gough *et al.* (1991, 1993) and Julien *et al.* (1995a,b), who regard the oscillations as a superposition of waves represented by sinusoidal functions of phases that deviate only slowly from linearity, with amplitudes that also vary slowly. Variation of the background state is determined from analysis of the variation of the phases. Interference between parallel propagating waves can be eliminated by removing that part of the phase gradient that travels with the group velocity, leaving a remaining contribution whose spatial variation is stationary and which can in principle be inverted for horizontal velocity and structural inhomogeneity. However, interference between obliquely propagating waves travels with any velocity less than the group velocity and, with observations of necessarily limited duration, cannot easily be eliminated. Whether there are other means by which it can be removed is an issue of current research. But in the meanwhile it should be possible from observations of the entire solar disk to isolate at least the genuine sectoral modes, which propagate around the equator and are essentially parallel waves, and thereby infer longitudinal variation near the equator from the phase analysis. It might also be possible to consider oblique analogues, which are sectoral modes about axes other than that about which the sun rotates, and which rotate about the rotation axis at an appropriate rate, but that would entail accounting for the distortion of the modes out of the rotating planes perpendicular to those axes, brought about by the rotational shear. Phase analysis of even genuine sectoral modes has not yet been carried out on solar data.

A third method is called time-distance helioseismology (Duvall *et al.*, 1993), and was inspired by the delay-time measurements of seismic waves produced by earthquakes. Instead of analysing disturbances on the solar surface as sequences of images each taken at a given instant of time, disturbances at distant points, A and B, on the surface are cross-correlated with a time lag, which, for any given frequency, represents the travel time modulo the oscillation period both of the phase of the surface motion (which is an interference phenomenon) and of the phase of the actual wave which asymptotically travels along the ray path (at the phase speed). Because there are many similar ray paths, waves on adjacent paths interfere, inducing phase jumps at the caustic surfaces on which the turning points lie, and which must be taken into account in the analysis. By varying the positions of A and B, the correlation measurement enables the phase patterns to be mapped out over the surface of the sun, and thereby in principle the eigenfunctions of normal modes are measured. But there is a greater potential use of this method. By concentrating attention on local variations, as in the previous two local methods, local variations in the structure of the sun can be inferred directly. For examples, by comparing cross-correlations with both positive and negative time lags, which represent waves travelling in opposite directions, Duvall *et al.* (1995) have evidence for downflow in the vicinity of sunspots. This was obtained by placing point A above the downflow and point B far from it. Such a detection would have been extremely difficult, if not impossible, by normal-mode frequency analysis. Procedures to invert such measurements to obtain the three-dimensional structure of the flow have not yet been developed.

I conclude by pointing out that Fourier analysis is not the only means by which the temporal variation of the data need be analysed. For example, Duvall (personal communication) has found interesting features in temporal cross-correlations of projections of seismic data onto spherical harmonics. Although in principle these contain no new information, they may provide the means to extract some of the information more readily.

Acknowledgements

I thank A.G. Kosovichev for supplying Figures 8 and 11, T. Sekii for Figures 9, 16, 17 and 18, and M.J. Thompson for 'original' versions of Figures 10 and 15.

9 REFERENCES

Backus, G. and Gilbert, F. 1967, *Geophys. J. R. astr. Soc.*, **13**, 247-276.
Backus, G. and Gilbert, F. 1968, *Geophys. J. R. astr. Soc.*, **16**, 169-205.
Backus, G. 1970, *Proc. Nat. Acad. Sci.*, **65**, 281-287.
Backus, G. and Gilbert, F. 1970, *Phil. Trans. Roy. Soc.*, A **266**, 123-192.
Bachmann, K.T., Schou, J. and Brown, T.M. 1993, *Astrophys. J.*, **412**, 870-879
Barratt, R. 1993, *GONG 1992: Seismic investigation of thesun and stars*, (ed. T.M. Brown, ASP Conf. Ser., San Francisco), **42**, 233-236.
Basu, S. and Antia, H.M. 1994, *Astrophys. J.*, **426**, 801-811.
Baturin, V.A., Gough, D.O., Vorontsov, S.V. and Däppen, W. 1994, *The equation of state in astrophysics*, (ed. G. Chabrier and E. Schatzmann, CUP, Cambridge), *Proc. IAU Colloq.*, **147**, 545-549.
Berthomieu, G., Cooper, A.J., Gough, D.O., Osaki, Y., Provost, J. and Rocca, A. 1980, *Nonradial and nonlinear stellar pulsation*, (ed. M.A. Hill and W.A. Dziembowski, *Lecture notes in physics*, Springer, Heidelberg), **125**, 307-312.
Berthomieu, G., Morel, P., Provost, J. and Zahn, J-P. 1993, *Inside the stars, Proc. IAU Colloq.*, **137**, (ed. W.W. Weiss and A. Baglin, ASP Conf. Ser., San Francisco), **40**, 60-63.
Brown, T.M., Christensen-Dalsgaard, J., Dziembowski, W.A., Goode, P.R., Gough, D.O. and Morrow, C.A. 1989, *Astrophys. J.*, **343**, 526-546.
Cooper, A.J. 1981, Ph. D. dissertation, University of Cambridge.
Craig, J.J.D. and Brown, J.C. 1986, *Inverse problems in astronomy: a guide to inversion strategies for remotely sensed data*, (Adam Hilger, Bristol)
Christensen-Dalsgaard, J. and Gough, D.O. 1980, *Nature*, **288**, 544-547.
Christensen-Dalsgaard, J. and Gough, D.O. 1981, *Astron. Astrophys.*, **104**, 173-176.
Christensen-Dalsgaard, J., Duvall, T.L. Jr, Gough, D.O., Harvey, J.W. and

Rhodes, E.J. Jr 1985, *Nature*, **315**, 378-382.

Christensen-Dalsgaard, J. and Gough, D.O. and Thompson, M.J. 1989, *Mon. Not. R. astr. Soc.*, **283**, 481-502.

Christensen-Dalsgaard, J., Hansen, P.C. and Thompson, M.J. 1990, *Mon. Not. R. astro. Soc.*, **264**, 541-564.

Christensen-Dalsgaard, J., Schou, J. and Thompson, M.J. 1990, *Mon. Not. R. astr. Soc.*, **242**, 353-369.

Christensen-Dalsgaard, J. and Berthomieu, G. 1991, *Solar interior and atmosphere*, (ed. A.N. Cox, W.C. Livingston, M.S. Matthews, Univ. Arizona Press, Tucson), 401-478.

Christensen-Dalsgaard, J., Gough, D.O. and Thompson, M.J. 1991, *Astrophys. J.*, **378**, 413-437.

Christensen-Dalsgaard, J. and Pérez-Hernández, F. 1991, *Challenges to theories of the structure of moderate-mass stars*, (ed. D.O. Gough and J. Toomre, Springer, Heidelberg), *Lecture Notes in Physics*, **388**, 43-50.

Christensen-Dalsgaard, J. and Däppen, W. 1992, *Astron. Astrophys. Rev.* **4**, 267-361.

Christensen-Dalsgaard, J. and Pérez-Hernández, F. 1992, *Mon. Not. R. astr. Soc.*, **257**, 62-88.

Christensen-Dalsgaard, J., Hansen, P.C. and Thompson, M.J. 1993, *Mon. Not. R. astr. Soc.*, **264**, 541-564.

Christensen-Dalsgaard, J. and Pérez-Hernández, F. 1994a, *Mon. Not. R. astr. Soc.*, **267**, 111-124.

Christensen-Dalsgaard, J. and Pérez-Hernández, F. 1994b, *Mon. Not. R. astr. Soc.*, **269**, 475-492.

Christensen-Dalsgaard, J., Larsen, R.M., Schou, J. and Thompson, M.J. 1995, *GONG'94: helio-and asteroseismology from the earth and space*, (ed. R.K. Ulrich, E.J. Rhodes Jr and W. Däppen, ASP Conf. Ser., San Francisco), **76**, 70-73.

Däppen, W., Gough, D.O. and Thompson, M.J. 1988, *Seismology of the sun and sun-like stars*, (ed. E.J. Rolfe, ESA SP-286, Noordwijk), 505-510.

Däppen, W., Gough, D.O., Kosovichev, A.G. and Thompson, M.J. 1991, *Challenges to theories of the structure of moderate-mass stars*, (ed. D.O. Gough and J. Toomre, Springer, Heidelberg), *Lecture Notes in Physics*, **388**, 111-120.

Deubner, F.-L., Ulrich, R.K. and Rhodes, Jr, E.J. 1979, *Astron. Astrophys.*, **72**, 177.

Duvall, T.L. Jr, 1982, *Nature*, **300**, 242-243.

Duvall, T.L. Jr, Harvey, J.W. and Pomerantz, M.A. 1986, *Nature*, **321**, 500-501.

Duvall, T.L. Jr, Harvey, J.W., Libbrecht, K.G. and Popp., B.D. 1988, *Astrophys. J.*, **324**, 1158-1171.

Duvall, T.L. Jr, Jefferies, S.M., Harvey, J.W. and Pomerantz, M.A. 1993, *Nature*,

362, 430-432.
Duvall, T.L. Jr, D'Silva, S., Jefferies, S.M., Harvey, J.W. and Schou, J. 1995, in preparation.
Dziembowski, W.A. and Goode, P.R. 1989, *Astrophys. J.*, **374**, 540-550.
Dziembowski, W.A., Pamyatnykh, A.A. and Sienkiewiez, R. 1990, *Mon. Not. R. astr. Soc.*, **244**, 542-550.
Dziembowski, W.A. and Goode, P.R. 1992, *Astrophys. J.*, **394**, 670-687.
Dziembowski, W.A., Pamyatnykh, A.A. and Sienkiewiez, R. 1992, *Acta. Astr.*, **42**, 5-15.
Elliott, J.R. 1995, *Mon. Not. R. astr. Soc.*, to be submitted.
Ellis, A.N. 1988a, *Advances in helio-and asteroseismology, Proc. IAU Symp.*, **123**, (ed. J. Christensen-Dalsgaard and S. Frandsen, Reidel, Dordrecht) 147-150.
Ellis, A.N. 1988b, Ph. D. dissertation, University of Cambridge.
Golub, G.H., Heath, M. and Wahba, E. 1979, *Technometrics*, **21**, 215.
Gordon, C., Webb, D. and Wolpert, S. 1992, *Bull. Am. Math. Soc.*, **27**, 134-137.
Gough, D.O. 1982, *Europhys. News*, **13**, 3-5.
Gough, D.O. 1985, *Solar Phys.*, **100**,65-99.
Gough, D.O. 1986, *Highlights in Astronomy* (ed. P. Swings), **7**, 283-293.
Gough, D.O. and Kosovichev, A.G. 1988, *Seismology of the sun and sun-like stars* (ed. E.J. Rolfe, ESA SP-286, Noordwijk), 195-201.
Gough, D.O. and Toomre, J. 1988, *GONG Newsletter*, (National Solar Observatory, Tucson), **9**, 20-60.
Gough, D.O. and Thompson, M.J. 1989, *Mon. Not. R. astr. Soc.*, **242**, 25-55.
Gough, D.O. 1990, *Progress of seismology of the sun and stars*, (ed. Y. Osaki and H. Shibahashi, Springer, Heidelberg), *Lecture notes in physics*, **267**, 283-318.
Gough, D.O. and Thompson, M.J. 1990, *Mon. Not. R. astr. Soc.*, **242**, 25-55.
Gough, D.O., Merryfield, W.J. and Toomre, J. 1991, *Challenges to theories of the structure of moderate-mass stars*, (ed. D.O. Gough and J. Toomre, Springer, Heidelberg), *Lecture Notes in Physics*, **388**, 265-270.
Gough, D.O. and Thompson, M.J. 1991, *Solar interior and atmosphere*, (ed. A.N. Cox, W.C. Livingston and M. Matthews, Univ. Arizona Press, Tucson), 519-561.
Gough, D.O. 1993, *Astrophysical fluid dynamics*, (ed. J.-P. Zahn and J. Zinn-Justin, Elsevier, Amsterdam), 399-560.
Gough, D.O., Merryfield, W.J. and Toomre, J. 1993, *GONG 1992: Seismic investigation of the sun and stars*, (ed. T.M. Brown), *ASP Conf. Ser.*, **42**, 257-260.
Gough, D.O. and Sekii, T. 1993, *GONG 1992: Seismic investigation of the sun and stars*, (ed. T.M. Brown, ASP Conf. Ser., San Francisco), **42**, 177-180.
Gough, D.O. and Toomre, J. 1993, *GONG Report*, (National Solar Observatory,

Tucson), **11**

Gough, D.O. and Kosovichev, A.G. 1995, *Solar Phys.*, **157**, 1-15.

Gough, D.O., Sekii, T. and Stark, P.B. 1995, *Astrophys. J.*, in press.

Gough, D.O. and Vorontsov, S.V. 1995, *Mon. Not. R. astr. Soc.*, **273**, 573-582.

Hansen, P.C. 1992, *Inverse Problems*, **8**, 849.

Hansen, P.C., Sekii, T. and Shibahashi, H. 1992, *SIAM J Sci. Stat. Comp.*, **13**, 1142-1150.

Harvey, J.W. and Duvall, T.L. Jr, 1984, *Proc. Conf. Solar seismology from space*, (ed. R.K. Ulrich, J.W. Harvey, E.J. Rhodes Jr and J. Toomre, JPL Publ. 84-84, Pasadena), 165-172.

Hill, F. 1988, *Astrophys. J.*, **333**, 996-1013.

Iglesias, C.A., Rogers, F.J, and Wilson, B.G. 1987, *Astrophys. J.*, **322**, L45-L48.

Iglesias, C.A., Rogers, F.J, and Wilson, B.G. 1992, *Astrophys. J.*, **397**, 717-728.

Julien, K.A., Gough, D.O. and Toomre, J. 1995a, *GONG'94: helio-and asteroseismology from the earth and space*, (ed. R.K. Ulrich, E.J. Rhodes Jr and W. Däppen), *ASP Conf. Ser.*, **76**, 196-199.

Julien, K.A., Gough, D.O. and Toomre, J. 1995b, *Proc. Fourth SOHO Workshop: Helioseismology*, (ed. B. Battrich, ESA, Noordwijk), in press.

Keller, J.B.. 1958, *Ann. Phys.*, **4**, 180-188

Kosovichev, A.G. and Parchevskii, K.V. 1988, *Pis'ma. Astron. Zh.*, **14**, 473-480.

Kosovichev, A.G., Christensen-Dalsgaard, J., Däppen, W., Dziembowski, W.A., Gough, D.O. and Thompson, M.J. 1992, *Mon. Not. R. astr. Soc.*, **259**, 536-558.

Lanczos, C. 1961, *Linear differential equations*, Van Nostrand, London.

Lavely, E.M. and Ritzwoller, M.H. 1992, *Phil. Trans. R. Soc.*, A**339**, 431.

Lavely, E.M. and Ritzwoller, M.H. 1993, *Astrophys. J.*, **403**, 810-832.

Ledoux, P. and Walraven, Th. 1958, *Handbh der Physik*, **51**, 353-604.

Libbrecht, K.G., Woodard, M.F. and Kaufman, J.M. 1990,*Astrophys. J. Suppl.*, **74**, 1129-1149.

Lynden-Bell, D. and Ostriker, J.P. 1967, *Mon. Not. R. astr. Soc.*, **136**, 293-310.

Marchuk, G.I. 1977, *Methods in computational mathematics*, Nauka, Moscow.

Masters, G. 1979, *Geophys. J. R. astr. Soc.*, **57**, 507.

Mihalas, D., Däppen, W. and Hummer, D.G. 1988, *Astrophys. J.*, **331**, 815-825.

Monteiro, M.J.P.F.G., Christensen-Dalsgaard, J. and Thompson, M.J. 1993, *Inside the stars, IAU Colloq.*, **137**, (ed. W.W. Weiss and A. Baglin, ASP Conf. Ser., San Francisco), **40**, 557-559.

Murawski, K. and Roberts, B. 1993a, *Astron. Astrophys.*, **272**, 595-600.

Murawski, K. and Roberts, B. 1993b, *Astron. Astrophys.*, **272**, 601-608.

Oldenberg, D.W. 1976, *Geophys. J. R. astr. Soc.*, **44**, 413-431.

Patrón, J., Hill, F., Rhodes, E.J. Jr, Korzennik, S.G. and Cacciani, A., *GONG'94: helio-and asteroseismology from the earth and space*, (ed. R.K. Ulrich, E.J. Rhodes Jr and W. Däppen), *ASP Conf. Ser.*, **76**, 208-211.

Pijpers, F.P. and Thompson, M.J. 1992, *Astron. Astrophys.*, **262**, L33-L36.
Pijpers, F.P. and Thompson, M.J. 1994, *Astron. Astrophys.*, **281**, 231-240.
Provost, J. and Berthomieu, G. 1986, *Astron. Astrophys.*, **165**, 218-226.
Rao, C.R. and Mitra, S.K. 1971, *Generalized inverse of matrices and its applications*, Wiley, New York.
Ritzwoller, M.H. and Lavely, E.M. 1991, *Astrophys. J.*, **369**, 557-566.
Rogers, F.J. 1986, *Astrophys. J.*, **310**, 723-728.
Rosenthal, C.S. and Gough, D.O. 1994, *Astrophys. J.*, **423**, 488-495.
Rosenthal, C.S., Christensen-Dalsgaard, J., Nordlund, A. and Trampedach, R. 1995, *Proc. Fourth SOHO Workshop: Helioseismology*, (ed. B. Battrick, ESA, Noordwijk), in press.
Roxburgh, I.W. and Vorontsov, S.V. 1993, *GONG 1992: Seismic investigation of the sun and stars*, (ed. T.M. Brown, ASP Conf. Ser., San Francisco), **42**, 169-171.
Schou, J. 1991, *Challenges to theories of the Structure of moderate-mass stars* (ed. D.O. Gough and J. Toomre, Springer, Heidelberg), *Lecture Notes in Physics*, **388**, 93-100.
Schou, J. 1993, Ph. D. dissertation, Arhus Universitet.
Schou, J., Christensen-Dalsgaard, J. and Thompson, M.J. 1994, *Astrophys. J.*, **433**, 389-416.
Sekii, T. and Shibahashi, H. 1988, *Seismology of the sun and sun-like stars*, (ed. E.J. Rolfe, ESA SP-286, Noordwijk), 521-523.
Sekii, T. 1990, *Progress of seismology of the sun and stars*, (ed. Y. Osahi and H. Shibahashi, Springer, Heidelberg), *Lecture Notes in Physics*, **367**, 337-340.
Sekii, T. 1991, *Publ. Astron. Soc. Japan*, **43**, 381-411.
Sekii, T. 1993a, *GONG 1992: Seismic investigation of the sun and stars*, (ed. T.M. Brown, ASP Conf. Ser., San Francisco), 237-240.
Sekii, T. 1993b, *Mon. Not. R. astr. Soc.*, **264**, 1018-1024.
Sekii, T. and Gough, D.O. 1993, *Inside the stars*, *Proc. IAU Colloq.*, **137**, (ed. W.W. Weiss and A. Baglin, ASP Conf. Ser., San Francisco), **42**, 97-100.
Sekii, T. 1995, *Mon. Not. R. astr. Soc.*, submitted.
Tassoul, M. 1980, *Astrophys. J. Suppl.*, **43**, 469-490.
Thompson, M.J. 1995, *Inverse problems*, **11**, 709-730.
Tikhonov, A.N. and Arserun, V.Y. 1977, *Solutions of ill-posed problems*, (Winston, Washington DC).
Toutain, T. and Fröhlich, C. 1992, *Astron. Astrophys.*, **257**, 287-297.
Unno, W., Osaki, Y., Ando, H., Saio, H. and Shibahashi, H. 1989, *Nonradial oscillations of stars*, (Univ. Tokyo Press, Tokyo)
Vorontsov, S.V., Baturin, V.A. and Pamyatnykh, A.A. 1991, *Nature*, **349**, 49-51.
Zahn, J-P. 1970, *Astron. Astrophys.*, **4**, 452-461.

Global Changes in the Sun

J. R. Kuhn

National Solar Observatory, Sunspot NM, and
Michigan State University, E. Lansing, MI, USA

1 INTRODUCTION

This is almost an impossible task, to summarize the subject of *Global Changes in the Sun* so I must apologize in advance for limiting the scope of these lectures to issues that have been choosen, in part, because of personal interests. I hope that the references provide the reader with footpoints from which to explore a larger set of questions which bear on this subject.

Here we will not discuss the very long, evolutionary, timescales over which the sun changes, nor will we explore the fast changes associated with flares and other transient phenomena. While these discussions depend on some results from MHD models of the solar magnetic cycle, we will not be concerned with the MHD mechanism. These lectures will not address the questions needed to understand local physical models that describe, for example, granulation. On the other hand we will describe some of the physical problems that connect the small-scale behavior of the sun to its global properties. By "global property" I mean an observable that is connected by physically important timescales to the entire sun: limb shape and brightness, largescale magnetic field, oscillation frequencies, solar luminosity, and solar irradiance are all examples of global properties.

Here we are interested in understanding the deviations of the sun from some standard one-dimensional static stellar model. This is a subject we can hardly approach for other stars, and for the sun it is difficult because the physics of magnetic fields and convection are linked over a wide range of spatial and temporal scales. Of course this fact makes the problem rich, interesting, and difficult.

A new set of tools is coming of age with the helioseismic, photometric, and numerical experiments that are now, or may soon be, possible. I am excited by the prospects of using these to make definite progress against some of the theoretical scenarios. I think we stand a chance of uncovering how and why the sun is changing on human timescales - a question that should have more than academic interest to a broader community than our scientific peers.

It is impossible to be exhaustive or even to do justice to most of the models that have been proposed to explain global solar changes. I will not try, in part because I haven't found one which is complete and consistent with all of the observational constraints. Although the criticism of incompleteness can also be applied to the scenarios I will discuss in these lectures - they will serve to illustrate some of the important issues, observations, and questions that bear on the solar cycle problem.

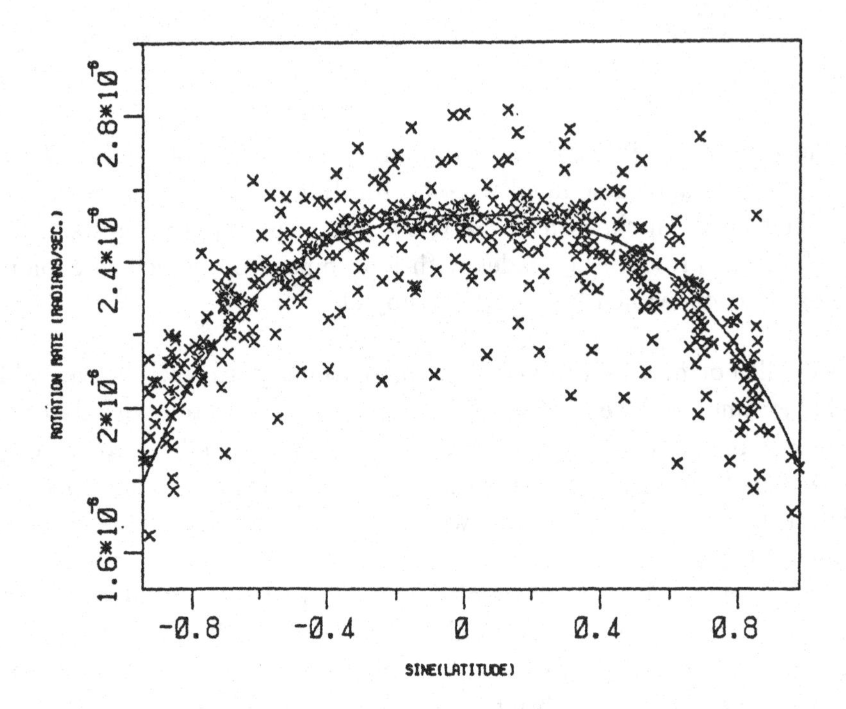

Figure 1: *Solar Rotation Profile*

2 SOME INTERESTING NON-ACOUSTIC OBSERVABLES

2.1 Global Observations

Most of these results should be in the nature of a review to you, but I'll collect some graphs and figures to remind you of the major puzzles yet to be quantitatively described. We don't have a complete model that explains all of these observables but we have many of the pieces of the complete picture.

We should appreciate that the sun is nearly perfectly spherically symmetric. The most obvious departure from spherical symmetry is the solar rotation. With an equatorial rotation period of 27d the rotational "asphericity", the ratio of the squares of the equatorial rotation velocity and the escape velocity, is a small number, 10^{-5}. The angular differential rotation profile is also a puzzle (Fig. 1). An expansion in Legendre polynomials gives quite significant $l = 2$ and $l = 4$ terms with

$$\Omega(\theta) = 2.53 \times 10^{-6} \, (1 - 0.187 P_2(\cos\theta) - 0.036 P_4(\cos\theta)) \tag{1}$$

It seems that the temporal variation in the surface rotation profile is small (cf.

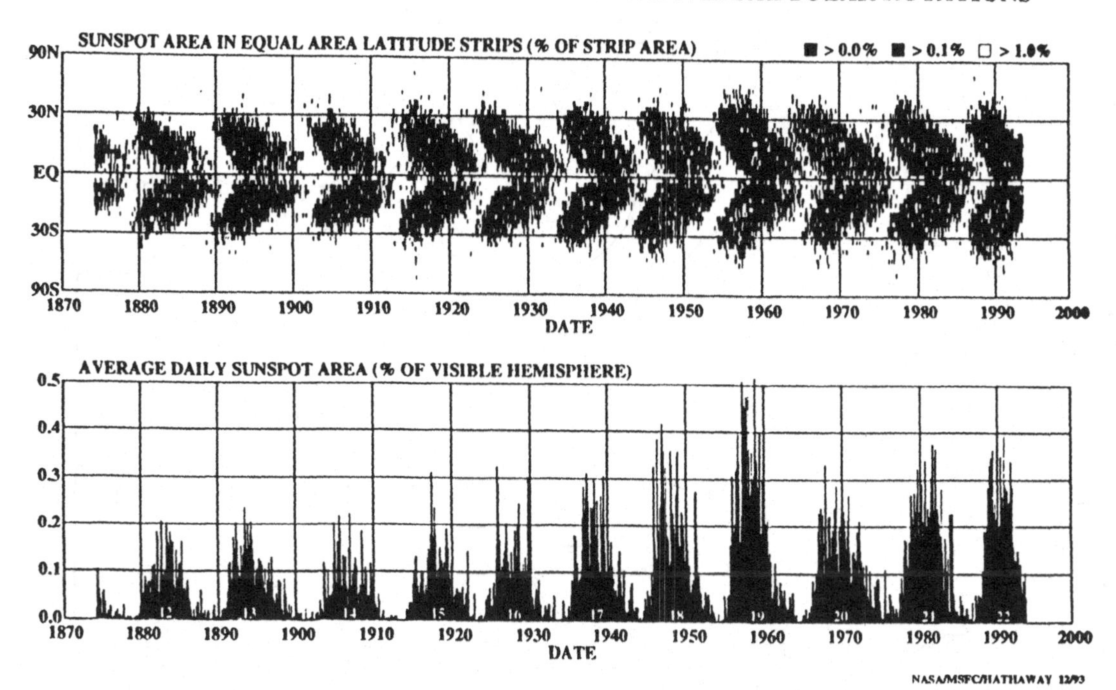

Figure 2: *Sunspot Cycle Variations*

Howard 1990) – probably no larger than 1% over the solar cycle. Other large-scale velocity fields have been observed, like "torsional waves" but they have a tiny velocity amplitude of 5 m/s or less (cf. Snodgrass 1992). The evidence for meridional fields is inconsistent (cf Hathaway 1991). Steady north-south vertical photospheric flows are probably less than 100m/s if they exist at all. Recall the well established facts; that the photospheric velocity field includes granulation convection (characterised by a velocity scale of 1km/s, spatial scale of 1000km and lifetime of 10min), and the "supergranular motions" (characterised by a horizontal velocity of 100m/s, spatial scale of 30,000km and a lifetime of 40 hours). Observational evidence for larger-scale convective flows is inconsistent.

Of course the solar magnetic field is both aspherical and time variable. Fig. 2 (Hathaway, personal communication) shows how the sunspot number and latitudinal distribution changes during the 11 yr sunspot cycle. It seems that sunspots provide a reasonable proxy for the solar magnetic field evolution. Magnetograph data provide a more quantitative measure of the temporal variation of the mean longitudinal magnetic field (or flux). For example, from near solar minimum (1986) to near maximum (1989), the average flux density magnitude varied between 8 and 18 G (as measured from Kitt Peak magnetograms). A Legendre polynomial decomposition of the mean field also yields low order terms that are of order 10G.

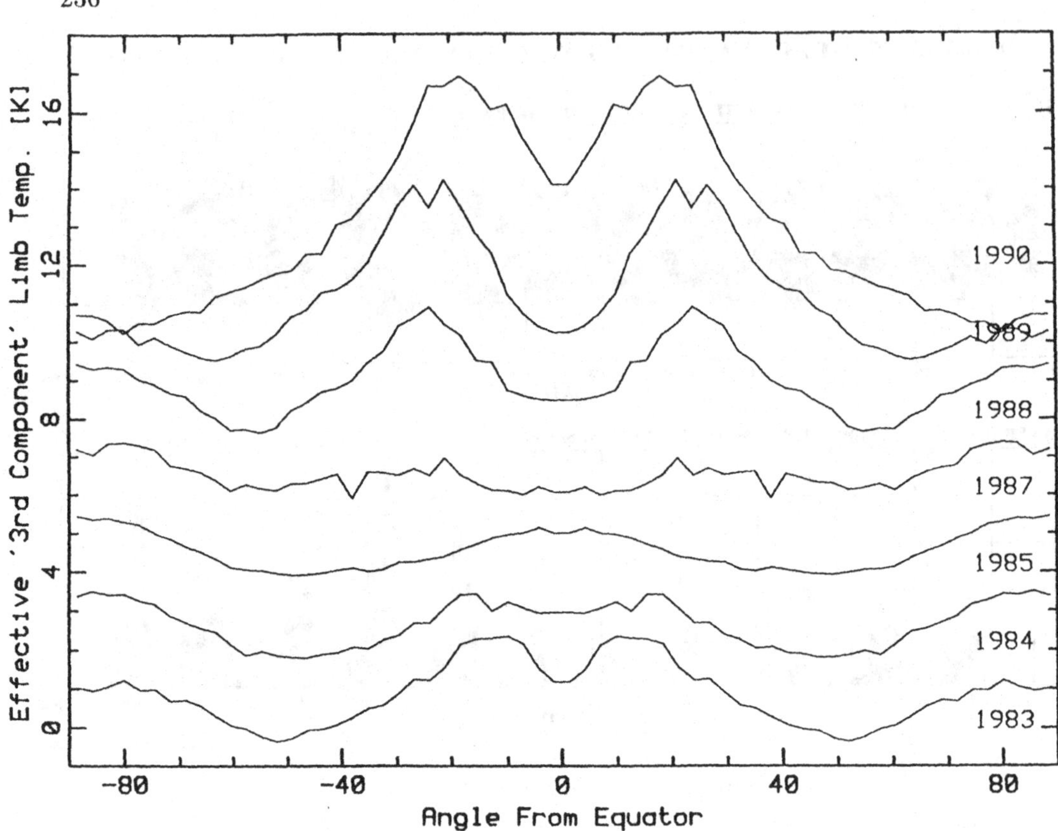

Figure 3: *Solar Limb Brightness Variations*

Two observables I am particularly interested in are the solar shape and surface brightness distributions. Both of these measurements are obtained against a large background (the mean solar brightness) and are obtained using specialized limb photometers (Kuhn,Libbrecht, Dicke 1988). Measurements of the limb brightness and shape obtained between 1983 and 1989 indicate a changing asphericity in the solar limb brightness and a small, probably static, solar asphericity (Kuhn 1990). The difference between the equatorial and polar solar radii during these years was approximately 0.010 arcseconds.

Figure 3 shows how the limb brightness, expressed as a color temperature, changed during the last solar cycle. The only ground-based measurements of solar shape asphericity are of the solar oblateness (the $l = 2$ Legendre harmonic) and these data allow the possibility of only a small change in the oblateness during a 3 year interval (1983-1985) (Dicke et al. 1987). It is notable that the fractional change in the solar brightness (for example, parametrized by the $l = 2$ component of the average photospheric temperature distribution, T_2) is much larger than limits we have on the $l = 2$ (δr_2 oblateness) change in the shape of the limb. We know that $\delta r_2/r \leq 3 \times 10^{-3} T_2/T_\odot$

Recently satellite observations from Yohkoh (the white light data) have been used to measure the limb shape and brightness. The absence of an atmosphere to look through allowed very sensitive limits on changes in the limb brightness and shape. The time series is very short (only a few hours) so we cannot measure solar-cycle timescale changes, but short period variations (in the 5-min band to slightly longer periods) also are dominated by brightness variations that are much larger than shape changes. The Yohkoh data provide limits on the rms variation of $l = 2$ to $l = 64$ shape harmonics that are less than a few milliarcseconds (Kuhn et al. 1994).

There is also no good evidence that the solar radius is changing on yearly timescales, although there is controversy over whether or not the sun's radius may change over longer timescales (Ribes et al. 1991). The upper limit on solar radius variations, $\delta r/r$, over timescales of a year from photoelectric measurements (cf. Ribes et al. 1991) obtained at HAO is 5×10^{-5}.

Finally, the last global property we'll mention here that changes on cycle timescales is the solar irradiance. This is the mean energy flux in the ecliptic plane (which should not be confused with the solar *luminosity*). The ACRIM and NIMBUS/ERB satellites have been measuring this quantity now for about one solar cycle (cf Willson 1991). Fig. 4 shows a plot of how the irradiance has changed since 1980 (Pap, personal communication). Note that the scale of the variation is 10^{-3} of the mean, and the phase of the irradiance change is such that it is a maximum near sunspot maximum.

2.2 Solar Neutrino Variations?

The solar neutrino problem will be more conclusively addressed by other lecturers during this school. Here I only want to briefly address the question of whether *variations* in the measured neutrino flux from the ^{37}Ar experiment are likely to tell us much about the solar structure. It has been suspected for some time (cf. references in Bahcall and Press 1991) that there is apparently an anticorrelation between the ^{37}Ar production rate and the solar activity. Figure 5 plots the ν capture rate on Chlorine against the corresponding average daily sunspot number. Numerous statistical tests (cf. Bieber et al. 1990, and Bahcall and Press 1991) have been aimed at this correlation, but it seems to be statistically robust. Thus some have even argued (Krauss 1990) that we may learn about the solar structure from these data. The leading theoretical basis for interpreting these flux changes centers on the possibility that electron and muon ν's could be coupled by the interaction of a significant ν magnetic dipole moment with the solar magnetic field.

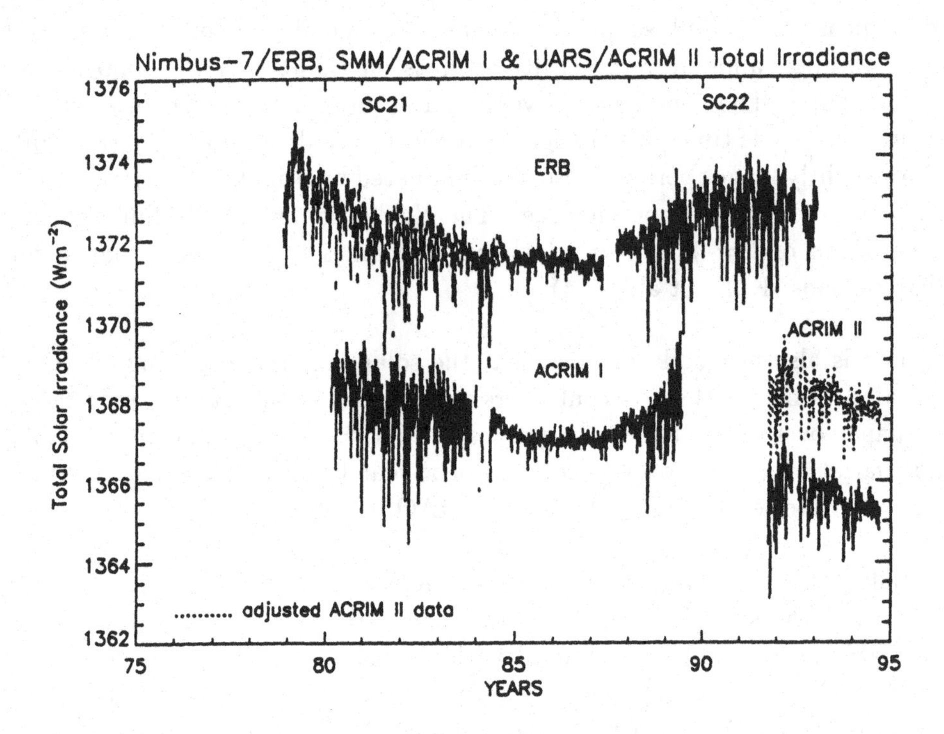

Figure 4: *Solar Irradiance Variations*

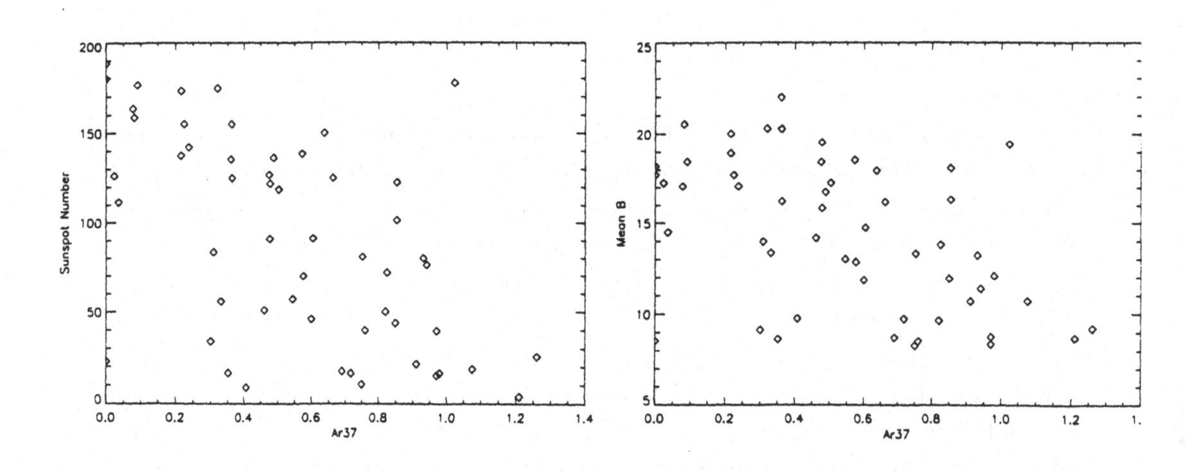

Figure 5: *Measured Neutrino Flux vs Sunspot number and Average Magnetic Field*

There are least three concerns one should have with this interpretation. First, if the physical mechanism is related to the interaction of the neutrino with the solar magnetic field then we might expect to find a larger correlation between the ν flux and a better proxy of the solar interior field near the plane of the ecliptic. For example, I used the Kitt Peak full disk magnetic field measurements to arrive at a mean photospheric field estimate, and also plotted these against the ν flux in Fig. 5. The correlation coefficient is not stronger but is in fact *weaker* than the ν correlation with sunspots.

A second concern is related to the cosmic ray background. Much work by Davis and collaborators has been done to pin down the expected background signal due to cosmic ray muons. The Davis (1978) extrapolation from measurements at over 2 orders of magnitude higher Ar production rates, yields a background correcton of 0.08 Ar atoms/day (these are the flux units in Fig. 5). In the absence of deeper background measurements these are the best estimates possible. Now we know that the cosmic ray flux (except at high energies) at the surface of the earth shows large variations during the solar cycle. In the Gev energy range the proton flux (cf. Jokipii 1991) can vary by orders of magnitude. Since it is the interaction of the solar and terrestrial magnetic fields that "shields" the earth from these cosmic rays the correlation is negative, i.e. when the sun is most active (large sunspot number) the terrestrial cosmic ray flux is smallest. Thus an underestimate of the cosmic ray background or some other mechanism that couples the ^{37}Ar count rate to cosmic rays could produce an apparent negative correlation with sunspots. At solar maximum the actual background is smaller so that the apparent ν flux appears to be larger. For an alternate opinion see Bieber et al. (1990), who also apparently recognized this possibility but discounted it.

The last concern is that the Kamiokande II measurements of neutrino flux (which have completely different background problems) see no evidence of semi-annual ν flux variations (Hirata et al. 1991)

In the absence of more conclusive measurements I am highly skeptical of a solar *interior* origin to the apparent ν flux variation seen in the Homestake mine experiment.

2.3 Another Solar Mode?

The puzzle is now to interpret these deviations from spherical symmetry and solar-cycle temporal variations. Of the global properties we've described, the aspherical parts (subscripted with a 2 for the $l = 2$ component) are: $B_2/B \approx 1$,

$\Omega_2/\Omega \approx 0.2$, $F_2/F \approx 10^{-3}$, and $r_2/r \approx 10^{-5}$. Here B, Ω, F, and r, refer to the global magnetic field, surface rotation, flux, and solar radius. We have also noted that there is a time dependence to the aspheric quantities during a solar cycle (here δ signifies the temporal variation) given as: $\delta B/B \approx 1$ and $\delta F/F \approx 10^{-3}$. limits on the rotation variation are about 10^{-2} and on the oblateness, $\delta r_2/r \leq 10^{-5.5}$.

Can we relate the asphericity, or the temporal variations of any of these quantities? Unlike the solar acoustic oscillations these temporal variations aren't described by a Sturm-Liouville problem, e.g. we can't write down the equations that relate the amplitude of the surface magnetic field to the solar flux changes. This several year-timescale "mode" we think of as the solar cycle is not at all well understood.

We might begin by asking what physical solar property is most directly tied to the solar cycle variation? The largest dimensionless changes are in the magnetic field. Thus the global solar magnetic fields should be key to understanding variations in the solar structure. Many would argue that the basic MHD mechanism for the cycle is an $\alpha-\omega$ dynamo, perhaps a "shell" dynamo near the base of the convection zone will solve the observational problems (see the review by Rosner and Weiss 1992). I think the dynamo model problem is very hard, and by no means clear. We won't address it except to note the apparent consequence that a toroidal field builds up in the solar interior over an 11 year timescale. Its regeneration and dissipation are questions that we may speculate on, but they remain very interesting problems yet to be understood.

Now we start the job of interpreting the photospheric structural observations and to try relating them using some simple "toy" models. From there we'll consider the subsurface data – both helioseismic and numerical simulation results.

3 INTERPRETING VELOCITY AND SHAPE DATA

3.1 Lowest Order: A Hydrostatic Sun

Limb observations measure local conditions in the solar atmosphere. The density scale height near the photosphere is about $h = 200$km so the density varies as $\rho(r) = \rho_0 e^{-z/h}$. Here z represents the height above some reference point (say the $\tau_{500} = 1$ surface at the limb). At temperatures of a few thousand degrees we approximate the opacity by $\kappa \approx K\rho^{1/2}T^5$. Now integrating $d\tau = \kappa dr$ at an impact parameter z through the atmosphere gives $\tau(z) = \sqrt{1.5\pi h(R_\odot + z))}K'e^{-1.5z/H}$. We can obtain the intensity from the source function $S(\tau)$ using $dI/d\tau = S(\tau)e^{-\tau}$. If S is slowly varying then $I(z) = S(z)(1 - e^{-\tau(z)})$ so that the intensity near the limb is a rapidly falling (exponential of an exponential) function of z. Thus limb measurements probe the solar atmospheric source function over a vertical scale less than the density scale height. Recalling the Eddington-Barbier relation we note that the intensity near the limb gives a reasonable estimate of the local source function near $\tau = 1$ and in some cases the temperature in the solar atmosphere near the limb. Note that by observing surfaces of constant intensity we can determine the shape of surfaces of constant temperature near the photosphere.

Important information on the solar structure is contained in limb observations. If we assume, to a first approximation, that the solar surface satisfies hydrostatic equilibrium then $\nabla p = -\rho \nabla \phi$. Taking the curl of this equation leads immediately to the conclusion that surfaces of constant density and gravitational potential are coincident (Von Zeipel's theorem). If density and potential surfaces are coincident then from the hydrostatic equation pressure surfaces must also be coincident. Finally if the fluid satisfies an equation of state of the form $P = f(\rho, T)$ where f is any function of only density and temperature then it follows that surfaces of constant temperature must also be coincident with pressure, density, and potential surfaces.

We can use this formalism (cf. Dicke 1970) to relate the solar rotation to the measured limb shape. If rotation is constant along cylinders, i.e. $\omega = \omega(R, \theta)$ with $R = r\sin\theta$ then we can replace the gravitational potential by an "effective" potential that includes a centrifugal term so that $\phi(R,\theta) = \phi' - \int_0^R R'\omega^2(R')dR'$. Allowing for the gravitational potential to contain a quadrupole term we write $\phi'(r) = \frac{-GM}{r}(1 - J_2(r_0/r)^2 P_2)$. Here $P_2 = 1.5\cos^2\theta - 0.5$ and J_2 is a numerical coefficient that quantifies the solar gravitational quadrupole moment. If we further assume uniform rotation then we can relate the difference between measured polar and equatorial radii in the limb shape (which must also describe the shape of ϕ) to the rotation and J_2. Thus,

$$\frac{(r_{eq} - r_p)}{r} = 1.5J_2 + \frac{\omega^2 r^3}{2GM}$$

For a surface rotation of 27d (with $J_2 = 0$) we obtain $\Delta r/r = 9.3 \times 10^{-6}$. We observe a fractional radius difference of 0.01"/960" $\approx 10^{-5}$ so that the apparent limb shape is well accounted for by a rigidly rotating sun with a rotation rate of 27d.

3.2 Mean Flows

Of course von Zeipel's theorem is violated by static or variable flows in the photosphere, and the Lorenz forces associated with magnetic fields. Recall the Euler equation

$$\rho \partial \vec{v} / \partial t + \rho(\vec{v} \cdot \vec{\nabla}) \cdot \vec{v} \;=\; -\nabla p - \rho \nabla \phi$$

This equation, supplemented with a magnetic force term, contains most of the physics we need to describe structural changes, although exact solutions to this equation are not common.

As an example, let's estimate the magnitude of the steady meridional flow needed to generate the observed oblateness in a non-rotating sun without significant gravitational oblateness ($J_2 = 0$). In this case we take gravitational potential surfaces to be spherical and let p, ρ, and T surfaces have the small oblateness given by $\epsilon(1.5\cos^2\theta - 0.5)$ with $\epsilon = 10^{-5}$. Thus $\vec{\nabla} p$ and $\rho\vec{\nabla}\phi$ are no longer parallel. The angle between these vectors is $\alpha = 1.5\epsilon \sin 2\theta$ and must be balanced by the $\hat{\theta}$ term of the LHS of the Euler equation. The RHS of the Euler equation is approximately $\rho\nabla\phi \cdot \alpha\hat{\theta}$. Solving this for the variation in v_θ^2 between the pole and equator we get $\Delta v_\theta^2 = 1.5 g R \epsilon$ with the maximum meridional velocity at the pole. Here g is the photospheric gravitational acceleration. The direction (poleward or equatorward) of v_θ is indeterminate in this simple analysis and the magnitude of the flow is large (not surprisingly), 1.7km/s.

3.3 Turbulent Flows

The turbulent motions in the sun have a very definite contribution to the mean asphericity that is not apparent from the form of the Euler equation above. A standard procedure is to separate the velocity field v into mean and fluctuating components, $\vec{v} = \vec{U} + \vec{u}$ where $\overline{\vec{v}} = \vec{U}$ and $\vec{U}$ is the mean flow. Here the overline represents a time or space average of the instantaneous velocity. Substituting for $\vec{v}$ and averaging each term of the Euler equation yields

$$\rho(\frac{\partial \vec{U}}{\partial t} + (\vec{U} \cdot \nabla)\vec{U}) \;=\; -\nabla p - \rho\nabla\phi - \nabla \cdot \mathbf{t} \quad . \tag{2}$$

All of the thermodynamic quantities above now refer to corresponding means and the tensor $t^{ij} = \overline{\rho u^i u^j}$ is known as the turbulent Reynolds stress. We have also assumed that the fluctuating velocity satisfies $\nabla \cdot \vec{u}\rho = 0$ (the anelastic approximation).

The quantity $\mathbf{t}$ is a turbulent momentum flux density tensor (cf Landau and Lifshitz 1959). Thus, the momentum flux, $d\vec{f}$, (or force) through a surface of area da characterised by surface normal $\hat{n}$, and due to the turbulent fluid motion, is $d\vec{f} = da\mathbf{t} \cdot \hat{n}$. Written in this form we see that each of the terms on the RHS of eq. (2) provide a driving force for the mean flow on the LHS. The diagonal terms of t^{ij} contribute to the force like a pressure, and this part of the tensor is often called the turbulent pressure.

A complication, hidden by eq. (2), is that the stress tensor $\mathbf{t}$ is often a function of the mean flow. Thus the correlation between fluctuating velocity components in general depends on the mean velocity field, $\vec{U}$. It is common then to write (cf. Durney and Spruit 1979)

$$\mathbf{t} = \mathbf{t_0} + \mathbf{T} \qquad (3)$$

where $\mathbf{t_0}$ includes the velocity correlation in the absence of a mean flow (usually a diagonal, but not necessarily isotropic, matrix), and $\mathbf{T}$ includes the contribution to $\mathbf{t}$ due to a non-zero mean flow, $\vec{U}$. $\mathbf{T}$ is a matrix which generally includes both diagonal and off-diagonal components, and satisfies $\mathbf{T}(\vec{U} = 0) = 0$. It is usually called the turbulent viscosity.

A mean flow can generate significant anisotropy or off-diagonal terms in the effective stress tensor in the presence of what is otherwise isotropic turbulence. An expression for $\mathbf{T}$ in terms of $\mathbf{t_0}$ and $\vec{U}$ was developed using the formalism of "molecular" turbulence by Wasiutynski (1946) and Elsasser (1966). They consider fluid elements (like "eddies") to have a distinct momentum-carrying lifetime of τ in the presence of a mean flow field $\vec{U}$. It is not difficult to prove that when $\vec{U}$ is not uniform a microscopically isotropic velocity distribution becomes "biased" by the non-uniform flow. A simple example of this is illustrated in fig. 6.

We start with a fluid at rest (or in uniform motion) and assume that fluid elements have an isotropic velocity distribution with speed v_0. Consider a point P in the fluid. At a given time all fluid elements arriving at P (on average) started out a time τ earlier and originated from a spherical surface S centered on P with a radius of τv_0. We assume all velocity directions are equally likely (isotropic velocity distribution in the absence of a mean flow) so that the expectation of $v^i v^j$ at P is $\delta_{ij} v_0^2/3$. Now consider this problem when there is a velocity gradient at P. Assume the left half-space ($y < 0$) flows upward with velocity $u\hat{z}$ and the right half-space flows downward with velocity $-u\hat{z}$. The average velocity at P is still 0. In this case the fluid elements from $y < 0$ which arrive at P originated from a hemisphere of radius τv_0 displaced to $z = -\tau u$. Similarly those elements from $y > 0$ originate from a hemisphere displaced vertically upward from P by τu. The resulting fluctuating velocity distribution at P now consists of a

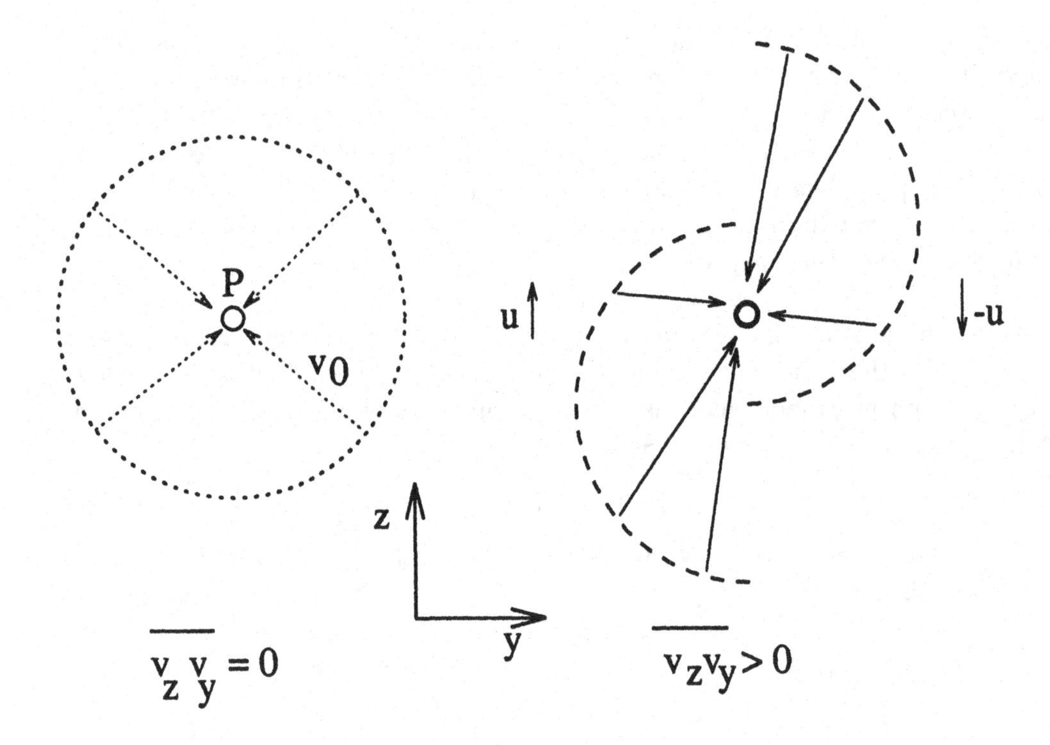

Figure 6: *A gradient in the mean flow induces turbulent stress.*

uniform distribution of velocity vectors from each hemisphere displaced by $\pm u\hat{z}$ (see fig. 6). Thus, the expectation of $v^i v^j$ is no longer an isotropic matrix. The vertical offset of each hemispherical velocity distribution also generates off-diagonal (stress) terms in the velocity correlation. Notice that in this case $\overline{v^z v^y} > 0$, while the remaining off-diagonal terms remain zero. Elsasser (1966) derived an expression for the tensor **T** resulting from a general mean flow, $\vec{U}$, (repeated indices are implicitly summed in the following equation),

$$T^{ij} = -\tau(t_0^{ik}\partial_k U^j + t_0^{jk}\partial_k U^i + U^k \partial_k t_0^{ij}) \quad (4)$$

Equation 4 gives the viscous stress when the velocity shear u, is infinitesmal, i.e. assuming finite velocity derivatives. Note that this equation quantitatively describes our example from fig. 6. We now have, in principle with eq. 2-4, what we need to relate mean velocity flows to the random turbulent fluid motion in the convection zone.

3.4 Generating Mean Flows

Equations 2-4 are quite difficult to solve. I am not aware of any published self-consistent, and realistic solution for the global solar velocity field. Kippenhahn

(1963) does solve eq. 2 (with $\mathbf{T}$ in spherical coordinates even) for an idealized non-isotropic viscous fluid. He proves that all steady state solutions when $\mathbf{t_0}$ are not isotropic require some meridional circulation. He also generates approximate solutions for the induced differential rotation. A more realistic approximate solution was constructed by Durney and Roxburgh (1971).

Here we will gain some appreciation for the physical effect of turbulence on the mean mass motion by isolating the mean flow, $\vec{U}$, from the turbulent motion. We'll approach the meridional flow problem by assuming the measured surface turbulence is characteristic of the $\mathbf{t_0}$ tensor, then ask how the angular differential rotation should be related to a meridional circulation. In this case, if the derived meridional circulation is smaller than our initially imposed velocity field, we can hope that a self-consistent solution for $\vec{U}$ may be constructed by some iterative process.

The tensor quantities are more complex in spherical symmetry (but see below): we will consider the cartesian approximation where a point on the photosphere characterised by r, θ, and ϕ has local $\hat{z}$ parallel to $\hat{r}$, $-\hat{\theta}$ is parallel to $\hat{y}$, and $\hat{\phi}$ is parallel to local $\hat{x}$. The angular differential rotation leads to nonzero $\partial u_x/\partial y$.

We will see later from numerical experiments that the dominant term in $\mathbf{t_0}$ is $t_0^{rr} = t_0^{zz}$. This may have interesting consequences for the radial derivative of the rotation rate. Consider that the dominant off-diagonal term from eq. 4 is $T^{r\phi} = T^{zx} = -\tau t_0^{zz}\partial_z u^x$. Consider then a surface with normal in the $\hat{z}$ direction. The vector $\mathbf{T}\cdot\hat{n}$ yields a flux density (force per area) which is $-\hat{x}\partial u_x/\partial z$. While we haven't solved the full equation this force term represents a viscous force that opposes any radial (depth) variation in the rotation rate, $u_x(z)$. Thus, in the absence of other forces the rotation rate should evolve toward a constant ($\partial u_x/\partial z = 0$) in the vertical direction. This fact does seem to be borne out by current helioseismic observations of the vertical rotation profile, as you will learn from other lectures during this school.

Consider the $\hat{x}$ component of the LHS of eq. 2 for steady flows. Let's estimate the induced meridional flow u_y from the granulation convection turbulence. If we assume vertical gradients are small and that the flow is axisymmetric we get

$$\rho u_y \frac{\partial u_x}{\partial y} = t^{yy}\frac{\partial^2 u_x}{\partial y^2}$$

At a distance y from the equator, and using eq 1 to estimate the derivatives, we get $u_y \approx v_y^2\tau/y$. Letting the granulation velocity scale be, v_y=1km/s, with a lifetime τ=500s at midlatitudes we expect a meridional flow of only about 2m/s poleward. Larger convective motions, or longer flow lifetimes can produce stronger meridional flows. It is evident from these estimates that the turbulent meridional flow can have little to do with the measured limb shape distortion. Note that to reverse the direction of the meridional flow the equatorial rotation rate should be less than at higher latitudes.

Durney and Spruit (1979), and Durney (1989) have developed these equations further in order to quantitatively describe the effect of a more general circulation pattern (approaching the scale of the depth of the convection zone) on global circulation. The form of eq. 4 suggests that the observed latitudinal differential rotation (if it results from mean flows interacting with isotropic turbulence) is a consequence of radial variations in the rotation rate (in contrast to our above assumptions); or is due to anisotropy in the turbulent viscosity (cf. Kippenhahn 1963); or is the result of large-scale flow patterns like those investigated by Durney and collaborators.

4 INTERPRETING BRIGHTNESS DATA

4.1 Turbulent Heat Flow

Large-scale variations in the heat transport through the convective zone may be related to the observed effective photospheric temperature variations we noted above. To see this recall that the convective heat transport, F, depends on the convective velocity, $\vec{v}$, the specific heat c_p, and the temperature excess of the convected fluid element, Θ, as $\vec{F} = \vec{v}\rho c_p\Theta$. The temperature difference depends on the superadiabatic temperature gradient, $\nabla\Delta T$, as $\Theta = \tau\vec{v}\cdot\nabla\Delta T$. A suitable average over fluid elements shows that the convective flux then also depends on $\mathbf{t}$ as

$$\vec{F} = \kappa\cdot\nabla\Delta T = c_p\mathbf{t}\cdot\nabla\Delta T \tag{5}$$

If we assume a diagonal $\mathbf{t_0}$ but with $t_0^{rr} = a$ and $t_0^{\theta\theta} = t_0^{\phi\phi} = b$ then the components of $\mathbf{T}$ can be transformed using the cartesian tensor, $\mathbf{T}$, and the spherical metric tensor (cf. Durney and Spruit 1979)

$$T_r^r = -2a\tau\partial_r U_r$$

$$T_r^\theta = -\frac{b\tau}{r^2}(\partial_\theta U_r + r\partial_r U_\theta - U_\theta) - \frac{\tau(a-b)}{r^2}\partial_r(rU_\theta) = \frac{1}{r^2}T_\theta^r$$

$$T_r^\phi = -b\tau\partial_r(U_\phi/r) - \frac{\tau(a-b)}{r^2}\partial_r(rU_\phi) = \frac{1}{r^2\sin^2\theta}T_\phi^r$$

$$T_\phi^\theta = -\frac{b\tau}{r}\sin^2\theta\partial_\theta U_\phi = \sin^2\theta T_\theta^\phi$$

$$T_\phi^\phi = -\frac{2b\tau}{r}(U_r + \cot\theta U_\theta)$$

$$T^\theta_\theta = -\frac{2b\tau}{r}(\partial_\theta U_\theta + U_r)$$

This equation closely follows the expression used by Kippenhahn (1963). The superadiabatic gradient is a covariant vector (indices down). If we assume a purely radial local temperature gradient then the resulting turbulent heat flux has the form

$$(F_r, F_\theta, F_\phi) = c_p|\nabla\Delta T|(a + T^r_r, T^r_\theta, T^r_\phi)$$

Let's continue to assume that anisotropy in the convection zone tends to minimize radial gradients in U^i. Then the magnitude of the heat flux at the pole and at the equator are $F_{e,p} \propto \sqrt{a^2 + (\tau(a-b)U_{\phi,e,p})^2/r^2}$. Thus the fractional pole-equator change in the convective flux is

$$\Delta F/F \approx (\frac{\tau\delta U_\phi(a-b)}{aR})^2/2$$

We will argue later that $a > 10b$, so we will assume strongly anisotropic convection here also. We use $\delta U_\phi = 1$km/s (the rotation velocity) and pick a timescale of 10^3s (corresponding roughly to granulation timescales). We find that the scale of the pole-equator flux variation is about 10^{-3}. Of course we are only estimating the magnitude of a single term here. Other steady flows could change the actual flux variation, and changes in the superadiabatic temperature gradient have been ignored. Nevertheless the calculation illustrates how turbulent anisotropy may lead to a pole-equator flux variation. The magnitude of this variation is small, but notable. More complete (but still non-realistic) solutions (cf Durney and Roxburgh 1971) which include effects due to a meridional motion also tend to produce a hotter equator but with a flux excess that may be a factor of 10 or so higher.

The pole-equator flux difference has its origin in the coupling between rotation and the strong radial turbulent velocities. The natural scale for this flux, or effective temperature, excess is about 1^oK which agrees in magnitude with the measured polar temperature excess (fig. 3), but disagrees in sign.

4.2 How Many Irradiance Components?

Satellite observations (Willson 1991) established conclusively that the solar irradiance is not constant. The irradiance data in Fig. 4 are changing on all scales between the 5min oscillations out to the solar cycle timescale. The variability on day to year timescales is of most interest to us here. These irradiance changes have a large amplitude and are directly related to changes occuring at the photosphere and in the solar interior. One of the first questions we could ask about these longer periods is how many distinct physical solar phenomena contribute to the irradiance? We started these lectures by noting that the magnetic field seems to dominate the asphericity and temporally changing observables. How does the magnetic field affect the irradiance? Figure 7 shows how the mean Kitt Peak magnetic field measurements correlate with the irradiance. The bottom panels in this figure show the mean daily irradiance and magnetic field residuals after subtracting a 150d running mean. This establishes that the high frequency variability in the irradiance is anticorrelated with the mean solar magnetic field. The top panels show that the correlation before removing the slower variability is positive, i.e. on timescales longer than one month months the irradiance has just the opposite magnetic dependence and increases with increasing magnetic field.

¿From these observations, even before looking at a picture of the sun, we conclude that there are at least two physical components to the solar irradiance. The simplest interpretation is just that the negative correlation comes from sunspots (dark), while the positive correlation results from faculae (bright regions). Of course this is not entirely satisfying since the difference in timescale between the correlated and anticorrelated behavior is unexplained. Could it be that faculae/active regions evolve over very different timescales than spots, or that the ratio of faculae to spots is a strong function of the average number of sunspots – there being, fractionally, very many more facular regions near sunspot maximum? Schrijver (1988) looked for systematic differences in the size distribution of active regions during a solar cycle. He found no evidence that the phase of the cycle was correlated with changes in the facular distribution. The alternative conclusion is that the irradiance depends on more than the spot and facular brightness contributions.

Sunspots are nearly invisible in limb observations, but faculae increase in constrast as they approach the limb and are easily identified there. The oblateness experiment did a good job of measuring the bolometric flux within 20" of the limb with an angular resolution of about 1.5°. Small brightness variations were recorded and a large statistically useful dataset was obtained from about 100d per yr during 7 years between 1983-1990. The brightness signal showed an intermittent ("facular") contribution and a latitudinal variation in the residual mean limb brightness (Kuhn and Libbrecht 1991). Figure 8 shows the clear signature for the average limb brightness due to faculae. The seasonally constant, but solar-cycle variable, limb brightness change was plotted in Fig. 3. This brightness distribution is quite distinct, in latitudinal and temporal variation, from the facular signal and represents a "third" component, in addition to sunspots and active region faculae, that contribute to the solar irradiance. It is worth noting that the correlative studies using proxy indicators of faculae (cf.

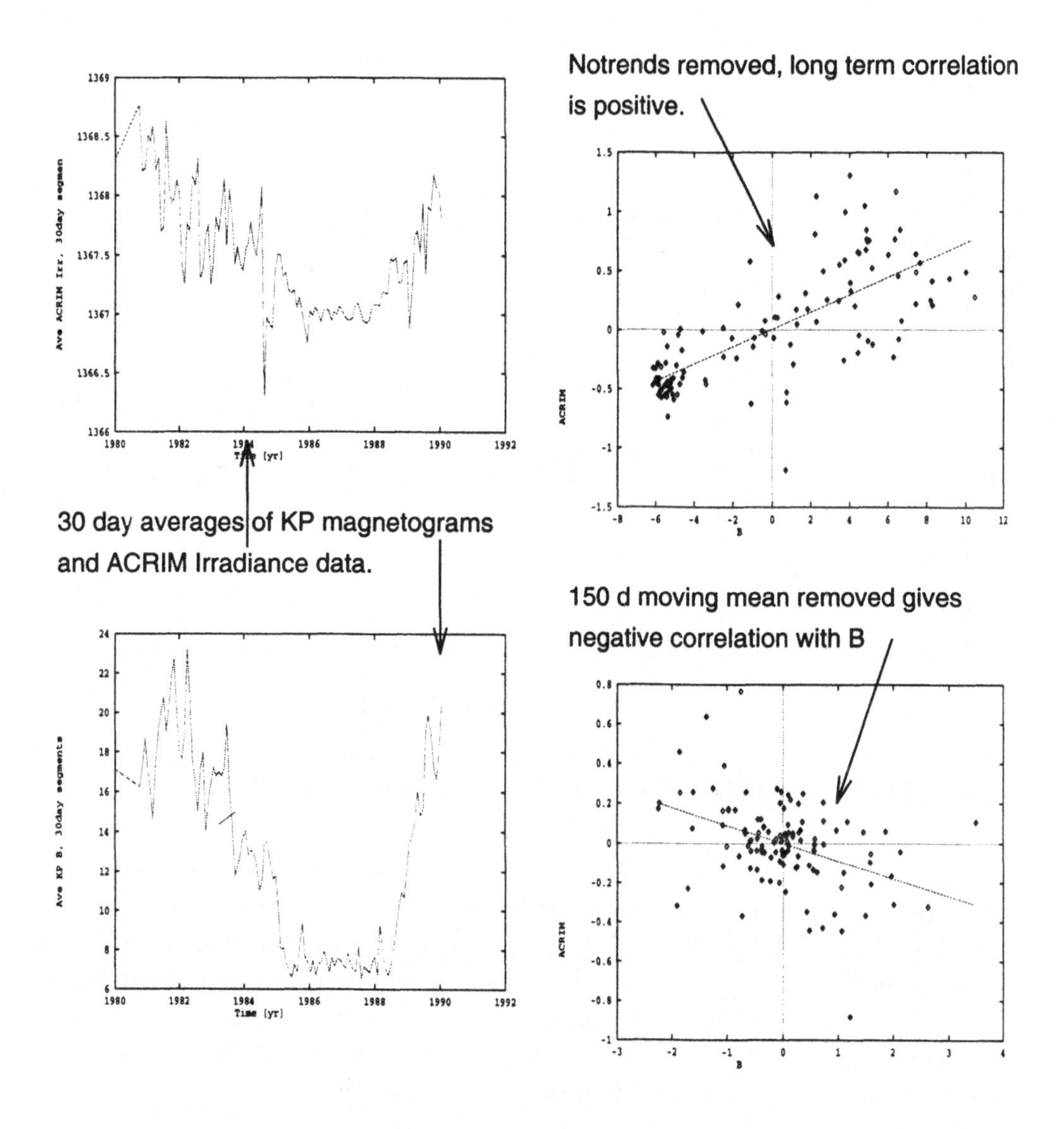

Figure 7: *Correlated and Anticorrelated Behavior of Irradiance and Magnetic Fields*

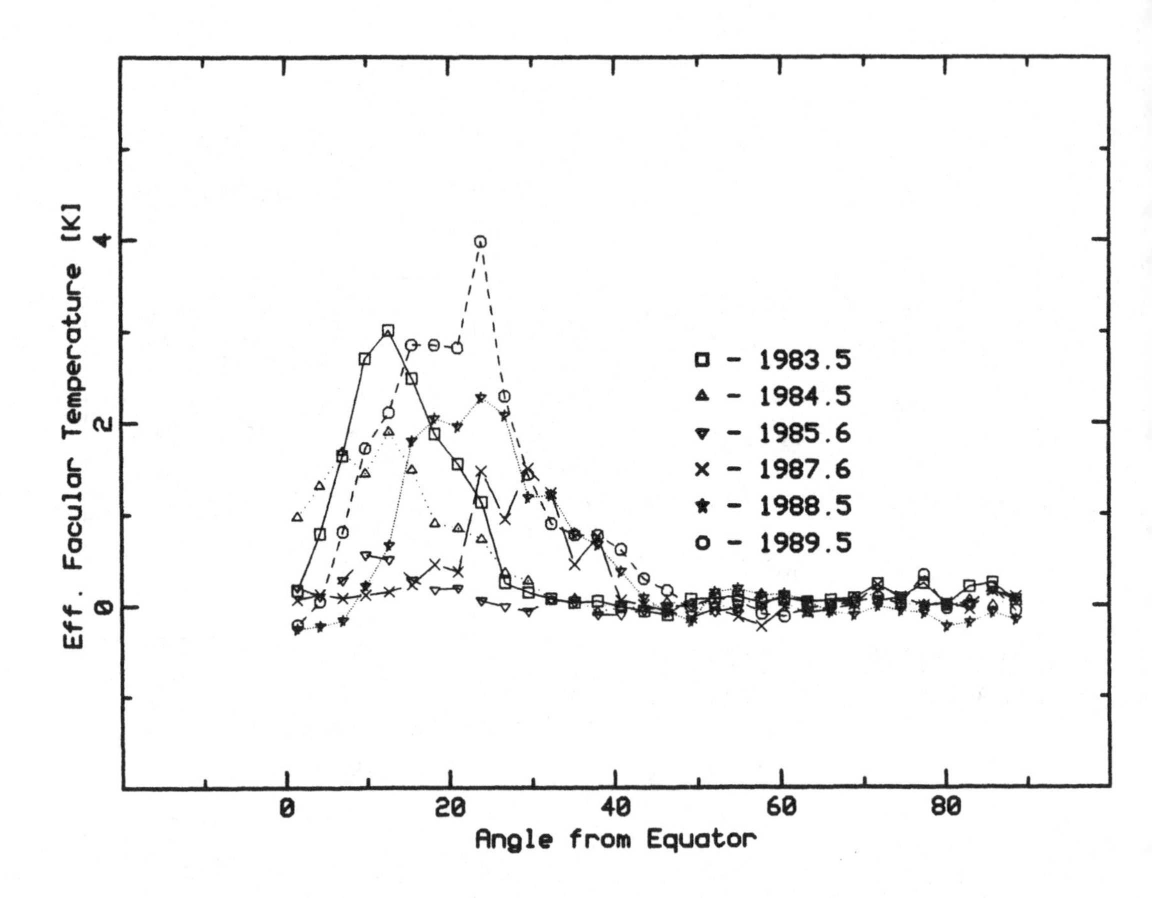

Figure 8: *The Mean Facular Contribution to the Limb Brightness*

Foukal and Lean 1990) were also unable to reproduce the long-term irradiance changes without a third brightness component. This other brightness signal has not been identified. Foukal and Lean (1990) postulated that it may be due to changes in the continuous "network" facular distribution associated with supergranular cell boundaries. Unfortunately the limb observations, which directly measure this brightness component, do not have sufficient spatial resolution to identify the surface morphology of the residual limb brightness.

4.3 Irradiance or Luminosity Changes?

Is the total energy output from the sun changing or are the irradiance variations the result of a changing nonisotropic radiation field at the photosphere? Although the ERB and ACRIM data are corrected to give the mean flux at 1AU – with only one vantage point to measure the sun's irradiance from this is not a simple question to answer. Note that magnetic fields, either as sunspots

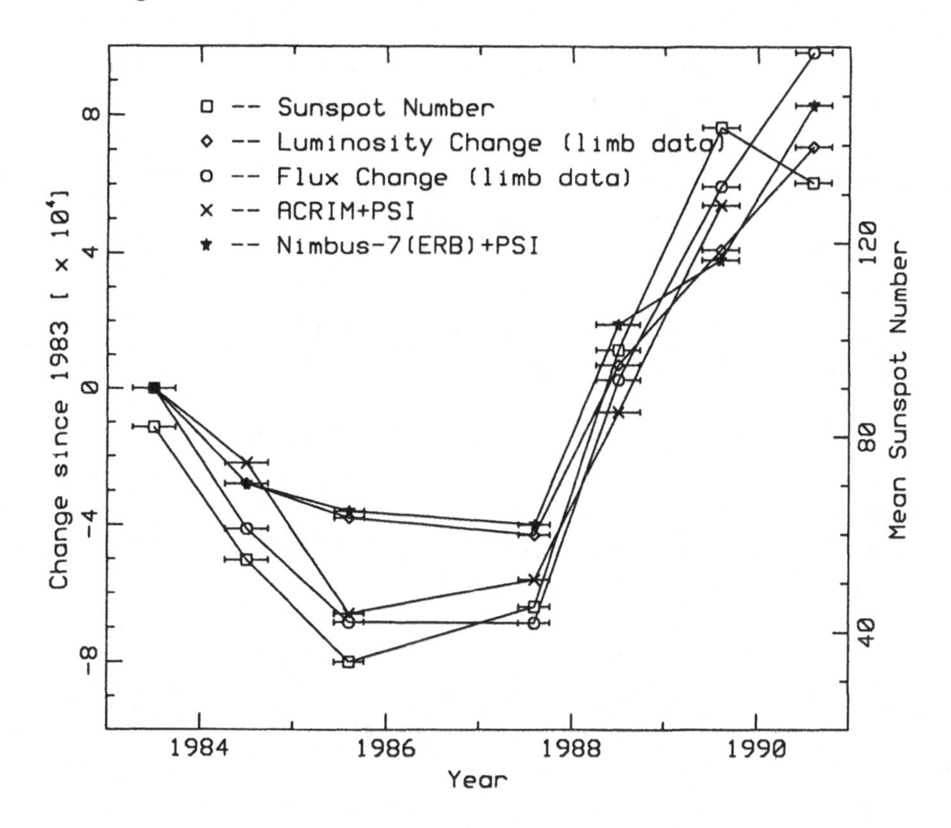

Figure 9: *Solar Cycle Changes of the Mean Solar Irradiance and Luminosity*

or faculae, have the effect of redistributing the emergent intensity so there is more energy at large angles from the local photospheric normal, i.e. sunspots are dark and more apparent near the disk center, while faculae are bright and more visible near the limb. Thus, on average the solar irradiance will change just because the latitudinal distribution of magnetic fields varies with the solar cycle.

On timescales longer than the rotation time we can compute the solar luminosity or irradiance by integrating the limb brightness measurements. These observations measure the average limb brightness over several rotations. If we use a measured facular contrast function (cf. Muller 1975) we can compute the full disk brightness change due to faculae from the facular data in fig. 8. The residual limb temperature (from fig. 3) can also be integrated over the full disk to get the third irradiance component. The sunspot contribution to the irradiance is tabulated as a "photometric sunspot index" (cf. Hudson et al. 1982) and is the final ingredient in the total brightness change calculation. The resulting fractional irradiance and luminosity change are then normalized to the 1983 solar values and have been plotted in figure 9. Notice that the average of the ERB and ACRIM irradiance change tracks the irradiance variation computed from the limb observations. It is also clear that the facular signal alone is more than a factor of two too small to explain these long term brightness changes. The sunspot number is also plotted and evidently has a slightly different phase (it peaked before the irradiance by about a year or two). Finally we can use the same data to compute the *luminosity* change (assuming axial symmetry and averaged over several rotations). The calculations and data demonstrate that the solar irradiance and luminosity are changing with the same phase and relative amplitude, and that the dominant contribution to this variation is *not* from active region faculae.

What about the faster irradiance changes? Does the solar luminosity change on timescales of a few days or less? This is a harder question to answer, and we really don't have the data to solve this problem yet. We can show that the short term irradiance data are consistent with *no* luminosity fluctuations, but can't quite rule out the possibility of any short term changes.

We will assume that evolutionary effects in sunspots and active regions are dominated by the effect of solar rotation on an active region contrast function. As the photospheric features that cause the variation in the satellite irradiance records rotate across the solar disk we can hope to see systematic changes in the irradiance caused by the changing line-of-sight direction through the feature. For example, a sunspot causes the largest negative dip in the irradiance when it is near disk center. If spots/active regions have redistributed that flux to larger angles then we could expect to find an *increase* in the irradiance record 1/4 of a rotation before or after the negative dip. A single observation or even a few timeseries records may not tell us much because of the large "noise" associated with the averaging effect of many photospheric features which contribute to each single irradiance measurement. The approach I took was to look at the autocorrelation of the irradiance data after filtering the slow (yearly) changes. If the flux redistribution is significant we can expect a *negative* dip in the autocorrelation function corresponding to 1/4 of a rotation, or about 6-7 days. The negative autocorrelation arises because dips in the irradiance should, on average, be accompanied by a flux excess 6-7 days before or after the negative fluctuation. Of course what we measure in the irradiance record is the collective effect of many photospheric features, and we should think of this procedure as yielding an estimate of the contrast function for the largest features or active regions. Figure 10 shows the autocorrelation from two years of ACRIM date obtained during the rising phase of the solar cycle. The negative peak at 1/4 rotation is apparent.

It is a non-trivial inversion problem to obtain the short term irradiance contrast function from the autocorrelation data, but a simple forward calculation

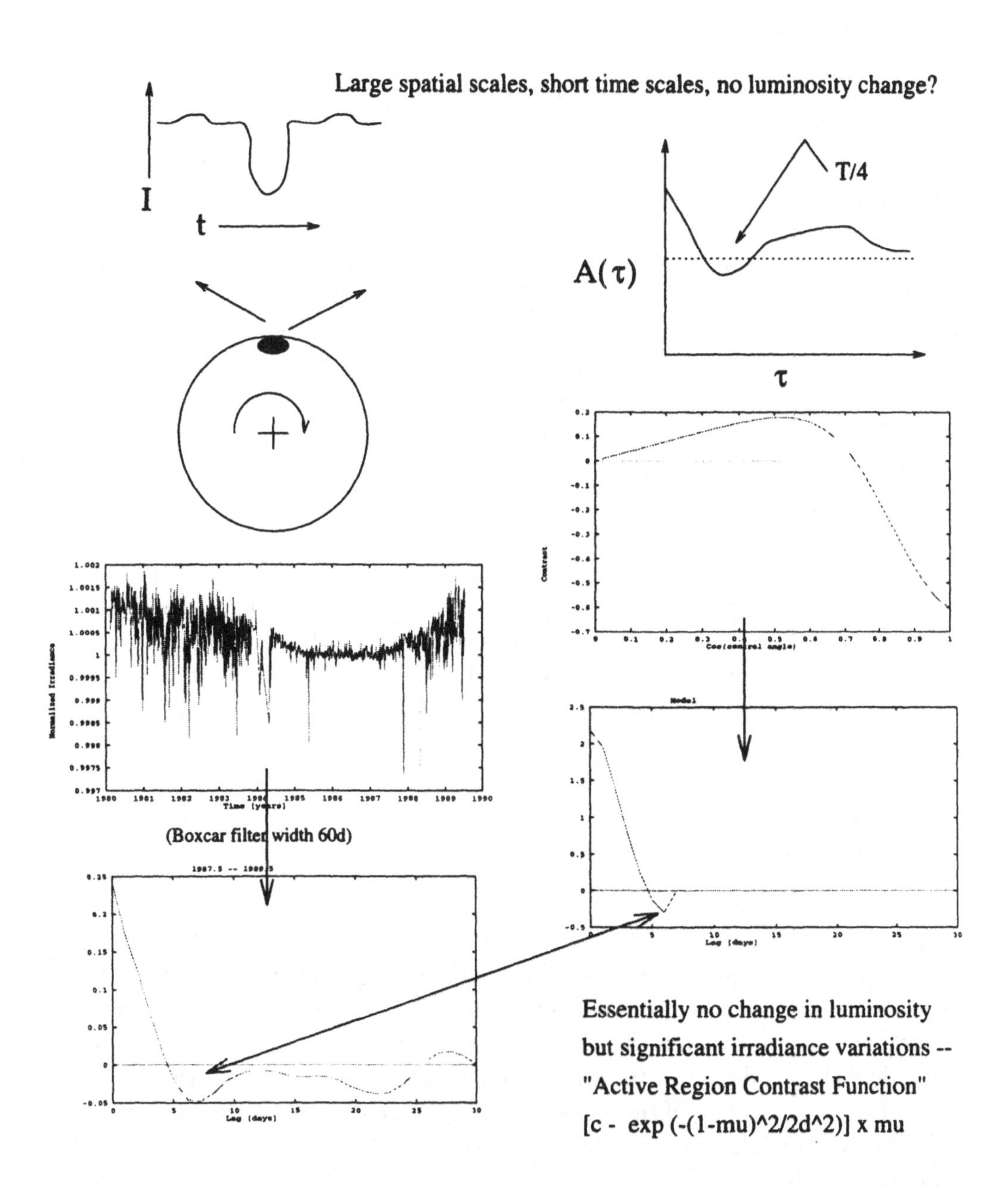

Figure 10: *ACRIM Autocorrelation Analysis*

will illustrate an important point. We arbitrarily choose a two parameter contrast function of the form $dI(\mu)/I = (c - \exp\{-\frac{(1-\mu)^2}{2\mu_0^2}\}) \times \mu$ where the parameter c and μ_0 are to be determined. Figure 10 plots an example of such a function and its resulting autocorrelation assuming a 27d solar rotation rate. The choice of $c = 0.4$ and $\mu_0 = 0.3$ yields an autocorrelation function that has the same shape as the data plotted in fig. 10. Notice that this contrast function exhibits an irradiance decrease near disk center ($\mu = 1$) and falls to zero at the extreme limb. More interestingly the integral of this function over 2π steradians is essentially zero, i.e. it implies no luminosity change. Thus, we may conclude that the short term irradiance fluctuations do not imply any change in the solar luminosity on rotation timescales. Refining this conclusion requires precise *full-disk* photometry from a long (several rotation periods) timeseries. Then a statistical measure of active region *luminosity* changes could be extracted from the rotationally modulated surface brightness records. To compare one timeseries record with the next, it is critical that the local active region brightness be known with respect to the solar surface brightness far from the region, i.e. global (full-disk) photometry is needed. Since ground-based, higher resolution observations, cannot measure the brightness normalization (the total irradiance) because of uncertain atmospheric extinction corrections, the satellite irradiance record is crutial for interpreting the resolved datasets (e.g. note that the interpretation of the limb data depended critically on the comparison with ACRIM or ERB timeseries).

4.4 Timescales

The problem of describing heat transport in the convection zone (CZ) is difficult because many phenomena are linked by timescales that vary dramatically between the top and the bottom of the convection zone. For example, the timescale for the upper few thousand kilometers of the solar CZ to radiate its stored energy is less than a day, and the time to convect this energy is less than a few hours while from the base of the convection zone these timescales are respectively 2×10^5 yr and one month. There are many scenarios by which solar irradiance or luminosity changes could be generated by magnetic fields within the CZ – read Spruit (1991) or Parker (1994) for a sampling of some of the important issue in idealized problems. Our approach here will be to stay close to the data while trying to sort through some of the brightness mechanisms.

A fundamental issue in this problem is whether entropy perturbations near the photosphere can be coupled to the entire convection zone. The heat capacity of the CZ is large so if localized entropy changes (like the perturbing influence of sunspots) are coupled to the CZ then there is some belief (cf. Spruit 1991) that the perturbed flux from below the region goes toward a small temperature change in the convection zone, and an insignificant contribution to the solar luminosity on observable timescales. This belief originates from a mixing-length calculation of the "effective turbulent thermal conductivity." In effect then a sunspot must change the luminosity (and irradiance) of the sun by blocking flux from below, which is not reradiated on timescales shorter than the thermal timescale of the CZ (10^5 years). In this picture a sunspot is a near-perfect "plug" for thermal radiation. This scenario must be modified if sunspots are very shallow, not extending below the superadiabatic layer, but this is entirely unphysical given the good evidence that sunspot magnetic fields are organized by the global scale solar magnetic field (i.e. they are deep). In fact, as we discussed above, the irradiance data provide no evidence that sunspots change the solar luminosity at all. It is not that the missing flux appears as a bright ring around the spots, but that it may be redistributed in angle.

The rather surprising expectation that the entire CZ participates in entropy changes near the photosphere runs into obvious problems with the observation of largescale photospheric temperature variations. This effective bolometric temperature change with latitude and time could not be supported if the surface entropy changes are coupled to an isotropic CZ. The apparent escape is to confine the changes to the extreme outer superadiabatic region of the CZ, thereby decoupling them from the efficient turbulent mixing from below. In this case, as we argue below, there should be no luminosity change on long timescales, in contradiction with previous observational conclusions.

Let's examine the suggestion that small flux tubes protruding through the superadiabatic layer could provide thermal "leaks" which increase the solar luminosity. The first problem we must solve is to carefully avoid blocking the convective flux from below, as is the case for the larger flux bundles in sunspots, because this would tend to *decrease* the luminosity. The more significant problem is that if we decouple the energy output of the superadiabatic layer from the CZ and increase its radiative efficiency in localized regions by using magnetic fields to decrease the local gas density, then we do not increase the overall luminosity of the region very much. The outer atmosphere is exponentially stratified in both photon free-path-length and heat capacity. If we make a small "antiplug" so that the local radiation resistance is decreased the excess energy radiated from the "antiplug" comes predominantly from *horizontal* layers near the effective radiating surface. Thus if we consider the total energy radiated by a box around the "antiplug" there is only a small increase, since the local excess results in a radiation deficit in the surrounding region.

Figure 11 summarizes a simple numerical model that illustrates this point.

We start with a diffusive 2-d model that has exponentially increasing thermal conductivity with height, and exponentially decreasing heat capacity with height (in analogy with the solar atmosphere). We solve for the temperature using $dT/dt = ke^{z/h}\nabla \cdot e^{z/h}\vec{\nabla}T$, where h is the scale height and k is an arbitrary constant. The calculation displayed in this figure used a 64×64 grid with a scale height that was 6 grid units. As indicated in fig. 11 an antiplug with variable depth was inserted (adjusting the density to be 10 percent of the unperturbed density) in the region and the flux through the indicated surface was computed as a function of time. Dimensional variables were scaled to match mean solar conditions near the photosphere and the equilibrium condition was solved for using fixed temperature boundary conditions on the top and bottom surface and periodic conditions on the horizontal surfaces. The graph in fig. 11 shows how the antiplug initially increases the total luminosity from the box, but after several hours the luminosity drops back to nearly the inital value after the surrounding layers have adjusted to the perturbation (the arrow on the graph). The greyscale figures show the temperature changes soon after the perturbation, and after it equilibrates. The horizontal adjustment in temperature decreases the outward emergent luminosity. By making the antiplug longer and deeper the luminosity change could be increased, but not by much. With an antiplug between 1 and 4 scale heights deep we could increase the luminosity by only 0.3 to 4.7% of the initial flux per pixel.

One final point should be made of the delay between the maximum in the observed photospheric magnetic flux and the maximum luminosity. The sunspot number and the Kitt Peak full-disk magnetograph data are well correlated so that the sunspot number seems to be a good proxy for the total magnetic field. It is puzzling then that the sunspot number (and the facular brightness contribution) peaked before the solar luminosity (fig. 9) by a year or two. If the luminosity change is driven by photospheric "leaks" then the maximum luminosity change should occur immediately after the magnetic flux appears at the photosphere (note that the luminosity will start to decay just a few minutes after the field appears, fig. 11). A similar behavior has been observed in a few stars (D. Grey, personal communication) where the peak brightness change occurs after the chromospheric (magnetic) proxies.

The observational evidence is against a model of the solar CZ where entropy perturbations are well mixed, only to reappear at the photosphere after timescales of 10^5 years. The reason is that our mixing-length model, while it is adequate for describing equilibrium stellar properties, does not describe the *transport* properties of the CZ, as we will see below when we experiment with small regions of the CZ using numerical simulations.

Exponential Diffusion Model: Flux = K exp(z/h) grad T

dT/dt = k exp (z/h) div Flux

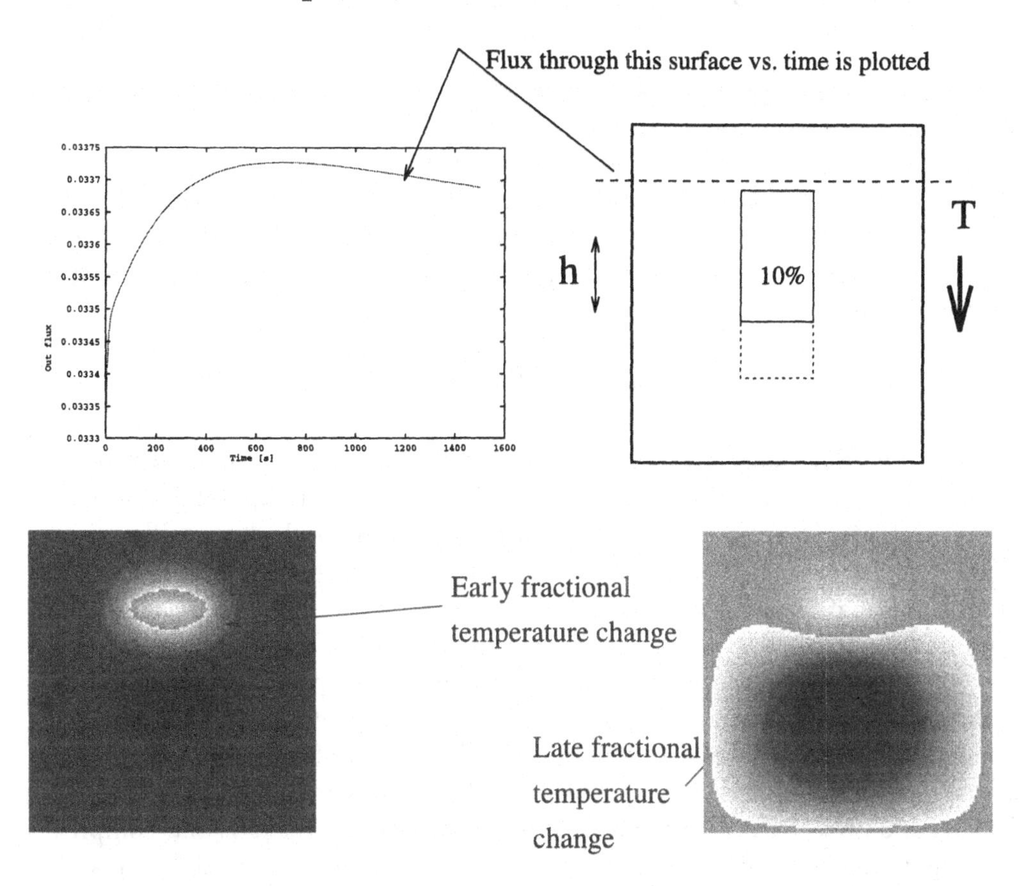

Late time flux change is between 0.3 and 4.7% of the initial flux per pixel

luminosity contrast is attenuated

Figure 11: *Summary of Diffusion Calculation for Surface Magnetic Field*

5 EXPERIMENTING BELOW THE PHOTOSPHERE

5.1 Helioseismic Tools

It should be obvious by now that the acoustic oscillations are a wonderful probe of the solar interior. Small variations in the oscillation frequencies have been observed and are a direct measure of structural changes occuring within the sun. You will see the oscillation equations several times during this school, but I am fond (Kuhn 1989) of a particular formalism I learned from Dicke (1978) which is geometry independent and compact. We start from a fluid Lagrangian density written in the form

$$L = \frac{1}{2}\rho g_{jk}\dot{\xi}^j\dot{\xi}^k - \frac{1}{2}\rho c^2(\nabla\cdot\vec{\xi})^2 + \frac{1}{2}\xi^k\phi_{,k}\xi^j\rho_{,j} + \rho\xi^k\phi_{,k}\nabla\cdot\vec{\xi}$$

Here ξ is the Eulerian fluid displacement, comma's represent partial differentiation with respect to a spatial index, over-dots represent temporal derivatives, $\mathbf{g} = diag(1, r^2, r^2\sin^2\theta)$, ϕ is the gravitational potential (which we will assume is not changed by the oscillation, ξ^j. This expression is geometry independent, in that the matrix $\mathbf{g}$ contains all the information we need to know about the coordinate system, as long as we use a covariant formalism. Note then that $\nabla\cdot\vec{\xi} = \frac{1}{\sqrt{G}}(\sqrt{G}\xi^i)_{,i}$ where $G = r^4\sin^2\theta$ is the determinant of the metric tensor $\mathbf{g}$ and indices are raised and lowered by multiplying by corresponding diagonal elements of the metric tensor. The physical derivation of each term in this expression can be found in Tolstoy (1963), although you may recognize that the first term on the RHS is the kinetic energy density, the next is the compressional potential energy density, while the last two terms include the bouyancy potential energy density. With ξ as our canonical variable we obtain from the Euler-Lagrange equation the following equation for the equation of motion of $\xi^1 = y(r)/r^2$

$$y'' + y'\frac{d}{dr}\ln\{\frac{\rho c^2\omega^2}{r^2(\omega^2 - c^2k^2)}\} + y\{\omega^2/c^2 - k^2 - \frac{g}{c^2}\frac{d}{dr}\ln\{\frac{\omega^2 g}{(\omega^2 - c^2k^2)r^2}\} + \frac{k^2N^2}{\omega^2}\} \tag{6}$$

Here $N(r)$ is the Vaisala bouyancy frequency $(\sqrt{-g(\rho_{,1}/\rho + g/c^2)}$, $k^2(r) = l(l+1)/r^2$ and we have assumed spherical harmonic angular dependence to the $j = 1-3$ components of ξ. Thus the covariant $j = 2,3$ components are given by

$$\xi_j = \frac{gy - c^2y'}{r^2(\omega^2 - c^2k^2)}\frac{d}{dx^j}Y_{lm}$$

The primes refer to radial derivatives and g here refers to the local gravitational acceleration.

If we restrict our attention to modes with periods near 5 min and with $l < 100$ then $\omega^2 > c^2k^2$. Also $\omega^2 > N^2$ and we can simplify Eq. 6 to

$$-c^2y'' - y'(c^2/H + c^{2\prime} - 2c^2/r) + y(c^2k^2 - 2g/r + g') = \omega^2 y \qquad (7)$$

Here $H(r)$ is the density scale height. Our solution for the radial part of the eigenmode now depends on only three functions of r, $c^2(r)$, $H(r)$, and $g(r)$. If we go to the standard solar model to get these functions then it is straightforward to convert Eq. 7 into a tridiagonal matrix eigenvalue problem (since it is an explicit 2nd order differential equation). Note that solutions are degenerate in spherical harmonic index m, and that $y_{nlm}(r,\theta,\phi) = y_{nl}(r) \cdot Y_{lm}(\theta,\phi)$. Following Cox (1980) we impose boundary conditions on y_{nl} so that ξ^1 is zero near $r = 0$. To satisfy the zero pressure derivative boundary condition at the top we also require $y'_{nl} = 0$ at the surface. We will not explicitly display the difference equations corresponding to Eq. 7 but see Press et al. (1986) and Dubrelle (1970) for a general discussion of efficient solution techniques for such tridiagonal eigenvalue problems. Note that each choice of l leads to a single eigenvalue problem and all values of n are naturally obtained for a given angular harmonic index l. Then the finite difference eigenvalue problem takes the form $\mathbf{M}^{(l)}\vec{y}_l = \omega^2_{(l)}\vec{y}_l$. In general the solution space includes eigenvalues ω^2_{nl} and eigenvectors $\vec{y}_{nl}$, $n = 1, \ldots, N$. Notice that the eigenvectors satisfy the integral orthogonality relation $\int_0^R y_{pl}(r)y_{ql}(r)\rho r^{-2}dr = \delta_{pq}$.

5.2 Asphericity

How does an aspherical sun affect these equations? Let's consider perturbations of the form

$$c^2(r,\theta) = c^2(r)\{1 + \sum_i c_i f_i(r) P_i(\cos\theta)\}$$

and

$$g(r,\theta) = g(r)\{1 + \sum_i c_i f_i(r) P_i(\cos\theta)\},$$

where P_i are Legendre polynomials and we require $d\ln f_i/dr \ll 1/R$ We normalize the relative perturbations so that $f_i(R) = 1$. Notice that we have explicitly assumed that the sound speed and gravitational acceleration asphericities are equal. This is accurate to the extent that the local temperature determines c^2 and von Zeipel's theorem holds. In any case, for high degree 5-min modes the terms in Eq. 7 that depend on g are small compared to the the c^2 terms.

Substituting the perturbed sound speed and acceleration into Eq. 7 and using the integral orthogonality relation yields an equation for the perturbed frequency difference $\delta\omega_{nlm} = \omega_{nlm} - \omega_{nl}$,

$$\delta\omega_{nlm} = \sum_i c_i R_{iln} S_{ilm} \omega_{nl}/2 \qquad (8)$$

where $S_{ilm} = \int Y^*_{lm} Y_{lm} P_i d\Omega$ and $R_{iln} = \int f_i(r) y^2_{nl}(r) \rho / r^2 dr$. Observationally the frequency splittings are well described by a Legendre polynomial expansion in m/l of the form

$$\delta\omega_{nlm} = 2\pi \sum_i b_i P_i(-m/l)$$

and what I have called the b_i coefficients (cf. Kuhn 1988) are related to the measured a_i coefficients of, for example, Duvall et al. (1986) by $b_i \equiv l a_i$ where l is the spherical harmonic of the observed mode (or some appropriate average)

There is another important simplification of Eq. 8 if we work on S_{ilm}. Dziembowski (1988) displays some useful recursion relations for the spherical harmonic moments of $\cos\theta$. Using these we can show that for large l, $S_{2lm} \to -\frac{1}{2} P_2(m/l)$ and $S_{4lm} \to \frac{3}{8} P_4(m/l) \ldots$. In general the coefficient of $P_{2s}(m/l)$ is $(-1)^s \frac{(2s-1)!!}{(2s)!!}$. Thus, an asphericity of angular harmonic form $P_{2s}(\cos\theta)$ generates frequency splittings with an m dependence proportional to $P_{2s}(m/l)$. The radial dependence of the asphericity appears in the l and n dependence of the splittings. For example, suppose the asphericity had no depth dependence so that $f_i(r) = 1$, then the relationship between splittings and asphericity for all n and l is simply

$$b_{2nl} = -0.04 c_2 \omega_{nl}, \quad b_{4nl} = 0.03 c_4 \omega_{nl}, \quad \ldots$$

5.3 The Asphericity Data and Splitting Coefficients

The formalism developed in the previous section simply uses knowledge of the perturbed sound speed assuming that all other relevant thermodynamic quantities are unchanged. For example, a concern one might have is that the location of the outer boundary condition (limb shape) is also perturbed along with the sound speed. In fact, from our previous discussion of the limb data, we know that limb shape changes are two and a half orders of magnitude smaller than the limb temperature and thus irrelevant to the discussion of the splittings. Our approach then will be to use the limb temperature data to estimate the sound speed asphericity at the photosphere. We will not address the question here of how the sun manages to perturb its surface brightness distribution (without changing its shape), it is enough to know simply that *that is what the sun does.*

The simplest comparison between brightness observations and the helioseismic data is to average splittings over the 5-min band in both l and n. Then

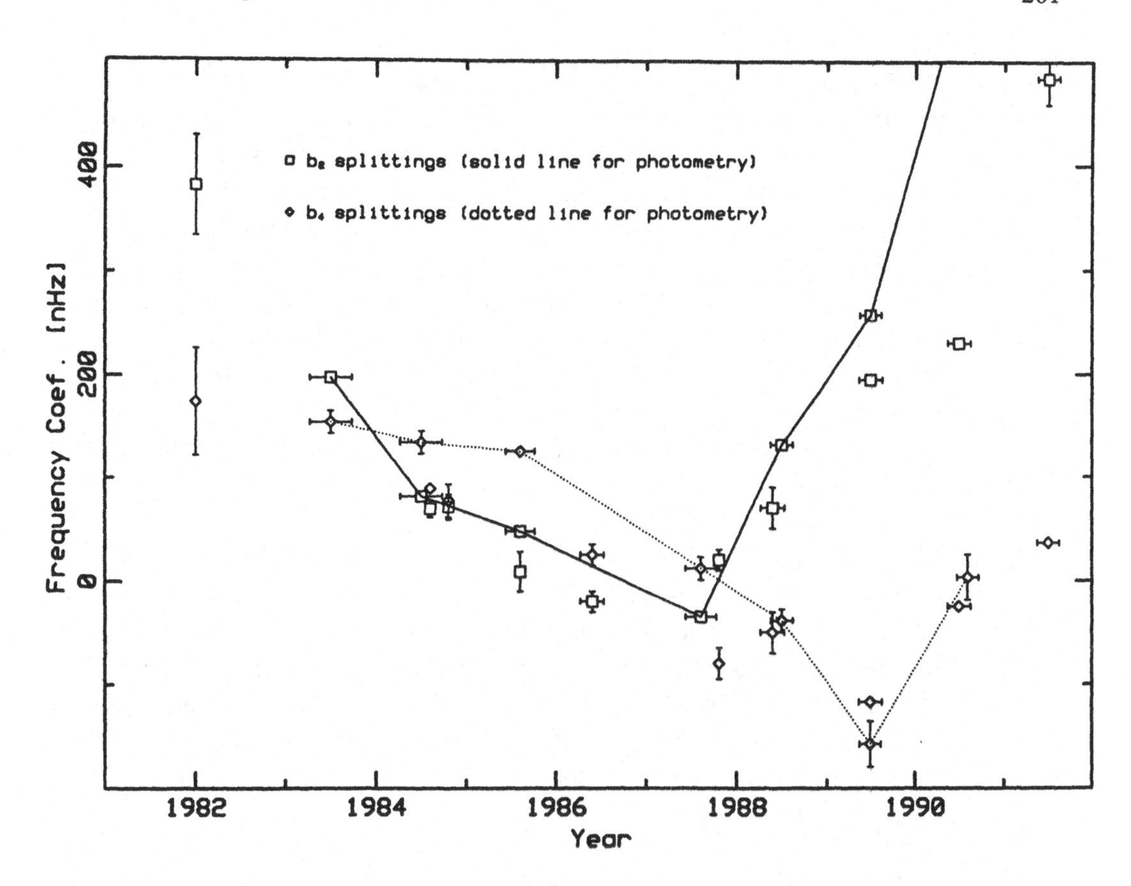

Figure 12: *Photometric-Helioseismic Splitting Comparison*

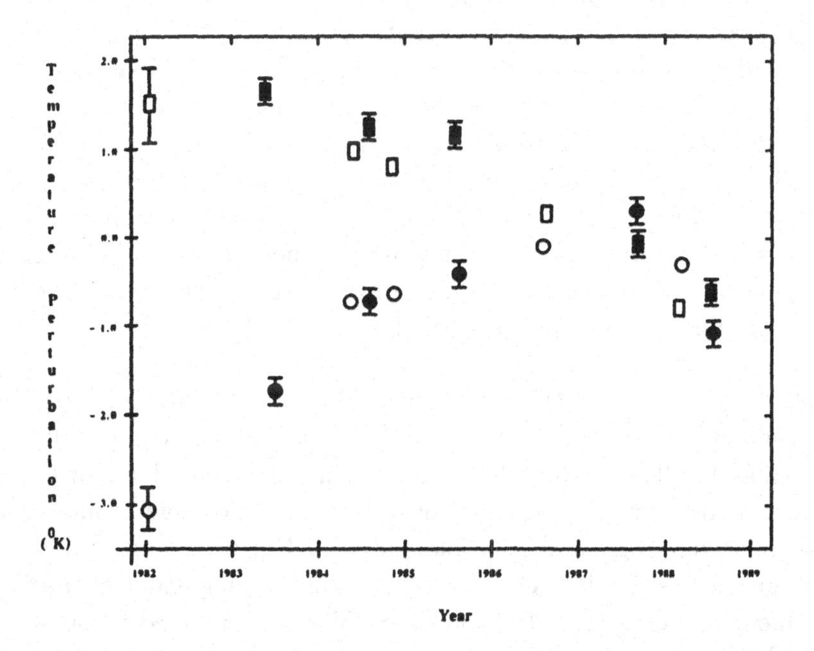

Figure 13: *Rotation Corrected Splitting-Photometry Comparison*

if we assume a constant asphericity $f(r) = 1$ we might expect to make an order of magnitude comparison. In this case a Legendre polynomial expansion of the surface temperature yields coefficients (t_{2s}) and we have $b_{2s} = K_{2s}t_{2s}$ with $K_0, K_2, \ldots = 175, -140, 110$ nHz/K. Figure 12 shows the splitting data plotted with temperature measurements (converted to equivalent splittings in this simple approximation). I found the agreement in trend with time and magnitude to be striking (cf Kuhn 1990). Goode and Kuhn (1989) corrected the splitting data for the contribution due to rotational stretching and found even better agreement (fig. 13). Here the filled in symbols refer to the photometric splitting predictions and the open symbols are helioseismic results (see Goode and Kuhn 1989 for the references to the original data). The circles refer to $l = 2$ and the squares refer to $l = 4$.

In our model the mean frequency shifts are determined by the $s = 0$ even coefficients of the temperature decomposition (through a calculation no different than for the other even coefficients). Physically it is the angular mean temperature change that is responsible for the centroid frequency shifts. Figure 14 compares the low-l helioseismic data with the $s = 0$ temperature coefficients (see Kuhn 1991 for references to the data sources). Again the comparison is reasonable considering that no free parameters and an "un-tuned" solar model went into this calculation.

The radial kernel in the calculation of R_{iln} is essentially the kinetic energy density of the mode. Since the 5-min p-modes have their energy density concentrated near the surface, the asphericity, $f(r)$, for $r < 0.9$ has little effect on the observed splittings. Figure 15 shows how the required surface amplitude drops as a function of the assumed depth of the perturbation ($f(r) = 0$ below the depth plotted on the horizontal axis). The data in this figure were obtained in late 1984 (see Goode and Kuhn 1989 for references). Varying the depth of the perturbation changes the required temperature asphericity from about 1-4K for the $l = 2, 4$ components (the measured asphericity was about ± 1K). The surface photometry data agrees in magnitude and variation with the splittings without fine-tuning the surface asphericity.

I am not aware that a structural inversion for the asphericity has been accomplished yet. In general the noise in the even-order splitting coefficients has been too high to allow meaningful inversions, as have been done for the internal rotation. The frequency dependence of the frequency change in modes between times of relatively high and low solar activity has been analysed (cf. Kuhn 1990). Figure 16 shows BBSO measurements of the magnitude of the frequency shift of modes observed in 1986 and 1988. The lines plotted through the data show the best fitting asphericity for several depth assumptions. It is quite clear

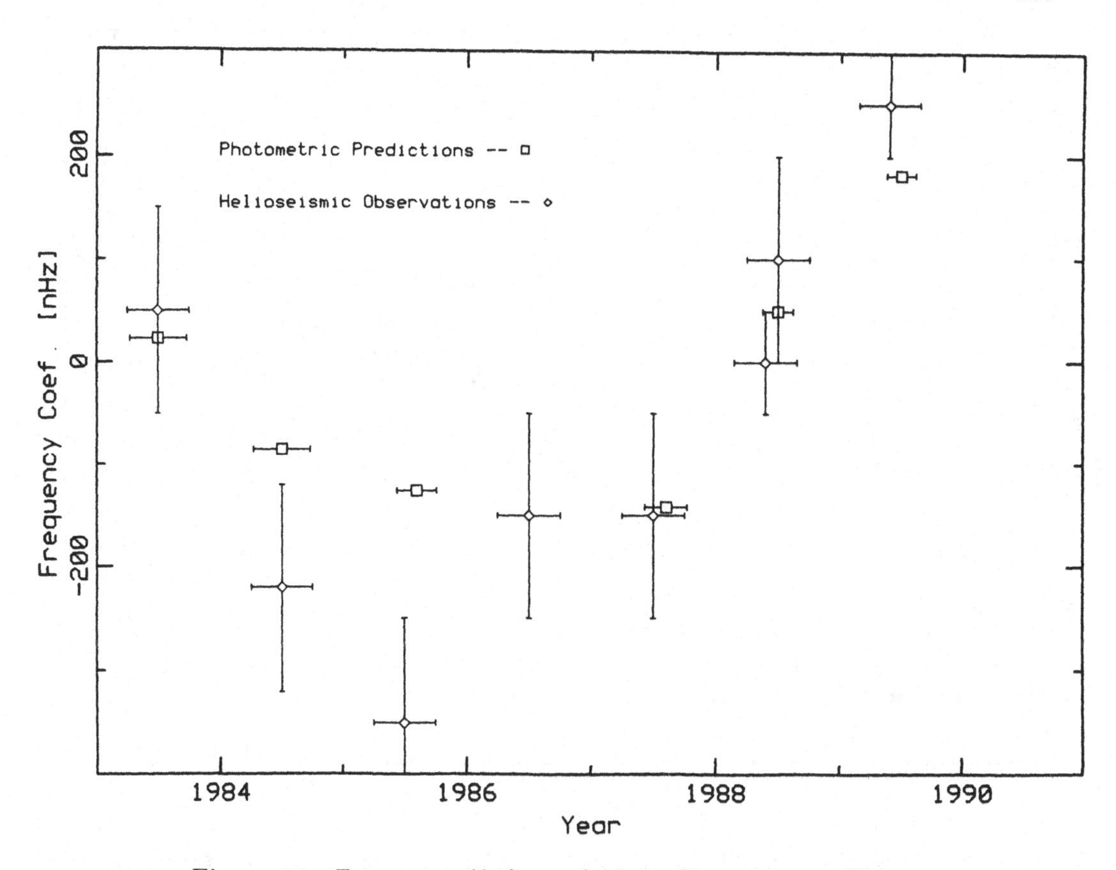

Figure 14: *Frequency Shifts and Mean Temperature Changes*

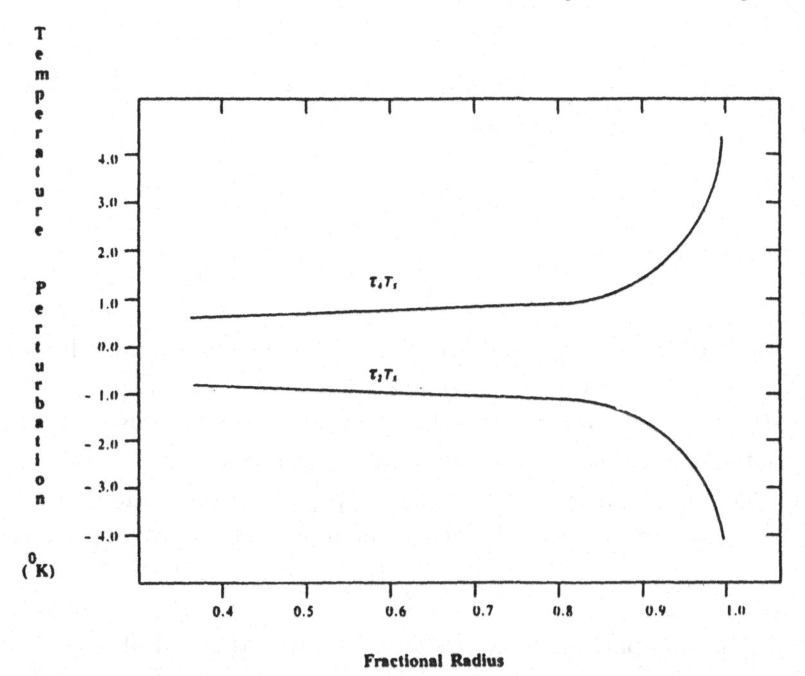

Figure 15: *Surface Temperature Amplitude Dependence on Depth of Asphericity*

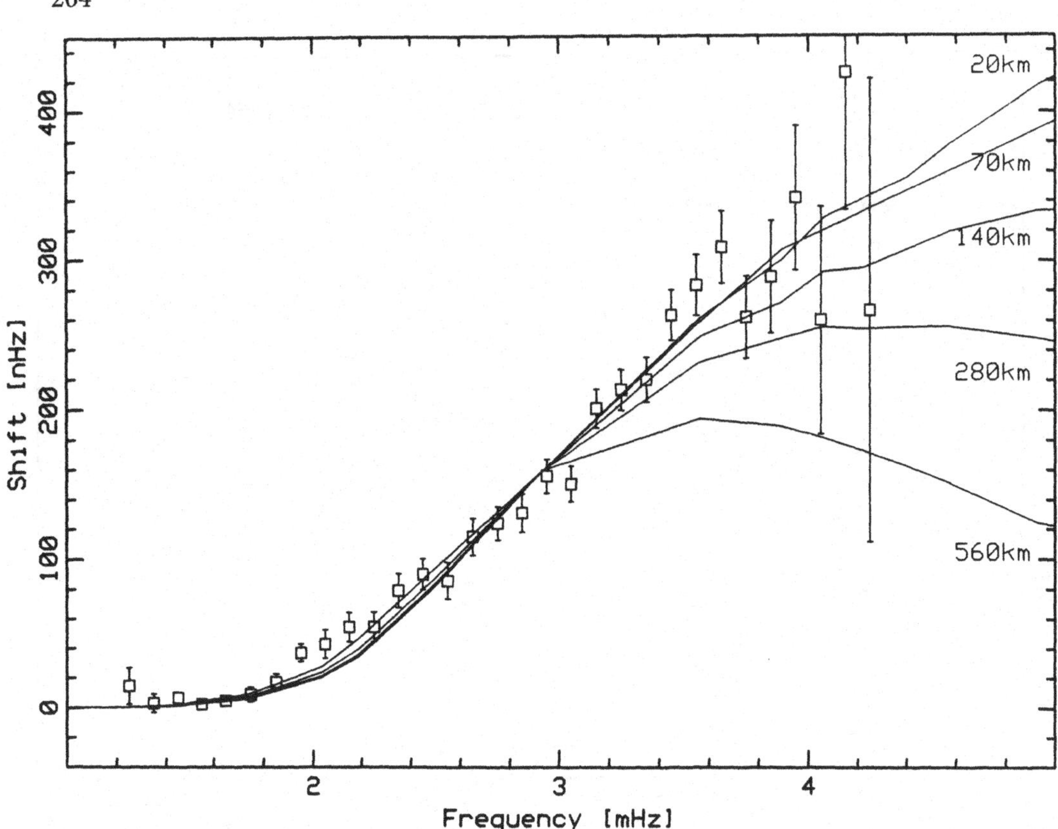

Figure 16: *Estimating the Depth of the Asphericity from Frequency Dependent Frequency Shifts*

that at high frequencies these data decline sharply and that the asphericity is no deeper than a few hundred kilometers.

5.4 Mechanisms

How are these surface changes generated? Our first conclusion must be that since changes in the oblateness are not coincident with the latitudinal temperature variations there must be mean flows that generate the Reynolds stresses needed to maintain the limb shape against temperature changes. The observed magnetic field is predominantly coincident with the temperature excess and doesn't appear to have the global coherence at the photosphere needed to provide this restoring stress.

Perhaps the comparison of limb temperature data to full disk irradiance observations, and the prediction of sound speed variations from the limb temperature are fortuitous. Each of these comparisons depend on the assumption that our measurement of the color temperature near the limb yields the local effective thermodynamic temperature. The effect of unresolved flux tubes (like the network magnetic fields) might generate a latitudinal and solar-cycle dependent temperature-like flux variation. Note that the helioseismic frequency shifts mimic the solar activity cycle. Couldn't then the frequency data be due to the variation in the mean magnetic field near the photosphere? This mechanism has been suggested by several (cf Gough and Thompson 1988; Goldreich et al. 1992; Jain and Roberts 1993) calculations aimed at being "self consistent." These are hard problems so it is not unexpected that no model includes Reynolds stresses, and describes the full frequency dependence of the shifts, and confronts the irradiance/luminosity constraints. They do generalize a point Bogdan and Zweibel (1987) made that the direct dynamical effect of fibril magnetic fields near the photosphere could generate significant frequency shifts. On the other hand if the outer envelope of the sun is permeated by a fibril field structure which is responsible for the cycle dependence in the frequency shifts, why does this frequency shift appear to originate in the outer superadiabatic layer? As we show below there are good physical reasons for the temperature perturbation to be largest in the superadiabatic region.

A more serious problem with the magnetic field calculations is that they require field strengths far in excess of observational constraints. For example, Goldreich et al. (1991), require an rms photospheric magnetic field strength of 250G! Now the mean flux density even during solar maximum is only 18G, and we know directly from observations that this field is predominantly radial. Since the sound speed perturbation depends on B^2, a truly realistic magnetic frequency shift calculation misses the observed magnitude of the helioseismic shifts by nearly two orders of magnitude (too low)! The observational constraints on the "true" magnetic field strengths have gotten even better from recent IR observations. For example, Lin (1994) measured the intranetwork and network field strength directly (not the flux) and confirmed that field strengths between 500-1400G are confined to small flux tubes with an average filling factor of about 3×10^{-3}. Average photospheric fields clearly cannot be anywhere near as large as 250G.

In contrast, the situation for turbulent flows is quite different. The temperature perturbation at the photosphere does not show a corresponding limb shape change. This implies that the fractional density perturbation (from the spherical sun) is much smaller than corresponding temperature and pressure perturbations. Since the gravitational potential change depends on the density we can simplify eq. 2 for steady flows to get $\nabla\delta P + \vec{\nabla}\cdot\mathbf{t} = 0$ where δP is the pressure perturbation due to the observed temperature excess. If $\mathbf{t}$ is dominated by $t_{rr} = \rho v^2$ and the equilibrium density and pressure scale heights define the dominant terms in the derivatives, we obtain $v^2 = c^2\delta T/T$ where c is the sound speed. For the observed temperature excess of $\delta T \approx 2°$K and $c = 6$km/s we get flow velocites of about 100m/s – more than an order of magnitude smaller than typical granular velocities and of the same magnitude as some meridional flow observations. It seems that no consistent model of the cycle variations in the outer part of the sun can afford to ignore the Reynolds stresses.

Braun et al. (1992) have shown that photospheric magnetic fields in the form of sunspots have a direct effect on p-modes. They measured the attenuation of the acoustic wave flux, but could not detect a change in p-mode frequencies within the spot. Could sunspots affect the global frequency measurements? Since global frequencies increase near sunspot maximum it is not likely. These local measurements show that the largest effect of the magnetic fields is to dissipate acoustic energy, and adding dissipation to an oscillator *decreases* its frequency.

There are alternative explanations, for example Gough and Thomson's (1988) model incorporating latitudinal or temporal variations in the convective mixing length, but it is fair to say that none of these formulations are consistent with all of the solar data (helioseismic and photometric). The interpretation of the limb observations as a thermodynamic temperature is attractive in that it describes both the helioseismic splittings (although requiring additional Reynolds stresses that are usually ignored), *and* the irradiance/luminosity changes (which are not addressed by any of the alternative scenarios).

6 NUMERICAL EXPERIMENTS, DEEP QUESTIONS?

6.1 Numerical Experiments

The literature is full of calculations aimed at explaining small parts of the solar cycle puzzle. A common thread to all of these models is the uncertainty of how to express the largescale effects of the CZ on mean field, or global solar properties. Times are changing and we now have the computational power to do useful *experiments* on small pieces of a realistic solar convection zone. Such numerical experiments should greatly clarify these theoretical uncertainties.

Consider the problem of the outer convection zone and how excess entropy may be transported. Figure 17 shows a snapshot of a slice through a simulation done by Nordlund and Stein (cf Stein et al. 1988). Their calculations include a realistic three dimensional treatment of radiation and a complete equation

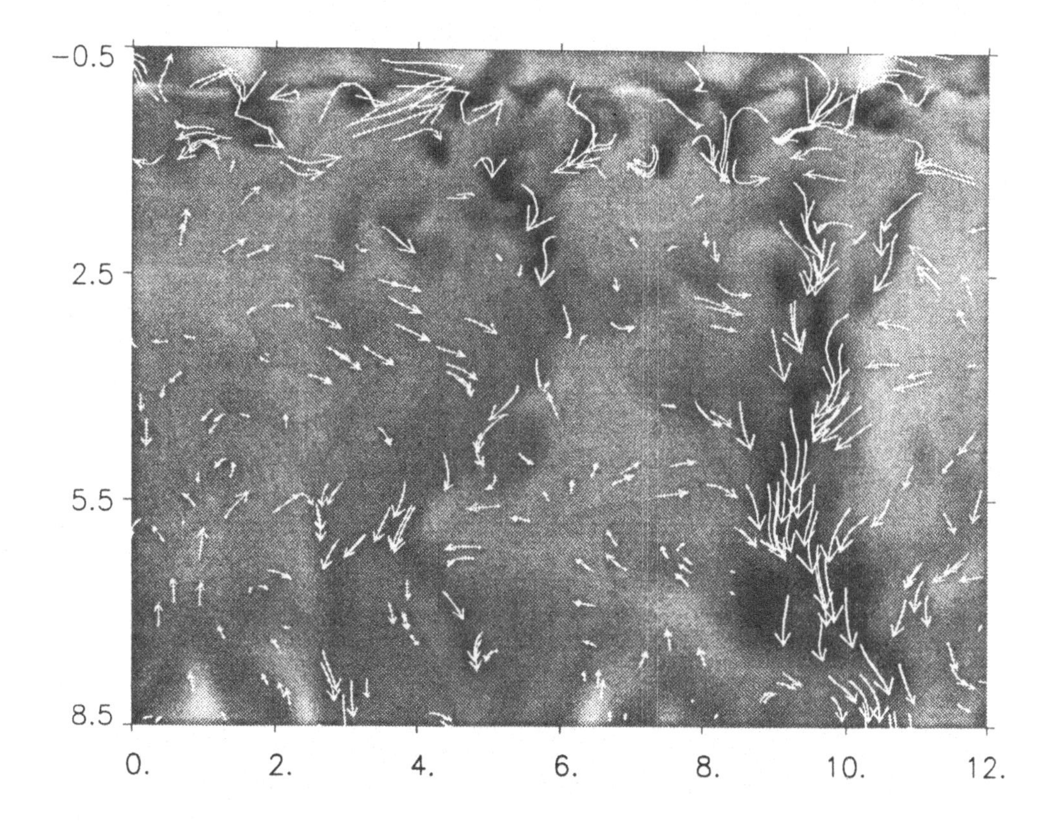

Figure 17: *Solar Convection Exposed by Numerical Experiments*

of state. The vertical scale of the simulation is 9000km and it is 12000km wide. The scale height is much smaller - about 200km. The image shows the instantaneous temperature excess superimposed on vectors that show the fluid motion. It is striking that the temperature and velocity structure have a very long correlation length in the vertical direction (much longer than the scale length) but the horizontal correlation length is considerably shorter, comparable to the scale length. We may conclude from this (and other numerical work) that turbulent convection in the outer convection zone is certainly not isotropic.

One of the interesting questions related to the luminosity puzzle is whether entropy fluctuations deep in the CZ are mixed by the turbulent fluid motion before appearing as excess heat at the photosphere. We performed some experiments with a smaller volume approximately 3100km on a side. On the left half of this cube we increased the entropy of incoming fluid at the bottom by 1%, while on the opposite side we decreased it by the same factor (Kuhn 1991). After waiting for about three rise times (90 minutes of solar time) we checked to see how the entropy perturbations were mixed toward the top of the volume. We were surprised to learn that the mean temperature difference (in horizontal planes) between the hot and cold sides of the box actually *increased* toward

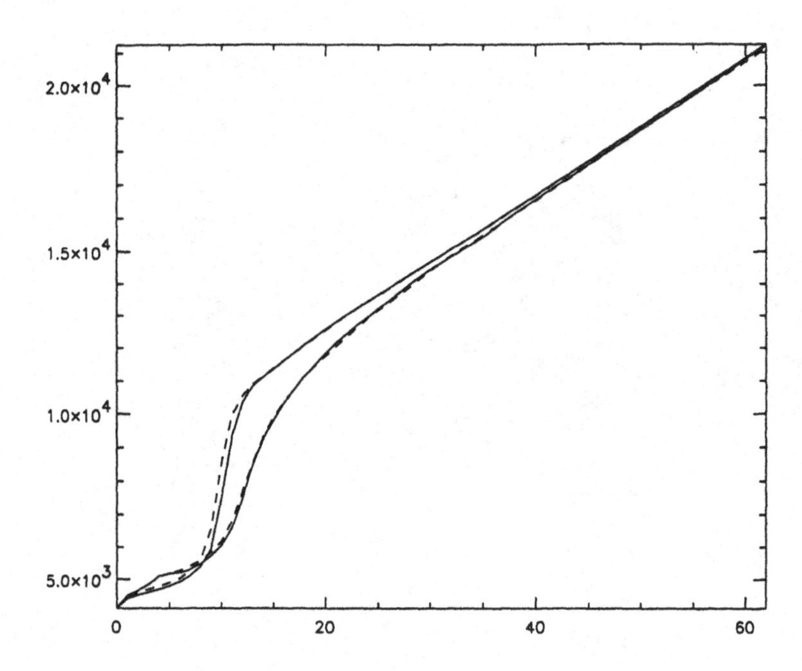

Figure 18: *Vertical Variation in Mean Temperatures of Convection Experiment*

the top of the box (Fig. 19). Figure 18 shows how the temperature varies with height between the two sides of the box and between upflowing and downflowing gas. Each grid cell is 50km. The rapidly falling temperature between cells 10 and 15 (a region about 300km deep) clearly shows the effect of the superadiabatic temperature regime, where the atmosphere is becoming transparent. The upper(lower) pair of lines show the temperature of upflowing(downflowing) gas and the dotted(solid) lines show the temperature of gas from the hotter(cooler) half of the volume.

Not only is the horizontal convective mixing generally ineffective, but in the superadiabatic region near the photosphere (where the convective efficiency is low) the temperature difference between hot and cold regions is amplified. There is some horizontal mixing in the lower 2000km, but as the atmosphere becomes transparent a larger temperature gradient is needed to carry the convective flux. In a linear diffusion model the vertical temperature difference should be exponentially damped. From the weak decline in dT near the lower region one can estimate that the horizontal "conductivity" is at least 10 times smaller than the vertical conductivity. The visual anisotropy of Fig. 17 is borne out by this simple estimate of the conductive anisotropy.

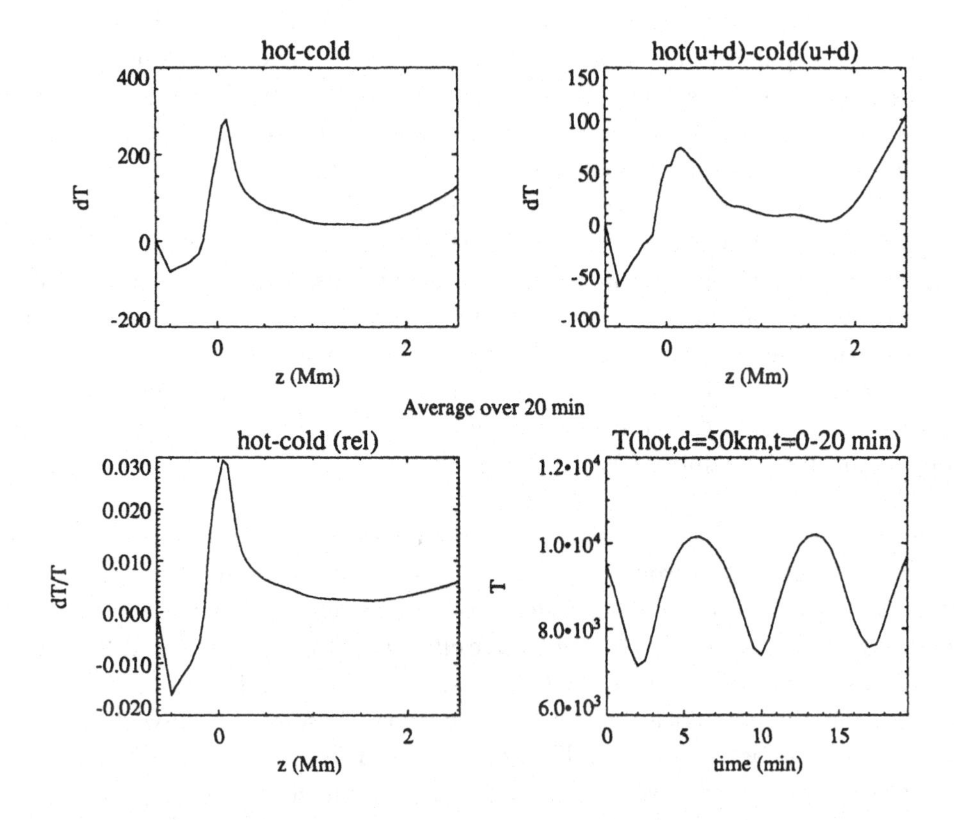

Figure 19: *Horizontal Hot-Cold Temperature Differences in the Convection Experiment*

6.2 Deeper Mechanisms?

An important observational result that restricts the possible mechanisms responsible for solar cycle variations is the ratio of the fractional radius change to the fractional luminosity change (the so-called W parameter). On yearly timescales we have seen that the fractional radius change must be less than 5×10^{-5} while the luminosity change is about 10^{-3}, and so $W < 0.05$. The $l = 2$ component of the shape and brightness change is even smaller ($W_2 < 0.005$) but we don't have theoretical models to compare with, since none of the structural equilibrium calculations are two-dimensional (nor do they include Reynolds flows needed to understand how the shape relaxes in response to the perturbation). Endal et al. (1985) summarize much of the work that has been done to estimate how magnetic perturbations, or changes in the convective efficiency would alter a one dimensional star. In particular from their simple mixing length calculation we must conclude that changes in the convective efficiency occur in the outer few thousand kilometers of the CZ in order to accomodate such low values of W. Deeper perturbations to the convective efficiency lead to larger relative radius changes (and larger values of W).

Endal et al. (1985) also considered how an extra magnetic pressure term (as a function of depth) would affect the equilibrium structure of a star. They find that the radius changes are much larger (in magnitude) than the relative luminosity perturbations due to convective efficiency changes for a magnetic field perturbation anywhere in the solar CZ. In anticipation of a discussion below we note that they found that a magnetic field near the base of the convection zone could be as large as a 10^6G, would generate a radius change smaller than the observational limits, and would produce an insignificant luminosity variation. Magnetic field strengths larger than this could (depending on the field geometry) produce observable radius variations.

There are no thermal timescales in the outer few thousand kilometers of the sun as long as the solar cycle, yet the largest cycle variations occur in just the outer few hundred kilometers. How does the sun store and release energy on an 11 year timescale? Since magnetic fields protruding through the photosphere do little to change the net luminosity it is likely that the mechanism for luminosity change lies deeper – near the base of the convection zone in the magnetic field generating region. This is where the largest shear velocities exist and where magnetic field lines get wrapped up to form toroidal bundles.

We suppose that a toroidal flux sheath is generated just below the bottom of the convection zone during the course of the solar cycle. Within the flux bundle the magnetic field contributes to the pressure equilibrium with the surroundings so that $\frac{B_i^2}{8\pi} + P_i = P_e$, where B_i, P_i, and P_e are the internal and external magnetic field strength and pressure of the flux sheath. To maintain equilibrium the density inside the magnetised region must decrease to diminish the gas pressure. Thus on short (acoustic) timescales the density in the magnetised region decreases. The geometry is summarized in Fig. 20. The temperature is decreasing outward but starts out the same for the magnetised and unmagnetised regions. Now, because the density in the magnetised region is smaller that fluid experiences a larger heat flux and the top of the flux sheath gets hotter than the corresponding unmagnetised region. Since the top of the sheath is hotter than nearby regions the temperature gradient *above* the sheath must also increase, in comparison to the nearby unmagnetised fluid. Thus, not only does the magnetised fluid have a higher entropy than surrounding latitudes outside of the toroidal activity band, but the magnitude of the temperature gradient above this region is enhanced. The timescale for the temperature to change due to this radiative process should be

$$\tau \approx \frac{3\pi C_p \kappa \rho l}{16\sigma T^2 \Delta dT/dz} \tag{9}$$

Here σ and C_p are the Stefan Boltzmann constant and specific heat, κ is the mean opacity, l is the vertical scale of the magnetised region and Δ is the fractional change in density due to the magnetic field. This approximates the decay time assuming radiative diffusion and a fixed magnetic field.

A numerical solution for the temperature increase of magnetised fluid near the top of the radiative zone is displayed in Fig. 21. The upper panels in this figure show; the change in temperature with time of the top end of the magnetised fluid box (a region where the magnetic field decreased the gas density by 10%), and the geometry of the calculation. Initial conditions were chosen to match a region of approximately one density scale height below the base of the convection zone using a 64×64 grid. A magnetic pressure contribution was added in the region indicated and the thermal response of the system was computed. After 11 years the fluid temperature has increased fractionally by 0.07% . The lower two panels show greyscale plots of how the temperature of the fluid and the temperature gradients have changed after 11 years. In the stable region near the CZ boundary a small increase in the magnitude of the temperature gradient can cause the stable fluid to become convectively unstable. Thus, it is interesting that the effect of the magnetised region may be to locally increase the depth of the convection zone (since the temperature gradient is increased above the field region) while increasing the specific entropy of the magnetised fluid. Our numerical calculation used a thick magnetic sheath to minimize the required numerical resolution. Consequently the magnetic field (about 30×10^6 G) is unphysically large and probably ruled out by the observed limits on changes in the solar radius. A better calculation here would scale down the sheath thickness

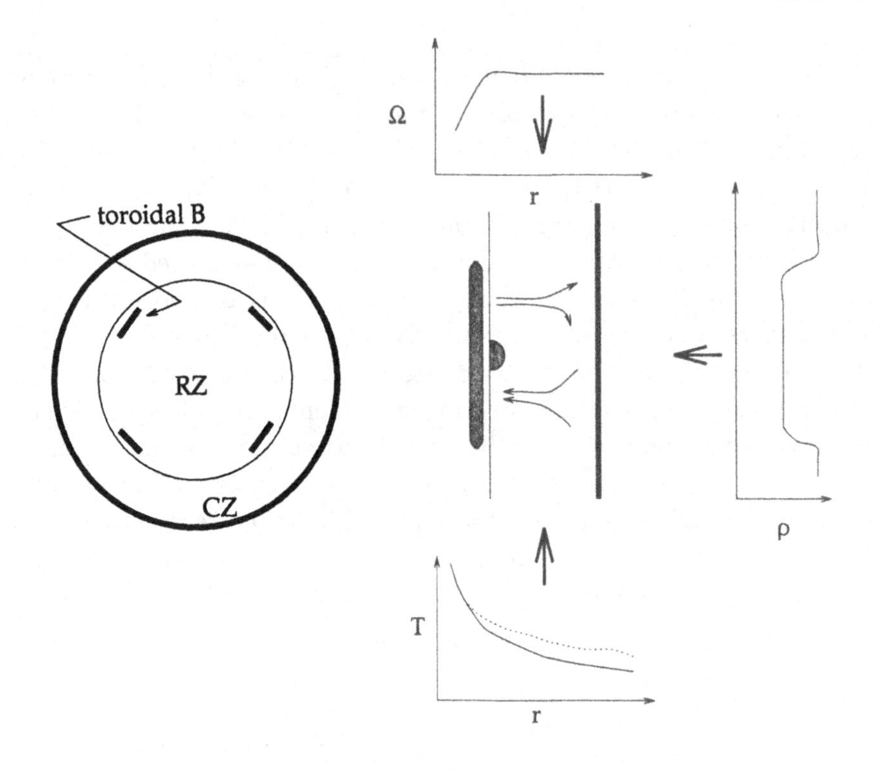

Figure 20: *Growing Entropy Perturbations and Hot Magnetic Fields*

and the magnetic field by one or two orders of magnitude. This must be checked, but if Eq. 9 is accurate this will not change our conclusions; that the growth of a toroidal field immediately below the CZ may be limited by a radiative process that enhances the convective instability of the fluid above the field region, and the magnetised fluid that is advected into the convection zone carries excess entropy compared to similar regions outside the toroidal activity bands.

The ingredients to an interesting picture may now be in hand. During the course of the solar cycle we generate a toroidal field below the convection zone where the velocity shear is largest. The shear is small in the CZ because of the effects of the strong anisotropy of the turbulent convection there. As the toroidal magnetic field is generated it experiences radiative heating and in the latitude bands where the field is strongest the convection zone drops a little to provide a mechanism for pulling the field out of the radiative zone. The magnetised fluid is of order 10^{-3} hotter than fluid from other latitudes and this entropy eventually appears at the photosphere. Since the CZ is anisotropic the entropy excess at the base of the CZ is not well mixed as it is advected to the surface. The surface active latitude bands, where the magnetic field appears, should then coincide with the regions of highest photospheric temperature. Because the excess heat must diffuse along the flux tubes the temporal correlation between magnetic field and luminosity increase may not be perfect. Perhaps the diffusion process

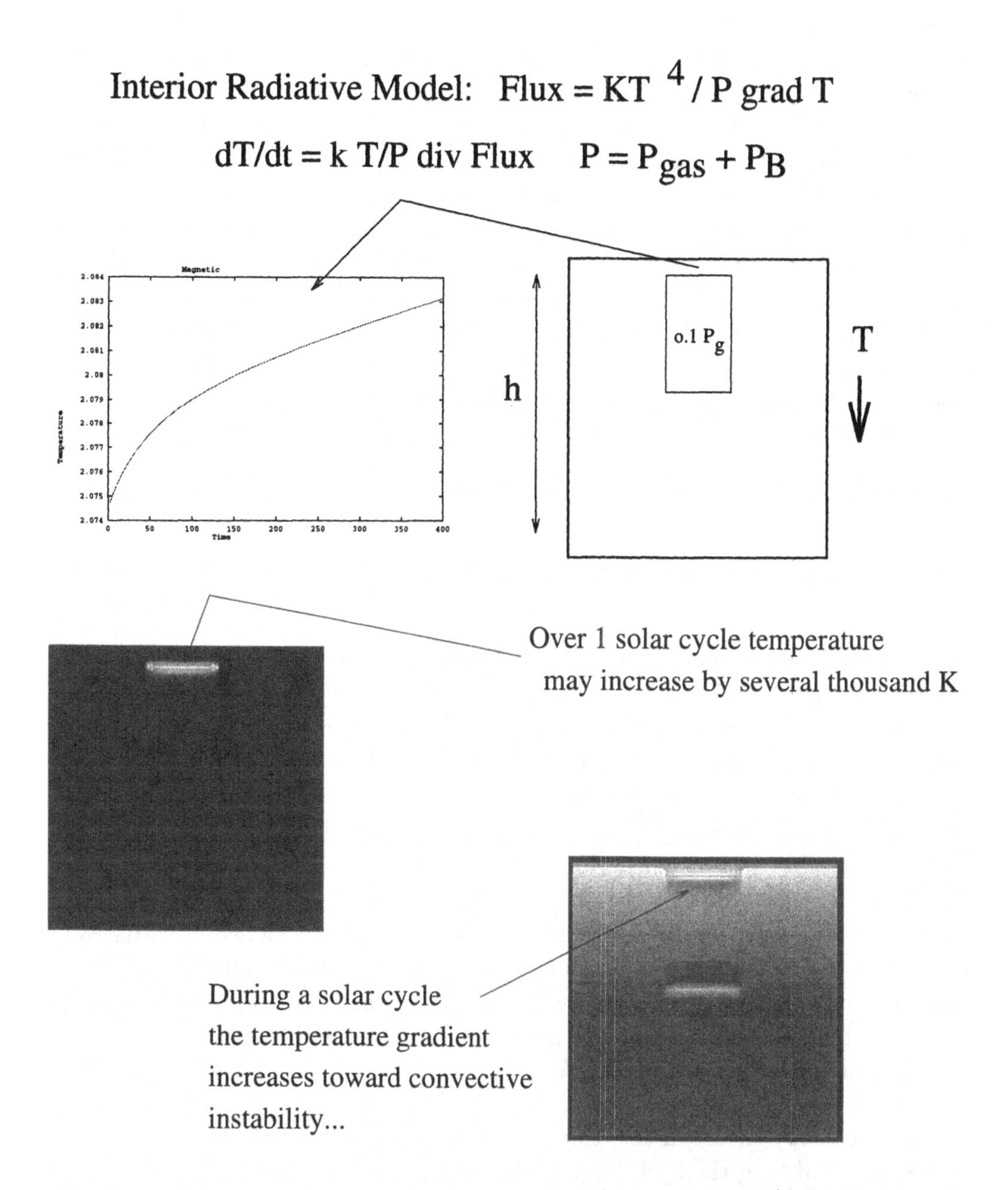

Figure 21: *Numerical Solution for Heated Magnetic Field Regions*

leads to a lag between when the magnetic field emerges from the photosphere and when the excess entropy appears (which is distributed along the flux tube). In any case we expect the solar luminosity to increase with increasing magnetic field. The effect of surface magnetic fields may have a compensating effect on the irradiance though. It is quite possible that the primary effect of magnetic fields protruding through the photosphere (in the absence of excess entropy from below) is to redistribute emergent energy away from the surface normal direction. This means that the net effect of emerging magnetic fields depends on their latitudinal distribution and angle with respect to the observer. This picture might be tested against the brightness variability of other solar-like stars. The surface temperature distribution also has a static latitudinal variation. This must reflect the effects of the latitudinal differential rotation in the presence of the strong turbulent anisotropy.

6.3 Magnetic or Thermal Variations?

For magnetic field strengths of order 1MG the thermal effects described in the previous section will dominate the dynamical effect of the field on helioseismic observations. Such measurements should be sensitive to weaker fields because of the induced thermal changes in the solar model. Consider the numerical results we obtained for a region with a vertical scale of 10^5km (the height of the magnetised region in fig 21). We found that a fractional temperature excess of 10^{-3} was possible over a 10yr period. The field strength used in that calculation was 30MG. Following eq. 9 we should expect roughly the same temperature excess over the same time period in a region 100km thick with a field of 1MG. We can't make a region much thinner than this because the equilibrium temperature gradient times the thickness scale would be smaller than the temperature excess we're inducing near the colder top of the magnetised volume (the photon free path in this region is less than a cm).

Is the entropy excess in this thin sheath sufficient to explain the 10^{-3} luminosity change during the cycle? Using the CZ model computed by Spruit (1974) we compute e, the ratio of the excess stored energy to the time-integrated excess luminosity, as

$$e = \frac{\rho C_p 4\pi r^2 dr \Delta T}{10^{-3} L_\odot \times 10yr} = 40$$

Note also that the magnetic flux contained with this region ($\approx 10^{24}$Mx) is still much larger than the flux in a typical active region, 10^{20}Mx (cf Hoeksema 1994).

I believe that this fractional temperature perturbation may be responsible for the observed surface temperature perturbations of the same order, that appear to cause acoustic frequency shifts during the solar cycle of about 300nHz (fig 14). The magnitude of the thermal-induced frequency changes are one part in 10^4. The same 100km region near the base of the convection zone has a smaller direct effect on the mode frequencies. The magnetic phase velocity, v_a is found from $v_a^2 = c^2 + \frac{B^2}{8\pi\rho}$ so that the relative change from the unperturbed sound speed is $\delta v/c = 0.02$ for a magnetic field of 10^6G at the bottom of the convection zone. The acoustic travel time is changed by $\Delta\tau = l\delta v/c^2$ and the relative frequency shift is just $\Delta\tau/300\text{s} = 10^{-5}$ – an order of magnitude smaller than the shift due to the effect of the thermal stratification.

We should stress that there are many ways that the preceding scenario could be wrong. Magnetic bouyancy, for example, has been ignored. How does diffusion from the hotter magnetic fluid elements describe the excess luminosity? Can we build an entirely self-consistent model that includes the needed circulation currents that yield such small surface shape changes?, etc. Nevertheless I hope these discussions serve to emphasize that it will soon be possible to put together the tools of helioseismology, numerical simulations, and precise photometry that will lead to an overconstrained, and perhaps even a correct picture of the mechanisms of solar-cycle variability.

6.4 Top 10 List of Needed Data and Calculations Relevant to These Problems

10. Measure the angular radiation distribution (contrast function) for spots, active regions and network elements.

9. Measure the *luminosity* change due to sunspots and faculae.

8. Spatially resolve the 3rd photospheric luminosity component.

7. Explain the rapid drop in cycle dependent helioseismic frequency shifts at frequencies above 4.3mHz.

6. Confirm or refute the neutrino flux variation as terrestrial background problem

5. Use a realistic model to compute the magnetic-thermal effects for a *thin* magnetic field near the bottom CZ boundary

4. Convincingly measure the meridional circulation and other flows associated with the latitudinal temperature variation.

3. Convincingly measure and separate the magnetic and thermal effects in the helioseismic data.

2. Measure the latitudinal corrugation (due to thermal flux instability?) at the base of the convection zone.

1. Simulate a realistic convecting fluid region large enough to confirm the turbulent viscosity fundamentals.

7 ACKNOWLEDGEMENTS

I'd like to thank the IAC for its support and generous hospitality during the VI Canary Island Winter School.

8 REFERENCES

Bahcall, J.N., Press, W.H.: 1991, *Ap. J.* **370**, 730.
Bieber, J.W., Seckel, D., Stanev, T., Steigman, G.: 1990, *Nature* **348**, 407.
Bogdan, T.J. and Zweibel, E.G.: 1987, *Ap. J.* **318**, 888.
Braun, D.C. et al.: 1992, *Ap. J. Lett.* **391**, L113
Cox, J.P.: 1980, *Theory of Stellar Pulsation*, (Princeton Univ. Press, Princeton).
Davis, Jr, R.: 1978, *Proc. Informal Conf. on Status and Future of Sol. Neutrino Research* (G. Friedlander, ed) (Brookhaven National Lab, Upton NY) 1.
Dicke, R.H.: 1970, *Ap. J.* **159**, 1.
Dicke, R.H.: 1978, *Mon. Not. Roy. Ast. Soc.* **182**, 303.
Dicke, R.H., J.R. Kuhn, and K.G. Libbrecht: 1987, *Ap. J.* **318**, 451.
Dubrelle, S.: 1970, *Num. Math.* **15**, 450.
Durney, B.R.: 1989, *Ap. J.* **338**, 509.
Durney, B.R., and Roxburgh, I.: 1971, *Sol. Phys.* **16**, 2.
Durney, B. R., and Spruit, H.C.: 1979, *Ap. J.* **234**, 1067.
Duvall, T.L., Harvey, J.W. and Pomerantz, M.A.: 1986, *Nature* **321**, 500.
Dziembowski, W.: 1988, *Seismology of the Sun and Sun-Like Stars* (ESA), 259.
Elsasser, K.: 1966, *Zeits. fur Astrophys.* **63**, 65.
Endal, A.S., Sofia, S. and Twigg, L.W.: 1985, *Ap. J.* **290**, 748.
Foukal, P. and Lean, J.: 1990, *Science* **247**, 505.
Goldreich, P., Murray, N., Willette, G., Kumar, P.: 1991, *Ap. J.* **370**, 752.

Goode, P.R. and Kuhn, J.R.: 1990, *Astrophys. J.* **356** 310.
Gough, D.O., Thompson, M.J.: 1988, *Advances in Helio- and Asteroseismology* (Christensen-Dalsgaard, J. and Frandsen, S. eds) (Reidel, Dodrecht), 175.
Hoeksema, J.T.: 1994, *IAU Colloquium 143, The Sun as a Variable Star: Solar and Stellar Irradiance Variations*, J. Pap, C. Frohlich, H.S. Hudson and S. Solanki, eds. (Cambridge University Press), 138.
Hathaway, D.: 1991, *Solar Phys.* **137**, 13.
Hirata, et al.: 1991, *Phys. Rev. Lett.* **66**, 9.
Howard, R.H.: 1990, *Solar Phys.* **126**, 299.
Hudson, H. d'Silva, S., Woodard, M. and Willson, R.C.: 1982, *Sol. Phys* **76**, 211.
Jain, R., and Roberts, B.: 1993, *Ap. J.* **414**, 818.
Jokipii, J.R.: 1991, *The Sun In Time* (University of Arizona Press: Tucson), 205
Kippenhahn, R.: 1963, *Ap. J.* **137**, 664.
Krause, L.: 1990, *Nature* **348**, 403.
Kuhn, J.R.: 1988, *Ap. J. Letts.* **331** L131.
Kuhn, J.R.: 1989, *Sol. Phys.* **123**, 1.
Kuhn, J.R.: 1990, *Progress of Seismology of the Sun and Stars* (Osaki, Y, and Shibahashi, H, eds.) 197.
Kuhn, J.R.: 1991, *Adv. Space. Res.* **11**, 171.
Kuhn, J.R.: 1994, *IAU Colloquium 143, The Sun as a Variable Star: Solar and Stellar Irradiance Variations*, J. Pap, C. Frohlich, H.S. Hudson and S. Solanki, eds. (Cambridge University Press), 130.
Kuhn, H. Hudson, J. Lemen, T. McWilliams, P. Milford: 1994 (to be submitted to Ap. J.)
Kuhn, J.R. and K. G. Libbrecht: 1991 *Ap. J. Letts.* **381**, L35.
Kuhn, J.R, K.G. Libbrecht, and R.H. Dicke: 1988, *Science* **242** 908.
Kuhn, J.R., K.G. Libbrecht, and R.H. Dicke: 1985 *Ap. J.* **290**, 758.
Kuhn, J.R. and Libbrecht, K.G.: 1991, *Ap. J. Letts.* **381**, L35.
Landau, L.D. and Lifshitz, E.M.: 1959, *Fluid Mechanics* (Pergamon Press, NY), 12.
Lin, H.: 1994, *Ap. J.* (in the press).
Muller, R.: 1975, *Sol. Phys.* **45**, 105.
Parker, E.N.: 1994, *IAU Colloquium 143, The Sun as a Variable Star: Solar and Stellar Irradiance Variations*, J. Pap, C. Frohlich, H.S. Hudson and S. Solanki, eds. (Cambridge University Press), 264.
Press, W.H., Flannery, B.P., Teukolsky, S.A. and Vetterling, W.T.: 1986 *Numerical Recipes*, (Cambridge University Press, Cambridge).
Ribes, E., B. Beardsley, T.M. Brown, P. Delache, J.R. Kuhn, F. Laclare, and N. Leister: 1991, *The Sun In Time* (University of Arizona Press: Tucson), 59.
Rosner, R. and Weiss, N.O.: 1992, in K. Harvey (ed.), *The Solar Cycle, Proc.*

of 12 NSO/Sac Peak Workshop, (PASP), 511.

Schrijver, C.J.: 1988, *Ast. Astrophys.* **189**, 163.

Snodgrass, H.: 1992, in K. Harvey (ed.), *The Solar Cycle, Proc. of 12 NSO/Sac Peak Workshop*, (PASP), 205.

Spruit, H.C.: 1974, *Sol. Phys.* **34**, 277.

Spruit, H.C.: 1991, *The Sun In Time* (University of Arizona Press: Tucson), 118.

Stein, R.F., A. Nordlund, and J.R. Kuhn: 1988, *Seismology of the Sun and Sun-Like Stars* (ESA) 529.

Tolstoy, I.: 1963, *Rev. Mod. Phys.* **35**, 207.

Wasiutynski, J.: 1946, *Astrophysica Norvegica* **4**, 1.

Woodard, M.F, J.R. Kuhn, N. Murray, and K.G. Libbrecht: 1991, *Ap. J. Letts* **373**, L81.

Willson, R.C. and Hudson, H.S.: 1991, *Nature* **351**, 42.

Solar Interior and Solar Neutrinos

John N. Bahcall

Institute for Advanced Study, School of Natural Sciences
Olden Lane, Princeton, NJ 08540, USA

1 NUCLEAR FUSION REACTIONS IN THE SUN

This first lecture is designed to introduce you to the basics of nuclear fusion reactions in the sun. I hope you will learn enough from this introduction to appreciate the significance of the role of nuclear reactions in helping to determine the characteristics of the sun. You will then, I believe, be in a position to understand what is at stake for astronomers in the ongoing and future solar neutrino experiments. First, we need to do some order of magnitude estimates.

Let us estimate the Coulomb barrier between two charged nuclei, the barrier that prevents nuclei fusing at low temperatures. Let the charges on the nuclei be Z_1 and Z_2, where the charges (atomic numbers) are (usually) small integers. The Coulomb barrier is

$$E_{\rm Coulomb} = Z_1 Z_2 e^2/r \ \approx \ Z_1 Z_2 \ {\rm MeV}/A^{1/3}. \tag{1}$$

You can check the arithmetic used in Eq. (1) since I have only inserted $e^2/(2 * a_{\rm Bohr})$ ≈ 13.6 eV (familiar to you from atomic physics), the size of the Bohr radius, 5×10^{-8} cm and the size of the nuclei, which is characteristically $A^{1/3} \times 10^{-13}$ cm.

How does the calculated energy compare with the temperature at the center of the sun? We can estimate the temperature directly from the equation of hydrostatic equilibrium,

$$\nabla P = -\rho \nabla \phi, \tag{2}$$

where P, ρ, and ϕ, are, respectively, the pressure, density, and gravitational potential. For this estimate, we can assume a perfect gas equation of state: $P = (\text{number density}) \times \text{kT} = \rho \text{kT}/\text{m}_\text{H}$. The effect of the gradient operator can be crudely approximated by multiplying by $R_\odot^{-1}$. We then obtain, canceling the factor of $\rho/R_\odot$ that appears on both sides of Eq. (2),

$$kT \sim GM_\odot m_H/R_\odot = 1 \text{ keV}. \tag{3}$$

This temperature is at least a factor of a thousand less than the characteristic Coulomb repulsion between charged nuclear particles.

Why is the temperature so much lower than the potential barrier that must be penetrated in order for two nuclei to fuse? Suppose that nuclei had temperatures in the center of the sun that were as large or larger than the characteristic energy, MeV, calculated above for Coulomb barriers. Then the characteristic luminosity that would be generated is

$$L_{\rm characteristic} \sim \epsilon * N * E_{\rm release}/\tau_{\rm characteristic}, \tag{4}$$

where ϵ is the fraction of the total number, N, of solar nuclei participating in the fusion process, and $\tau_{\text{characteristic}}$ is the characteristic lifetime to fusion of a nuclear particle in the sun. The lifetime is

$$\tau_{\text{characteristic}} = [n\sigma v]^{-1}, \tag{5}$$

where n is the number density of nuclei, σ their typical nuclear cross section for interaction, and v their characteristic relative velocities. We know n crudely from the mass and radius of the sun and we know that nuclear reactions above the Coulomb barrier normally have cross sections measured in milibarns $\sim 10^{-27}\text{cm}^{-2}$. The velocities are certainly much less than the velocity of light; they are of order 10^{9}cm s^{-1} (the precise value does not matter for our purposes). Inserting plausible numbers in Eq. (5) (the particle density in the solar core is about 10^{2} the average density, which also does not make any difference in this calculation), we find a characteristic lifetime to fusion of only 10^{-8} s. This is much, much shorter than the lifetime of the sun (10^{17} s). Inserting the characteristic fusion lifetime in Eq. (4), and taking N to be of order 10^{57} particles, E_{release} of order 10 MeV, and ϵ to be of order 0.01, we find a huge luminosity

$$L_{\text{characteristic}} \sim 10^{20} L_{\odot}. \tag{6}$$

This calculation shows us why stellar temperatures are below the Coulomb barrier. Otherwise, the stellar luminosities would be so large that the stars would blow themselves apart. The stellar temperatures must be limited by atomic processes, most importantly (for our purposes) the radiative opacity of the material.

It is convenient and conventional to represent nuclear cross sections at low energies in the following way,

$$\sigma(E)\frac{S(E)}{E}exp(-2\pi\eta), \tag{7}$$

where $S(E)$ is known as the low-energy cross section factor and η is

$$\eta = Z_1 Z_2(e^2/\hbar v). \tag{8}$$

Over the past 25 years, increasingly precise measurements of the nuclear reaction cross sections have been carried out in laboratories around the world, in large part in an attempt to clarify the implications of the disagreement between the predicted and the observed solar neutrino event rates. The cross section factors, $S(E)$, are usually quoted in units of keV-barns; the extrapolated value at zero energies is generally denoted by S_0. The expression for the nuclear cross section given in Eq. (7) must be convolved with a Maxwell-Boltzmann distribution at the local solar interior temperature in order to determine the rate of a given reaction under the stellar conditions.

Part of the way the sun keeps the average cross sections from being too large is to make the number of participating particles only a small fraction of the total number of nuclear particles, N. For example, the fraction of the mass in the form of ^{3}He nuclei is only of order 10^{-5} in the solar interior. The fact that the abundance is so low corresponds to the fact that the nuclear cross section for the destruction of ^{3}He is relatively high at solar temperatures. Of course, the rate of a given reaction is proportional to the number density of each participating particle times the cross section.

The sun shines, we believe, by fusing four hydrogen nuclei, protons, into helium nuclei, alpha particles. The schematic way this is achieved is:

$$4\mathrm{p} \rightarrow \alpha + 2\mathrm{e}^{+} \ + \ 2\nu_{\mathrm{e}} + 25 \ \ \mathrm{MeV}, \tag{9}$$

where 25 MeV is approximately the average energy release to the star (from the difference in rest mass energies of the initial and final nuclei). The particular way that this transformation is achieved depends primarily upon the individual nuclear cross sections and the stellar temperature. Table I shows the principal nuclear reactions that occur in the solar interior and that are believed to supply the luminosity of the sun.

The neutrino-producing reactions in Table I are of special interest. They are reactions 1a (the *pp* reaction), 1b (the *pep* reaction), 5 (the ^{7}Be electron-capture reaction), and 8 (^{8}B decay). [I began working on solar neutrinos in 1962 by calculating the rate of reaction 5 under solar conditions; lots of lessons, not all positive, can be inferred from the fact that almost none of you listening to the lecture today were even born then.] The ^{7}Be $p-\gamma$ reaction, number 7, gives rise to ^{8}B that almost immediately decays. The ^{8}B decay produces the high energy neutrinos that are expected to dominate the rates of the first two solar neutrino detectors to operate successfully, the chlorine (Homestake) and water (Kamiokande) experiments.

Table II shows the neutrinos produced in the CNO cycle in the sun. This cycle, which was originally suggested by Bethe to be the source of most of the solar luminosity, is now believed to supply about 1.6% of the energy generation in the sun. The neutrinos produced by ^{13}N and ^{15}O decay are believed to produce only a small contribution to the operating and planned solar neutrino experiments.

The calculated rates of solar neutrino experiments depend only on the ener-

Table I: *The pp chain in the Sun. The average number of pp neutrinos produced per termination in the Sun is 1.85. For all other neutrino sources, the average number of neutrinos produced per termination is equal to (the termination percentage/100).*

Reaction	Number	Termination† (%)	ν energy (MeV)
$p + p \rightarrow {}^2H + e^+ + \nu_e$	1*a*	100	≤ 0.420
or			
$p + e^- + p \rightarrow {}^2H + \nu_e$	1*b* (pep)	0.4	1.442
${}^2H + p \rightarrow {}^3He + \gamma$	2	100	
${}^3He + {}^3He \rightarrow \alpha + 2p$	3	85	
or			
${}^3He + {}^4He \rightarrow {}^7Be + \gamma$	4	15	
${}^7Be + e^- \rightarrow {}^7Li + \nu_e$	5	15	(90%) 0.861 (10%) 0.383
${}^7Li + p \rightarrow 2\alpha$	6	15	
or			
${}^7Be + p \rightarrow {}^8B + \gamma$	7	0.02	
${}^8B \rightarrow {}^8Be^* + e^+ + \nu_e$	8	0.02	< 15
${}^8Be^* \rightarrow 2\,\alpha$	9	0.02	

† The termination percentage is the fraction of terminations of the pp chain, $4p \rightarrow \alpha + 2e^+ + 2\nu_e$, in which each reaction occurs. The results are averaged over the model of the current Sun. Since in essentially all terminations at least one pp neutrino is produced and in a few terminations one pp and one pep neutrino are created, the total of pp and pep terminations exceeds 100%.

Table II: *The CNO cycle. The main reactions are shown for the carbon-nitrogen-oxygen cycle in the Sun.*

Reaction	ν energy (MeV)
${}^{12}C + p \rightarrow {}^{13}N + \gamma$	
${}^{13}N \rightarrow {}^{13}C + e^+ + \nu_e$	$\lesssim 1.199$
${}^{12}C + p \rightarrow {}^{14}N + \gamma$	
${}^{14}N + p \rightarrow {}^{15}O + \gamma$	
${}^{15}O \rightarrow {}^{15}N + e^+ + \nu_e$	$\lesssim 1.732$
${}^{15}N + p \rightarrow {}^{12}C + \alpha$	

gies of the neutrinos that are produced in the solar core, provided that nothing happens to the neutrinos after they are created in the solar interior. This "provided" is the origin of much excitement in particle physics today since it is possible that something does happen to the neutrinos before they get to their detectors on earth. If something does happen, it would be the first demonstration of particle physics beyond the standard electroweak model.

What is the flux of neutrinos of all types from the sun? The flux (making the very plausible assumption that neutrinos do not decay on the way to the earth) is

$$\text{Flux} \approx L_{\odot}/[4\pi \times (25\ MeV)(A.U.)^2] \sim 7 \times 10^{10}\text{cm}^{-2}\ \text{s}^{-1}. \tag{10}$$

In order to calculate the number of neutrinos of a definite energy that are produced in the sun (and expected to be detected by solar neutrino detectors), we have to calculate the relative rates of the nuclear reactions shown in Table I averaged over the conditions of the solar interior. For this, we need a precise solar model, whose interior characteristics must be calculated much more precisely than for any other problem in stellar interior research.

So what is at stake for astronomers in solar neutrino experiments? Why do solar neutrino experiments matter if you are an astronomer? The reason is that the measurement of the neutrino fluxes provide precise, quantitative tests of the theory of stellar evolution, a theory which all of us astronomers use implicitly nearly all the time when we discuss astronomical phenomena. In addition, the observed fluxes can be used to place strong phenomenological constraints on the conditions in the solar interior. We will begin to discuss solar neutrino experiments in the next lecture.

2 THE CHLORINE SOLAR NEUTRINO EXPERIMENT

In this lecture, we will begin discussing the operating solar neutrino experiments. We will concentrate on the implications of these experiments, i.e., what the results mean for physics and for astronomy. I will not discuss the details of how the experiments are done (for a general review see Bahcall 1989). I will also say relatively little about the details of the solar models (although that is presumably closer to your interests) that are used in interpreting the experiments. I have recently written about the solar models extensively elsewhere (see, for example, Bahcall and Pinsonneault 1995 and Bahcall 1994 and references therein).

As we shall see, the most important implications for physics are essentially independent of the details of the solar models, although the details are important

Table III: *Characteristics of the ^{37}Cl detector.*

Location	Homestake Mine, Lead, South Dakota
Depth	4850 ft
Tank	20 ft diameter $\times$ 48 ft long
Detector fluid	C_2Cl_4
Total weight of fluid	615 tons
Volume	3.8×10^5 liters
Threshold	0.814 MeV
Cl37 atoms	2.16×10^{30}
Half-life ^{37}Ar	35.0 days
Neutrino sensitivity (major)	^{8}B, ^{7}Be

for astrophysical and astronomical questions.

Before discussing the chlorine experiment, I should point out that there is one important way in which solar neutrino astronomy and helioseismology are similar (and different from many other fields of astronomy) and one important way in which helioseismology and neutrino astronomy are very different from each other. (I mean, of course, in addition to the fact that the sun is the source of the signal in both cases.) Both for $p-$modes and for neutrinos, it is essential to have a good theoretical model in order to decide if you have discovered something new or not. This is different from many other fields of astronomy in which you can make a discovery by purely observational means, just by looking (for example, quasars, x-ray sources, $\gamma-$ray bursters, infrared objects, etc.). Neutrinos and $p-$modes differ in that the data rate is much higher for your observations of the sun with photons. In the neutrino game, we currently get only a few events per month per experiment (although this may increase by about a factor of order 10^2 in the next set of experiments).

The first experiment that we shall discuss is the famous ^{37}Cl experiment of Ray Davis Jr., which has been operating since 1968 (see Davis 1964, 1978). The reaction that is used to detect solar neutrinos is the inverse of the laboratory decay of ^{37}Ar. The neutrino absorption reaction is:

$$\nu_e + {}^{37}\mathrm{Cl} \rightarrow e^- + {}^{37}\mathrm{Ar}, \quad E_{\mathrm{th}} = 0.814 \ \mathrm{MeV}. \tag{11}$$

The 0.8 MeV threshold detection energy permits the detection of all the major neutrino sources discussed in the previous lecture except the fundamental *pp* neutrinos, which have a maximum energy (0.4 MeV) that is less than the chlorine energy threshold. Some of the basics characteristics of the Davis chlorine detector are listed in Table III.

Before we go on to discuss the implications of the chlorine experiment, I

Table IV: *Prediction of Standard Model:* ^{37}Cl *Detector.*

Source	Capture rate (SNU)
p-p	0.0
pep	0.2
^{7}Be	1.2
^{8}B	7.4
^{13}N	0.1
^{15}O	0.4
Total =	$9.3^{+1.2}_{-1.4}$

1σ uncertainties

need to record for you what the latest standard solar model calculations predict. Table IV shows the contributions to the calculated capture rate from each of the solar neutrino branches. I am using here the results from the Bahcall-Pinsonneault (1995) model. This model includes helium and heavy element diffusion and is in excellent agreement with the results of a similar calculation by Proffitt (1994).

As you can easily see, about 80% of the predicted rate is due to the very rare, high-energy ^{8}B neutrinos. The next largest contribution, 13%, is calculated to come from the lower energy (0.86 MeV) ^{7}Be neutrinos. We will have a lot to say about both of these neutrino branches when we discuss the Kamionkande (pure water) and gallium experiments in the following lectures.

In calculating the uncertainties in the theoretical predictions, I have used the quoted 1σ experimental error for all quantities that are measured. For the theoretical errors, I have used generous estimates of the possible errors based upon the spread in theoretical values among different published calculations. For example, I have used one-third the difference between calculations that include both helium and metal diffusion and calculations that completely neglect diffusion. (Thus, the difference between full diffusion and no diffusion is regarded as a 3σ uncertainty.) Since neglecting diffusion causes the calculated event rates in solar neutrino experiments to decrease, this large uncertainty causes the error in the lower direction to be somewhat larger than the error in the upper direction. The uncertainty in the diffusion rate in the upper direction is taken from the published estimated uncertainty by Thoul, Bahcall, and Loeb (1994).

The quoted uncertainties are intended to approximate 1σ errors, although there is no particular reason why they should rigorously follow a normal distribution. On the other hand, the calculated results for the past 25 years have fluctuated up and down with amplitudes that are consistent with the error estimates that have been cited (see Figure 1.2 of Bahcall 1989).

Why are the calculated results so robust? There are two principal constraints: 1) the measured luminosity of the sun; and 2) the measured p−mode helioseismological frequencies. Several of the principal lecturers (and a number of the students) have discussed the constraints provided by helioseismology on the structure of the sun; therefore, I will not comment on these constraints. I will just give a schematic argument of how the measurement of the solar luminosity tightly constrains the predicted neutrino fluxes. The sun's luminosity is measured to an absolute accuracy of about 0.4%. Any solar model that derives its energy mostly from the pp chain has a luminosity that depends approximately on the central temperature, T_c, as $L \propto T_c^4$ (Bahcall 1989). Crudely speaking, one can also parameterize the dependence of the neutrino fluxes as a power-law dependence upon T_c, i.e., $\phi_\nu \propto T_c^n$. Therefore, we conclude that roughly speaking:

$$\frac{\Delta\phi}{\phi} \approx \left(\frac{n}{4}\right)\frac{\Delta T}{T}. \tag{12}$$

The strongest temperature dependence, the most uncertain neutrino flux, is the ^{8}B neutrino flux, for which $n \approx 18$ (Bahcall 1989). According to the schematic argument given above, the uncertainty in the calculated ^{8}B neutrino flux should therefore be less than 2%. I actually estimate a much larger error, about 17%, because the argument that led to Eq. (12) omits some of the important physics. But, you can see from this argument that the luminosity is a powerful constraint on the neutrino fluxes. The argument is especially powerful for the lower-energy neutrinos like ^{7}Be (for which $n \approx 8$) and the pp neutrinos (for which $n \approx -1.2$).

There are also other reasons why the predictions are robust. Each of the important input parameters have been determined a number of times in the last quarter of a century and their values are relatively well known. Also, there is a fairly large number of significant input parameters, about nine. Sometimes when the models are redone, a few parameters will change in a way to increase the predicted rates and some other parameters will change in a way so as to decrease the predicted rates. By the central limit theorem, the average change will be a small number.

The most recent result of the chlorine experiment is (Lande 1995):

$$\text{Rate Cl Obs} = (2.55 \pm 0.25), \tag{13}$$

where the experimental error quoted is the quadratic sum of statistical and systematic errors.

We are now in these lectures at the position that solar neutrino work was in after the first two decades of operation of the chlorine detector, 1968–1988. There is a well-defined conflict between the predictions (Table IV) of the standard solar model and the measured rate (Eq. 13). For the two decades mentioned, the conflict between calculation and observation was known as "The Solar Neutrino Problem."

We shall see in the next lectures that the conflict is refined and strengthened by the three other operating experiments.

3 CHLORINE VERSUS KAMIOKANDE

In the previous lecture, we saw that the observed counting rate in the chlorine solar neutrino detector of Ray Davis and his associates is 2.55 ± 0.25 SNU, which conflicts with the theoretical rate, $9.3^{+1.2}_{-1.4}$ SNU, calculated from a solar model that includes helium and heavy element diffusion. For many years, in fact for two decades, the conflict between the chlorine observation and the solar model calculation constituted "The Solar Neutrino Problem." The situation changed dramatically in 1989, when the first results of the Kamiokande solar neutrino experiment became available.

As we shall see in this lecture, the Kamiokande results, taken together with the chlorine experiment, constitute an even more stringent solar neutrino problem. This new problem arises when the chlorine and the Kamiokande experiments are compared with each other: they are supposed to be measuring the same thing (mostly ^{8}B neutrinos for chlorine; only ^{8}B neutrinos for Kamiokande), but they give different answers. The discrepancy with the standard solar model prediction is larger for the chlorine than for the Kamiokande experiment.

Many authors have suggested that the reason for this apparent paradox is that the two experiments are really not measuring the same thing, that the assumption based upon the standard model of electroweak interactions that nothing happens to the neutrinos after they are created is wrong. The apparent paradox would be resolved, as has been suggested, if the lower energy neutrinos to which the chlorine experiment is sensitive are depleted more strongly by a new physical process (the most popular one is the so-called MSW effect) than are the higher energy neutrinos to which the Kamiokande experiment is sensitive.

We will see in the last lecture that the gallium experiments constitute by themselves a separate solar neutrino problem. So, we will really have three solar neutrino problems. With these preliminary remarks out of the way, we can begin to discuss the Kamiokande experiment and its implications. For a more detailed discussion of this experiment see the references by the Kamiokande team (e.g., Hirata, et al. 1989, 1991; Suzuki 1995). The detector is a water Cerenkov detector that is located about 300 km west of Tokyo in a mine in the Japanese Alps. Neutrinos reveal themselves in this detector by scattering off electrons,

$$\nu + e \rightarrow \nu' + e'. \tag{14}$$

The experiment is at a depth of about 1 km (2700 m.w.e.) and has a fiducial mass of about 680 tons (about 10% more than the mass of the chlorine detector), which is shielded by a much larger mass of about 3 ktons. The large (approximately 50 cm) Cerenkov light detectors represent a major advance in this experiment; they make it possible to cover about 20% of the surface area of the fiducial region with light detectors.

The electron-scattering mechanism provides additional information that is not available for radiochemical experiments like the chlorine and gallium experiments. Most importantly, the electrons provide directionality. The electrons are primarily scattered in the forward direction since the incoming energy of the neutrinos is, for the Kamiokande experiment, much larger than the rest mass energy of the electrons. Therefore, the electrons predominantly move in the forward direction along the axis from the sun to the detector. Each event also provides a measure of the energy of the scattered neutrino since the Cerenkov detectors measure well the energy of the electrons. Thus the recoil electron energy spectrum is an additional quantity that can be compared with experiment; the recoil energy spectrum can be computed if one knows the spectrum of energies of the incoming neutrinos. If nothing happens to the neutrinos after they are created, then the neutrino energy spectrum is well-known. Different particle physics processes that affect the neutrino fluxes also affect the energy spectra and lead to predictable changes in the electron recoil energy spectrum. In addition, an electron scattering experiment provides information about the flavor (type) of the neutrinos. For neutrinos with typical energies like those observed in the Kamiokande experiment (~ 10 MeV), the scattering cross section is about six times larger for electron-type neutrinos than for muon or tau neutrinos. Finally, the electron scattering process allows one to record the exact time of arrival of each neutrino that is detected. In radiochemical experiments, one can only infer that the neutrino sometime during the exposure period (before a chemical extraction is made), which is typically of order weeks or months.

Just a word about the history of this wonderful experiment. It was originally designed to detect nucleon decay, which is the origin of the name of the experiment. The detector is located in the Kamioka mine and the last three letters of Kamiokande, "nde," denote "nucleon decay experiment." When the higher energy events corresponding to nucleon decay were not found in the detector, the experiment was upgraded with improved electronics to allow the lower energy solar neutrinos to be observed. Just months before the only supernova observed in recent history in a nearby galaxy exploded, the Kamiokande detector became capable of detecting the $\sim$ 10 MeV neutrinos from solar neutrinos or from a supernova. The most recent result of the Kamiokande experiment is (Hirata et al. 1989, 1991; Suzuki 1995).

$$\phi_K(^8\mathrm{B}) = \left(2.89^{+0.22}_{-0.21} \pm 0.35(\mathrm{syst})\right) \times 10^6 \ \ \mathrm{cm}^{-2}\,\mathrm{s}^{-1}, \tag{15}$$

which is only about 44% of the ^{8}B flux predicted by the standard solar model. The Kamiokande result for the ^{8}B flux confirms the deficit of high-energy ^{8}B neutrinos originally inferred from the chlorine experiment. In addition, the Kamiokande team has demonstrated the directionality of the signals: the scattered electrons move in the opposite direction from the sun, along the line from the sun to the detector. This same detector also shows that the recoil spectrum gives no hint of neutrinos with energies in excess of the $\approx$ 14 MeV end point of the ^{8}B neutrino spectrum. Finally, the arrival times are consistent with a neutrino flux that is constant in time. So, this one detector gives crucial information on the energy spectrum, the time dependence, and the absolute value of the ^{8}B neutrino spectrum.

We will now show how the comparison of the Kamiokande and the chlorine experimental results presents evidence for a second solar neutrino problem, a problem that is largely independent of the detailed results of the standard solar model (for more details of this argument see Bahcall 1994 and Bahcall and Bethe 1990, 1993).

The capture rate in the chlorine experiment just from ^{8}B neutrinos observed in the Kamiokande experiment is, using the known relative cross sections:

$$\phi_K(^8\mathrm{B})\sigma_{\mathrm{Cl}} = \left(3.21^{+0.24}_{-0.23} \pm 0.39(\mathrm{syst})\right) \ \mathrm{SNU}. \tag{16}$$

The only assumption used in deriving Eq. (16) is that the shape of the ^{8}B neutrino spectrum is the same for ^{8}B neutrinos produced in the sun as it is for ^{8}B neutrinos produced in a laboratory. This assumption can be shown to be correct to an accuracy of one part in 10^5 provided standard electroweak theory is correct (see Bahcall 1991).

If nothing happens to the neutrinos once they are created in the center of the sun, then the ^{8}B neutrinos expected from the Kamiokande experiment would cause a counting rate in the chlorine experiment in excess of the total counting rate in the chlorine experiment.

Subtracting Eq. (16) from the observed rate in the chlorine experiment for the sum of the rates due to ^{7}Be, CNO, and pep neutrinos:

$$\text{Rate Cl Obs } \left(\text{pep } + {}^7\text{Be} + \text{CNO}\right) = (-0.66 \pm 0.52) \text{ SNU}, \tag{17}$$

if standard electroweak theory is correct.

The total spread in the calculated capture rates for pep neutrinos is, for a set of 12 recently-published standard solar models (Bahcall 1994),

$$(\phi\sigma)_{\text{Cl}}(\text{pep}) = \ (0.22 \pm 0.01) \text{ SNU, standard models,} \tag{18}$$

where much of the dispersion is due to round-off errors in the published *pep* neutrino fluxes. The accuracy with which the *pep* flux is calculated reflects the fact that it is closely related to the basic *pp* reaction, which is ultimately responsible for nearly all of the solar luminosity. Subtracting Eq. (18) from Eq. (17), we find that

$$(\phi\sigma)({}^7\text{Be}) + (\phi\sigma)(\text{CNO}) \leq \ 0.46 \text{ SNU, } 95\% \text{ conf. limit Cl expt.} \,. \tag{19}$$

The result given in Eq. (19) is in strong disagreement with the entire range of values for the ^{7}Be neutrino fluxes calculated from 12 recently-published solar models:

$$(\phi\sigma)_{\text{Cl}}({}^7\text{Be}) = \ (1.13 \pm 0.11) \text{ SNU, standard models.} \tag{20}$$

The discrepancy between Eq. (19) and Eq. (20) is a quantitative statement of the solar neutrino problem that results from the comparison of the chlorine and the Kamiokande experiments. The upper limit on the sum of the capture rates from ^{7}Be and CNO neutrinos is significantly less than the lowest value predicted for any of the recently-published solar neutrino models. (I did not subtract the CNO neutrinos to isolate the ^{7}Be neutrinos and make the conflict even stronger, because it is already very strong and because the flux of CNO neutrinos is somewhat uncertain in standard solar models.) We conclude that if both solar neutrino experiments are correct, then something happens to the neutrinos after they are created in the solar interior (i.e., standard electroweak theory must be modified).

The flux of ^{7}Be neutrinos is independent of uncertainties in the measurement of the $^7\text{Be}(p,\gamma)^8\text{B}$ capture cross section. We have not used the calculated value of the ^{8}B neutrino flux anywhere in the above argument. In fact, one would have to increase the value of the $^7\text{Be}(p,\gamma)^8\text{B}$ cross section by more than two orders of magnitude (the currently estimated uncertainty is 10%) in order to affect significantly the calculated flux of ^{7}Be neutrinos.

We conclude that the comparison of the chlorine and Kamiokande experiments leads to a solar neutrino problem that is independent of details of solar models. In the next lecture, we will discuss a solar neutrino problem that is associated with the gallium experiments.

4 GALLEX AND SAGE

In the previous two lectures, we discussed the solar neutrino problem that arises because of the conflict between the observed chlorine and the calculated solar model results and a second solar neutrino problem that arises because of the apparent difference in the ^{8}B fluxes that are seen in the chlorine and the Kamiokande (pure water) experiments. Of course, both of these conflicts exist only if one assumes that nothing happens to the neutrinos after they are produced in the center of the sun.

In this lecture, we will discuss a third solar neutrino problem that exists because the measured event rates in the two gallium experiments are less than the rate calculated from standard solar models.

There are two operating gallium solar neutrino experiments (Anselmann et al., 1992, 1993, 1994; Abazov et al., 1991; Abdurashitov, et al., 1994), the Russian-American gallium experiment, SAGE, and the European-American-Israeli gallium experiment, GALLEX. The reaction they use to detect neutrinos is

$$\nu_e + {}^{71}\mathrm{Ga} \rightarrow \mathrm{e}^- + {}^{71}\mathrm{Ge}, \ E_{\mathrm{th}} = 0.2332 \ \mathrm{MeV}. \tag{21}$$

The low threshold energy makes possible the detection, for the first time, of *pp* neutrinos. The radioactive ^{71}Ge decays by electron capture, the inverse of Eq. (21), in 11.43 days. Both experiments count ^{71}Ge atoms produced by neutrino capture and use the ^{71}Ge atoms to measure the neutrino event rate. The GALLEX experiment utilizes 30 tons of an aqueous GaCl-HCl solution in the underground laboratory at Gran Sasso. The SAGE experiment uses 60 tons of metallic gallium in a special laboratory built underneath a tall mountain in the North Caucus region of Russia.

Table V shows the rate predicted for the gallium experiments by a standard solar model that includes helium and heavy element diffusion. More than half of the predicted capture rate arises from the basic *pp* neutrinos, which are respon-

Table V: *Prediction of Standard Model:* ^{71}Ga *Detector.*

Source	Capture rate (SNU)
p-p	70
pep	3
^{7}Be	38
^{8}B	16
^{13}N	4
^{15}O	6
total =	137^{+8}_{-7}

1σ uncertainties

sible for most of the luminosity of a standard solar model and whose flux can, I believe, be predicted to an accuracy of better than 1%. The next most important contributor to the predicted event rate, according to the standard solar model, is the ^{7}Be neutrino flux. The ^{7}Be neutrinos are expected to contribute about 30% of the total event rate. Note that the rare, high-energy ^{8}B neutrinos which dominate the expected event rates in the chlorine and Kamiokande experiments, contribute less than 15% to the expected capture rate for gallium. The theoretical uncertainties are relatively small for the gallium experiment because the dominant $p-p$ and ^{7}Be neutrino fluxes can be calculated with high accuracy from a standard solar model. The results from 12 recently-published different solar models (Bahcall 1994) constructed using different equations of state, opacities, nuclear parameters, chemical composition, luminosity and age values, and different assumptions about diffusion (from no diffusion to full diffusion) all yield predicted event rates for the gallium experiments of 74 ± 1 SNU from the $p-p$ and pep neutrinos and 34 ± 4 SNU from the ^{7}Be neutrinos. The theoretical uncertainties that I quote below are asymmetrical because of the contributions of transitions between the ground nuclear state of ^{71}Ga and excited nuclear states of ^{71}Ge (see Bahcall 1989).

Before we begin to discuss the experimental results, we need to know one other theoretical result. I state the result in the form of a theorem (Bahcall 1989). The lowest event rate that can occur in a gallium solar neutrino experiment is 80 SNU, provided only that: 1) the sun is currently generating energy in its interior by nuclear fusion reactions at the rate it is loosing energy by photons from the surface; and 2) nothing happens to the neutrinos after they are produced. This minimum is achieved by terminating all of the $p - p$ reactions by the $^3\mathrm{He} -^3 \mathrm{He}$ reaction, which means that only the $p - p$ and pep neutrinos are produced and none of the higher energy neutrinos are generated.

The GALLEX and the SAGE experiments have reported (Anselmann, et al., 1992, 1993, 1994; Abazov, et al., 1991; Abdurashitov, et al., 1994) consistent neutrino capture rates (respectively, 79±12 SNU and 73±19 SNU). The weighted average of their results is

$$\text{Rate Ga Obs} = (77 \pm 10) \text{ SNU}. \tag{22}$$

Subtracting the accurately known rate for $p - p$ and pep neutrinos (quoted above), we find

$$\text{Rate Ga}(^7\text{Be} +^8 \text{B}) = (3 \pm 10) \text{ SNU}. \tag{23}$$

We must also subtract the ^{8}B flux that is measured in the Kamiokande experiment.We can do this provided we know the shape of the solar ^{8}B neutrino spectrum. If we assume that standard electroweak theory is valid, then we know the shape of the spectrum; the shape of the solar spectrum is the same as the shape of the laboratory spectrum. This gives a contribution of ^{8}B neutrinos to the gallium capture rate of

$$\phi_K \sigma_{\text{Ga}}(^8\text{B}) \;=\; 7.0^{+7}_{-3.5} \text{ SNU}, \tag{24}$$

where the quoted errors are dominated by the uncertainties in calculating the transitions from the ground state of gallium to various excited states of germanium. Subtracting the rate from ^{8}B neutrinos from the combined rate for ^{7}Be plus ^{8}B neutrinos, one again finds a negative best-estimate flux for ^{7}Be neutrinos,

$$(\phi\sigma)_{\text{Ga}}(^7\text{Be}) = (-4^{+11}_{-12}) \text{ SNU}. \tag{25}$$

The negative flux for the ^{7}Be neutrinos that is inferred from the results of the GALLEX and the SAGE experiments is the simplest statement of the gallium solar neutrino problem. Following the conservative procedure described in the Particle Data Book for handling negative best-estimates of positive-definite quantities, one finds the following upper limit on the ^{7}Be contribution to the gallium capture rate,

$$(\phi\sigma)_{\text{Ga}}(^7\text{Be}) \;\leq\; 19 \text{ SNU, } 95\% \text{ conf. limit.} \tag{26}$$

This result is much less than the calculated contribution of 34 ± 4 SNU. We again find that the ^{7}Be neutrinos must be depleted relative to the calculated value.

It is important to note that we have not used anywhere in the above argument the calculated value of the ^{8}B neutrino flux. Instead, we used the value measured in the Kamiokande experiment and assumed that the *shape* of the energy spectrum was unchanged (i.e., electroweak theory is valid). Thus the argument is completely independent of uncertainties in the measured $^{7}Be(p,\gamma)^{8}B$ cross section.

Now it is time to sum up. We have spent a lot of our time discussing how the solar neutrino experiments are in conflict with the combined predictions of the standard solar model and the standard electroweak model. However, it is worth noting, in fact it is worth emphasizing, that the original goals of the experiments have been wonderfully satisfied!

The agreement between the astronomical predictions and the observations is much better than I, or anyone else I know, would have anticipated back in 1964 when this work began. The Kamiokande experiment shows that the extraterrestrial neutrinos we observe at earth (to currently achieved levels of sensitivity) come from the sun; the electrons are scattered in the water detector in the forward direction from the sun to the earth. The neutrino energies are predicted to lie between 0 and 14 MeV, and the Kamiokande observations are consistent with this prediction also. The standard solar model predicts that the neutrino fluxes are constant in time (except for seasonal variations due to the changing earth-sun distance) and, with low statistical accuracy, this result also seems to be correct. Moreover, all four of the experiments yield observed rates that are of the same order as the predicted rates, i.e., the observed and the predicted rates differ by only a factor of at most a few.

If we were wrong about the predominant source of energy and the sun shone instead via the CNO cycle, the event rate for the chlorine experiment would have been 28 SNU (instead of the observed 2.6 SNU), the event rate for the gallium experiments would have been 610 SNU (instead of the observed 77 SNU), and the event rate for the Kamiokande experiment would have been identically zero (all of the CNO neutrinos are below threshold for the Kamiokande experiment).

We have evidence that the existing solar neutrino experiments are pointing toward new physics. Additional experiments are required to determine whether or not this inference is correct. Three new experiments, SNO, Superkamiokande, and BOREXINO will test the need for new physics in the next few years. I would have loved to discuss those experiments with you, but tomorrow is my wife's birthday (and I am going to join her), so I will have to cut these lectures short.

I want to close these remarks with a look backward. As a community, I believe we can already take pride in the joint effort, based upon the work of hundreds of chemists, physicists, astronomers, and engineers collaborating over several decades, experimentalists and theorists, that has successfully brought to a conclusion an inquiry begun in the middle of the 19th century. The solar neutrino experiments reviewed in these lectures demonstrate directly that the sun shines by nuclear fusion reactions among light elements; the neutrinos produced by those reactions deep in the solar core have been observed directly.

5 REFERENCES

Anselmann, P., et al.: 1992, *Phys. Lett. B* **285**, 376; 1993, **314**, 445; 1994, ibid, submitted.

Abazov, A. I., et al.: 1991, *Phys. Rev. Lett.* **67**, 3332.

Abdurashitov, J. N., et al.: 1994, *Phys. Lett. B* **328**, 234.

Bahcall, J. N.: 1989, *Neutrino Astrophysics* (Cambridge University Press, Cambridge, England).

Bahcall, J. N.: 1991, *Phys. Rev. D* **44**, 1644.

Bahcall, J. N.: 1994, *Phys. Lett. B* **338**, 276.

Bahcall, J. N., and Bethe, H. A.: 1990, *Phys. Rev. Lett.* **65**, 2233; 1993, *Phys. Rev. D* **47**, 1298.

Bahcall, J. N., and Pinsonneault, M. H.: 1995, *Reviews of Modern Physics*, submitted.

Davis, R., Jr.: 1964, *Phys. Rev. Lett.* **12**, 303; 1978, in *Proceedings of the Informal Conference on Status and Future of Solar Neutrino Research*, G. Friedlander (ed.), (Brookhaven National Laboratory: Upton), Report No. 50879, Vol. 1, p1.

Hirata, K. S., et al.: 1989, *Phys. Rev. Lett.* **63**, 16; 1991, *Phys. Rev. D* **44**, 224.

Lande, K., et al.: 1995, *Proc. 16th Int. Conf. on Neutrino Physics and Astrophysics*, Eilat, Israel, May 29–June 3, 1994, to be published.

Proffitt, C. R.: 1994, *Astrophys. J.* **425**, 849.

Suzuki, Y.: 1995, *Proc. 16th Int. Conf. on Neutrino Physics and Astrophysics*, Eilat, Israel, May 29–June 3, 1994, to be published.

Thoul, A. A., Bahcall, J. N., and A. Loeb, A.: 1994, *Astrophys. J.* **421**, 828.

The Solar Magnetic Field

Eugene N. Parker

Enrico Fermi Institute. U.S.A.
University of Chicago. 5640 South Ellis Av.
Chicago, Illinois 60637. U.S.A.

1 INTRODUCTORY THEORETICAL DYNAMICS OF ASTROPHYSICAL PLASMAS AND FIELDS

Much of theoretical astrophysics is concerned with the dynamics of ionized gases in the presence of magnetic fields. There are many excellent textbooks on the subject and the reader is undoubtedly familiar with the basic equations. However, there are divergent ideas on the proper fromulation of the momentum and induction equations that leave one wondering how early the foundation of the equations is grasped. One point of view begins with the fact that charged particles respond only to the electric field $\vec{E}$ in their own frame of reference. Therefore the electric field is the fundamental entity, providing the current $\vec{j}$ that creates the magnetic field $\vec{B}$. For instance in magnetospheric physics it is asserted that the electric field $\vec{E} = -\vec{v} \times \vec{B}/c$ of the solar wind that is the prime mover, actively penetrating into the magnetosphere to produce the observed activity. There is also a growing practice of declaring an equivalent governing electric circuit for the basic magnetohydrodynamic motions of a system of plasma and magnetic field arising from the concept that $\vec{E}$ and $\vec{j}$ are the basic physical quantities. Then one hears that the familiar hydrodynamic momentum equation cannot be applied to a collisionless gas because the concept of pressure is undefined in the absence of collisions and because the "mirror force" is neglected. The "mirror force" refers to the reflection of spiralling ions and electrons from regions of strong field. The conventional Lorentz force $(\vec{\nabla} \times \vec{B}) \times \vec{B}/4\pi$ has no component parallel to $\vec{B}$, so clearly it ignores the "mirror force".

The purpose of this lecture is to go back to the equations of Maxwell and Newton to establish the theoretical foundations of the large-scale dynamics of plasmas. As we shall see, the familiar magnetohydrodynamic (MHD) equations

$$\frac{\partial \vec{B}}{\partial t} = \vec{\nabla} \times (\vec{v} \times \vec{B}) + \text{dissipations terms}$$

$$\rho \frac{d\vec{v}}{dt} = -\vec{\nabla} p + \frac{(\vec{\nabla} \times \vec{B}) \times \vec{B}}{4\pi}$$

are nothing more than Faraday's induction equation and the applicability of Newton's equation of motion to the center of mass of any aggregate of particles. They are direct consequence of the basic laws of physics and cannot be avoided. The concerns and objections are without merely questions that are easily answered in any specific case content of Maxwell's and Newton's equations.

To begin the exposition, note that for nonrelativistic velocity $\vec{v}$ ($<< c$) of a particle of mass m and charge q, Newton's equation of motion can be written conveniently as

$$m\frac{d\vec{V}}{dt} = q\vec{E}' \tag{1}$$

where $\vec{E}'$ is the electric field in the frame of reference of the moving particle. In terms of $\vec{E}$ and $\vec{B}$ in the laboratory frame,

$$\vec{E}' = \vec{E} + \vec{v} \times \vec{B}/c \tag{2}$$

neglecting terms second order in v/c compared to one. We use the electrostatic system of units in which $\vec{E}$ is measured in statvolts per cm (1 statvolt = 300 volts) and $\vec{B}$ is in gauss (1 gauss = 10^{-4} tesla) in order to avoid the grotesque distortion of the natural symmetry of $\vec{E}$ and $\vec{B}$ in Maxwell's equations,

$$\frac{\partial\vec{B}}{\partial t} = -c\vec{\nabla} \times \vec{E} \tag{3}$$

$$4\pi\vec{j} + \partial\vec{E}/\partial t = c\vec{\nabla} \times \vec{B} \tag{4}$$

arising in SI units. Thus, $\vec{E}$ and $\vec{B}$ have the same dimensions $(m^{\frac{1}{2}}/l^{\frac{1}{2}}t)$. They are numerically equal when their energy densities and pressures $E^2/8\pi$ and $B^2/8\pi$ are equal. The only asymmetry between $\vec{E}$ and $\vec{B}$ is the general absence of magnetic charge in nature, although the open end of a long thin closely wound solenoid serves that purpose quite well for many laboratory purposes. The current density is in electrostatic units (esu) in which the unit of charge is $\frac{1}{3} \times 10^{-9}$ coulomb, so that the charge on an electron is $e = 4.80 \times 10^{-10}$ esu.

In addition to equations (3) and (4) there is the condition

$$\vec{\nabla} \cdot \vec{E} = 4\pi\delta, \tag{5}$$

where δ is the electric charge density. Equation (5) combined with the divergence of equation (4) dictates charge conservations, just as the assumption of charge conservation applied to the divergence of equation (4) yields equation (5). There is also the condition that

$$\vec{\nabla} \cdot \vec{B} = 0, \tag{6}$$

which tells us that magnetic field lines have no ends. Equation (5) asserts that the total integrated flux of $\vec{E}$ from a positive charge q is $4\pi q$. Electric charge q is defined in terms of the electrostatic force $F = q_1q_2/r^2$ between two charges q_1 and q_2 separated by a distance r. The electric field $\vec{E}(\vec{r},t)$ at a point $\vec{r}$ at time t is defined as the force, per unit charge exerted on a charge q, so that $\vec{E} = \vec{F}/q$. It must be understood that $\vec{E}$ is a real physical force field and not merely a mathematical device for representing action at a distance. There is real energy density and stress density at every position within a nonvanishing $\vec{E}$.

The magnetic field can be similarly defined in terms of the force per unit magnetic charge g. For a long thin solenoid of radius R with N turns of current I per unit length, the field throughout the interior is $B = 4\pi NI/c$ and the total flux of B in the interior is $\phi = \pi R^2 B$. The flux per unit charge is 4π so the open end of the solenoid is equivalent to a magnetic charge, $g = \phi/4\pi = \pi R^2 NI/c$. This field spreads out radially from the open end at distances in excess of R but substantially less than the total length L of the solenoid. So for $R \ll r \ll L$ the open end is essentially a magnetic monopole.

A magnetic field $\vec{B}$ is also defined by the Lorentz force $\vec{I} \times \vec{B}/c$ exerted on a specified current $\vec{I}$, and $\vec{B}$ can be defined by the Lorentz transformation (2) between known fields $\vec{E}$ and $\vec{E}'$ in two frames of reference with relative velocity $\vec{v}$. Again it must be understood that $\vec{B}$ is a physically real stress field in space and not merely a mathematical abstraction. The simplest experimental proof of the physical reality of $\vec{E}$ and $\vec{B}$ is human vision, which senses photons with frequencies of the order of 6×10^{14} cycles/s. The photons are composed of nothing but $\vec{E}$ and $\vec{B}$, of course.

We are interested in applying Maxwell's equations to the dynamics of a fluid with high electrical conductivity, i.e. a continuum with a well defined local velocity $\vec{v}(\vec{r}, t)$ that is uncapable of supporting any significant electric field $\vec{E}'$ in its own frame of reference. Such media are made up of free ions and electrons with electric charge neutrality over any dimension in excess of the Debye radius $R_D = (kT/4\pi Ne^2)^{1/2}$.

Consider the equation of motion for any arbitrary volume V of the plasma. If Z_i denotes the total electromagnetic and gravitational force exerted on the plasma of mass M in V, then the momentum equation can be written

$$M\frac{dv_i}{dt} + \int dS_j p_{ij} = Z_i$$

where v_i is the motion of the center of mass of the volume V and p_{ij} is the thermal momentum flux Mv_i transported by thermal motions u_j. In terms of the thermal velocity distribution function $f(x_k, u_k, t)$ the tensor p_{ij} is defined as

$$p_{ij}(x_k) = \int d^3u_k f(x_k, u_k, t) m \; u_i \; u_j$$

for particles with mass m. Since not all particles have the same mass M, it must be understood that p_{ij} is the sum over all particle species. Use Gauss's theorem to convert the surface integral to a volume integral, so that the equation of motion can be written,

$$M\frac{dv_i}{dt} + \int dV \frac{\partial p_{ij}}{\partial x_j} = Z_i$$

This applies to every volume V with dimensions large compared to the Debye radius, R_D. Then if the scale l of variation of N, v_i, and p_{ij} is large compared to R_D, it follows that there exists an intermediate scale λ such that $R_D << \lambda << l$. For volumes with dimension λ it follows to good approximation that N, v_i, p_{ij} etc, at the center of V, are closely equal to the volume average over V. Thus if F_i is the force per unit volume, we have $Z_i = VF_i, M = NmV$, etc, so that the moment equation becomes

$$NM\frac{dv_i}{dt} = -\frac{\partial p_{ij}}{\partial x_j} + F_i \tag{7}$$

where now v_i represents the local mean bulk velocity of the fluid. It is well known that the same momentum equation follows from the first velocity moment of the collisionless Boltzmann equation, where the same p_{ij} is refered to as the pressure tensor. The pressure tensor is also the momentum flux tensor.

The force on an individual particle of mass m and charge q is indicated by equations (1) and (2). Summing over all the particles in a volume V of dimenssion λ, charge neutrality decrees that $\sum q = 0$ while the sum of $q\vec{v}$ gives the electric current since the electrons and ions have slightly different mean bulk velocities. Hence, it follows from the sum that the force per unit volume is $\vec{j} \times \vec{B}/c$, where $\vec{j}$ is the current density. The result is the equations of motion

$$Nm(\frac{\partial v_i}{\partial t} + v_j\frac{\partial v_i}{\partial x_j}) = -\frac{\partial p_{ij}}{\partial x_j} + \frac{\delta_{ijk}\, j_j B_k}{c} \tag{8}$$

to which gravitational forces, viscosity, etc. can be added if desired. This can be expressed in terms of B_i alone using Ampere's law to eliminate j_i,

$$Nm(\frac{\partial v_i}{\partial t} + v_j\frac{\partial v_i}{\partial x_j}) = -\frac{\partial p_{ij}}{\partial x_j} - \frac{\partial}{\partial x_i}\frac{B^2}{8\pi} + \frac{B_j}{4\pi}\frac{\partial B_i}{\partial x_j}$$

Now, how do we treat Maxwell's equations? The medium cannot support an electric field $\vec{E}'$ in its own frame of reference. That is to say, a very small $\vec{E}'$ produces an enormous current density $\vec{j}$ which quickly neutralizes whatever charge separation caused $\vec{E}'$. In fact the instantaneous introduction of an $\vec{E}'$ produces electron plasma oscillations in this way, with a frequency $\omega_p = (4\pi Ne^2/m)^{1/2}$, equal to $6 \times 10^4 N^{1/2}$ radians/s, where N is the number of electrons per cm^2. The oscillations are subject to Landau damping and the bottom line is that the macroscopic (on any scale larger than R_D) electric field $\vec{E}'$ is suppressed to an exceedingly small level ($| \vec{E}' | << | \vec{v} \times \vec{B}/c |$). It follows from equation (2) that

$$\vec{E} \cong -\vec{v} \times \vec{B}/c. \tag{9}$$

That is to say, the component of $\vec{E}$ perperdicular to $\vec{B}$ is small compared to $\vec{B}$ by the factor v/c. The component $\vec{E}_{\|}$ parallel to $\vec{B}$ is generally even smaller, except for brief transients (shocks, etc.) and the special circumstance in which Ampere's law requires a current density $\vec{j}_{\|}$ along the field in a plasma so tenuous that the electron conduction velocity must be as large or larger than the ion thermal velocity.

So if the principal electric field is given by equation (9), it follows from equation (3) (Lundquist, 1952; Elsasser, 1954) that

$$\frac{\partial \vec{B}}{\partial t} = \vec{\nabla} \times (\vec{v} \times \vec{B}) \tag{10}$$

This familiar induction equation is merely a consequence of Faraday's induction equation and the condition that $\vec{E}'$ is small compared to $\vec{v} \times \vec{B}/c$. The equation asserts that the magnetic field is transported bedily with the bulk velocity $\vec{v}$. If the characteristic scale of the fields is l, then it follows from equation (5) that the charge density δ is of the order of E/l, which is $(v/c)B/l$. The electrostatic force on δ is, in order of magnitude,

$$E\delta = E^2/l = (v^2/c^2)B^2/l \tag{11}$$

which is small to second order in v/c compared to the Lorentz force B^2/l exerted by the magnetic field. Hence it can be neglected.

Consider equation (4). The right hand side is of the order of cB/l. On the left hand side, $\partial/\partial t$ is of the order of v/l so that

$$\left| \frac{\partial \vec{E}}{\partial t} \right| \sim \frac{v^2}{c^2} \frac{cB}{l} \tag{12}$$

and can be neglected as second order in v/c. The result is Ampere's law

$$4\pi \vec{j} = c\vec{\nabla} \times \vec{B} \tag{13}$$

which states that there is always an electric current associated with a deformation $\vec{\nabla} \times \vec{B}$ of a magnetic field. The Lorentz force $\vec{j} \times \vec{B}/c$ can be written $(\vec{\nabla} \times \vec{B}) \times \vec{B}/4\pi$ on the right hand side of the momentum equation (8), providing the familiar MHD momentum equation.

It is at this point that confusion arises as to what field variables are fundamental and which are only secondary. As already noted, it is common practice in magnetospheric physics to declare that $\vec{E}$ and $\vec{j}$ are the fundamental quantities. But $\vec{E}$ is small $O(v/c)$ compared to B, so in the non relativistic limit $(v/c << 1)\vec{E}$, hardly ranks as a primary physical quantity representing stress and energy that is small $O(v^2/c^2)$. In any case, try writing the momentum equation (8) in terms of $\vec{E}$ and $\vec{j}$, using the Biot-Savart integral

$$\vec{B}(\vec{r}) = \frac{1}{c} \int d^3\vec{r'} \frac{\vec{j}(\vec{r'}) \times (\vec{r} - \vec{r'})}{|\vec{r} - \vec{r'}|^3} \tag{14}$$

over all space to replace $\vec{B}$ with $\vec{j}$. The equations become hopelessly clumsy in all but the simplest circumstances.

Part of the motivation for using $\vec{E}$ and $\vec{j}$ is concern over charge conservation, concern that the Newtonian mechanics of the electron and ion motions may produce a $\vec{j}$ that does not satisfy Ampere's law, and the basic belief that $\vec{j}$ is the cause of $\vec{B}$, so that $\vec{j}$ is more fundamental. Taking these concerns one at a time, take the divergence of equation (4) and then use equation (5) to eliminate $\vec{\nabla} \cdot \vec{E}$. The result is the statement of conservation of charge

$$\vec{\nabla} \cdot \vec{j} + \frac{\partial \delta}{\partial t} = 0 \tag{15}$$

Given the smallness of δ, this reduces to $\vec{\nabla} . \vec{j} = 0$, which is the result of the divergence of equation (14), of course. So charge conservation is automatic and cannot be violated in a system described by Newtonian mechanics in which particles are conserved.

Suppose, then, that $\vec{j}$ momentarily fails to satisfy Ampere's law. That is to say, suppose that $\vec{j}$ falls short of $c\vec{\nabla} \times \vec{B}/4\pi$ by some small amount $\vec{\Delta} j$. It follows from equation (4) that

$$\frac{\partial \vec{E}}{\partial t} = 4\pi \Delta \vec{j} \tag{16}$$

which states that an electric field grows in the direction of $\Delta \vec{j}$. The growth rate is enormous and in view of Newton's equations of motion for the individual ions and electrons, the growing $\vec{E}$ soon forces $\Delta \vec{j}$ to zero (in a characteristic time given by the electron plasma oscillation period), so that equation (4) reduces to equation (13). This is the fundamental compatibility of Newton and Maxwell about which more will be said later.

So far as the idea that $\vec{j}$ causes $\vec{B}$, note that the term "cause" refers to the energy source. In the laboratory the practice is to drive electric currents through conductors with an applied emf, thereby causing magnetic fields. The current and the emf together clearly cause the magnetic field. But in the Sun there are no applied emf and the energy that causes $\vec{j}$ in opposition to the slight inertia and frictional drag (resistivity) of the electrons is the magnetic field. The electric field that compels compliance with Ampere's law is induced by $\partial B/\partial t$ in the manner described by equation (3). So the idea that $\vec{j}$ is fundamental is without basis. It is the magnetic field that is the cause of $\vec{j}$ via a weak electric field, in astrophysical settings.

This brings us to Poynting's theorem which establishes in a general way the compability of the energy and momentum concepts of Newton and Maxwell. It is convenient to describe the individual particles by the continuous density $\rho(\vec{r},t)$ of matter in space. An electron is represented by a localized peak in ρ with a characteristic radius of the order of 10^{-13} cm (the classical electron radius) and a proton by a similar but much higher peak. An ion becomes much broader with a central peak with a radius of 10^{-13} cm and the broad low wings spreading out to 10^{-3} cm. The individual peaks move with the velocity v_i of each particle, of course. The charge density is similarly represented by $\delta(\vec{r},t)$ with moving peaks coinciding with the peaks in ρ, but the peaks for the electrons are negative holes of finite depth instead of peaks. The velocity v_i is defined only at each particle. The equation of motion (1) is extended to all particles, having the form

$$\rho\frac{d\vec{v}}{dt} = \delta(\vec{E} + \vec{v}\times\vec{B}/c)\ , \tag{17}$$

which is applicable at every point in space. Note, then, that the electric current density is $\vec{j}(\vec{r},t) = v\ \delta(\vec{r},t)$. Use equation (5) to eliminate δ_i, use equation (6) to justify adding the term $\vec{B}\vec{\nabla}\cdot\vec{B}/4\pi$ to the right hand side, and use equation (4) to eliminate $\vec{j}$. The result contains the term

$$\frac{\partial\vec{E}}{\partial t}\times\vec{B} = \frac{\partial}{\partial t}(\vec{E}\times\vec{B}) - \vec{E}\times\frac{\partial\vec{B}}{\partial t} = \frac{\partial}{\partial t}(\vec{E}\times\vec{B}) + c\vec{E}\times(\vec{\nabla}\times\vec{E})\ .$$

The final form can be rearranged to

$$\rho\frac{d\vec{v}}{dt} + \frac{\partial}{\partial t}\frac{\vec{E}\times\vec{B}}{4\pi c} = \frac{\vec{E}\vec{\nabla}\cdot\vec{E} + \vec{B}\vec{\nabla}\cdot\vec{B} + (\vec{\nabla}\times\vec{E})\times\vec{E} + (\vec{\nabla}\times\vec{B})\times\vec{B}}{4\pi} \tag{18}$$

Use the vector identity $(\vec{\nabla}\times\vec{A})\times\vec{A} = -\nabla\frac{1}{2}A^2 + (\vec{A}\cdot\vec{\nabla})\vec{A}$ and add the vanishing quantity $v_i(\partial\rho/\partial t + \partial\rho v_i/\partial x_j) = 0$, obtaining

$$\frac{\partial\rho v_i}{\partial t} + \frac{\partial}{\partial x_j}\rho v_i v_j + \frac{\partial}{\partial t}\frac{P_i}{c^2} = \frac{\partial M_{ij}}{\partial x_j}\ , \tag{19}$$

where d/dt has been written as $\partial/\partial t + v_j\partial/\partial x_j$ and where the Poynting vector P_i is defined as $P_i = c\epsilon_{ijk}E_jB_k/4\pi$, i.e. :$\vec{P} = c\vec{E}\times\vec{B}/4\pi$, while the Maxwell stress tensor is defined as

$$M_{ij} = \frac{E_iE_j + B_iB_j}{4\pi} - \delta_{ij}\frac{E^2 + B^2}{8\pi}\ . \tag{20}$$

The form of equation (19) makes it clear that M_{ij} is indeed a stress tensor while P_i/c^2 represents the electromagnetic momentum density. Since $\rho = 0$ in the space not occupied by a particle, if follows that

$$\frac{\partial}{\partial t}\frac{P_i}{c^2} = \frac{\partial M_{ij}}{\partial x_j}\ , \tag{21}$$

everywhere except at a particle. The essential point is that $\vec{E}$ and $\vec{B}$ together represent momentum and individually they exert stresses in space, accelerating both P_i and the particles.

With the electric field of equation (9), it follows that the electric contribution to M_{ij} is small to second order in v/c compared to $\vec{B}$, so that M_{ij} reduces to

$$M_{ij} \cong -\delta_{ij}\frac{B^2}{8\pi} + \frac{B_iB_j}{4\pi} \tag{22}$$

representing an isotropic pressure $B^2/8\pi$ and a tension $B^2/4\pi$ along the field B_i. Thus the electric field is a secondary quantity . It plays no significant role in the nonrelativistic dynamics of the plasma. The stress resides in $\vec{B}$. The Poynting vector is

$$\vec{P} = c\frac{\vec{E}\times\vec{B}}{4\pi} = -\frac{(\vec{v}\times\vec{B})\times\vec{B}}{4\pi} = \frac{B^2\vec{v}-(\vec{v}\cdot\vec{B})\vec{B}}{4\pi} = \vec{v}_\perp B^2/4\pi \ , \tag{23}$$

where $v_\perp$ is the component of $\vec{v}$ perpendicular to $\vec{B}$. The Poynting vector represents the convection of magnetic enthalpy density $B^2/4\pi$ and the equation of motion (19) is nothing more than the usual MHD momentum equation.

Consider, then, the equation for the kinetic energy of the particles

$$\frac{d}{dt}\frac{1}{2}\rho v^2 = \vec{v}\cdot\vec{E}\cdot\delta = \vec{j}\cdot\vec{E} = (c\vec{\nabla}\times\vec{B} - \frac{\partial\vec{E}}{\partial t})\cdot\frac{\vec{E}}{4\pi} =$$

$$= -\frac{\partial}{\partial t}\frac{E^2+B^2}{4\pi} + \frac{\vec{B}}{4\pi}\cdot\frac{\partial\vec{B}}{\partial t} + c\frac{\vec{\nabla}\times\vec{B}\cdot\vec{E}}{4\pi} = \frac{\partial}{\partial t}\frac{E^2+B^2}{4\pi} + \frac{c}{4\pi}(\vec{\nabla}\times\vec{B}\cdot\vec{E} - \vec{B}\cdot\vec{\nabla}\times\vec{E})$$

Then since $\vec{\nabla}\cdot(\vec{E}\times\vec{B}) = \vec{B}\cdot(\vec{\nabla}\times\vec{E}) - \vec{E}\cdot(\vec{\nabla}\times\vec{B})$ this can be written

$$(\frac{\partial}{\partial t} + v_j\frac{\partial}{\partial x_j})\frac{1}{2}\rho v^2 + \frac{\partial}{\partial t}\frac{E^2+B^2}{8\pi} = -\frac{\partial P_j}{\partial x_j} \tag{24}$$

at every point in space. In the space between particles, where $\rho = 0$, the result is just

$$\frac{\partial}{\partial t}\frac{E^2+B^2}{8\pi} = -\frac{\partial P_j}{\partial x_j} \ . \tag{25}$$

Obviously the energy density of the electromagnetic field is $(E^2+B^2)/8\pi$ while the flux density of electromagnetic energy is represented by the Poynting vector P_i. In the MHD limit.

$$\frac{\partial}{\partial t}\frac{B^2}{8\pi} + \vec{\nabla}\cdot(\vec{v}_\perp\frac{B^2}{4\pi}) = 0$$

In the present case the energy of the electric field is small to second order in v/c and can be neglected. The magnetic field, then, posseses the energy density $B^2/8\pi$, which is transported by $\vec{P}$. So it is the magnetic energy that drives the particles and hence causes the electric current flow. It is clear, then, that in any system where the medium cannot support an electric field $\vec{E}'$ in its own frame of reference, the primary variables are $\vec{B}$ and $\vec{v}$. The current $\vec{j}$ and $\vec{E}$ are secondary, playing no direct role in the dynamics, and can be computed, if needed, from equations (13) and (9), respectively. The quantities $\vec{E}$ and $\vec{j}$ are peripheral and the dynamical equations (10) and (8) cannot be formulated in terms of $\vec{E}$ and $\vec{j}$ in any useful form. Note, then, that the momentum equation (8) can now be written

$$\rho(\frac{\partial v_i}{\partial t} + v_j \frac{\partial v_i}{\partial x_j}) = -\frac{\partial p_{ij}}{\partial x_j} + \frac{\partial M_{ij}}{\partial x_j} ,$$

with M_{ij} given by equation (22).

As a final example, note that if the individual particle motions in $\vec{B}$ are computed in the guiding center approximation (in which the cyclotron radius $Mv_\perp c/qB$ is small compared to the scale l of $\vec{B}$), then the current density can be obtained by summing $\vec{v} \cdot \delta$ over all the particles. The calculation is straightforward in principle but care must be taken to include all the geometric factors. Substituing the expression for the current density into equation (4), the result can be written

$$\frac{\partial \vec{E}_\perp}{\partial t} = -\frac{4\pi c}{B^2} \vec{B} \times \{\rho \frac{d\vec{v}}{dt} + \vec{\nabla}(\vec{p}_\perp + \frac{B^2}{8\pi}) - \frac{[(\vec{B}\cdot\vec{\nabla})\vec{B}]}{4\pi}[1 + \frac{p_\perp - p_\parallel}{B^2/4\pi}]\} ,$$

where $\vec{v}$ is the electric drift velocity $c\vec{E} \times \vec{B}/B^2$ of the individual particles, and $p_\parallel$ and $p_\perp$ represent the particle pressures parallel and perpendicular to $\vec{B}$. Since $\vec{E}_\perp$ is given by equation (9), and $\partial \vec{E}_\perp/\partial t$ is of the order of v^2/c^2 times $c\vec{B}/l$, it follows that the left hand side of this equation is small to second order in v/c and may be put equal to zero. Hence the perpendicular (to $\vec{B}$) component of the quantity in braces on the right hand side must vanish, providing the equation

$$\rho \frac{d\vec{v}}{dt} = -\vec{\nabla}_\perp(\vec{p}_\perp + \frac{B^2}{8\pi}) + \frac{[(\vec{B}\cdot\vec{\nabla})\vec{B}]_\perp}{4\pi}[1 + \frac{p_\perp - p_\parallel}{B^2/4\pi}],$$

(Parker, 1957). The term $(p_\parallel - p_\perp)$, represents the centrifugal force of any anisotropic thermal motions around the curvature $[(\vec{B}\cdot\vec{\nabla})\vec{B}]_\perp/B^2$ of the field, so that the centrifugal force of the thermal motion opposes the magnetic tension $B^2/4\pi$ around the same curvature. Microscopic plasma instabilities are usually effective in pushing the thermal motions toward isotropy so that $p_\parallel - p_\perp$ can generally be neglected.

The essential point is that Newton's equations of motion automatically provide the electric currents required by Ampere's law, accomodating the condition that $\partial E/\partial t$ is small, $O(v^2/c^2)$. Then putting $\partial E/\partial t = 0$ in Maxwell's equation (4), that equation gives back Newton's equation (8) for the bulk motion $\vec{v}$ of the collisionless plasma. There need be no concern, then, about conservation of charge and current.

The dynamics of astrophysical plasmas and magnetic fields is a push and pull contest between the transport P_{ij} of thermal momentum, the transport of bulk momentum represented by the Reynolds stress tensor $R_{ij} = -\rho v_i v_j$ and the magnetic stress M_{ij} in the field, as well as gravity, of course. The basic physical quantities are $\vec{B}$ and $\vec{v}$. No physics is omitted by failing to mention $\vec{j}$ or $\vec{E}$ any more than the internal quantum mechanics of the biological molecules that make up the arm of a baseball pitcher need to be mentioned in working out the Newtonian mechanics of throwing a baseball.

The electric current becomes interesting only in cases where the ionization level is so low or the gas so tenuous that there are not enough free electrons to carry the current density required by Ampere's law without a significant $\vec{E}_{\parallel}$. In such cases the "frozen in" field condition indicated by the induction equation (10) breaks down (cf Schindler, Hesse and Birn, 1991) and interesting particle acceleration sometimes occurs. That generally occurs first in the thin current sheets that arise spontaneously in large-scale deformed fields (Parker, 1972, 1979, 1994), which are the topic of the fourth of these five lectures.

2 THE ORIGIN OF THE MAGNETIC FIELD OF THE SUN

The magnetic field at the visible surface of the Sun is made up of many separate small intense bundles of magnetic flux that have risen from somewhere below. The mean field (averaged over dimensions of 10^{10} cm in latitude and longitude) shows a dipolar character within about 35° of each pole and an intensity of the order of 10 gauss. In the wide equatorial band of ± 55° latitude the mean pattern is usually made up of three to eight head to tail bipolar regions, each with an approximately east-west orientation, and having opposite orientation on opposite sides of the equator. This head to tail bipolar pattern is interpreted as indicating a general azimuthal magnetic field somewhere below the surface, with opposite sign in the northern and southern hemispheres. The individual bipolar regions represent individual Ω - loops of that field, driven to the surface by their magnetic buoyancy (Parker, 1955 a). The entire magnetic system alternates in sign with varying amplitude on a period that fluctuates about a mean value of 22 years. Observations of other stars show similar magnetic cycles with periods of 2-20 years or more, so the oscillatory magnetic behavior of the Sun is typical of most stars.

It is interesting to note that the azimuthal magnetic field (i.e. the bipolar magnetic regions at the surface) first appears at the surface at middle latitudes ($\sim 40°$) and migrates toward the equator over the next decade (at about 1m/s) where it disappears, presumably cancelling across the equator with the opposite azimuthal magnetic field from the other hemisphere. The dipole, or poloidal, component on the other hand, reverses near the peak of the magnetic activity, when the azimuthal fields are at latitudes of the order of $\pm 15°$. The reversal appears to be a rotation of the dipole moment through 180° over a period of a couple of years, rapidly passing undiminished through the perpendicular orientation (relative to the spin axis of the sun) on its way to the new opposite orientation.

One can imagine that there may still be some remnant magnetic fields trapped in the high electrical conductivity of the radiative core of the Sun from the time the Sun was formed some 4.6×10^9 years ago. But such fields are not so strong as to provide observable evidence of their existence. So the problems posed by the observed magnetic field of the Sun concerns the rapid creation and destruction of magnetic field over scales of the solar radius $R_\odot(7 \times 10^{10}$ cm) in times of the order of a decade. That is to say, we are dealing with a magnetohydrodynamic dynamo, which implies that the action takes place in the convective zone. Present theoretical models of the interior of the Sun indicate that the convective zone extends downward from the visible surface to a depth of about 2×10^{10} cm (cf Spruit, 1974) with perhaps some overshoot for a distance $\sim 3 \times 10^9$ cm into the stable radiative zone below the region of convective instability.

The strongest magnetic inductive effect is the interaction of the non-uniform rotation of the sun $\omega(\tilde{\omega}, z)$ with the poloidal magnetic field $B_{\tilde{\omega}}, B_z$ (in cylindrical polar coordinates $(\tilde{\omega}, \varphi, z)$). In the simple case of rotational symmetry $(\partial/\partial\varphi = 0)$, the magnetohydrodynamic induction equation is

$$\frac{\partial B_\varphi}{\partial t} = \tilde{\omega}\frac{\partial}{\partial_z}\omega B_z + \frac{\partial}{\partial\tilde{\omega}}\tilde{\omega}\omega B_{\tilde{\omega}}$$

It is convenient to express the poloidal field in terms of an azimuthal vector potential $A(\tilde{\omega}, z)$ as $B_{\tilde{\omega}} = \frac{1}{\tilde{\omega}}\frac{\partial A}{\partial z}$, $B_z = -\frac{1}{\tilde{\omega}}\frac{\partial A}{\partial \tilde{\omega}}$. The field lines are given by $A(\tilde{\omega}, z) = \lambda$ and the induction equation becomes

$$\frac{\partial B_\varphi}{\partial t} = \frac{\partial\omega}{\partial\tilde{\omega}}\frac{\partial A}{\partial z} - \frac{\partial\omega}{\partial z}\frac{\partial A}{\partial\tilde{\omega}} \, .$$

A dissipative effect is essential if the field is to be created and destroyed, so we add a uniform scalar resistive diffusion coefficient η as a convenient expression of dissipation, obtaining finally,

$$[\frac{\partial}{\partial t} - \eta(\nabla^2 - \frac{1}{\tilde{\omega}^2})]B_\varphi = \frac{\partial\omega}{\partial\tilde{\omega}}\frac{\partial A}{\partial z} - \frac{\partial\omega}{\partial z}\frac{\partial A}{\partial\tilde{\omega}} \,. \tag{26}$$

Cowling's theorem asserts that there are no self-sustanining solutions to the MHD equations for a magnetic field with a rotationally invariant topology, so the generation of the poloidal field, described by A, must involve localized motions (i.e. $\partial/\partial\varphi \neq 0$), presumably individual convective cells. The rising and sinking convective cells provide both upward and downward pointing Ω-loops in the azimuthal field. The convective cells are cyclonic as a consequence of the Coriolis force in the rotating Sun. The sinking fluid is located around the periphery of the rising cells and so gives less rotation. The result is that the Ω-loops are rotated so that they have non vanishing projection on meridional planes. That is to say, the rotated Ω-loops provide magnetic circulation in the meridional planes, and that represents a net poloidal field (Parker, 1955b, 1957a, 1970b), equivalent to generating azimuthal vector potential. Diffusion destroys the small-scale structure of the individual Ω-loops and the result is a mean poloidal field described by the mean azimuthal vector potential A. One can think of the projection on the meridional plane of each rotated Ω-loop as contributing a local azimuthal A. The individual clumps of A merge throuhg diffusion to form the macroscopic A describing the large-scale poloidal field. The generation of A can be described by

$$[\frac{\partial}{\partial t} - \eta(\nabla^2 - \frac{1}{\tilde{\omega}^2})]A = \Gamma(\tilde{\omega}, z)B_\varphi \,, \tag{27}$$

where the generation coefficient Γ is comparable to the rms cyclonic velocity of the convective cells multiplied by their filling factor. The quasi-linear approximation provides Γ, conventionally written as α , in terms of the mean small-scale helicity $< \vec{v} - \vec{\nabla} \times \vec{v} >$ of the convection (Steenbeck, Krause, and Rädler, 1966; Steenbeck and Krause, 1966; Krause and Rädler, 1980). The short sudden approximation (Parker, 1955b, 1970a, b; 1979; Bachus, 1958; Lerche, 1991) provides α or Γ in terms of the helical structure of the individual convective cells (cf. Moffatt, 1978).

There are higher order effects, involving the interaction of the cyclonic convection with the poloidal field, the gradient of Γ, η, B_φ, etc. (cf. Krause and Rädler, 1980; Parker, 1979). But attention is restricted here to the basic dynamo effect described by equations (26) and (27), the so called $\alpha\omega$-dynamo.

To obtain some idea of the magnitudes of Γ, η, and $G = \mid \nabla\omega \mid$, it is sufficient to consider the dynamo in plane geometry, in which the spherical shell of the convective zone is flattened into an infinitely broad slab and the thickness of the slab becomes large without limit. Let the y-direction correspond to azimuth, z

to the vertical, and x to latitude. In this simplified scenario, the magnetic field is a plane wave with wave vector k in the x-direction with the idealization that η, Γ, and G are constants and the principal shear ($G = dv_y/dz$) is vertical. Then let $B_z = \frac{\partial A}{\partial x}$ with $B_x = 0$, so that the dynamo equations become

$$(\frac{\partial}{\partial t} - \eta \frac{\partial^2}{\partial x^2}) B_\varphi = G \frac{\partial A}{\partial x} \tag{28}$$

$$(\frac{\partial}{\partial t} - \eta \frac{\partial^2}{\partial x^2}) A = \Gamma B_\varphi \ . \tag{29}$$

Solutions of the form $exp(\sigma t + ikx)$ yield the dispersion relation,

$$\sigma = -\eta k^2 \pm (\frac{k\Gamma G}{2})^{\frac{1}{2}}(1+s) \ . \tag{30}$$

The upper sign gives the regenerative mode, with

$$Re(\sigma) = (\frac{k\Gamma G}{2})^{\frac{1}{2}} - \eta k^2 \ , \tag{31}$$

$$Im(\sigma) = (\frac{k\Gamma G}{2})^{1/2} \ , \tag{32}$$

if k, Γ, and G are positive quantities.

Consider what it takes to sustain such a dynamo wave. For constant amplitude the growth rate $Re(\sigma)$ must vanish, from which if follows that

$$\Gamma G = 2\eta^2 k^3 \ . \tag{33}$$

The observed 22 years period for the solar dynamo would require $Im(\sigma) \sim 10^{-8}$/s . The bands of azimuthal field have a latitudinal extent of about 35° or 4×10^{10} cm, representing half a wavelength π/k. Hence $k \cong 0.8 \times 10^{-10}$cm^{-1}. The result from equation (32) is $\Gamma G \cong 2.5 \times 10^{-6}$cm/s^2. Then equation (33) gives $\eta \sim 1.6 \times 10^{12}$cm^2/s, in order of magnitude.

The nonuniform rotation at the surface of the Sun involves a 25 day sidereal period for low latitudes (equatorial velocity $\omega R_\odot \sim$ 2Km/sec) and a 35 day period at the poles. This suggests that there may be a velocity difference of as much as 4×10^4 cm/s across the 2×10^{10} cm depth of the convective zone, providing $G = 2 \times 10^{-6}$/sec. In that case Γ assumes the modest value of 1 cm/s. This is to be compared with the characteristic cyclonic velocity ωl across a correlation length or mixing length l, where $\omega \sim 2 \times 10^{-6}$/s represents the mean angular velocity of the Sun. Then if $l \sim 10^9$ cm, if follows that $\omega l \sim 2 \times 10^3$cm/s.

It is curious that the necessary effective mean cyclonic motion Γ is so small compared to ωl, we might have expected that $\Gamma \sim 0.1\omega l$.

Consider, then, the physical basis for a diffusion coefficient η as large as 10^{12} cm^2/s and a Γ or α of the general order of 1cm/s. The resistive diffusion coefficient in ionized hydrogen at 10^6 K is 0.5×10^4cm/s. The requirement of 10^{12} cm^2/sec can be met only with turbulent diffusion. Dimensional analysis applied to the turbulent mixing of scalar fields suggests an eddy diffusion coefficient $\eta \sim 0.1lv$, where v represents the velocity of the dominant eddies, of characteristic scale l. There is no comprehensive theory for the diffussion of vector fields, so the scalar concept is taken over into MHD. The mixing length models of the convective zone associate l with the pressure scale height and provide convective velocities of such a magnitude that $\eta \sim 10^{12}$ cm^2/sec. For instance, the granules at the visible surface show $v \sim$1 km /s across correlation lengths $l \sim 5 \times 10^7$cm, providing $\eta \sim 0.5 \times 10^{12}$ cm^2/s. At a depth of 10^5 km, halfway down through the convective zone, the temperature is 0.9 $\times10^6$ K, the mean molecular weight is 0.62, and the gravitational acceleration g is approximately 3.6 $\times$ 10^4cm/s. The scale height is 0.9 $\times$ 10^6 K, the mean molecular weight is 0.62, and the gravitational $kT/\mu Mg$ is 3.4 $\times$ 10^9 cm. Spruit (1974) gives a turbulent velocity of 3.4×10^3 cm/s. Equating the characteristic mixing length l to the scale height yields $\eta \sim 1.2 \times 10^{12}$ cm^2 /s. So the numbers appear to satisfy the needs of the theoretical solar dynamo in a natural way, and it is easy to believe that an $\alpha\omega$ dynamo in the convective zone is a plausible theoretical basis for the observed oscillatory magnetic fields of the Sun. Indeed, there is no known alternative. A variety of detailed kinematical dynamo models have been constructed within the spherical shell occupied by the convective zone (cf. Steenbeck and Krause, 1969; Köhler, 1973; Yoshimura, 1975, 1977; Stix, 1976; Krause and Rädler, 1980; Priest 1982; Soward, 1983; Choudhuri, 1984, 1990 and references therein) showing that the detailed behaviour of the magnetic field of the Sun can be fitted quite well with suitable choice of η, Γ, G, etc.

However, a closer examination of the situation leads to serious problems. The first difficulty arises with estimates of the mean azimuthal magnetic field, based on the formation of large bipolar magnetic regions at the surface. The large bipolar magnetic regions develop through successive eruptions of many small flux bundles, or Ω-loops, in the same site over a period of months. Gaizauskas, et al. (1983) studied a long lived bipolar region over a period of about a year, during which time the total magnetic flux at the surface increased to $6-8\times10^{22}$ Maxwells through a succession of eruptions of $10^{21}-10^{22}$ Maxwells at irregular intervals of the order of a week. Assuming that all the flux remained connected to the azimuthal field in the deep convective zone, it follows that the total azimuthal flux is not less than 8 $\times$ 10^{22} Maxwells. It is reasonable to suppose that the successive eruptions of flux in the local site that produced the bipolar region at the surface came from a restricted band of latitude, with a width of, say, 10° or less, ($< 10^{10}$ cm). One presumes that the azimuthal field is confined to the lower half (10^{10} cm) of the convective zone (Parker, 1975, 1977). Thus we estimate that the 8×10^{22} Maxwells or more had a cross section of 10^{20} cm^2. It follows that the mean azimuthal field $< B >$ in the lower convective zone is at least as large as 8×10^{22} Maxwells divided by the area of 10^{20} cm^2, yielding $< B >> 8 \times 10^2$ gauss. In fact the mean field may be substantially larger than this lower limit. But even 800 gauss is enough to stifle the essential turbulent diffusion and the cyclonic α-effect in regions with such large magnetic Reynolds numbers ($> 10^{10}$).

The essential point is that the turbulent mixing and diffusion of a mean large-scale vector field $< \vec{B} >$ involves the displacement and stretching of the individual flux bundles into long thin ribbons whose thickness is so small that ordinary molecular diffusion can dissipate the field in the allotted time. The thinning of the flux bundles is associated with comparable lengthening since the volume of each flux bundle is conserved. The result is a field intensity (within each ribbon) that is enormously larger than the mean field. But turbulence with a mean square velocity $< v^2 >$ provides Reynolds stresses of the order of $\rho < v^2 >$ which cannot stretch tightly packed flux bundles with larger magnetic tension $B^2/4\pi$. That is to say, the total rms field $< B^2 >^{1/2}$ is not expected to exceed the equipartition value $4\pi\rho < v^2 >^{1/2}$, and that is estimated from the standard mixing length models (cf. Spruit, 1974) to have a maximum of about 3×10^3gauss in the lower convective zone, falling rapidly to zero through the bottom of the convective zone into the overshoot region below as $< v^2 >$ declines to zero.

To be more quantitative, the stretching of a flux bundle in turbulence with characteristic eddy velocity v over a scale l increases the length $L(t)$ of a line (carried with the fluid) at an exponential rate given in order of magnitude by $L(t) = L(0)exp(t/\tau)$, where τ is of the order of l/v, essentially the correlation time for the turbulence. Thus a flux bundle with initial diameter l is dram into a long thin ribbon with length $L(t)$, width of the same general order as l, and thickness of the order of $\delta(t)$ where conservation of volume requires that $\delta(t)L(t) \cong lL(0)$ in order of magnitude. Hence $\delta(t) \cong l\ exp(t/\tau)$. The extended ribbons are interwired like the individual noodles in a bowl of spaghetti. The characteristic diffusion time in which one ribbon becomes blended into its neighbours as a consequence of the atomic resistive diffusion η is of the order of $\delta(t)^2/4\eta$. The flux bundle loses its identity when it becomes so thin that the diffusion time falls to the growth rate τ. This occurs when $exp\ 2t/\tau \cong \delta(0)^2/\eta\tau \cong lv/\eta \equiv R_m$, where R_m is the characteristic magnetic

Reynolds number. Conservation of magnetic flux within the flux bundle requires $B(t)\delta(t) \cong B(0)\delta(0)$ in order of magnitude, from which the familiar result that $B(t) \cong B(0)R_m^{1/2}$ in order of magnitude at the time that resistive diffusion is becoming important. With $R_m \cong 10^{10}$ in the solar convective zone, it follows that the field within the individual flux bundle would approach $10^5 B(0)$ before being dissipated. There is no way that the convection could manipulate so strong a field to form long thin ribbons.

Suppose, then, that the destruction of flux ribbons by diffusion and reconnection with neighboring ribbons is carried out by rapid reconnection between contiguous ribbons at the maximum theoretical rate (Petschek and Thorne, 1967) in which one ribbon cuts across a neighboring (nonparallel) ribbon at a speed comparable to the Alfven speed $B(t)/(4\pi\rho)^{1/2}$ in the ribbon field multiplied by a factor of the order of 0.1. The characteristic cutting time is then of the order of $10(4\pi\rho)^{1/2}[\delta(t)/B(t)]lnR_m$ across the thickness $\delta(t)$. The flux bundle loses its identity when the cutting time falls to τ, at which point

$$exp\ 2t/\tau = 10\frac{v\ lnR_m}{C(0)}\ ,$$

where $C(0)$ is the Alfven speed $B(0)/(4\pi\rho)^{1/2}$ in the mean field $B(0)$. The field within the individual flux bundle has increased to $B(t) = B(0)(10\ \frac{v}{C}\ lnR_m)^{1/2}$, in order of magnitude. If the equipartition field is 3000 gauss and the mean-field is 800 gauss, it follows that $v/C \sim 0.25$. The width $R_m = 10^{10}$, it follows that the field intensity $B(t)$ with the individual flux bundle would grow to at least $11B(0)$ in the processes of intermixing flux bundles to the point of dissipation. The equipartition field is not more than $4B(0)$, so at $11B(0)$ the magnetic stresses would be about 11 times the Reynolds stress and there is no reason to believe that such intense fields could be stretched out by the convection. This is only a lower limit on the field of the individual flux bundle, of course, because there is no compelling reason to believe that the reconection proceeds as fast as $0.1C(t)/lnR_m$ (in the Petschek mode). It may instead progress as slowly as $C(t)/R_m^{1/2}$ (Parker, 1957b, 1979; Sweet, 1958a, 1969). In that case, at the time when resistive diffusion takes over

$$exp\frac{2t}{\tau} = \frac{\delta(0)}{C\tau}R_m^{1/2}, \text{and } B(t) = B(0)(\frac{v}{C}R_m^{1/2})^{1/2} \cong B(0)(\frac{1}{5}R_m^{1/2})^{1/2} \cong 10^2 B(0)\ .$$

This is very much stronger than the equipartition field ($\sim 4B(0)$).

In any case, the mean field of 800 gauss or more appears to be too much to allow significant dissipation of the individual flux bundles. So turbulent diffusion is greatly suppressed, although it is not possible to state by how many factors of ten.

The cyclonic α-effect suffers the same limitations and it is clear that the original estimates of η and Γ or α, that provided simple plausible dynamo models, are much too optimistic in the face of mean azimuthal magnetic fields of 800 gauss or more.

The resolution of the difficulty is suggested by the result inferred from helioseismology (Dziembowski, Goode, and Libbrecht, 1989; Brown et. al., 1989; Thompson 1990; Schou and Brown, 1994) that the velocity shear, and hence B_φ, lie below the base of the convective zone. It follows that the cyclonic α effect up in the convective zone occurs where there is very little B_φ. This suggests the theoretical possibility that the turbulent diffusion η is relatively unsuppressed in the convective zone by a weak azimuthal field B that leaks up from the shear layer in where the azimuthal field B is so strong as to suppress the eddy diffusivity to some very small value $n = \mu^2\eta$ where $\mu^2 << 1$. Some small amount of poloidal field created in $z > 0$, is punched down into the shear layer by concentrated downdrafts, providing the small vector potential a in $z < 0$. The dynamo equations in plane geometry become

$$[\frac{\partial}{\partial t} - \eta(\frac{\partial^2}{\partial x^2} + \frac{\partial^2}{\partial z^2})]B = 0 \ , \ [\frac{\delta}{\partial t} - \eta(\frac{\partial^2}{\partial x^2} + \frac{\partial^2}{\partial z^2})A = \Gamma B$$

in the convective zone ($z > 0$) in the absence of any shear G. In the shear layer ($z < 0$), where there is no significant cyclonic motion Γ, the dynamo equations are

$$[\frac{\partial}{\partial t} - \mu^2\eta(\frac{\partial^2}{\partial x^2} + \frac{\partial^2}{\partial z^2})]b = G\frac{\partial a}{\partial x} \ , \ [\frac{\partial}{\partial t} - \mu^2\eta(\frac{\partial^2}{\partial x^2} + \frac{\partial^2}{\partial z^2})]a = 0$$

The simultaneous solution of these four equations must satisfy the boundary conditions that the fields vanish at $z = \pm\infty$ and match at $z = 0$ where

$$a = A \ , \ b = B \ , \ \frac{\partial a}{\partial z} = \frac{\partial A}{\partial z} \ , \ \mu^2\eta\frac{\partial b}{\partial z} = \eta\frac{\partial B}{\partial z}$$

The result (Parker, 1993) is a dynamo surface wave propagating along the interface $z = 0$ and declining exponentially in both directions away from $z = 0$. The horizontal profile of the azimuthal field intensity is shown in Fig. 1 with the phase throughout one cycle indicated on each curve for the special case that $\mu^2 = 0.01$.

The surface wave is a less efficient dynamo than the conventional one where G and Γ are superposed, because the fields created by G and Γ have then to be transpoted to the regions of Γ and G, respectively. For comparison note that a 22 year period with $k = 0.8 \times 10^{-10}$/cm again requires $\eta \sim 1.4 \times 10^{-10}\text{cm}^2$/s but now $\Gamma G = 1.7 \times 10^{-3}$ cm/s^2, larger by a factor of about 700. With $G = 10^{-5}$/s

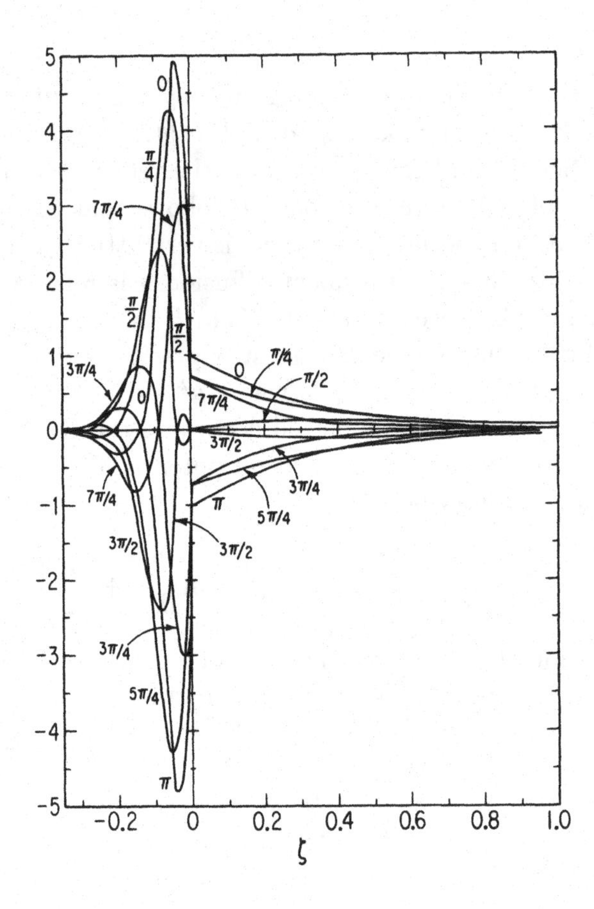

Figure 1: *The horizontal (north-south) profile of the azimuthal field at the successive phases indicated for each curve.*

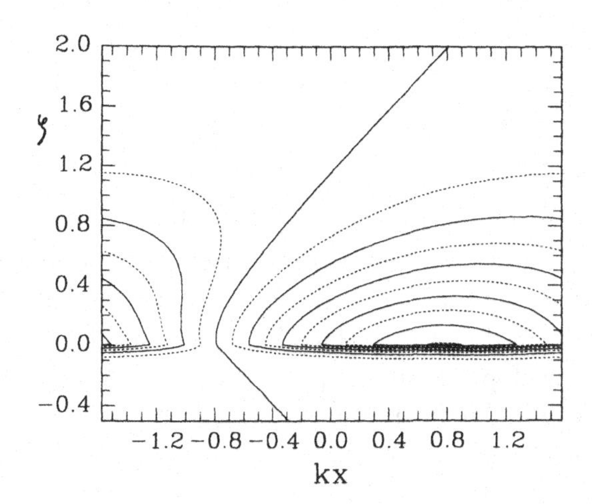

Figure 2: *A plot of the field lines at equal flux intervals of the poloidal field, where kx represents the phase in latitude*

again, it means that $\Gamma \sim 2 \times 10^2$ cm/s. This is more nearly what one might expect for cyclonic velocities in the large convective cells ($l \sim 10^9$ cm) in the lower convective zone.

So it would appear that the scenario suggested by helioseismology may perhaps provide a workable dynamo. The big theoretical unknown remains the suppression of η and Γ by the azimuthal magnetic field and the actual effective upward transport of azimuthal field into the region of cyclonic turbulence Γ and the downward transport of A into the region of shear G . The simple dynamo surface wave indicates that the suppression of the eddy diffusivity by the factor μ^2 provides an azimuthal field in the convective zone that is reduced by the factor $1/\mu$ below the field in the shear layer.

It is essential that this dynamo surface wave be tested both theoretically and observationally. For this, and for many other problems in MHD turbulence, we should continue to push for a quantitative theory of the turbulent diffusion of magnetic fields. Unfortunately this question is confused by the observational fact of the intense fibril state of the magnetic field at the surface of the Sun, suggesting the possibility that the magnetic field is in a fibril state throughout the convective zone, so the physics of the fibril field must also be addressed. It is a daunting task, but we cannot claim to understand any stellar MHD dynamo until we have solid reasons for believing that we fully understand the solar dynamo.

3 300000 GAUSS AZIMUTHAL MAGNETIC FIELDS

3.1 The Azimuthal Field

It is commonly believed that the head to tail bipolar magnetic regions that extend around the Sun at low latitudes are a consequence of large scale azimuthal fields at some depth in the Sun. It is presumed that the magnetic buoyancy of the azimuthal field far below the surface is responsible for the upward bulges (Parker, 1955) that develop into the shape of an Ω-loop as they rise to the surface to form the observed bipolar magnetic regions. However, theoretical studies in recent years have shown that the rise of a buoyant flux bundle is not the simple phenomenon it was once thought to be. In particular, Choudhuri and Gilman, (1987) have showed that the Coriolis force strongly deflects a rising Ω-loop, causing it to move nearly parallel to the spin axis of the Sun thereby failing to provide the bipolar regions at low latitude. The importance of the Coriolis force can be seen directly from the fact that the time of rise is of the same order of magnitude as the rotation period of the Sun. Now helioseismology suggests that the principal gradient in the angular velocity of the Sun lies immediately below the bottom of the convective zone, in the uppermost layers of the stably stratified radiative zone. One infers that the azimuthal magnetic field lies in much the same stratum, where it is generated by the interaction of the gradient in the angular velocity with the poloidal field of the Sun (see the previous lecture on the solar dynamo). It is interesting to note, then, that the works of Spruit and Van Ballegooijen (1982), and Schüssler et al (1993) show that the magnetic buoyancy of the azimuthal magnetic field below the base of the convective zone is strongly suppressed by both the subadiabatic temperature gradient and the strong non uniform rotation in that region. They estimate that the magnetic buoyancy does not overcome these stabilizing effects at low latitude until the field strength reaches something of the order of 10^5 gauss.

It is interesting to note, then, that recent numerical simulations of the buoyant rise of an Ω-loop from below the convective zone show that the flux bundles come nearly radially upward and arrive at the surface with the observed small east-west tilt of the bipolar regions if, and only if, the initial individual flux bundle at the base of the convective zone has an intensity of the general order of 10^5 gauss (Fisher, McClymont and Chou, 1991; D'Silva, 1993; D'Silva and Choudhuri, 1993; Fan, Fisher and De Luca, 1993). With such strong field the buoyant rise is rapid, thereby moving nearly radially, while the magnetic tension is sufficient to suppress the large overall cyclonic rotation of the apex of the Ω-loop. The dynamics of Ω-loops and bipolar regions appears to follow naturally, then, for azimuthal flux bundles of the general order of 10^5 gauss. That is to say, it seems reasonable to conclude that the azimuthal field of the Sun contains flux bundles with intensities of the order of 10^5 gauss. Such large field strengths are astonishing, being larger by at least one factor of ten than any relevant kinetic energy density of wich we are aware.

Incidentally, one finds (Parker, 1967, 1979; Matsumoto et al 1993) that the rising flux bundle is unstable to shredding into many thinner flux bundles in the process of floating up to the surface, explaining perhaps why the emergence of magnetic flux at the surface is in the form of many individual flux bundles (of the order of 10^{19} Maxwells or less) sometimes appearing at a rate of 10^2 or more per day during a time of active emergence. It is during such times of emergence of fresh flux that the separate flux bundles show the peculiar tendency to cluster together to form pores and sunspots.

The essential goal for the present narrative is to understand why there are magnetic flux bundles of 10^5 gauss in the shear layer below the bottom of the convective zone. The magnetic pressure of 10^5 gauss is 4×10^8 dynes/cm^2. The magnetic tension $B^2/4\pi$ is twice as large. The ambient gas pressure is 6×10^{13} dynes/cm^2 so the plasma $\beta \equiv 8\pi p/B^2$ is 1.5×10^5. On the other hand, the Reynolds stress ρv^2 associated with the nonuniform rotation arises from a total velocity difference Δv of no more than about 4×10^4 cm/s across a vertical scale h of perhaps 4×10^9 cm. This yields a vertical shear $G = dv_\varphi/dz \cong 10^{-5}$s.

The first question is whether the nonuniform rotation is fast enough to produce 10^5 gauss in a few years at the onset of each activity cycle. Another question is whether the nonuniform rotation is driven hard enough by the meridional circulation to overcome the Maxwell stress in the combined azimuthal and poloidal fields B_φ and B_ρ, respectively. To answer these two questions, note that the azimuthal field is generated from the vertical component of the poloidal field at the rate

$$\frac{\partial B_\varphi}{\partial t} = GB_p$$

It follows for a constant B_p that after a time t, $B_\varphi = GtB_p$. Make the optimistic assumption that the vertical B_p down in the shear layer below the convective zone is as large as 10 gauss. Then with $Gt = 10^4$ and $B_\varphi = 10^5$ gauss we find that $t = 10^9$s=30 years. This is approximately ten times slower than what is needed. We might cut the time to 15 years by assumming that the shear occurs over a height of only 2×10^9 cm, but even so it is still too slow. Consider, then, the energy requirements. The energy density of 10^5 gauss is 4×10^8 ergs/cm^3, whereas the kinetic energy density of the rms Δv across a uniform shear larger is $\frac{1}{6}\rho(\Delta v)^2$, equal to 0.5×10^8 ergs/cm^3 where $\rho = 0.2$ g/cm^3. Relative to the midplane of the shear layer the mean kinetic energy density is only one quarter of this amount, or 1.2×10^7 ergs/cm^3. The nonuniform rotation would have to be maintained, in opposition to the enormous Maxwell stress, by a characteristic acceleration time of only a few solar rotation periods.

However, there is really no observational requirement to fill the shear layer with a field of 10^5gauss. Observations of flux emergence at the surface of the Sun imply only that there is a minimum of about 10^{23} Maxwells at 10^5 gauss. The cross sectional area of this field would be about 10^{18} cm^2, or 0.04 of the cross section of the region of depth 4×10^9cm and latitudinal width 10^{10} cm which might be occupied by the azimuthal field. The mean magnetic energy is then 1.6×10^7 ergs/cm^3. If the nonuniform rotational energy of the shear layer could be closely coupled to the production of the concentrated azimuthal magnetic flux bundles, it might be adequate for a couple of 11-year cycles. In this case, then, the formation of the restricted shear layer must have a characteristic time of no more than a few decades.

This raises the question of how the nonuniform rotation could produce isolated azimuthal flux ropes with a filling factor of only 4×10^{-2}. The obvious conjecture is that the poloidal field is in the form of isolated azimuthal flux ropes each one of which is sheared to provide a single azimuthal flux rope. Thus, a filling factor of 4×10^{-2} for a vertical poloidal field with flux bundles of 250 gauss provides a mean poloidal field of 10 gauss and requires only that $Gt = 4 \times 10^2$ or $t \cong 1$ year to provide 10^5 gauss in the azimuthal bundles. If the flux in the individual poloidal bundle is 10^{19} Maxwells, the diameter of the bundle is approximately 2×10^8 cm. It is not clear why or whether the poloidal field has an intense fibril form in the shear layer, of course, but in our present state of ignorance it remains a theoretical possibility. The azimuthal flux bundle has the same total flux as the poloidal bundle from which it is produced (since there is no reason to think that the symmetry of the nonuniform rotation is so perfect that an azimuthal bundle closes on itself after being extended a whole turn around the Sun). The cross sectional area of the azimuthal bundle would be only 10^{14} cm^2 or 10^2 km diameter.

In summary, if the hydrodynamic forces that create the restricted nonuniform rotation layer are strong enough to form the layer in a characteristic time of a couple of decades ($\sim$ 200 rotation periods), then it might be possible to understand the 10^5 azimuthal flux bundles based on a poloidal field made up of intense magnetic fibrils with a filling factor of 4×10^{-2}. However, it appears that there is a more likely alternative.

3.2 The Barometric Origin of the Intense Field

It has been pointed out that the 10^5 gauss azimuthal field intensity may be related to the observed fact that the azimuthal flux bundles form buoyant Ω-loops that rise to the surface of the Sun (Parker, 1994). Imagine, then, that the azimuthal field has a mean value of the order of 10^3 gauss (based on the assumption that the shear layer has a thickness of 4×10^9 cm) in order that 10^{23} Maxwells is available. This mean azimuthal field follows for $Gt = 10^2$ from a vertical poloidal field of 10 gauss, i.e. for $t \sim 10^7$ s or about four solar rotations, (longer if $B_p << 10$ gauss). The azimuthal field is known to form Ω loops first at middle latitudes where the shear is relatively weak and azimuthal fields of 10^3 gauss have a chance to form Ω-loops which billow upward into the convective zone. The Ω-loops may, or may not, appear at the surface. The essential point is that the process of forming succesive Ω-loops in a superadiabatic temperature gradient (i.e. in the convective zone) provides powerful barometric forces that concentrate the azimuthal flux bundles in the shear layer below. It has been pointed out (Spruit, Title and van Ballegooijen, 1979; Wilson, McIntosh and Snodgrass, 1990) that Ω-loops very likely reconned across their base after a time, leaving a free O-loop and restoring the field below the O-loop to its original azimuthal form, sketched in Fig. 1. We have no idea how quickly the reconnection may take place, of course. One might identify the reconnection with times as short as the period of a month associated with the rise time of Ω-loops from the bottom of the convective zone, or with the several months in which a large bipolar region may appear or disappear.

We are inclined towards the longer periods of many months because the reconnection proceeds at only a small fraction of the Alfven speed, which is 10^3 cm/s in a field of 10^3 gauss in the lower convective zone ($\rho = 0.2$ g/cm^3). It is theoretically possible that the reconnection proceeds at increasing rates as the flux bundles become more intense, of course. But for the moment we make no other assumption but the Ω-loops are pinched off across their base in times of 1 year or less to form O-loops, restoring the azimuthal field connection. Hence over an 11 year activity cycle, an azimuthal flux bundle forms many successive Ω-loops at several different active longitudes along its length. Whether an individual azimuthal flux bundle maintains its identity for 11 years, or only for a few years at a fixed latitude is not known of course, the answer depending upon the degree to which meridional circulation, as distinct from the dynamo creation and destruction of azimuthal field, is responsible for the equatorward migration of the location of azimuthal field.

Consider, then, the effects that arise in the formation of an Ω-loop. The first thing that comes to mind is the lengthening of the flux tube from its original azimuthal form. Cutting off the Ω-loop at its base to restore the azimuthal field means that the gas in the azimuthal flux bundle has been reduced by the amount carried away in the free O-loop. In the simple case that the temperature T of the gas is the same inside and outside the azimuthal flux bundle, the hydrostatic equilibrium determines the field strength B within the flux bundle as $\frac{B^2}{8\pi} = 2kT(N - N_i)$ for ionized hydrogen, where N is the ambient number density of hydrogen atoms and N_i is the number density within the flux bundle where the magnetic field is B. The energy that goes into increasing B comes from the convective forces that raise the Ω-loop, and the next question is how strong are those forces. The answer is that the gas moves nearly adiabatically in the rapid upwelling ($\sim 10^4$ cm/s) associated with a rising Ω -loop. So by the time it rises 10^{10} cm and reaches the middle of the convective zone it is hotter by $\Delta T \sim 1$ K (cf. Spruit, 1974) where $T = 0.8 \times 10^6$K. Another 5×10^9 cm and $\Delta T = 5$K where $T = 3.5 \times 10^5 K$ at a depth of 5×10^9 cm. At a depth of 2×10^9cm, $\Delta T = 14$ K where $T = 1.5 \times 10^5$ K.

These temperature enhancements above the ambient T provide strong upward buoyancy, causing the fluid to billow upward both within and around the rising Ω-loop. The essential point is that the upward convective force $NMg\Delta T/T$ in the upper part of the Ω-loop reduces the gas pressure within the lower part of the Ω-loop, so that gas streams from the azimuthal flux bundle into the Ω-loop. To put it differently, there is approximate horizontal pressure equilibium at each level across the updraft, and the field is so expanded at the upper end of the Ω-loop that $p_i \cong p$. With the elevated temperature $T + \Delta T$ of the adiabatically rising gas within the vertical legs of the Ω-loop the gas pressure increases downward less rapidly than in the ambient fluid. Integrating downward within the flux blundle from $p_i \cong p$ at the apex, it follows that $p_i(0)$ at the base of the Ω-loop is substantially less than the ambient p. The result is the reduction of p_i in the azimuthal flux bundle connected to the base of the Ω-loop, so that the ambient pressure compresses the flux bundle to 10^5 gauss or more. One imagines that the full 10^5 gauss is achieved in the azimuthal flux bundle only after several successive Ω-loop have been spawned and converted to O-loops. But this is only a guess, and the dynamics of the problem has to be treated quantitatively before the degree to which the development of a single Ω-loop evacuates and concentrates the azimuthal magnetic field can be shown. Here we can only outline the static barometric effect that drive the process.

The effect can be treated quantitatively for a fixed Ω-loop. The equation of motion for the internal motion along the static field B is

$$\frac{1}{p_i}\frac{\partial p_i}{\partial s} + \frac{M}{2k(T + \Delta T)}(v\frac{\partial v}{\partial s} + g cos\theta) = 0 \ , \tag{34}$$

where θ is the angle by which the local field direction tilts away from the vertical,

v is the flux velocity along the field, p_i is the internal gas pressure ($2N_ikT$ for ionized hydrogen), and s represents distance measured along the field lines. This equation integrates to

$$p_i(s) = p_i(0)exp(-\int_o^z \frac{dz}{\wedge + \Delta\wedge})exp(-\frac{1}{2g}\int_o^{s(z)} \frac{ds}{\wedge + \Delta\wedge}\frac{\partial v^2}{\partial s}) \tag{35}$$

upon noting that $dz = ds\ cos\theta$ and writing the scale height as

$$\wedge = 2kT/Mg,\ \Delta\wedge = 2k\Delta T/Mg \tag{36}$$

with g taken to be uniform. Note then, that $\Delta\wedge/\wedge$ and $g^{-1}\partial v^2/\partial s$ are both small compared to one, so that

$$p_i(z) \cong \pi(z)[1 + I_1(z) - I_2(z)] \tag{37}$$

neglecting terms second order in $\Delta T/T$ and $g^{-1}\partial v^2/\partial s$, where the ambient pressure varies as

$$p(z) = p(0)exp[-\int_o^z \frac{dz}{\wedge(z)}] \tag{38}$$

while

$$k(z) \equiv p_i(0)p(z)/p(0) \tag{39}$$

$$I_1(z) \equiv \int_o^z \frac{dz\Delta\wedge(z)}{\wedge^2(z)} << 1 \tag{40}$$

$$I_2 \equiv \frac{1}{2g}\int_o^{s(z)} \frac{ds}{\wedge(z)}\frac{\partial v^2}{\partial s} << 1\ . \tag{41}$$

The ambient barometric pressure $p(z)$ exceeds the pressure $p_i(z)$ by the amount

$$p(z) - p_i(z) = p(z)\{1 - \frac{p_i(0)}{p(0)}[1 + I_1(z) - I_2(z)]\}\ . \tag{42}$$

In the vicinity of the apex of the Ω-loop at some level $z = \lambda$ the billowing field is so weak that $p \cong p_i$. If follows that

$$p(0) = p_i(0)[1 + I_1(\lambda) - I_2(\lambda)]\ . \tag{43}$$

In this limiting case, the magnetic field at the base ($z = 0$) of the Ω-loop is evacuated to the intensity

$$\frac{B^2(0)}{8\pi} = p(0) - p_i(0) \cong p(0)[I_1(\lambda) - I_2(\lambda)]\ . \tag{44}$$

The Bernoulli effect, represented by $I_2(\lambda)$, is small as the upward billowing fluid is dispersed, leaving

$$B(0) \cong [8\pi p(0)I_1(\lambda)]^{\frac{1}{2}} \tag{45}$$

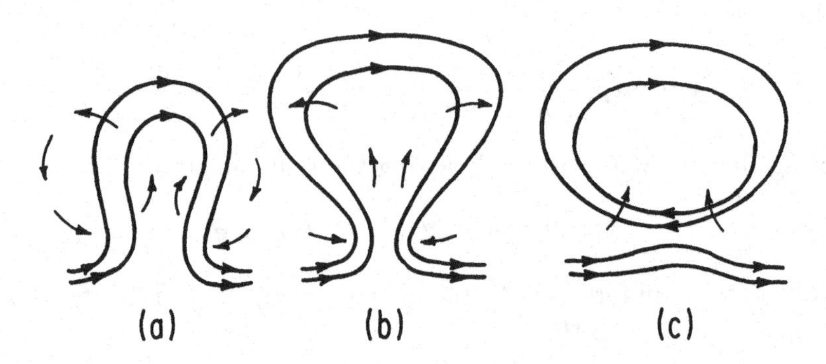

Figure 3: *A schematic drawing of the successive phases (a) and (b) of forming an Ω-loop and pinching off the Ω loop by reconnection across its base to form a O-loop in (c)*

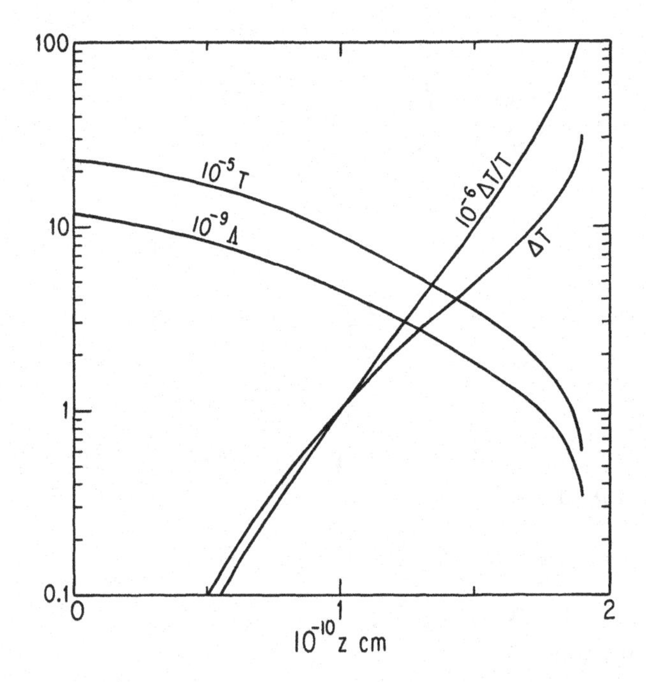

Figure 4: *A plot of $\triangle T, T, \triangle T/T$ and $\wedge$ as a function of height z above the bottom of the convective zone, based on Spruit's (1974) model of the convective zone.*

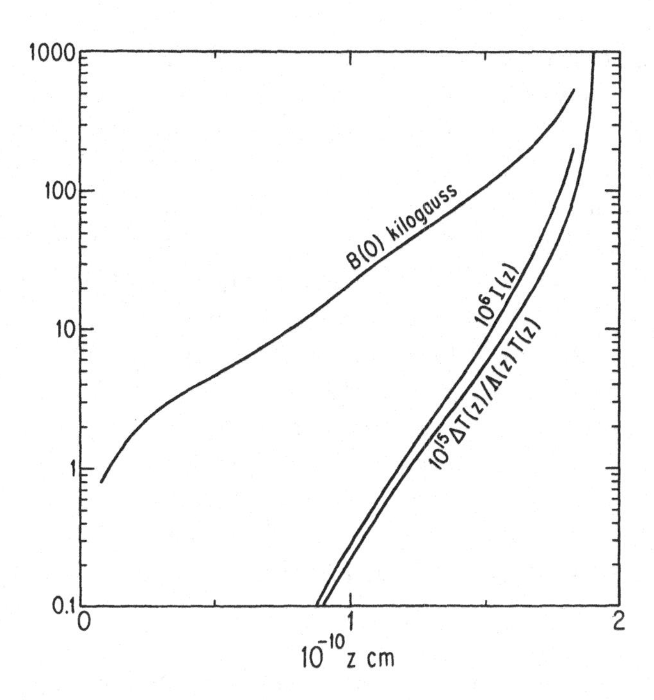

Figure 5: *A plot of* $\Delta T/T and \Lambda$ *as a function of height* z *above the bottom of the convective zone, together with* $I_1(z)$.

Fig. 4 provides $\Delta T, \Delta T/T, T$ and Λ as a function of height z above the base ($z = 0$) of the convective zone, while Fig. 5 provides the integrand $\Delta T/T\Lambda$ and a plot of $I_1(z)$ as a function of z, with $B(0)$ shown in kilogauss for $p(0) = 5 \times 10^{13}$ dynes/cm^2. Note that $B(0)$ has a value of 10^5 gauss for $\lambda \cong 1.5 \times 10^{10}$ cm. That is to say, the convective forces that arise when the upwelling reaches a depth of 5×10^9 cm ($z = 1.5 \times 10^{10}$ cm) are sufficient to evacuate the azimuthal flux bundles at $z = 0$ to an intensity of 10^5 gauss. At a depth of 2×10^9 cm ($\lambda = 1.8 \times 10^{10}$ cm) the limiting field intensity is up around half a megagauss.

We presume that several Ω-loops, successively pinching off to O-loops, are required to approach the asymptotic limit of equation (12), so the calculations here show only the limiting possibilities. We have no clear idea as to the effective value of λ, except that the stationary ($\partial/\partial t = 0$) approximation of equation (34) probably breaks down in the vigorous convection and small pressure scale height above $\lambda = 1.8 \times 10^{10}$ (depth of 2×10^9 cm). It is clear that the effect is just one small part of the turbulent hydrodynamics of an upwelling Ω-loop. The unique feature of an Ω-loop rising from the bottom of the convective zone to the visible surface is the coherence of the vertical wake created by the rising apex of the Ω-loop over many scale heights.

One may ask, of course, to what extent the upwelling of the Ω-loop is driven by the magnetic buoyancy and to what extent the upwelling is driven by the convective forces both inside and outside the flux bundle once the Ω-loop, is on its way up. Then to what extent is the wake created by the upward passage of the Ω-loop mantained by the convective forces, and to what extent do the convective forces cause the wake to overrun the Ω-loop ? Whatever the correct quantitative scenario, there is at least a temporary coherence in the updraft over a major portion of the height of the convective zone. Indeed, the emergence of individual flux bundles or Ω-loops at the surface of the Sun suggest a rapid succession of rising Ω-loops, i.e. a ladder of Ω-loops at any given time. As we shall see in the next lecture the coherence of the updraft appears to be the major cause of the increase in solar brightness during the years of high activity.

4 BRIGHTNESS VARIATION OF THE SUN

Absolute self-calibrating radiometers have viewed the Sun from spacecraft above the atmosphere of Earth since 1978. They have established the astonishing fact (Willson and Hudson, 1991; Hoyt et al 1992; Donnelly, 1993) that the mean brightness, or irradiance, of the Sun varies substantially with the changing level of solar activity. The NIMBUS 7 instrument began recording the brightness of the Sun in 1978. The brightness increased with the increasing activity to a maximum of about 1374.0 watts/m^2 in 1979 and thereafter declined along with the general level of solar activity to a broad minimum at about 1371.3 watts/m^2 in 1986, increasing thereafter into the activity maximum of 1990. The decline in brightness from the activity maximum in 1979 to the minimum in 1986 is approximately 0.2 percent, or two parts in 10^3. The ACRIM instrument, launched into space in 1980 gave an initial figure of 1368.4 watts/m^2, which declined to 1367.0 watts/m^2 in 1986. The readings from NIMBUS at these same times were 1372.5 and 1371.3, respectively. Thus, while the absolute intensities provided by the two instruments disagree by 4.3 watts/m^2, they agree on the decline from 1980 to 1986, with NIMBUS 7 providing 1.2 and ACRIM providing 1.4 watts/cm^2 i.e. a decline of one part in 10^3.

No one was surprised by the transient dips of about 1 watt/m^2 associated with the appearance of large spot groups (cf Spruit, 1977). But no one anticipated the 0.2 percent variation of the mean brightness with the 11-year activity cycle. How can a star change its brightness over a period of a few years? Foukal and Lean (1988) showed that a major part of the variation of brightness arises from the increased photospheric faculae and chromospheric plages during the years of magnetic activity. The faculae etc. more than offset the darkening effect of sunspots, on the average. Unfortunately the physics of faculae and plages is still a matter of conjecture, so it is not clear why the Sun brightens. A small contribution also comes from the enhanced UV and X-rays in the years of high solar activity. Finally, there may be a change ΔT in the photospheric temperature ($\Delta T \sim 1$ K) in nonfacular regions, although it is extremely difficult to establish this effect through direct observation (Kuhn and Libbrecht, 1991).

The first question, then, is the physics for the delivery of energy to the radiative surface of the Sun at a rate that varies with the level of magnetic activity. The manner in which the energy is radiated away is another question for later consideration. The delivery must be from a depth in excess of 10^9 cm, because there is not enough thermal capacity above that level to carry over an enhanced brightness for a period of years. For instance, $\Delta L_\odot = 4 \times 10^{30}$ ergs/s amounts to 6×10^{38} ergs over a period of 5 years. An understanding of the physical cause of faculae and plages, which occur where the meanfield is in excess of 20 gauss, might well shed some light on the problem. But in the absence of that knowledge, we consider only the gross thermodynamics of the energy transport.

The thermal energy U per unit area of the gas above a given level z in the convective zone can be characterized at z by the product of the pressure scale height $kT(z)/\mu(z)Mg(z)$ and the thermal energy density $\rho(z)c_p(z)T(z)$ where μ is the mean molecular weight, M is the mass of a hydrogen atom, g is the acceleration of gravity (2.7×10^4 cm/s^2 at the radiative surface), c_p is the specific heat per gram at constant pressure and ρ is the density (cf. Spruit, 1974). It follows that $U = 3 \times 10^3 \rho c_p T^2/\mu$ ergs/cm^2 in the region immediately below the visible surface. The characteristic relaxation time τ is given by U/I where I is the mean energy flux, equal to 6.3×10^{10} ergs/cm^2 s at the surface. Fig. 6 is a plot of τ in years for the depth (in units of 10^9 cm). The essential point is that $\tau \sim 4$ years at a depth of 1×10^9 cm, rising steeply at greater depth to 27 years at 1.55×10^9 cm and to 130 years at 2.5×10^9 cm. It follows, then, that the uppermost 10^9 cm must be supplied with heat from depths of 2×10^9 cm or more. The lower convective zone is a sufficient reservoir, with $\tau \sim 2 \times 10^4$ years at the depth of 10^{10} cm and $\tau \sim 2 \times 10^5$ years at the bottom (depth 2×10^{10} cm). In fact, the lower convective zone has characteristic thermal times comparable to those associated with the ice ages that intermittently beset the planet Earth. Unfortunately knowledge of the internal hydrodynamics of the Sun is insufficient to say anything further.

Consider, then, how the brightness of the Sun may vary with the 11-year magnetic activity cycle. The convection is believed to be responsible for the generation of the magnetic fields through an $\alpha\omega$ -dynamo (see section 1). The magnetic field generated by the dynamo involves a substantial amount of energy,

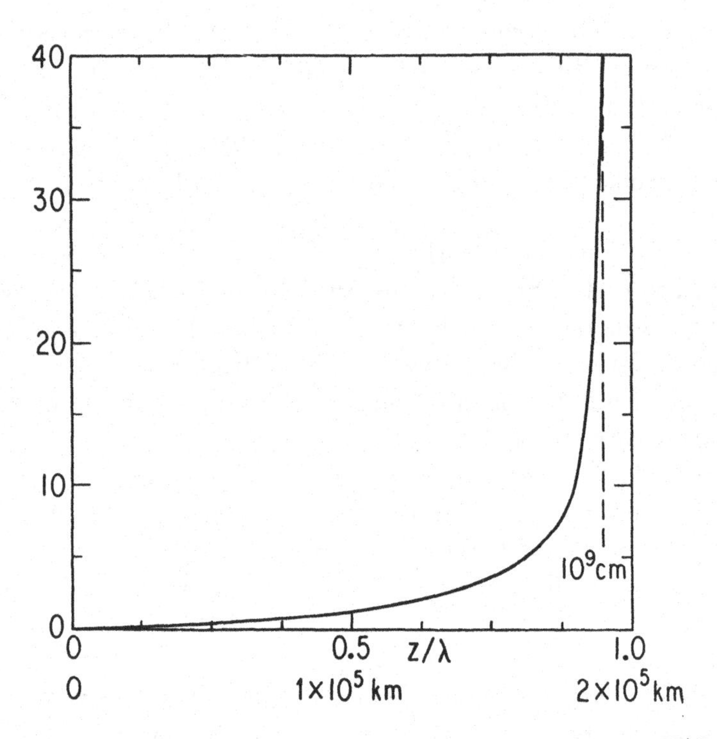

Figure 6: *A plot of the characteristic thermal response time* $\tau = U/I$ *(years) as a function of depth below the surface (in units of* 10^9 *cm)*

and that energy comes from the convective motions working against the Maxwell stresses of the magnetic field. So one might expect the Sun to be somewhat fainter during the years of increasing magnetic activity. A rough lower limit for the magnetic energy can be established, noting that the large bipolar magnetic regions observed at the surface of the Sun indicate 10^{23} Maxwells of azimuthal magnetic field in each hemisphere (Gaizauskas et al, 1983). It is believed that this field is in the form of separate intense magnetic fibrils of 10^5 gauss residing in the shear layer below the bottom of the convective zone. (Fisher, McClymont, and Chou, 1991; D'Silva and Choudhuri, 1993a,b; Fan, Fisher, and Deluca 1993). It follows that the magnetic energy density is 4×10^8 ergs/cm^3 and the cross sectional area is 10^{18} cm^2, so that there are 4×10^{26} ergs/cm along the azimuthal field. The length of the flux bundles extending around the Sun is 3×10^{11} cm, so the total magnetic energy is of the order of 2×10^{39} ergs, and perhaps more because we have no guarantee that the fields that emerge through the visible surface represent all of the flux bundles at the bottom of the convective zone.

The figure of 2×10^{38} ergs is somewhat smaller than the 6×10^{38} ergs, associated with $\Delta L_\odot$ over a five year period. So the diversion of energy into magnetic field does not appear to be a significant factor, nor is there any reason to think that the inflation of the Sun by the magnetic pressure has significant effects. The magnetic pressure of 10^5 gauss is only 0.6×10^{-5} of the gas pressure at the bottom of the convective zone, and the magnetic fibrils may occupy no more than about 4×10^{-2} of the volume. Finally, as noted above, the magnetic energy has the wrong phase to account for the enhanced luminosity at the time of maximum magnetic activity and field energy.

The kinetic energy of the nonuniform rotation of the convective zone is not sufficiently large, nor is it known to vary to any significant degree. For instance, the mass of the convective zone is approximately $0.02M_\odot = 4 \times 10^{31}$ g. The differential rotation involves peak velocity differences of about 2×10^4 cm/s across the shear layer below the convective zone at low latitude, and rather less elsewhere. So if half the mass of the convective zone has a velocity of 2×10^4 cm/s relative to the rest of the Sun, then the kinetic energy is 4×10^{39} ergs. No significant changes over the activity cycle in this energy are observed or expected. So evidently no significant fraction of the nonuniform rotation energy is available. The kinetic energy in the convective motions is too small. The mean kinetic energy density of the convection has a broad maximum of about 4×10^5 ergs/cm^3 across the lower half of the convective zone (so that the equipartition magnetic field has a broad maximum of about 3×10^3 gauss). The volume of the lower half of the convective zone is 4×10^{32} cm^3, so the total kinetic energy of the convection is of the order of 1.6×10^{38} ergs comparable to the estimated energy of the azimuthal field.

When we recall that the convection is the principal heat transport to the surface of the Sun, the smallness of $\Delta L_\odot$ indicates that the convection varies but little, if at all. Hence its kinetic energy is nearly constant in time and not available to provide an energy reservoir for $\Delta L_\odot$. In summary, it appears that we are limited to the thermal energy and its transfer to the surface to account for the observed variation in brightness of the Sun. Now the transfer of thermal energy through the convective zone to the surface is accomplished mainly by convective mixing, involving stepwise transfer of small temperature differences δT handed upward one eddy diameter at a time. The eddy diameter, or characteristic mixing length, is believed to be comparable to the local pressure scale height because of the expansion and consequent dynamical break up of a convective updraft. The elementary mixing length theory of turbulent heat transport indicates that the transport is proportional to the mixing length (eddy diameter) l to the power 3/2, because the convective velocity is proportional to $l^{\frac{1}{2}}$ while the transport length is l. The essential point is the substantial increase in convective heat transport with increasing vertical coherence length l.

Consider, then, the present concept that the bipolar magnetic regions appearing at the surface are a consequence of the direct rise of azimuthal flux bundles from below the bottom of the convective zone. The rise times, at something of the order of 10^4 cm/s, are estimated as 2×10^6 s over 2×10^{10} cm. The upward motions of the azimuthal flux bundles are propelled by the magnetic buoyancy and probably also by convective forces in the superadiabatic temperature gradient (Parker, 1979, 1988). The essential point is that the passage of a flux bundle from the bottom to the top of the convective zone establishes a wake, or updraft of some form with an initial coherence length comparable to the full depth of the convective zone. This process automatically provides a strong enhancement of the upward heat transports. Observations (Gaizauskas et al., 1983; Brants, 1985) find that the magnetic flux of a bipolar magnetic region emerges over an extended time in the form of many modest ($10^{18} - 10^{19}$ Maxwells) flux bundles or Ω-loops from the infered azimuthal magnetic field at the bottom of the convective zone. As much as 2×10^{21} Maxwells may emerge in a week in the form of 10^2 separate successive flux bundles. This suggests that at anyone time there may be as many as 10^2 Ω-loops on their way up from the bottom of the convective zone. The site of the emergence is quite small ($< 2 \times 10^9$cm), suggesting that the concentrated updraft is a royal road to the surface, with many separate flux bundles simultaneously in transit, probably accelerated upward by the convective forces in addition to the magnetic buoyancy, and continually reasserting the coherence of the associated updraft.

Only a small fraction of the total volume is involved, but the convective heat transport in the coherent wake is immense. The characteristic transport time is of the order of the rise time of 2×10^6 s. We suggest (Parker, 1994a,b, 1995a,b) that this may be the major effect in modulating the brightness of the Sun. One imagines that during the years of peak solar activity, when there are ten or more large bipolar active regions distributed around the Sun, each showing emergence of fresh azimuthal magnetic flux at intervals of a few days, there is perhaps a fraction $\epsilon \cong 10^{-3}$ of the total surface associated with emerging flux tubes. Then, if the convective heat transport is doubled over the fraction ϵ of the total surface, the solar luminosity is enhanced by the factor $1 + \epsilon$, i.e. the brightness increases by something of the order of one part in 10^3. The observed fact of the continuing emergence of flux bundles through the visible surface and the inference that the flux bundles have come all the way up from the bottom of the convective zone indicates that some such enhanced energy transport is inescapable.

It is not possible at the present time to be more precise. The hydrodynamics of a stratified convective zone is complicated and is not yet fully in hand, while the hydrodynamics of the wake of a rising flux bundle presents some special problems of its own (Parker, 1995 a,b). In particular, updrafts expand and break up so that numerical simulations show general convective overturning with vertical dimensions comparable to the scale height, not unlike the conventional mixing length concept, whereas the simulations show concentrated downdrafts that maintain their coherence through many scale-heights.

The wake of a rising flux bundle is neither of these, because it combines the expansion of an updraft with the coherence of the wake created by the rising flux bundle. Thus however much the expansion may divert fluid out the sides of the updraft, there is the coherence of the central core of the wake created by the continuing upward passage of successive flux bundles. The wake is more or less adiabatic, so that the superadiabatic temperature gradient provides powerful buoyant forces, over long vertical coherence length. Indeed, the buoyant forces may soon develop to the point that the wake propels the flux bundle to some degree (convective propulsion, already mentioned). Unfortunately this takes place at Reynolds numbers so large as to render precise numerical simulation intractable. About all that can be said is that the updraft associated with a rising flux bundle starts from the bottom of the convective zone and extends upward in a nearly adiabatic state to somewhere near the surface where the pressure scale height becomes smaller than the characteristic width of the updraft. We might guess that the updraft loses its coherence at a depth of the order of 10^9 cm (where the scale height is 3×10^8 cm) if we ignore the complications of the convective forces. The additional heat then spreads broadly before reaching the surface (Spruit, 1977). The effect of the convective forces is to accelerate the updraft, causing it to expand rapidly and billow outward from the initial slender wake established by the rising flux bundle. As already noted, the hydrodynamics of the wake and the subsequent convective updraft initiated by a rising flux bundle are not subject to either complete analytical or numerical simulation, as a consequence of the combined-large Reynolds number and strong density stratification. We can, however, examine indiviudal facets of the problem (Parker, 1995b,c). We provide here a single illustration of a two dimensional updraft in a polytropic atmosphere.

For a polytropic atmosphere ($p \sim \rho^{\alpha}$) in the presence of a uniform gravitational acceleration g the pressure, density, and temperature are given by

$$p(z) = p(0)(1 - z/\lambda)^{\alpha/(\alpha-1)}, \ \rho(z) = \rho(0)(1 - z/\lambda)^{1/(\alpha-1)}, T(z) = T(0)(1 - z/\lambda),$$

where $\lambda = \Lambda\alpha/(\alpha-1)$ with $\Lambda = kT(0)/Mg$. Then if y represents horizontal distance, the two dimensional steady flow $\rho v_y = \partial\psi/\partial z, \rho v_z = -\partial\psi/\partial y$, expressed in terms of the stream function

$$\psi = \frac{-\rho(0)v}{k} tanh\ ky(1 - \frac{z}{\lambda})^{\beta+\alpha/(\alpha-1)},$$

has the form

$$v_y = \frac{v}{k\lambda}[\beta + \alpha/(\alpha-1)] tanh\ k_y(1 - z/\lambda)^{\beta}, \quad v_z = v\ sech^2 k_y(1 - z/\lambda)^{\beta+1}.$$

For $-1 < \beta < 0$ the vertical velocity declines relatively slowly toward the surface $z = \lambda$, and then drops off abruptly to zero as z reaches λ while v_y increases without bound in the same limit. With $\beta = 0$ the vertical velocity declines linearly to zero while v_y remains bounded. Fig. 7 is a plot ($\alpha = 3/2$) of the stream lines ψ = constant for $\beta = 0$, showing the concentrated core of the updraft and the general horizontal expansion of the fluid away from the core (Parker, 1995a).

There are several curious aspects of an extended coherent updraft. For instance, as already noted, the updraft is itself nearly adiabatic, in view of the rapid rise of the fluid. Hence the fluid within the updraft is not subject to the ambient convective overturning instability. Thus, it appears that the surrounding fluid may be more turbulent than the fluid within the core of the updraft. The eddy viscosity, then, is a minimum within the updraft and we have the phenomenon of a relatively inviscid updraft confined by more rigid (viscous) but fluctuating fluid on both sides. The expected upward velocity of 10^4 cm/s exceeds the ambient rms turbulent velocity all the way from the bottom of the convective zone, where it is estimated to be 4×10^2cm/s, to the level at $z = 1.9 \times 10^{10}$ cm, only 10^9 cm below the visible surface. So the ambient eddy viscosity may help to confine the updraft over most of its long path to the surface. A simple analytical illustration of the effect appears in Parker, (1994a, 1995b), and the stability of a concentrated updraft is treated in Parker (1995c). It is clear that a definitive treatment of the problem must approach the dynamics from first principles, without the convenient parametrization of an eddy viscosity. The effect of strong turbulence cannot be represented by a single parameter. Unfortunately neither analytical nor numerical methods can handle the task at the large Reynolds numbers ($\sim 10^{14}$) that arise in the convective zone of the Sun.

The basic physics of the coherent updrafts feeding magnetic flux into the bipolar magnetic regions at the surface of the Sun is that they are coherent over the entire depth of the convective zone because they are associated with the rise of Ω-loops in the azimuthal flux bundles from the bottom of the convective zone. It would appear that the updraft is continually reaffirmed by the successive upward passage of many Ω-loops. Given the extended vertical coherence length l of the nearly adiabatic updraft, the convective heat transport is proportional to $l^{3/2}$, and is enormously enhanced. The arbitrary doubling of the heat transport in the foregoing illustration may be a serious underestimate.

Now we expect that the coherence is destroyed by the rapidly declining scale height close to the surface. The heat transport through the last 10^9 cm may be less coherent, with the increased thermal energy spread horizontally outward

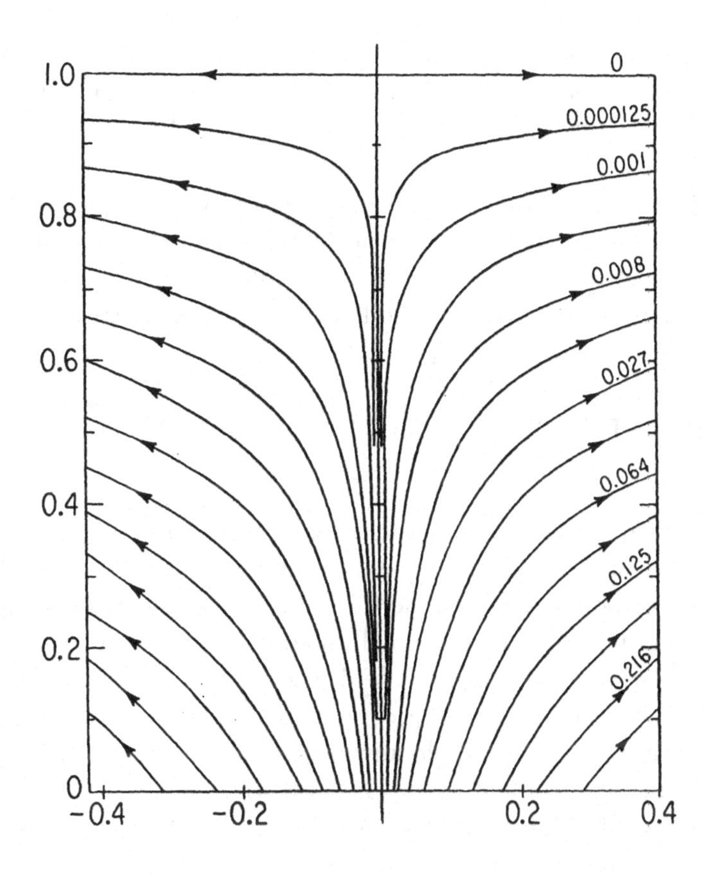

Figure 7: *A plot of the stream lines* ψ = *constant for* $\beta = 0, \alpha = 1.5$, *with the value of the mass flux* $\rho(0)v/k$ *indicated on each curve.*

over 10^9 cm or more. Observationally we would expect to see the active latitudes, defined by the sites of active flux emergence, slightly hotter (~1K) than the higher latitudes where the only flux emergence appears to be small disoriented bipolar magnetic regions. Unfortunately, in the presence of faculae and plages, network fields and supergranules, et al., it proves difficult to define a unique temperature and to show that there are large-scale temperature differences (over dimensions of 3×10^{10} cm) (cf. Kuhn and Libbrecht, 1991).

Meanwhile, the dynamics of coherent wakes of objects passing individually or successively upward through a stratified superadiabatic atmosphere is an interesting subject to contemplate. It is also interesting to consider what other physical effects may contribute to modulating the brightness of the Sun.

5 SOLAR AND STELLAR X-RAY EMISSION

The outer atmosphere of the Sun -the corona- has a temperature of the order of 2×10^6 K, with the result that it extends far into space. In regions where the magnetic field is too weak to confine the coronal gas, the gas expands outward, reaching several hundred km/s beyond about $10R_{\odot}$ to form the solar wind. In active regions, where the magnetic field is strong ($\sim 10^2$gauss), the magnetic stresses overpower and confine the coronal gas. The field maintains a bipolar form, with both ends of the bipole anchored in the dense convecting gas below the photosphere. The coronal gas is enclosed with the bipolar field and cannot escape in spite of its high temperature. The gas builds up to a density $\sim 10^{10}$ atoms/cm^3 such that the energy input, estimated at 10^7 ergs/cm^2 (Withbroe and Noyes, 1977), is consumed by downward thermal conduction and principally outward thermal radiative emission. The thermal radiation is in the form of soft X-ray, easily observed from a distance of 1 a.u.. A particularly interesting feature of the X-ray emitting bipolar active regions is that their surface brightness is essentially independent of their length over the range of 2×10^9 to 2×10^{10} cm. This curious fact has important implications for the coronal heat source. For comparison, the expanding coronal gas in the weak open field regions (~10 gauss) accumulates only to a density of about 10^3 atoms/cm^3. Hence there is little or no detectable X-ray emission so that the regions of open field are dark in X-ray pictures of the Sun and are called "coronal holes". The faster solar wind streams (600-800 km/s) evidently issue from the coronal holes. Withbroe (1988) estimates that the coronal holes are maintained by a heat input of the order of 10^6 ergs/cm^{2}s with the important requirement that most of the heat input is within 1-2 $R_{\odot}$ of the surface of the Sun in order to maintain the temperature in the presence of continuing expansion. However, the high velocity of the expansion at large radial distance indicates that some fraction of the heat is introduced beyond the sonic point in the wind.

We may suppose that the Sun is typical of most late main sequence stars, so that if we can work out the coronal physics for the Sun, we will also understand the X-ray emission and stellar winds of other stars. X-ray astronomy is limited to taxonomy until the physics of coronal heating is properly understood. The fundamental question is how the Sun, with a surface temperature of 5600 K, transmits energy to the corona that is 500 times hotter at 2.3×10^6 K. The basic principle was pointed out decades ago by Biermann (1948), Alfven (1949), and Schwarzschild (1948) (see also Osterbrock, 1961 and Whittaker, 1963). The convective zone is a heat engine, subject to the theoretical limitations of the Carnot cycle. But the mechanical work done by the heat engine can be converted directly into heat in a medium of any temperature. The problem then, is to understand how the work is transmitted to the corona and how the conversion to heat takes place in the corona. The early suggestions were based on the generation of sound waves, Alfven waves, internal gravity waves, etc. in the convective zone, with propagation up into the corona where the waves are dissipated into heat. It was imagined that the chromosphere, along with faculae and plages, was heated in the same way.

Technical difficulties began to appear with the construction of quantitative models of wave generation and propagation. In particular, sound waves and internal gravity waves steepen rapidly as they propagate up from their origin in the convection, so that they dissipate in the cromosphere and do not survive into the corona. Alfven waves on the other hand, being purely transverse steepen but little, so they are much less dissipative. They survive through the chromosphere and reach the corona without difficulty. However, being relatively nondissipative, they provide no significant heat as they propagate along the large-scale magnetic fields in the corona.

There is another difficulty with Alfven waves besides their nondissipative transit through the corona and that is the long wavelength for the expected wave periods of 10^2 s or more. The Alfven speed V_A is of the order of 2×10^8 cm/s in the strong bipolar X-ray emitting regions ($B \sim 10^2$ gauss, $N \sim 10^{10}$ H atoms/cm^3) and in the weak fields of the coronal holes ($B \sim 10, N \sim 10^8$ H atoms/cm^3). The strongest source known for generating waves is the granules at the surface of the Sun, with velocities of 1-2 km/s, diameters of 10^3 km, and characteristic correlation times of the order of 300 s. Granules would be expected to produce wakes predominatly with periods of the order of 300 s, corresponding to wavelengths of 6×10^5 km in the corona. Propagating out along the open field lines of a coronal hole, such waves might dissipate their energy over a characteristic distance of 10 $R_\odot$ or more. Hence, they may contribute to the high speed of the solar wind from coronal holes (Parker, 1991). But there is no known way by which they heat the near corona. Nor is there any way that they could be fitted into the typical X-ray emitting bipolar magnetic fields with length $0.2 - 2 \times 10^5$ km.

Short period Alfven waves ($\tau \sim 1$ s) are needed and it has been speculated (Martin 1984, 1988; Porter et al 1987: Porter and Moore, 1988; Parker, 1991) that strong short period waves may be emitted by the microflares in the small magnetic features in the boundaries of supergranules (the network fields). These waves dissipate in distances of $10^{10} - 10^{11}$ cm and could supply the heat for the near coronal holes. However, it remains to be shown that there is sufficient energy 10^6 ergs /cm^{2}s in the microflares to do the job. The unfortunate fact is that we still are not sure as to the energy source for the solar wind. The observed temperatures provide the wind as a consequence of hydrodynamic expansion. But the cause of the high temperature ($\sim 1.5 \times 10^6$ K) is still a matter of guess work. The larger energy requirement of the X-ray corona (10^7 ergs/cm^{2}s) rules out this possibility as a major contributor.

Now it is to be expected that the granules, with Reynolds numbers of the order of 10^{10}, are turbulent with small eddies of short period. In a Kolmogoroff spectrum the statistical velocity difference $v(l)$ over a length l is proportional to $l^{1/3}$. Hence the correlation time of motion on a scale l is $\tau(l) \sim l/v(l) \sim l^{2/3}$. It follows that the kinetic energy density $\frac{1}{2}\rho v(l)^2$ is proportional to $\tau(l)$. So the energy density and energy flux in Alfven waves falls off in proportion to $\tau(l)$. Waves with periods of 20 s ($\frac{1}{2}\lambda = 2 \times 10^9$ cm) can be fitted into the smaller bipoles, but they arise from eddies with a kinetic energy density much smaller ($\sim 10^{-1}$) tha the principal motion at $\tau = 300$ s). It has been suggested that Alfven waves with wavelength λ equal to twice the length of a bipole might set up a resonance, building oscillations to enormous amplitude (cf. Kuperus, Ionson, and Spincer, 1981; Lee and Roberts; Davila, 1987; Hollweg, et al, 1990, and references therein). But no such motions ($> 10^3$ cm/s) are observed, and the nearly uniform surface brightness of X-ray emission over bipoles with lengths of 2×10^9 cm to 2×10^{10} cm does not look like a resonance phenomenon, which one would expect to favor one length strongly over another.

These considerations leave us with only limited possibilities. For instance there is the recent suggestion that the X-ray bright points representing bipolar magnetic fields with lengths of the order of 2×10^9 cm, light up as a consequence of flares at their base where perhaps two small bipoles are in a prolonged collision. In contrast, the observations of larger bipoles show the apex of the emitting X-ray filaments to be brighter and hotter than the lower portions near the feet, while simple models of coronal magnetic loops with a fixed arbitrary distribution of heat input show that the loop is thermally stable only if the heat input is concentrated to some degree toward the apex. So we concentrate here on these larger bipoles, evidently not heated by flaring at their bases. The heat input is apparently strongest in the vicinity of the apex. Note, then, that the bipolar magnetic fields that contain significant X-ray emission almost always have intensities of about 10^2 gauss at the photosphere, the only exceptions being bipolar fields that intersect the photosphere near sunspots and occasionally in sunspot penumbrae (but never in umbrae, evidently).

The observational rule is that magnetic fields of the order of 10^2 gauss or more imply heat input of the order of 10^7 ergs/cm^{2}s, giving rise to the filamentary X-ray corona. Prof. Uchida showed many beautiful examples in his lectures. The theoretical problem is to understand how magnetic fields of 10^2 gauss with characteristic bipolar lengths of the general order on 10^{10} cm, can provide significant dissipation of magnetic energy when embedded in a tenuous gas of $2 - 3 \times 10^6$ K for which the resistive diffusion coefficient η is of the order of 10^3cm^2/s. The characteristic resistive dissipation time $s(l)$ over a scale l is of the order of $l^2/4\eta$. So if $l = 10^8$ cm, it follows that $s \sim 3 \times 10^{12}$ s $\cong 10^5$years.

In fact the characteristic dissipation time of the magnetic free energy is evidently only a few hours. For instance, if the fraction f of the magnetic energy $B^2/8\pi$ over a height h is free to be dissipated into heat, the available heat is sufficient to mantain the heat input of 10^7ergs/cm^2 for a time of 1.6×10^4 s (5 hours). The energy density of the gas ($\sim 3NkT$) is about 10 ergs /cm^3, which is 4×10^{10} ergs/cm^2 over the same height h. This is equivalent to the heat lost over 4×10^3 s or about an hour. Thus the gas cools radiatively with a characteristic time of an hour and there is enough magnetic free energy to maintain the high temperature of the gas (if it can be converted into heat) for a period of several hours. It is not surprising, then, to observe a continual shifting of the small-scale filamentary details of the X-ray emission in an hour or so, with large changes in structure over many hours, indicating a continually shifting heat supply.

It has been pointed out by several authors that the heating can be accomplished only if the electric currents in the field gradients are concentrated into thin sheets with thicknesses of the order of 40 m or less. The magnetic fields are rooted in the subphotospheric convection, which suggests that the footpoints of the bipolar field are shuffled and intermixed on scales comparable to the 500 km correlation length of the granules. Fig.8 is a sketch of the topology of a bipolar magnetic field after the field lines have been interwoven for a time by random continuous mapping of footpoints of the field. Now if the winding and intermixing of the footpoints proceeds far enough, one can imagine that the various elemental flux bundles that make up the whole field become increas-

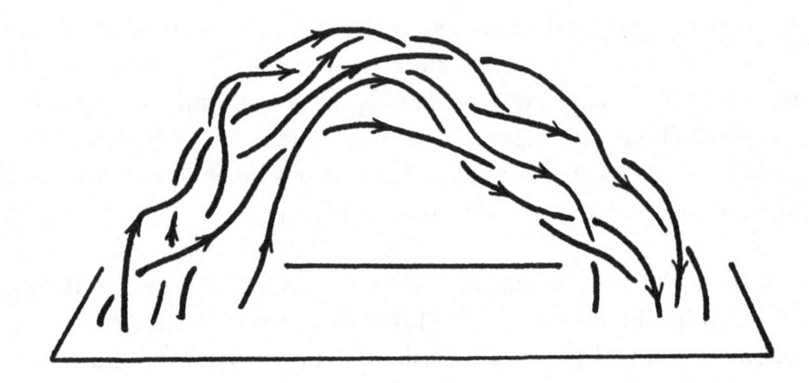

Figure 8: *Schematic drawing of the interwoven field lines of a bipolar magnetic region*

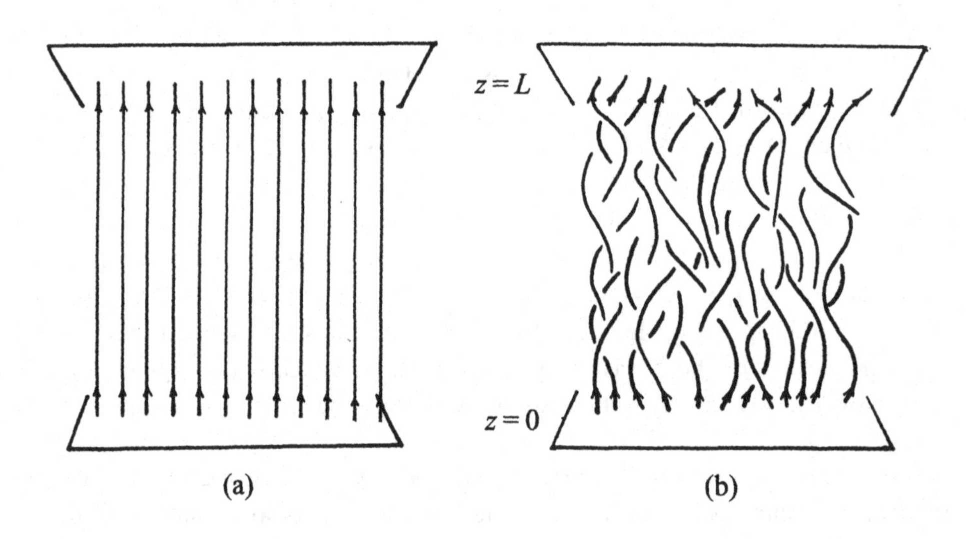

Figure 9: *A schematic drawing of the initial uniform magnetic field B_o (a) extending from $z=0$ to $z=L$ through an infinitely conducting fluid and the interwoven field (b) resulting from the continuous field motion described by equation(46)*

ingly misaligned and twisted so that individual flux bundles loop and kink back on themselves to produce interrecurrent sheets. However, high resolution X-ray photographs (Golub et al., 1990) show no signs of such tightly twisted flux bundles. In fact strong twisting is not necessary. It appears that the dissipation arises from the spontaneous formation of current sheets, i.e. shearplanes, in magnetic fields with only modest internal interweaving (Parker, 1972, 1981, 1983, 1986a,b, 1994a).

Consider a magnetic field in an infinitely conducting fluid. The basic theorem of magnetostatics assert that almost all field topologies result in the formation of internal tangential discontinuities when the field is allowed to relax to the lowest available energy state (i.e. relax to static equilibrium).

To fix ideas, consider the initially uniform magnetic field B extending in the z-direction from $z = 0$ to $z = L$ as sketched in Fig. 9(a). The space $0 < z < L$ is filled with an infinitely conducting fluid with uniform fluid pressure p maintained at the boundaries $z = 0, L$. The planes $z = 0, L$ are infinitely conducting and the plane $z = 0$ is held fixed. The fluid and the plane $z = L$ are then subjected to the smooth, bounded, continuous, n-times differentiable mapping described by the well behaved stream function $\psi(x, y, kzt)$, for which

$$v_x = kz\partial\psi/\partial y \quad ; \quad v_y = -kz\partial\psi/\partial x \quad , \quad v_z = 0 \tag{46}$$

This represents a progressive arbitrary winding introduced at $z = L$. The windings accumulate and compress progressively downward $z = 0$ as more winding is introduced at $z = L$. The magnetic field is readily shown (Parker, 1986a, b) to be

$$B_x = kt\ \partial\psi/\partial y \quad , \quad B_y = -kt\partial\psi/\partial x \quad , \quad B_z = B_o \tag{47}$$

as a consequence of the fluid motion. The fluid motion is switched off after some specified time t. The result is a bounded interwoven magnetic field shown schematically in Fig. 9(b). The field is continuous, containing no tangential discontinuties. The characteristic scale of transverse (x, y) variation of the field is comparable to the scale l of the applied fluid motions. Note that the field lines of the projection of $\vec{v}$ and $\vec{B}$ onto any plane $z = h$ $(0 < h < L)$ are given by the family of curves $\psi(x, y, kht) = \lambda$.

Suppose, then, that both boundaries $z = 0, L$ are held fixed while the dissipationless fluid throughout $0 < z < L$ is released so that the smooth continuous magnetic field described by equation (47) is free to relax to the lowest available energy state. It is convenient to introduce a small viscosity into the fluid to achieve this end. The basic theorem of magnetostatics states that for almost all choices of the well behaved function ψ the field develops internal tangential discontinuities (i.e. current sheets) as an essential part of the final static equilibrium. The tangential discontinuities can be avoided only by special choice of ψ, with special symmetry or without significant interweaving of the field lines. A close examination of the force-free field equation shows why this is so.

Consider the force-free field equation

$$\vec{\nabla} \times \vec{B} = \alpha \vec{B} \tag{48}$$

that describes the final equilibrium state of the field in the presence of a uniform fluid pressure. The curl and divergence of this equation yield

$$\vec{B} \times \vec{\nabla}\alpha = \nabla^2 \vec{B} + \alpha^2 \vec{B} \tag{49}$$

and

$$\vec{B} \cdot \vec{\nabla}\alpha = 0, \tag{50}$$

respectively. Equation (49) is a quasi-linear second order partial differential equation. The Laplacian operator indicates that the equation has two sets of complex characteristics. If that were the entire story, the system of equations would be fully elliptic and specification of the field on the boundaries $z = 0, L$ would determine the field uniquely throughout $0 < z < L$. That is to say, the topology of the field throughout $0 < z < L$, would be determined by the boundary conditions and viceversa. But the mapping described by equation (46) is a mechanical operation so that any desired winding pattern no matter how complicated, can be introduced. Then imagine that the field is interwoven through n different complex winding patterns along almost all lines of force. Let n and L increase without bound (so that L/n remains at some finite length λ). In the limit of large n the total topology along the unbounded length L becomes infinitely complicated with most of it located infinitely far from the boundaries. Radical changes of the field at the boundaries e.g. locally compressing or expanding the ends of the flux bundles, have no effect on the topology of the field throughout the distant deep interior of $0 < z < L$ where the tangled and interwired field lines are in local static equilibrium. That is to say, there is no direct relation between the field at the boundary and the topology of the field throughout the interior. Hence the force free field equation cannot be fully elliptic, for if it were, the interveawing of the field throught the deep interior would be completely determined by the field conditions at the boundary. This would be contrary to the fact of the infinitely extended sequence of random winding patterns throughout the deep interior that is not affected by radical changes at the boundary, such as expansion and compression of the flux bundles.

Then there is the obvious complication arising from equation (50), that the torsion coefficient

$$\alpha = \vec{B} \cdot \vec{\nabla} \times \vec{B} / B^2 \tag{51}$$

is rigorously constant along each field line. Yet in the extensive winding through n successive patterns an elemental flux bundle may wind both right and left about its neighbours at different location along the field.

The resolution of the dilemma arises from the fact, evident from equation (50), that the field lines constitute a set of real characteristics, i.e. a third set of characteristic (Parker, 1979, 1994a).

Now each set of characteristics of a partial differential equation represents a family of curves along which the solution to the equation is determined by the differential equation. Specification of the solution at any location on a characteristic determines the solution everywhere else along the characteristic. On the other hand, the solution on any given characteristic curve is independent of the solutions on the neighboring characteristics. It is sometimes said that the solution propagates along each characteristic curve. So there is no requirement that the solution is continuous from one characteristic curve to the next. Piecewise continuous is sufficient. A fully elliptic equation, e.g. Laplace's equation, has no real characteristics, so it allows no internal discontinuities. Hence, specification of the solution along any 2-D surface in real 3-D space provides a unique solution throughout the space. But the existence of a set of real characteristics means that the nature of the solution would have to be specified along some boundary cutting accross the real characteristics so as to determine the solution on each characteristic curve, if the boundary conditions were to provide a unique determination of the solution throughout the volume. In the present situation the surfaces $z = 0$ and $z = L$ intersect all the lines of force, but the only boundary condition is that $B_z = B_o$. There is nothing that requires continuity of B_x, B_y, and α at $z = 0$ and $z = L$. Indeed, if the field were continuous throughout $0 < z < L$, there would be no way to avoid the contradiction that the tension α is uniform along each characteristic while the elemental flux bundles wind both clockwise and counterclockwise at different locations along the field. This contradiction with equation (50) is avoided by the formation of surfaces of tagential discotinuity along the field lines. The magnitude B of the magnetic field is continuous across the surface of discontinuity, because the magnetic pressures on opposing sides of the surface are equal in static equilibrium. However, the direction of the field changes abruptly across the surface, say by some finite angle ϑ. Ampere's law requires that the surface of discontinuity is a current sheet, with a surface current density $J = (cB/2\pi) sin\frac{1}{2}\vartheta$.

To understand how a surface of tangential discontinuity permits right and left hand winding or torsion at different locations along a field line without violating equation (50), consider the nature of the tangential discontinuity. The essential point is that the tangential discontinuity represents the geometrical surface between two regions of continuous field and bounded torsion α. There is magnetic field of one direction on one side of this boundary surface and magnetic field of another direction on the other side, but there is no field within the surface because the surface has no interior. Hence equation (50) applies everywhere up to the surface on each side but not on the surface itself.

The torsion between the fields on opposite sides of the surface is undefined. One could, of course, say that the torsion is infinite in the sense that the jump $\triangle\vec{B}$ in the field direction across the surface is finite while the distance across the surface is zero. The right and left hand winding to be found at successive locations along the field is accommodated by the right and left hand "torsion" $\triangle\vec{B}$ across the surfaces of tangential discontinuity. Equation (50) does not apply because there is no field in the surface of discontinuity . So α remains rigorously constant along all field lines on each side. Thus the winding of the flux bundle about neighboring flux bundles may have different signs at different locations along the magnetic field with the difference between the torsion α within the flux bundle and the magnitude and sign of winding of the flux bundle appearing in the jump $\triangle\vec{B}$ across the surface of tangential discontinuity.

The surfaces of tangential discontinuity do not end anywhere in a force free field. This is obvious from the requirement that the surface current J is conserved without current flowing across field lines anywhere. It follows that the surfaces of tangential discontinuity extend all the way to the boundaries $z = 0, L$. Hence, the transverse field components B_x and B_y are made to be discontinuous at the boundaries by the discontinuities that appear in the interior.

The surfaces of tangential discontinuity are created in static equilibrium by the balance of the Maxwell stresses within the field, described by the stress tensor

$$M_{ij} = -\delta_{ij}\frac{B^2}{8_\pi} + \frac{B_iB_j}{4\pi} \tag{52}$$

representing an isotropic pressure $B^2/8\pi$ and a tension $B^2/4\pi$ along the field. This stress system can be in static equilibrium only if there are tangential discontinuities within the field (except in suitably symmetric topologies, where discontinuities can be avoided). So in retaining to the lowest available energy state the Maxwell stresses continually strive to produce a true mathematical discontinuity in $\vec{B}$.

Now, the finite resistivity of the fluids in the real physical world prevent the field from achieving a true discontinuity. The thickness of any transition is limited to some small but finite length δ where $\delta = (4\eta t)^{\frac{1}{2}}$, in order of magnitude, for a characteristic dynamical time t. There can be no static equilibrium within the transition thickness δ (equilibrium arises only for $\delta \rightarrow 0$) and the effect has been referred to as topological nonequilibrium (Parker, 1972). This is the basis to the familiar rapid reconnection, where in the Maxwell stress continually drives the field toward a discontinuity while the resistivity rapidly dissipates the magnetic energy in the steep field gradient that result. The consequence is a rapid dissipation of the magnetic free energy into heat, at rates characterized more by the local dynamical Alfven speed than by the local resistivity. Indeed, the extreme thinness of the current sheet may give rise to sufficiently high current densities as to invoke plasma turbulence and anomalous resistivity, further enhancing the dissipation rate. But in any case the dissipation proceeds relatively rapidly in any setting in the real physical world because the Maxwell stress perpetually drives the field toward large gradients in a continuing attempt to produce a tangential discontinuity. The effect continues until the topology of the field is reduced to so simple a form that static equilibrium no longer requires tangential discontinuities.

Returning to the X-ray corona of the Sun, the essential point is that the random walk of the footpoint of the bipolar magnetic fields in the photospheric convection zone has the effect of winding and mapping the field lines into nonsymmetric topologies. The winding and wrapping is slow, with the footpoints moving at something of the order of 1 km/s while the Alfven speed in the bipolar field at coronal levels is 2×10^3km/s. The bipolar field is in quasi-static equilibrium and the Maxwell stresses are busy driving the field toward internal surfaces of tangential discontinuity.

Now unfortunately one is not able in the unspecified local conditions in the corona to provide a formal calculation of the dissipation or reconnection rate across a thin layer of intense shear as the Maxwell stresses drive the field and tenuous plasma toward a tangential discontinuity. It is known that the speed with which the dissipation cuts across the field lines lies somewhere in the broad interval $V_A/N_R^{1/2}$ to V_A/lnN_R, depending upon the boundary conditions and the initial conditions. Here V_A is the Alfven speed computed for the discontinuous field component, and N_R is the Lundquist number computed for V_A and the breadth of the current sheet. On the other hand, it is relatively easy to relate the degree of winding of the field lines to the required dissipation rate (10^7 ergs/cm^2 s) and the velocity of the photospheric footpoint motions.

Suppose, for instance, that the footpoints of the magnetic field sketched in Fig. 9(b) are held fixed at $z = 0$ while those at $z = L$ move about in some continuous velocity field v_x, v_y. In particular, consider a single flux bundle whose footpoint on $z = L$ moves with constant velocity v along a random mean path among the footpoints of all the other elemental flux bundles. In the simple case that the other flux bundles remain straight, sketched in Fig. 10, the moving flux bundle forms the hypotenuse of a right triangle with altitude L and base $s = vt$ after a time t. The hypotenuse is inclined to the vertical by an angle $\theta = tan^{-1}vt/L$. If the vertical component of the field remains at its initial value B_o, the horizontal component is $B_o tan\theta$. The tension in the displaced flux bundle pulls backward against the advancing footpoint with a force F per unit are a given by

$$F = B_o^2 tan\theta/4\pi \tag{53}$$

so that the rate W at which F does work on the advancing footpoint of the field is

$$W = vF = vB_o^2 tan\theta/4\pi = \tag{54}$$

$$= (B_o^2/4\pi)v^2t/L \tag{55}$$

per unit area and time. Let $B_o = 10^2$ gauss and as a guess suppose that $v \sim 0.5$ km/s. Then with W set equal to the energy input rate of 10^4 ergs/cm^2 s for the active X-ray corona of the Sun, the result is $vt/L = 1/4$ i.e. $\theta \sim 14°$. It follows that if the winding and wrapping proceed to where the characteristic rms angular deflection of the field is 14°, then the random transport of the footpoints of the bipolar magnetic field introduces magnetic free energy at the rate of 10^7 ergs /cm^2 s. Under steady conditions the dissipation converts that free energy into heat at the same rate that the free energy is introduced by the footpoint motions.

The principal uncertainty in the calculation is the appropiate value for the intermixing velocity v, which we have taken to be 0.5 km/s as a consequence of the granule motions of 1 km/s. The observational problem is that while ground based telescopes detect the larger magnetic fibrils during periods of exceptionally good seeing, they do not resolve the fibrils and can say very little about the intermixing. Note, then, that if $v \sim 0.25$ km/s, we have $vt/L = 0.5$ and the mean inclination turns out to be 26.5°. On the other hand 1 km/s yields 7°.

The reader can easily repeat the calculation with the assumption that the magnitude B of the wandering flux bundle is restricted to B_o by the pressure of the undisplaced flux bundles through which the wandering flux bundle threads, instead of putting $B_z = B_o$ as we have done here. In that case $B_z = B_o cos\theta$ and horizontal component is $B_o sin\theta$. The Maxwell stress is $(B_o^2/4\pi)sin\theta cos\theta$ and the final result is $vt/L = 2 - \sqrt{3} = 0.27$ instead of 0.25 for $v = 0.5$ km/s.

It is important to note that for a given intermixing velocity v of the footpoints the rate at which work is done on the field, and hence the rate of heat production in steady state, varies inversely with the reconnection rate across the tangential discontinuities. To put it another way, $\theta = tan^{-1}vt/L$ increases with time until the transverse field component $B_o tan\theta$ becomes sufficiently large that the reconnection cuts across the regions of continuous field fast enough to keep up with the winding of the field by the motions of the footpoints in the photospheric convection. The larger is θ required to drive the reconnection at the necessary rate, the larger is the energy input W.

As an example, suppose that $L = 5 \times 10^9$ cm with $v = 0.5$ km/s again. Then $vt/L = 0.25$ when $t = 10^5$s. If the winding is on the granule scale $l \sim 500$ km, the speed v with which the reconnection progresses to maintain a steady state is of the order of $v \sim l/t \sim 1$ m/s $\sim 0.5 \times 10^{-6} V_A$, where $V_A \sim 2000$ km/s is the Alfven speed in the mean field. One would expect the reconnection to progress at a speed not less than the general order of the Alfven speed $V_A tan\theta$ in the transverse field component $B_o tan\theta$ divided by the square root of the Lundquist number $\ell V_A tan\theta/\eta$ (Parker, 1957, 1979; Sweet, 1958a, b, 1969), with $l = 500$ km, $V_A tan\theta = 500$ km/s and $\eta \cong 10^3 \mathrm{cm}^2/\mathrm{s}$ the result is a Lundquist number of 2×10^{12} and a reconection velocity no less than about 40 cm/s in order of magnitude. So the reconnection rate of 1m/s estimated for footpoint mixing at 0.5 km/s is not out of line with the theory of rapid reconnection. Indeed, if we had chosen $v = 0.25$ km/s, the result would be $t = 2 \times 10^5$ s and $v = 0.5$ m/s, in coincidence with the lower bound of 40 cm/s on the reconnection rate.

The idea is, then, that the random walk of the photospheric footpoints winds up the flux bundles of the bipolar field to the point where the number and intensity of the concentrated current sheets (developing discontinuities) reconnects the field across the sheets fast enough to keep up with the winding of the field lines by the random walk of the footpoints. The magnetic free energy is dissipated by the reconnection at the heating rate of $10^7 \mathrm{ergs/cm}^2$ s if the field is mixed at 0.5 km/s to an rms inclination of 14°.

Since the precise rate of reconnection lies beyond present calculations, the crucial step in testing these ideas is to observe the individual magnetic fibrils at the photosphere and determine the rate at which they wander among each other. This requires a telescopic resolution of 0.1", i.e. 75 km at the Sun. A 125 cm difraction limited telescope in a balloon or spacecraft would make the definitive determination of the motion. Specifically, with what velocity v do the footpoints of the fibrils mix among each other? With $B \cong 10^2$ gauss from magnetograms, and $W \cong 10^7$ ergs/cm^2, equation (54) allows $tan\theta$ to be estimated from the velocity v.

These considerations are fundamental to the X-ray astronomy of solitary stars. The Sun is the only star where the X-ray emitting regions can be resolved and studied. So it is essential to make the measurements that determine the nature of the heat source creating the hot dense X-ray emitting gas. Without that crucial question being answered, X-ray astronomy of other stars is limited

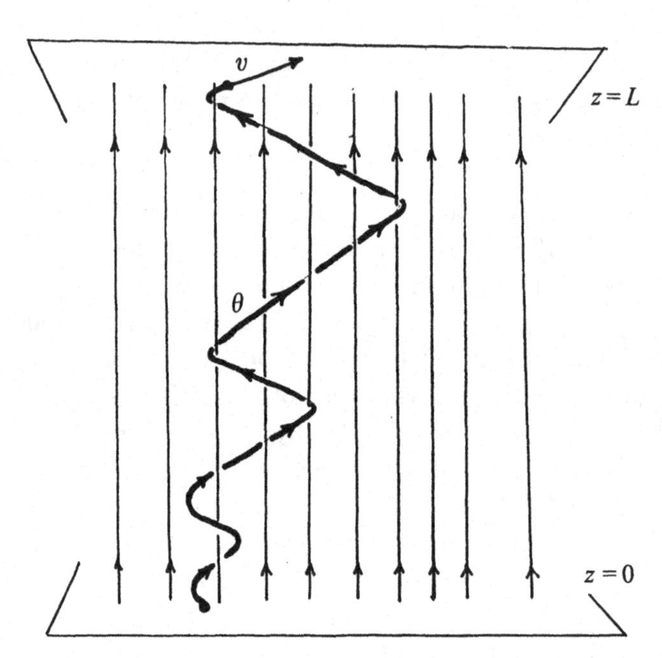

Figure 10: *A schematic drawing of a single flux bundle displaced from its initial position by the steady wanderning of its footpoint at z=L at a speed v among the other flux bundles that remained fixed*

to taxonomy and the physics of the X-ray emission from the Sun remains a matter of speculation. The problem is also of vital interest (Parker, 1994b) to the photochemistry and the general structure of the upper atmosphere of Earth where the solar X-rays are absorbed.

6 REFERENCES

Alfven, W.: 1947, *Mon. Not. Roy. Astron. Soc.*, **107**, 211.
Backus, G.E.: 1958, *Ann Phys.*, **4**, 372.
Biermann, L.: 1946, *Naturwiss*, **33**, 118.
Biermann, L.: 1948, *Z. Astrophys.*, **25**, 161.
Brants, J.J.: 1985, *Solar Phys.*, **98**, 197.
Brown, T.M., Christensen-Dalsgaard, J., Dziembowski, W.A., Goode, P. Gough, D.O. and Morrow, C.A.: 1989, *Astrophys. J.*, **343**, 526.
Choudhuri, A.R. and Gilman, P.A.: 1987, *Astrophys. J.*, **316**, 788.
Choudhuri, R.A.: 1984, *Astrophys. J.*, **281**, 846.
Choudhuri, R.A.: 1990, *Astrophys. J.*, **355**, 733.
D'Silva, S. and Choudhuri, R.A.: 1993b, *Astron. Astrophys.*, **272**, 621.

D'Silva, S. and Choudhuri, R.A.: 1993a, *Astrophys. J.*, **407**, 385.
Davila, J.M., 1987, *Astrophys. J.*, **317**, 514.
Donnelly, R.F.: 1993, *STEP Symp-Adv. Space Res.*, (in publication).
Dziembowski, W.A., Goode, P.R. and Libbrecht, K.G.: 1989, *Astrophys. J. Lett.*, **337**, L53.
Elsasser, W.M.: 1954, *Phys. Rev.* **95**, 1.
Fan, Y., Fisher, G.W. and DeLuca, E.E.: 1993, *Astrophys. J.*, **405**, 390.
Fisher, G.W., McClymont, A.N. and Chou, D.Y.: 1991, *Astrophys. J.*, **374**, 766.
Foukal, P. and Lean, J.: 1988, *Astrophys. J.*, **328**, 347.
Gaizauskas, V., Harvey, K.L., Harvey, J.W. and Zwaan, C.: 1983, **Astrophys. J.**, **265**, 1056.
Golub, L., Herant, M., Kalata, K, Lovar, I., Nystrom, G., Pardo, F., Spillar, E., and Wilczynski, J.: 1990, *Nature*, **344**, 842.
Hollweg, J.V.; Yang, G., Cadez, V.M. and Gakovic, B. 1990, *Astrophys. J.*, **349**, 335.
Hoyt, D.V., Kyle, H.L., Hickey, J.R.; and Maschhoff, R. W., 1992, *J. Geophys. Res.*, **97**, 51.
Khler, H.: 1973, *Astron. Astrophys.*, **25**, 467.
Krause, F. and Rädler, K.W.: 1980, *Mean-Field Magnetohydrodynamics and Dynamic Theory*, Oxford Pergamon Press.
Kuhn, J.R. and Libbrecht, K.G.: 1991, *Astrophys. J. Lett.*, **381**, L35.
Kuperus, M., Ionson, J.A. and Spicer, D.S.: 1981, *Ann. Rev. Astron. Astrophys.*, **19**, 7.
Lee, M. and Roberts, B.: 1986, *Astrophys. J.*, **301**, 430.
Lerche, I.: 1971, *Astrophys. J.*, **166**, 627.
Lundquist, S.: 1952, *Ark. f. Fys.* **5**, 297.
Martin, S.F., 1988, *Solar Phys.*, **117**, 243.
Martin, S.F.: 1984 in *Small-Scale Dynamical Processes in Quiet Stellar Atmospheres*, Ed. S.L. Keil (Sunspot, New Mexico: NSO/Sacramento Peak) p. 30.
Matsumoto, R., Tajima, T., Shibata, K. and Kaisig, M.: 1993, *Astrophys. J.*, **414**, 357.
Moffatt, W.K.,: 1978, *Magnetic field Generation in Electrically Conducting Fluids*, Cambridge University Press.
Osterbrock, D.E.: 1961, *Astrophys. J.*, **134**, 347.
Parker, E.N.: 1955a, *Astrophys. J.*, **121**, 491.
Parker, E.N.: 1955b, *Astrophys. J.*, **122**, 293.
Parker, E.N.: 1957, *Phys. Rev.* **107**, 924.
Parker, E.N.: 1957a, *Proc. Nat. Acad. Sci.*, **43**, 8.
Parker, E.N.: 1957b, *J. Geophys. Res.*, **62**, 509.
Parker, E.N.: 1967, *Astrophys. J.*, **149**, 535.

Parker, E.N.: 1970a, *Astrophys. J.*, **160**, 393.
Parker, E.N.: 1970b, *Ann. Rev. Astron. Astrophys.*, **8**, 1.
Parker, E.N.: 1970c, *Astrophys. J.*, **162**, 665.
Parker, E.N.: 1972, *Astrophys. J.* **174**, 499.
Parker, E.N.: 1975, *Astrophys. J.*, **198**, 205.
Parker, E.N.: 1975, *J. Geophys. Res.*, **62**, 509.
Parker, E.N.: 1977, *Astrophys. J.*, **215**, 370.
Parker, E.N.: 1979, *Astrophys. J.*, **232**, 282.
Parker, E.N.: 1979, *Cosmical magnetic Fields*, Oxford, Clarendon Press.
Parker, E.N.: 1981, *Astrophys. J.*, **244**, 631, 644.
Parker, E.N.: 1983, *Astrophys. J.*, **269**, 635, 642.
Parker, E.N.: 1986a *Geophys. Astrophs. Fluid Dyn.*, **34**, 243.
Parker, E.N.: 1986b, *Geophys. Astrophys. Fluid Dyn.*, **35**, 277.
Parker, E.N.: 1988, *Astrophys. J.*, **326**, 395.
Parker, E.N.: 1991, *Astrophys. J.*, **372**, 719.
Parker, E.N.: 1993, *Astrophys. J.*, **408**, 707.
Parker, E.N.: 1994, *Spontaneous Current Sheets in Magnetic Fields*, Oxford University Press.
Parker, E.N.: 1994a, *Astrophys. J.*, **433**, 867.
Parker, E.N.: 1994b, Proc. IAU Coll. n. 143, *The Sun as a variable star*, 20-25 June, 1993, Boulder, Colorado, Cambridge, Cambridge University Press, de. J.M. Pap, C. Frohlich, H.S. Hudon, and S. Solanki.
Parker, E.N.: 1994b, *J. Geophys. Res.*, **99**, 19155.
Parker, E.N.: 1995a, *Astrophys. J.*, 10 January.
Parker, E.N.: 1995b, *Astrophys. J.*, 20 March.
Parker, E.N.: 1995c, *Astrophys. J.*, submitted for publications.
Petscheck, H.E. and Thorne, R.M.: 1967, *Astrophys. J.*, **147**, 1157.
Porter, J.G. and Moore, R.L.: 1988, in *Proc. 9th Sacramento Peak Summer Symposium*, 1987, Ed. R.C. Altrock (Sunspot, USA).
Porter, J.G., Moore, R.L., Reichmann, E.J., Engrold, O. and Harvey, KL.: 1987, *Astrophys. J.*, **323**, 380.
Priest, E.R., 1982, *Solar Magnetohydrodynamics*, D. Reidel Publ. Co.
Schüssler, M., Caligari, P., Ferriz-Mas, A. and Moreno-Insertis, F.: 1994, *Astron. Astrophys.*, **281**, L69.
Schindler, K., Hesse, M. and Birn J.: 1991, *Astrophys. J.* **380**, 293.
Schou, J. and Brown, T.M.: 1994, *Astrophys. J.*, **434**, 378.
Schwarzschild, M.: 1948, *Astrophys. J.*, **107**, 1.
Soward, A.M.: 1983, *Stellar and Planetary Magnetism*, Gordon and Breach.
Spruit, H.C. and van Ballegooijen, A.A.: 1982, *Astron. Astrophys.*, **113**, 350.
Spruit, H.C., Title, A. M. and van Ballegooijen, A.A.: 1987, *Solar Phys.*, **110**, 115.
Spruit, W.C.: 1974, *Solar Phys.*, **34**, 277.

Spruit, W.C.: 1977, *Solar Phys.*, **55**, 3.
Steenbeck, M. and Krause, F.: 1966, *Z. Naturforsch A.*, **21**, 1285.
Steenbeck, M. and Krause, F.: 1969, *Astron. Nachr.*, **291**, 49, 271.
Steenbeck, M., Krause, F. and Rädler, K.W.: 1966, *Z. Naturforsch A.*, **21**, 369.
Stix, M.: 1976, *Astron. Astrophys.*, **47**, 243.
Sweet, P.A.: 1958a, *Electromagnetic Phenomena in Cosmical Physics*, Cambridge University Press. Ed. B.Lehnert, p. 123.
Sweet, P.A.: 1958b, *Nuovo Cim. Suppl.*, **8**, (10), 188.
Sweet, P.A.: 1969, *Ann. Rev. Astron. Astrophys.*, **7**, 149.
Thompson, M.J.: 1990, *Solar Phys.*, **125**, 1.
Whitaker, W.A.: 1963, *Astrophys. J.*, **137**, 914.
Whithbroe, G.L. : 1988, *Astrophys. J.*, **325**, 442.
Whithbroe, G.L. and Noyes, R.W.: 1977, *Ann. Rev. Astron. Astrophys.*, **15**, 363.
Wilson, P.R., McIntosh, P.S. and Snodgrass, N.B.,:1990, *Solar Phys.*, **127**, 1.
Willson, R.C. and Hudson, W. S.: 1991, *Nature*, **351**, 42.
Yoshimura, H.: 1975, *Astrophys. J. Suppl.*, **29**, 467.
Yoshimura, H.: 1977, *Solar Phys.*, **54**, 229.

Activity in the Solar Atmosphere as Observed by YOHKOH

Yutaka Uchida

Department of Physics, Science University of Tokyo,
Shinjuku-ku, Tokyo 162, Japan
and
Department of Astronomy, University of Tokyo,
Bunkyo-ku, Tokyo 113, Japan

ABSTRACT

The X-ray Solar Physics Satellite Yohkoh has provided us with a number of new findings about the high temperature and high energy processes occurring in solar flares, in active regions, and in the background corona. According to these new findings, hot and dense corona above active regions seem to be maintained, at least in part, with the injections of already heated mass along the magnetic loops from the footpoint below. The outermost loops of the magnetic structures of these active regions are expanding away almost continuously in the case of "active" active regions. These give us quite a different and lively picture about the active region corona compared with a previous static picture with steady heating that we had based on the previous low cadence observations. New clues to the mechanism of flares, which were hidden thus-far in the yet fainter and relatively short stages before the start of flares, have been revealed by the wide-dynamic range, high cadence observations with the scientific instruments aboard Yohkoh. Those preflare signatures and their changes containing essential information about the mechanism of flares, now allow us to pursue truer understanding about the flare mechanism. The same merits of Yohkoh (wide-dynamic range and high-cadence observations) have shown us for the first time in its full form the highly dynamical behavior of the faint background corona, together with the influence of the changes in active regions sometimes exerting overwhelming effects on the surrounding corona. These, together with the fact that the corona is an inhomogeneous system, suggest that the corona is not an atmosphere of the usual sense, but rather, is an entity essentially electromagnetic in character.

1 GENERAL INTRODUCTION

In my lecture series, some descriptions of the results of the X-ray observations from the X-ray Solar Physics Satellite Yohkoh (Ogawara et al. 1991) will be given. The scientific objectives of the satellite Yohkoh are to investigate the high temperature and high energy phenomena on the Sun in X-rays. The results obtained by Yohkoh about flares and the corona of the Sun have contributed in advancing our understandings of the physical processes occurring in these phenomena. In addition to this basic significance for the investigation of the Sun itself, these results also serve in providing us with (i) prototypes of cosmic magnetodynamical phenomena such as flaring and coronal activities on other active stars, and possibly on accretion disks in highly active cosmic objects, on one hand, and with (ii) vital information about the sole source of the disturbances in the terrestrial environment and interplanetary space on the other hand.

We describe the results and their significance in solar physics in Chapters II to V, but mention here only very briefly the above (i) and (ii). As for (i), there are various types of stars (dMe flare stars, RSCVn type close binary stars, young TTau stars, ...) showing activities of the similar kind with two or three orders of magnitude larger than those on the Sun in the amount of energy involved. When we say "similar kind", we mean the presence of emission of non-thermal character together with the thermal emission from extremely high temperature plasmas, ranging from γ-rays to radio. Such non-thermal photons are the characteristics of the activity involved with the action of electromagnetic processes in the source, and indicate the presence of magnetic field. Magnetic field sometimes serves as a channel of transferring energy from the powerful motion in the dense high-β (β is the ratio of gas pressure over magnetic pressure) part of the astronomical bodies to the outer rarefied low-β atmosphere, where the dissipation of the energy transferred along the magnetic field, if it occurs rapidly, possibly due to some magnetic instability, will lead to the production of an extremely high temperature plasma together with high energy particles. Among the stars showing such activities of electromagnetic character, the Sun is merely a relatively calm late type dwarf star, but is absolutely unique star for which the details of what is happening can be observed in a spatially resolved form. This is why the investigation of the solar activity research is astronomically important besides its significance in the understanding of the Sun itself. As for (ii), it is of direct importance in the solar-terrestrial physics as the Sun is the sole source of disturbances for the interplanetary space, and the interest in the observation by Yohkoh has increased in relation to the observations of the phenomena in the interplanetary space from the solar-polar mission Ulysses in the orbit, and from the quiet-Sun mission SOHO to be launched soon. The understandings of the interplanetary phenomena thus-far observed have been derived at the in-situ points of the satellites in the terrestrial environment or interplanetary space, but these are now supplemented by the imaging observations of the features causing them at their origin on the Sun. Increasing interests from the astrophysical side may also reside in the processes of mass-loss from stars in the form of a steady wind, and mass-loss of the form like CME's (coronal mass ejections), and other mass-loss processes like the active region expansion found by Yohkoh, and to be mentioned below.

We start from the brief explanation about the satellite and its scientific instruments in the following part of Chapter 1. Then we begin the discussion of the scientific outcome starting from the behavior of the active region corona in Chapter 2. We go on to the problem about flares in Chapter 3. Unexpected dynamical phenomena in the background corona are discussed in Chapter 4, and the global and solar-cycle variation of the various features are very briefly touched upon in Chapter 5. Conclusion will be given in Chapter 6.

Table 1. The Parameters of the Satellite Yohkoh

Size	100 cm x 100 cm x 200 cm
Weight	400 kg
Power	570 W
Data recorder	10 Mbytes bubble memory
Data rate	up to 32 kbps
Telemetry	265 kbps
Orbit	
Altitude	600 km
Inclination	31 deg
Period	97 min
Attitude control	
Absolute pointing	a few arcmin
Stability	1.2 arcsec/s
Ground stations	
Command	Kagoshima Space Center
Downlink	KSC + four NASA DSN stations

1.1 The Satellite YOHKOH

In order to make it clear what kind of instruments the results to be discussed below were obtained from, we first describe very briefly the Satellite Yohkoh itself and the scientific instruments aboard Yohkoh (Ogawara et al. 1991). The Satellite Yohkoh is a Japanese Scientific Satellite conceived by the consotium of the Japanse Institutes and Universities, and constructed by the Institute of Space and Astronautical Science (ISAS), Ministry of Education, Science and Culture, Japan, with international collaborations of the US and UK groups of scientists selected and supported by NASA and SERC in two of the four scientific instruments, the soft X-ray telescope, and the Bragg crystal spectrometer, respectively. The operation as well as the analyses of the scientific data are done by the combined efforts of the Team thus formed. The parameters of the satellite are given in Table 1. The satellite carries four scientific instruments, two imagers and two context instruments with instrumental parameters shown in Table 2 (Tsuneta et al. 1991, Kosugi et al. 1991, Culhane et al. 1991, Yoshimori et al. 1991).

Table 2. Scientific Instruments on Board Yohkoh

SXT (Soft X-ray Telescope)	
Instrument	Modif. Walter-I grazing incid. mirror + CCD
Wavelength	3 - 60 A
Angular resolution	2.5 arcsec
Field of view	Full solar disk
Time resolution	up to 0.5 s
SXA	Small optical telescope for coaligning
HXT (Hard X-ray Telescope)	
Instrument	64-element Fourier synthesis collimator
Energy bands	15-24, 24-35, 35-57, 57-100 keV
Angular resolution	5 arcsec
Field of view	Full solar disk
Effective area	70 cm^2
Time resolution	0.5 s
BCS (Bragg Crystal Spectrometer)	
Instrument	Bent crystal spectrometer
Spectral lines	SXV(5.0385A), CaXIX(3.1769A), FeXXV(1.8509A) FeXXVI(1.7780A)
Time resolution	up to 0.125 s
WBS (Wide Band Spectrometer)	
Instrument	Continuum spectrometers
Energy bands	2-30 , 20-400 keV, 2-100 MeV
Time resolution	up to 0.125 s

The satellite was successfully launched from the Kagoshima Space Center of ISAS on August 30, 1991. Data downlink has been performed with the assistance of NASA using its four Deep Space Network Stations, together with the Kagoshima Space Center, performing the data downlink as well as sending control commands for the programmed operations. The Satellite is obtaining data for over 3 years, and basically in good health, and is hoped to continue taking scientifically valuable data for some more years, possibly covering the solar activity minimum period.

1.2 Advantages of Yohkoh Observations

As already mentioned, the advantages of Yohkoh come from the simultaneous observations with co-aligned images in soft and hard X-ray ranges from (a) a wide-dynamic range (due to a high sensitivity CCD and a very low scattering mirror, with pictures of various thicknesses) grazing incidence soft X-ray telescope (SXT) which allows an effectively indefinite number of frame-taking, and therefore allows a very much higher cadence and continuous observations compared with its predecessor on Skylab, (b) a Fourier synthesis type 64-bigrid imagers in the hard X-ray range (HXT) with high spatial resolution, and free from the soft X-ray contaminations, (c) a set of four Bragg crystal spectrometers (BCS) for the spectral lines of SXV, CaXIX, FeXXV, and FeXXVI, covering several million to over one hundred million degrees, and (d) a wide-band continuum spectrometers covering 2 keV to 100 MeV. Combined use of these allows us to grasp the physics operating in the sources, and these are what have provided us with a number of new findings about the high temperature and high energy phenomena in the solar atmosphere.

The Imagers, SXT and HXT, identifying the locations, shapes, and their changes in time of the sources in the soft X-ray and hard X-ray energy ranges, respectively are coaligned with each other. This solved the otherwise difficult problem of determining the locations of the hard X-ray sources. A hard X-ray source appear at the time of the flare without any reference features in the background, and thus it was difficult to determine the exact location of the hard X-ray sources relative, for example, to the optically observable features. With the coalignment with SXT (and therefore with SXA, an attitude sensor, a small optical telescope installed along the axis of SXT), the HXT images can be compared not only with the images of SXT but also with any other images from the ground-based instruments by matching the locations of the sunspots and the limb in white light. Comparisons are made with images in H_α and radio, but more important in some cases, with the magnetograms. The possibility of exact overlay of the X-ray images with the ground-based data increased the trustability of the interpretation very much, especially for the hard X-ray sources.

The comparison of the imaging data with the spectrograph data is also important. For example, although BCS receives the whole-sun fluxes without spatial resolution, a method in which we associate it with the particular one of the SXT sources showing a time profile similar to that of BCS (especially, SXV, having a similar range of the temperature response) will give BCS a pseudo-spatial resolution, and the combined use of them turned out to be extremely useful in discussing the dynamics of the hot plasma in the source.

One of the aspects newly obtained by Yohkoh with the above instrumental advantage is the information about the behavior of active region in soft X-rays. Contrary to the previous expectation, the heated mass of the active region corona, at least an appreciable part, turned out to be injected from the footpoint of the loop (Shimizu et al.1992), and the outer shell of the active region magnetic structure expands away almost continuously (Uchida et al.1992). We describe the behavior of the active region corona in Chapter 2.

Another extremely important aspect made possible to discuss by this instrumental advantage of Yohkoh is the initial development of flares in the still faint preflare stages. It is not necessary to stress that the information about the configuration and its changes of the loop structures (=magnetic field structures) in the preflare stages is essential for the understanding of the flare mechanism, if it is due to the release of magnetic energy as generally believed. Observation by Yohkoh allowed us for the first time to see those occurring in the faint preflare phase before the "flare start" in the terminology used thus far. That faint preflare phase is an extremely important phase in which the mass and energy supply processes may be seen, and the combined use of SXT, HXT, and BCS is very important (Kosugi 1992, Tsuneta and Lemen 1992, Uchida 1993, 1995). We will come to this topic in Chapter 3.

Another aspect the above-mentioned merit of Yohkoh played an essential role is the finding of unexpectedly dynamical behavior of the background corona. Indeed, this finding that the background corona shows a highly dynamical behavior is one of the most important results of Yohkoh. The low scattering mirror (about ten times lower than the case of Skylab soft X-ray telescope), together with the high and regular cadence observations with CCD, revealed the subtle activity in the faint background corona with high clarity for the first time (Acton et al. 1992b, Uchida 1992). We come back to this problem in Chapter 4.

Lastly, the global characteristics of the coronal structure, or the location of active regions and flares and their changes with the cycle-phase, will be an important topic derivable from the Yohkoh data, but this is still a weak part in the analyses, partly due to the fact that Yohkoh experienced only a fraction of the solar cycle yet. Some suggestive results, however, like the finding of the non-rigid rotation of the coronal holes at high latitudes, have already been obtained (Takahashi 1993, Takahashi et al.1994), and it is expected that Yohkoh will provide us with many valuable information also in this area. Since the satellite is in quite a good health even 3.5 years after launch (by February 1995), it is hoped that it will cover a large enough fraction of the solar cycle to discuss the cycle phase-dependence of various phenomena. We briefly mention some of the results in this area in Chapter 5, but have to leave the full description for some later articles. Conclusion will be given in Chapter 6.

2 THE ACTIVE REGION CORONA REVEALED BY YOHKOH

2.1 Introduction – Previous Notions about Active Region Corona

Active region corona has been considered to consist of hot and dense plasma tied down statically to the photosphere by the strong sunspot magnetic field, and heated by some (unknown) steady heating mechanism. This is probably due to the information obtained from the previous low cadence observations. Before the pioneering imaging observation in X-rays from Skylab, major channels of information were the observations of the corona at eclipses and with coronagraphs on high mountains. These took images of Thomson-scattered white-light, or the line emissions from the ions whose ionization temperatures are in the coronal temperature ranges, and the structures were observed at the limb from sideways in heavy overlaps in small numbers of still pictures. In the case of observations from Skylab, the observation was made in the soft X-ray ranges from the top of the structure above the dark photosphere, and the cadence was considerably higher than the previous observations.The cadence, however, was still not high and regular enough to show the features found this time by Yohkoh and described below, because the pictures were taken manually on the photographic films by the astronauts.

2.2 New Observation of Active Region Corona by Yohkoh

2.2.1 Overview of the Yohkoh Observations of the Active Region Corona

One of the merits of Yohkoh-SXT observations mentioned above is that it can take effectively infinite number of pictures with high cadence because of the use of a CCD under the control of the computer on-board. Actually the number of frames taken exceeded one million frames which will never be attained with the previous type instruments. We here start from the description of the behavior of the high temperature corona above active regions which are so-to-say the source region of activity in the outer atmosphere of the Sun. These are not only the place where intense flares occur, but it turned out that the background corona is sometimes profoundly affected by the change in the active regions. Therefore, we deal with what have been found in active regions first, and then proceed to the problems about flares, and then to the problems of the background corona in the following chapters.

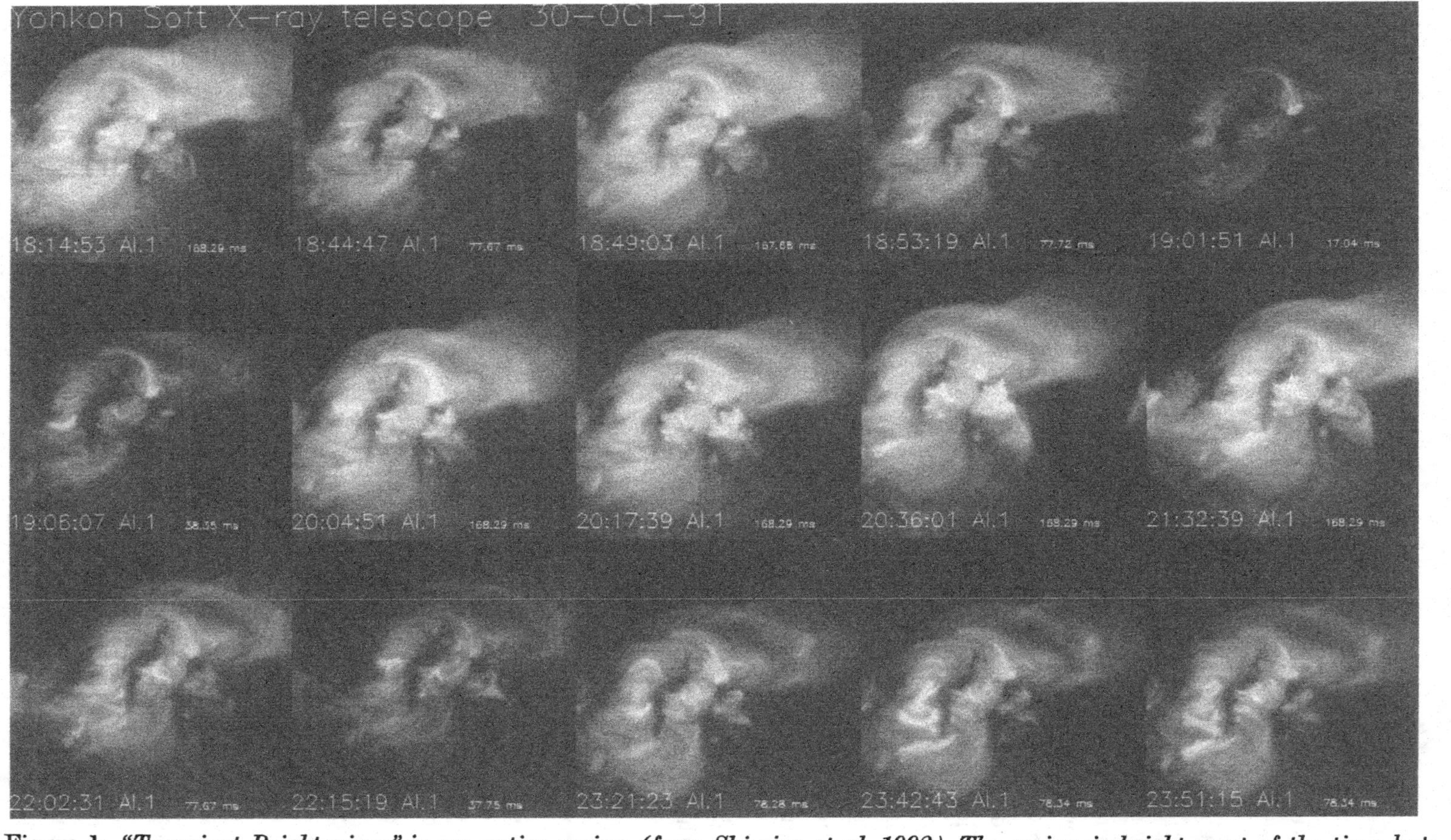

Figure 1: *"Transient Brightenings" in an active region (from Shimizu et al. 1992). The region is bright most of the time, but certain loops are bright at one time, and others are bright at other times. In some cases, the brightening seems to occur at the cross point of two loops, but in others, the loops are brightened from one footpoint and the bright region progresses along the loop with several hundred km/s. In some other cases, the loop brighten at once suggesting that a higher cadence observation is needed to resolve the fast time variation.*

2.2.2 Transient Brightenings of Active Regions

It was made clear for the first time that there occur frequent and intense brightenings of small loops inside active regions (these, however, are two to five orders of magnitude weaker in intensity than a typical flare) (Shimizu et al.1992). These are basically in the form of injections of *already heated* mass into the pre-existing magnetic loops inside active regions. This was not clearly noted in the Skylab time partly due to the observation with lower cadence picture-taking by using photographic films as mentioned above, and probably partly due to the saturation effect. This phenomenon of Transient Brightenings of the loops inside active regions was shown to have a typical life time of several to tens of minutes, and corresponds to a GOES class A to B tiny brightenings (Shimizu et al. 1992). The brightenings come up, in many cases, from one side of a pre-existing loop pattern, typically with a few hundred km/s velocity, with the density and the temperature on the average of the order of $(5-12) \times 10^9 \mathrm{cm}^{-3}$ and $(4-7) \times 10^6 \mathrm{K}$, with the estimated thermal energy of the order of 10^{26-29}ergs. In certain fraction of cases, they show some indication of interaction of two or more loops, but in the rest of the cases, they are single loops, and the phenomenon may most simply be described as non-steady injections of already heated mass into magnetic loops from one of the footpoints below (Figure 1).

These Transient Brightenings of active regions constitute a new mode of active region heating (and mass supply), thus far never thought of. The relation between the total output in the soft X-ray range from the relevant active regions and the frequency of such brightenings in them were examined by plotting a log-log diagram for a number of active regions by Shimizu et al, and it was found to have a slope of about 0.5. There must be, therefore, a tendency of creating larger brightenings in more active active regions, in order for the transient brightenings to be the source of heating of the whole active region. It may be more reasonable at this moment to say that we found a new component of active region heating, in addition to other heating mechanisms probably operating on the background in active regions (eg., Sakurai 1991).

Dynamic injection of the already heated mass from below requires special explanation because it markedly differs from what we expected from the steady heating models [by the slow dissipation of either DC current, or AC current (which corresponds to Alfven waves), with the subsequent evaporation of mass from both footpoints through downward heat conduction] discussed thus far. Possible alternative explanation may be that the mass is dynamically injected into the loop either through a magnetic reconnection which the loop makes with another (very) small emerging flux tube near one of the footpoints of such a loop low in the atmosphere, or through a mechanism like the dynamical relaxation

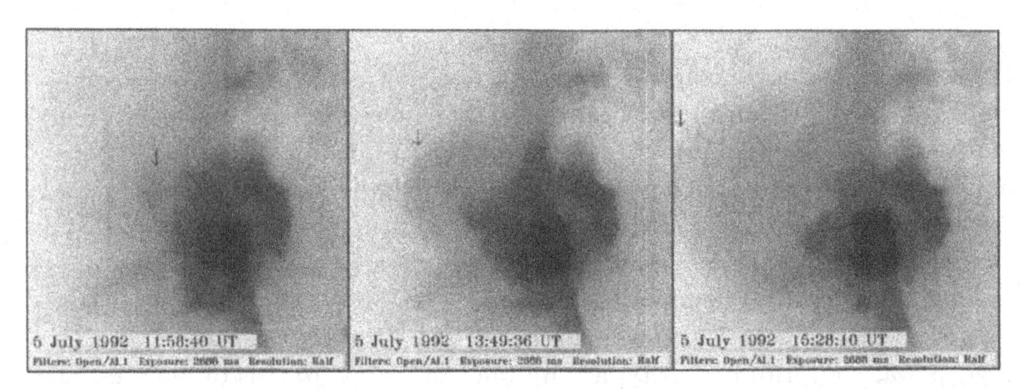

Figure 2: *An example of "Expanding Active Region Corona" (from Uchida et al. 1995a). The expanding fronts in many cases progresses keeping the smooth loop shape and with a velocity of the order of a few to a few tens km/s. It was found that the expansion is triggered by the occurrence of the "transient brightenings" mentioned above. It was found in the images of an expansion during the "off-point observation" that the expansion velocity increased with height to about 80 km/s at a height of 0.5* $R_\odot$ *in that particular event.*

of the magnetic twists which are produced below the photosphere by convective motions and released into the rarefied high atmosphere along the magnetic loop (Uchida and Shibata 1985, Shibata and Uchida 1986). [The twist can be released in the form of non-linear packets of torsional Alfven waves. This may be a fundamental process of non-thermal transfer of energy from the subphotosphere to the corona in the presence of a magnetic field. This process was considered in the loop flare model by Uchida and Shibata (1988), and may carry energy in the form of a self-closed coaxial current packets (= toroidal magnetic field produced by twisting of a portion of the magnetic loop), together with the kinetic energy of the gas carried with it. It was shown that the process also drives mass due to the "sweeping pinch mechanism" (Uchida and Shibata 1985). One of the remarkable points with this process is that the magnetic energy in the twists can be annihilated magnetohydrodynamically in a suitable situation even without any resistive effect which is usually assumed artificially large beyond the anomalous resistivity value].

2.2.3 Expansion of Active Region Corona

That the active region corona is almost always expanding, and not in a magnetohydrostatic equilibrium by being tied down by the strong magnetic field of the sunspot group, was another new finding by Yohkoh, giving considerable impact

(Uchida et al. 1992, 1995a) (Figure 2). This expansion is quite ubiquitous and occurs with most of the active regions, as if it were some intrinsic nature of active regions. The expansion occurs even without any reported flares or subflares. It should be remarked here that, although it has been imagined that a flare must affect the active region corona by its explosive action, what we actually found was that active region expansion also occurred at the times of flares, *but* flares led to only somewhat more pronounced version of the same type of slow expansion of the outermost loops as mentioned here, rather than a drastic blowing off of the active region corona. We concentrate ourselves here to the obviously unexpected case of expansion of active region corona *without* any flares or subflares of appreciable magnitude, in order to show this new phenomenon in its pure form.

The expansion is actually observed also when an active region is seen on the disk, but it is best seen, and its velocity can be measured most clearly, when the active region is at the limb. In many cases, the shape of the expanding active region loops is more or less retained similar during the expansion, and the velocity of expanding features measured by the time-lapse differencing of images ranges from several to a few tens km/s.

A question may arise on whether the observed expansion is a real physical expansion, or we are seeing some kind of propagating wave motion. We believe that it is a *real* expansion, because any wave (MHD fast wave, slow wave,...) should propagate with velocities one order of magnitude or more faster than that there. Another type interpretation, a successive filling up of outer loops by heated plasma, as in what is considered in the rising post-flare loops, would require some special situation like the "opening up - reclosing" type action of the magnetic field taking palce at such rather ordinary locations. We believe that this possibility can be excluded in our case because there are no such process of opening up seen to precede it.

We still need to find evidence for whether the motion actually continues out to interplanetary space, or the mass falls back to the Sun instead. There is supportive evidence for our expanding loops to reach interplanetary space obtained by a Japanese interplanetary mission *"Sakigake"*. *Sakigake* is one of the two Japanese Comet Halley Mission Satellites, and found peculiar interplanetary clouds, having a systematic out-of-ecliptic-plane magnetic field component, with systematically higher density, higher temperature, and slower velocity than the surrounding solar wind. These clouds, however, had *no solar surface counterparts* like flares or dark filament disappearances as in the case of other interplanetary features (Nakagawa et al. 1989). Recent examination late in the declining phase of the solar activity cycle (when smaller number of confusing active regions existed on the solar surface) indicated that actively expanding active region is nicely identified at or near the location on the solar surface to which the cloud motion is traced back with the observed cloud velocity (Nakagawa and Uchida 1995).

The roughly estimated mass flux and the magnetic flux in the expanding motion of the active region corona amount to $dM/dt \sim 10^{11}$ g/s and $dF/dt \sim 10^{16.5}$ Mx/s, respectively. This dM/dt amounts to be an appreciable fraction of the total mass loss rate by the solar wind partly because the density of the active region corona is higher, if it is established that the expanding active region loops actually come out to the interplanetary space, whereas dF/dt is of the same order as the rate of magnetic flux emerging into active regions from below (Zwaan 1987).

The expansion of active region corona, if established, is a newly recognized component of the mass loss from the Sun. The notion of the solar wind thus far, based on the solar wind picture by Parker (1963), is the thermal expansion wind accelerated along the field lines, whereas the mass motion in our expanding active region corona is basically accelerated by the Lorentz force, at least in its first phase of acceleration, perpendicular to the magnetic field. Thus the expansion of active regions we found may influence the notion of the mass-loss from the Sun, and therefore, the notion of mass-losses from other active stars, say RSCVn stars having gigantic starspots. There may be a stronger version of what we found in the solar active region also in such stars, which can possibly dominate the usual stellar wind.

Such expansions of the active region coronal structure requires some special explanations. Those newly found phenomenon of Transient Brightening of the active region loops in the core part of the active region below, and the expansion of the outermost part of the active region magnetic loop structures, seem to have some relationship. The transient brightenings occur in fixed closed loops below, and the expanding outer loops are clearly different from the brightened one. Possible explanation is that the heated mass injection into the brightened loop in the core part may accompany electric current (equivalent to the magnetic twists) as in the process to be discussed for the loop flare in Chapter 3, and the injection process raises the magnetic pressure inside the active region. The outermost part of the magnetic structure may be pushed outward due to the force imbalance, and start expanding. Once it starts expanding, the expansion continues because the expanding loop continues to be squeezed out along the gradient of the magnetic potential (similar to the so-called melon-seed effect).

Here it may be mentioned that the situation might have something in common with the case of the so-called CMEs (coronal mass ejections), although the scale is very different, in that the structure to be ejected goes with the smooth loop-like shape kept, indicating that it is not due to any localized strong impact by a flare blast, or flare ejecta, but rather, due to a global destabilization. It was seen by Yohkoh that a small brightening in active regions in some cases influences a large scale region of the surrounding corona, and as the result of this, a high latitude large arcade formation was caused. It is highly likely that a CME is launched above such a high latitude arcade formation. If this is the case, the CME launching is not caused by the flare effect like the dynamical impact, or blowing away, but through the destabilization of the equiliblium, and the rising may be due to the imbalance of the magnetic force. We come back to this problem in Chapter 4.

We next proceed to the description about flares which are the drastic energy release in active regions. Especially, one of the two (or three) types of flares, namely, the small simple loop type flares are of the similar geometrical property with the active region transient brightenings in active regions described above, though the energy released in loop flares amounts to values a few orders of magnitude greater. It is natural for us to suspect that there is some physical explanation to explain the relation between them. We come back to this point in the next Chapter.

3 NEW CLUES TO THE FLARE MECHANISM

3.1 Introduction – Previous Notion about Flares

Solar flares are powerful explosive energy liberation processes occurring in the active region corona. There had been various types of hypothesis to explain the basic mechanism of flares from early times, but gradually it became clearer that the source of energy must be magnetic field, and there should be a mechanism of storing energy together with a mechanism of liberating it quickly. The magnetic field is the only candidate having large enough energy stored in the region of flare occurrence, but of cource the energy of the magnetic field whose source is deep in the photosphere can not be annihilated in relation with the process occurring in the high atmosphere. The part of the magnetic energy raised by the effect of the local current can be annilated by the local process, and usable there. The magnetic field is good for this also because it has various kinds of instability. The local current systems considered for the present purpose have been (i) current system flowing along the magnetic loop, or (ii) the current system flowing in the current sheet. These basic current configurations corresponding to the configurations of loop flares and arcade flares, respectively, have been considered in the flare models. The main lines have been to consider mechanisms in which the current systems are brought up gradually, and the stored magnetic energy is liberated through some kind of fast instability inherent to the magnetic field. For example, a quenching of the steadily flowing current by its own non-linear nature (Alfven and Carlqvist 1969), or the occurrence of anomalous dissipation were considered for the current loop models for loop flares (Galeev et al. 1981). For the sheet current configuration, the neutral sheet magnetic reconnection model was proposed by Sweet (1956) and have been discussed by many people (Parker 1964, Petchek 1964, Yeh and Axford 1970). Those people dealt with the process by taking out the neighboring part of the X-type neutral point, and discussed the reconnection process under the simplifying assumption of a steady or quasi-steady flow in two-dimensional calculation. More recently, non-steady processes were treated by using two-dimensional numerical simulations (Sato and Hayashi 1979). These will be discussed a bit more in the respective sections.

The amount of heated mass in flares is not available in the corona where the energy source is believed to exist, and therefore, the mass somehow should come from the denser part of the atmosphere below, and this is another point of focus. In this context, the chromosphere is considered to provide mass, and thus the evaporation models have been treated in the last decade.

3.2 Observations of Flares by Yohkoh

3.2.1 Overview of the Observations of Flares by Yohkoh

The existence of at least two distinct flare types have been established, and their detailed behaviors have been revealed by Yohkoh. We discuss the findings about these two types of flares in the following: (A) Simple loop type flares: Small scale ($< 10^9$cm, showing a loop shape, but sometimes point-like showing an indication that it is a part of a loop), and shorter duration ($\sim$ 20 min) flares having a simple loop shape. (B) Arcade type flares: Large scale ($> (5-10) \times 10^9$ cm), long duration (more than a few to 10 hours) flares having an arcade type shape with relatively softer (and therefore weaker at high energies) hard X-ray emission. There are larger scale, fainter versions occurring outside active regions (scale $\sim (20-40) \times 10^9$ cm, duration $\sim$ 20 hours) having similar properties. Still larger ones with much fainter emission occurring in the circum-polar regions (scale $\sim 100 \times 10^9$ cm, duration $\sim$ (1-2) days) have been discovered by Yohkoh. [There may be a third category, the loop interaction type. This category, however, will not be discussed here since it has not yet been well confirmed.] We now describe what Yohkoh obtained about the (A) and (B) categories.

One of the important aspects obtained by Yohkoh with the above-mentioned instrumental advantage is the information about the initial development of flares in the still faint preflare stages. It is not necessary to stress that the information about the configuration and its changes in the magnetic field structure in the preflare stages is essential for the understanding of the flare mechanism. Observation by Yohkoh allowed us for the first time to see those occurring in the faint preflare period before the rise phase of flares, together with the behavior of the bright flares themselves.

3.2.2 Simple Loop Flares

3.2.2.1. *Previous Notion about Simple Loop Flares.* It was recognized from the time of Skylab that there are distinct types of flares, one being very small, having a loop shape, but often unresolvable with the Skylab spatial resolution (5 arcsec), and the other being larger in spatial scale and much longer in duration (Pallavicini 1975). It was clear even before the Skylab time that the amount of mass in the flare loops is larger than the mass pre-existing in the corona of that region, and can not be due to the process which collects the coronal mass. Therefore, it was considered that the mass supply from the high density chromosphere is necessary (chromospheric evaporation models), and two possible ways have been proposed for loop flares: one is to assume a heat source at the top of the loop, and consider that the heat conduction from the loop top source to the chromosphere "evaporates" the chromospheric mass into the loop. There were a number of treatments along this line. We call this type models as "heat conduction-evaporation models" (HCE models) in the following. The other way was to assume that there was a source of high energy particles at the top of the loop, and the particles streamed down into the chromosphere and heated the chromospheric mass by the bombardment, and "evaporates" it into the loop. We refer this type models as "electron bombardment-evaporation models" (EBE models) in the following. In both of them, the loop was assumed to have a fixed shape and a fixed cross-section, probably having a very strong straight (curved with the shape of the loop) magnetic flux tube in mind. A problem in both of them was the assumption of the source at the loop top without going into the physical discussion of the mechanism for that, while the investigation of this point is no doubt the most essential problem about the loop flares.

The points to be critically examined in the observational results will therefore be whether there is a source of heat conduction at the loop top having a large enough heat content before the appearance of blueshift in BCS lines, or the hard X-ray burst always precedes the blueshift in BCS lines. Doschek et al. (1983) and Tanaka et al. (1983), however, have already noted that the blueshift of the Ca XIX and Fe XXV lines strats in some flares earlier than the hard X-ray bursts. These points can best be examined by Yohkoh's observation with the combined use of SXT, HXT, and BCS.

3.2.2.2. *Typical Examples of Simple Loop Flares Observed by Yohkoh.* There are many examples of simple loop flares observed by Yohkoh, and some of them have been analysed and reported. Acton et al.(1992a), and Feldman et al.(1992) called attention to the highly localized soft X-ray source at the top of the loop in loop flares. Doschek et al.(1994) showed in four typical simple loop flares, Jan 13,1992, Feb 17,1992, Apr 17,1992, and Jul 11,1992 events, that both temperature and density of the loop top are higher compared with the leg part of the loop below. This is very significant from the mechanical (how can higher density mass be suspended above the less dense region along the legs of the loop, and what confines the high pressure gas at the loop top?), and thermal (how can the higher temperature region persist at the loop top without being cooled by the heat conduction loss, and without causing continued evaporation of the chromospheric gas *if* the heat conduction-evaporation model of a loop flare is correct ?) points of view. We discuss these later in section 2.2.3.

A more comprehensive analysis of the dynamical development has been given by Uchida et al. (1995b). A typical example treated in detail is the Feb 17, 1992 event (GOES M2-class). Some of the small but high energy flares of this type with smaller scale loop shape could have been classified in the Skylab observation as a "point-like flare" without time variation of its shape. Some of these might be due to insufficient spatial resolution or the cadence of the observation might not be high enough so that it failed to detect the preflare and the initial variation when the intensity was still low, and observed only the later phase in which the source was confined around the loop top. Such was indeed our initial impression about this type flares when the first Yohkoh data about this type flares were obtained. More detailed analysis of the Yohkoh-SXT observations, taking the fainter preflare stage with higher spatial resolution of 2.5 arcsec, wide dynamic range, and high enough cadence into proper account, however, revealed the thus far not known dynamical development in the initial part of this type flares, giving new clues to their origin.

In the Feb.17, 1992 event (Figure 3), the footpoints brightened first, and then the dynamical processes appeared in the high part of the loop, and the bright region that appeared high around the top of the loop expanded toward the footpoints (the expansion towards the northern footpoint was prominent, but the increase in the intensity indicates that the expansion also went down to the southern footpoint, too). The front of the expansion seemed to come down to the footpoints and then the hot gas seemed to shrink back to the loop

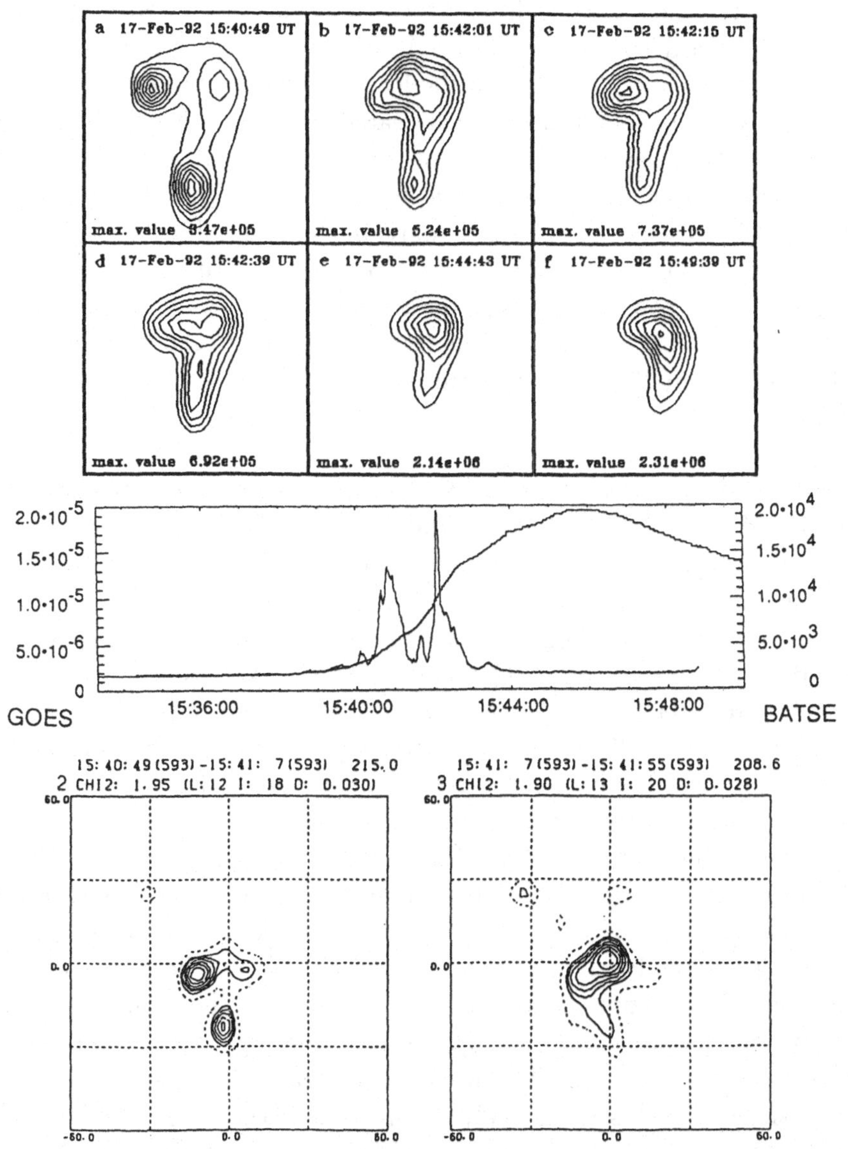

Figure 3: *Initial dynamical development of simple loop type flare of Feb. 17, 1992 which occurred at W81-N16, as seen by SXT (from Uchida et al, 1995b). (a) At 15:40:59, the two footpoints were bright, and the loop top started increasing the brightness. At 15:41:35, the loop top increased the brightness and the bright region started to expand. At 15:42:01, the bright knot progressed toward the northern footpoint. At 15:42:15, the bright knot went quite low near the footpoint and bounced back up by 15:42:31 and 15:42:43, respectively. The bright blob resided at the loop top, gaining brightness. The loop shape distorted into a question mark shape at around 14:43, but the distortion was relaxed before the flare maximum is reached as seen at 15:49:20. (b) Hard X-ray time profile from BATSE overlaid on GOES profile (courtesy of both group) showing the reference times in the event.*

top, and then the loop top brightness seemed to increase further by an order of magnitude toward the maximum brightness. It is remarkable that a quite appreciable *distortion* of the shape of the loop as a whole into a "question mark shape" occurred as the loop top source appeared, and this distortion disappeared by the time of the intensity maximum of the loop top source (Figure 3a). The "question mark shape" can be explained by a helical or kink distortion of that loop seen in projection. It is very remarkable that all these occurred in the period before the maximum phase of the flare, and after that the brightest part settled down at the top of the flare loop and did not show any marked motion. Therefore, observation with lower sensitivity and a lower cadence (eg. Skylab) would find it as a small point-like source suggesting loop-like shape not varying in time.

In contrast to the arcade type flares, the simple loop type flares have harder spectrum hard X-ray (HX) bursts around the start of the soft X-ray (SX) loop formation. The Feb. 17, 1992 flare had three HX impulsive bursts, two strong ones and a relatively weak one in between, together with much weaker background emission, prior, during, and after the bursts (Figure 3b). The sources of the first HX burst observed by HXT coincided with the footpoints of the loop (Sakao 1994, Masuda 1994, Uchida et al. 1995b). There was a less prominent second HX burst about 40 sec after the peak of the first one. This coincided with the time of the rapid increase of the SX loop top source and the start of its expansion toward the footpoints. The third HX burst, which occurred about 65 s after the peak of the first burst, had a very sharp spike component together with a slower and weaker component . The source of the latter component seemed to have relation with the SX loop top brightening, whereas the strong sharp spike in HX having a much harder spectrum came from the northern footpoint when the loop top bright region expanding downward along the loop touched down at the northern footpoint. This suggests that the high energy particles were contained in the expanding blob emitting soft X-rays. In other words, the high energy particles did not stream freely down to the lower layers but moved down along the loop in a blob with a velocity of the order of 400 km/s, and emitted the harder spectrum hard X-ray spike when the blob touched down the footpoint. It is probable that the bright blob may be the shock propagating downward. This is a new finding that differs from the processes of bombardment by individual free streaming high energy electrons as considered thus far.

Next, it is also to be remarked (Uchida et al. 1995b) that the creation of the high temperature plasma started *before* the first HX burst . The rise of the BCS lines started actually significantly earlier than the start of the first HX burst given by the BATSE/ Compton Gamma-ray Observatory data. BCS showed a remarkable blueshift in S XV before the start of the rise phase of the flare, and also clear red- and blueshifts in Ca XIX around the time of the appearance and the start of expansion of the loop-top source in the soft X-rays around the times of the second and the third HX bursts (Figure 3a,b). Since this loop flare occurred near the west limb, and the footpoints align almost on the same longitude, we do not expect to see appreciable Doppler shifts if the loop shape is a simple semi-circle because it was observed almost face-on. The fact that we actually saw Doppler shifts (blue first, and then red around the second weak HX burst) may mean either the loop performed a rolling motion, or the motion of the hot gas had velocity components along the line-of-sight from us. The latter is consistent with the distortion of the loop shape into a "question mark shape" in projection. The appearance of the redshifted component actually can be related to the occurrence of the flow of hot material.

Another example of Apr 22, 1993 loop flare (Uchida et al.1995c) also started from the two footpoints, and the bright source came up to the higher part of the loop in SX. This loop flare was observed almost top-on (E5, N12), and favorable for detecting the blueshift accompanying the mass rise along the legs of the loop. In the case of this flare, there occurred another small flare in the southern hemisphere, but its effect in the BCS spectrum could be nicely separated, and we could use the S XV line of BCS, as well as SXT and HXT images. It showed a blueshift of about 150 km/s, and the BCS lines rose earlier than the HX non-thermal impulsive burst, and it is found that a smooth thermal-like emission existed at that time in the HXT L-band (13-25 keV), suggesting the existence of a superhot gas before the non-thermal hard X-ray burst took place. It should, however, be added that the footpoint brightenings in the HXT L-band came up before this superhot source appeared at the loop top.

3.2.2.3. *Possible Interpretations of Loop Flares.* The observations described thus far raised several important new clues which posed several questions to the models discussed thus far. The models for the loop flares discussed in the last decades were models for the supply of the mass from the chromosphere to the loop, heated either via heat conduction from the heat source assumed at the loop top (heat conduction-evaporation model =HCE model), or via bombardment by high energy particles produced in a source *assumed* to exist at the loop top (electron bombardment-evaporation model =EBE model) (cf. Antonucci et al. 1986, and references therein), as mentioned in 3.2.2.1. The status about the models before the Yohkoh time is that some suitable choices of parameters in both models were claimed to give acceptable results for the formation of the observed soft X-ray loops, and therefore neither one of the models has been excluded in a definite way.

Now, we have Yohkoh observations providing better clues to decide what kind of models can withstand the new observations. The first question for the HCE model is whether there is a pronounced heat source having enough heat content at high part of the loop *before* the footpoint brightening in soft X-rays. One may argue that the "loop top source" may not be seen in SXT because the temperature was too high for it. It may be pointed out, however, that such a superhot source, if ever exists, can be detected in the lowest energy channel, L-band, of HXT. The absence of the continued evaporation flow in the post-maximum phase is another problem for the HCE model (unless the effects of an entangled magnetic field are introduced in a highly ad-hoc manner, but the HCE and EBE models are essentially non-magnetic models except for the assumption that the magnetic field provides the fixed loop shape with straight (loop shaped) field lines). The EBE model can withstand even if observation give no source at the loop top before the footpoint brightening if the density of the background medium at the loop top was low enough. The appearance of the loop top source in SX following the footpoint brightenings is compatible with these evaporation models, but these models need a "source" originating the sequence of processes to produce the footpoint brightenings high at the lop top, but that is to be verified. A problem about the EBE models arises from the Yohkoh observations of the behavior of the source of the sharp hard X-ray spike mentioned in the above. If this means that the moving blob had something to do with the transport of the high energy particles in a form contained in some moving region, it will require a considerable modification of the EBE models, because the packet carrying the high energy particles in confinement should be of magnetic character, and then the calculations without including the effect of the magnetic field might have been wrong.

A most basic question addressed to those evaporation models is the question of how such a source of large energy liberation can be suddenly provided at the top of a loop. Workers in the evaporation paradigm all *assumed* the presence of the source in an ad-hoc manner.

A magnetohydrodynamic model has been proposed in order to explain the origin of the energy, as well as the mass, in the loop top source (Uchida and Shibata 1988). They showed that the magnetic twist coming up from below the photosphere pinches the loop at its incidence to the upper part of the atmosphere, because the magnetic loop is supplied with a new toroidal component of the field in the form of twist. The incidence of the twist (non-linear torsional Alfven wave packet) therefore squeezes the material of that region and drives it upwards along the loop as it propagates with an Alfvenic velocity (the authors called this a "sweeping pinch mechanism"). Therefore such a process carries mass as well as magnetic energy in the form of the twist propagates into the loop. This process may be in operation in the so-called "active region transient brightening" discovered by Shimizu et al.(1992) in which the already heated mass in the active region may be injected from below. When, with much smaller probability, two of these magnetic twist packets are released into one and the same loop from both sides, the magnetic twist packets carrying the mass with Alfven velocity will collide at some high part of the loop, and convert the kinetic energy of the hypersonic motion and the magnetic energy in the twists into heat in shocks, and into the kinetic energy of the unwinding motion of the twists (the observed turbulence may be this if we can observe it in a resolved form) which may also be converted into heat eventually. This was the first proposal and simulation of magnetodynamic process in loop flares. Several simplifications of the situation like the use of 2.5D scheme, the adoption of symmetry with respect to the loop top to reduce computation time, etc., together with the assumption of the sudden release of the twist packets simultaneously into both footpoints may not have appeared realistic enough, and probably reduced the due impact of the model. The observational results of Yohkoh, however, suggest strongly that the origin of the loop flares may be highly magnetodynamic in character, and elaboration of this magnetodynamic model is called for. For example, the distortion and relaxation of the loop shape is just the property which is expected very naturally from their model if the simulation is performed in 3D, whereas such a distortion can never be explained by non-magnetic models (magnetic field was simply supposed to provide the fixed loop shape, and did not play any dynamic roles in the HCE and EBE models). Creation (not an assumption) of a loop top source with the bouncing shock due to the collision of the incoming magnetic-twist front is also what their 2.5D simulation could reproduce, and might explain the observed behavior of the hot source in the Feb. 17, 1992 event. The first hard X-ray emission with softer spectrum may be explicable by electrons accelerated in the betatron acceleration in the dynamical pinching, and the second and the third HX bursts may be produced through further acceleration of the betatron electrons accelerated to higher energies in between the approaching MHD shock fronts through the Fermi I mechanism.

We here add some more discussions about the observational results and interpretations about loop flares. The HXT imaging precisely aligned with the images from SXT performed by Masuda et al. (1994) indicated that a hard X-ray source exists right above the top part of some of the loop flares observed in SXT. They say that this may be favorable for the HCE model because the "above the top of the loop" source is interpreted as a superhot source (over 10^8 K) at the loop top. A question may arise as to why it is not in a part of the soft X-ray loop, but detached high from the loop and presumably outside of the loop itself. A possible answer by them might be that we are seeing a hard X-ray source before all the high energy electrons have been dumped to the footpoints in the bombardment. It may contribute to the production of another loop not yet heated just outside the one presently emitting soft X-rays. This reminded them of the classical arcade flare model in which the reconnection occurs in successively outer-lying shells in the once opened-up bipolar array field. Questions may still arise about some points. One is why an isolated single loop experiences the opening up and reclosing process similar to the arcade flares while their neighboring loops do not. The answer may be that the flaring loop may be the most bright one in the arrays of reconnecting loops. The other question is why the HX footpoint sources are on the SX footpoints and not outside, if the "above the loop top" source is outside the loop. There can be a time lag between the appearance of the superhot source and the SX loop because of the time taken by the "evaporation", but why the high energy particles are prevented from impinging into the footpoints of the loop on which the superhot source exists, namely, the loop above the SX loop.

Shibata et al.(1994), on the other hand, hinted from the ejection of a blob (heated part of the erupting dark filament) in arcade flares as described in 2.1.1., scrutinized the pre-flare images high above the simple loop flares, and they found some motion in the very faint structure in the high atmosphere in the region surrounding the active region in question. In some cases the faint moving object seems to be a blob-like, and in other cases it seems that the outer loops of active region surrounding the flare are expanding. They suggested that the process with loop flares might also involve the "opening up – reclosing process", but the situation reminds us of the "active region expansion" found by Yohkoh (Uchida et al. 1992) in which the outermost magnetic structure of active regions expands, most likely in relation to the so-called "active region transient brightening" (Shimizu et al. 1992) which itself is a phenomenon of mass injection into a closed loop near the central part of active regions, rather than an open magnetic structure leaking the hot gas from it.

We should, however, retain at this moment the possibility that the loop flares may be a part of the greater whole, although we may be able to say that it is now the time in which we should take the MHD models into consideration, rather than non-magnetic models which are dealing only with the evaporation of the gas alone in a fixed tube by assuming an ad hoc source of energy at its top, since what is occurring suggests its highly magnetohydrodynamic character.

3.2.3 Arcade Flares

3.2.3.1. *Previous Notion of Arcade Flares.* The other category of flares seen in X-rays, which had first been supposed to exist based on the information from Ha double-ribbon flare below and the post-flare loops seen to connect these double-ribbon above, is the arcade type flares. An X-ray arcade was actually seen for the first time by Skylab (Svestka 1976). However, the detail of their structure and their temporal behavior have been made clear first by the Yohkoh-SXT. The information obtained from the H_α observations was that the H_α double-ribbon were produced on both sides of a magnetic field polarity-reversal line in the photospheric field, and a dark filament lying along it disrupted and flew away sometime before the flare start. From its observation at the limb, it was known that a dark filament was a thin partition-type structure floating above the field polarity-reversal lines, and it was often confirmed that the dark filament had actually flown away upwards at the time of a flare. The scale size of this type flares was known to be larger than the size of loop flares, of the order of 5×10^9 cm $\times 10^{10}$ cm for the case occurring inside active regions. It was known that an "outside-active region version" of this type existed, having a larger scale of 10^{10}cm $\times$ 3×10^{10} cm, outside of active regions. Those were known as a quieter double-ribbon flares outside active regions (Michalitsanos and Kupferman 1974). The total energy involved in both of the active region type and "outside active region type" is around 10^{30-32}ergs, but the version outside active regions was less-energetic in the energy range of accelerated electrons is concerned.

It was interpreted by Sturrock (1966) that the magnetic arcade of loops were pulled outward by the rising dark filament, and the created oppositely-directed fields reclose through magnetic reconnection with each other, liberating magnetic energy there. and the reclosed field might explain the formation of the postflare loops. This energy is thought to be conducted down to the footpoints of the reconnected loops and caused the supply of the heated mass to explain the mass in the arcade flares. There were variations of this type models proposed by Hirayama (1974), Kopp and Pneuman (1976), and more recently by Priest and Forbes (1990), Forbes (1992) and others by making modifications.

3.2.3.2. *Typical Examples of Arcade Flares Observed by Yohkoh.* A typical one of this type first observed at the limb is the Feb. 21, 1992 flare (GOES class M3.2) (Figure 4). This flare was relatively low energy, large sized flare, belonging to the "long duration event" named after their long and gradual time profiles given already before Yohkoh. Tsuneta et al.(1992a) have given a detailed analysis of this flare, covering from its pre-flare state to the post-flare state: It showed a very impressive candle-flame type shape as it was observed on the east-limb, having a very neat dark tunnel below, and the size was gradually increasing with time before it faded and became invisible after several hours. The merit of SXT (wide dynamic range with high sensitivity, and high cadence of observation) played an important role in providing us with the information of what happened in the dark pre-flare state.

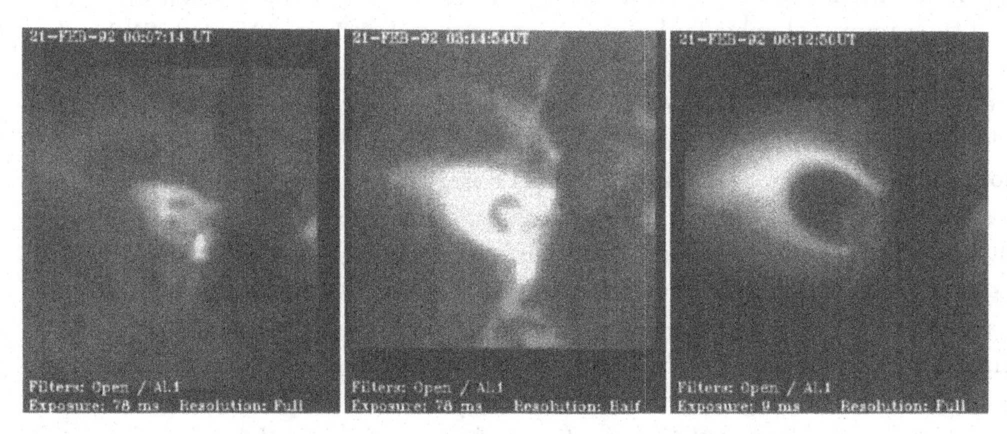

Figure 4: *Arcade type flare of Feb. 21, 1992 which occurred at the east limb (general description is referred to Tsuneta et al. 1992, and here we concentrate our attention to some intricate points which require some special notice, cf., Uchida 1995). Left pannel: Preflare "core" with faint structure connecting the top of the "preflare core" back to the photosphere. Also, we note a bright feature near the axis, and a vertical partition inside the region which grows into the dark tunnel in the flare maximum phase later. Middle pannel: Rise phase of the flare. The features mentioned continued to exist. Right pannel: A candle flame-like flaring with the dark tunnel below which is generally taken to be consistent with the classical model.*

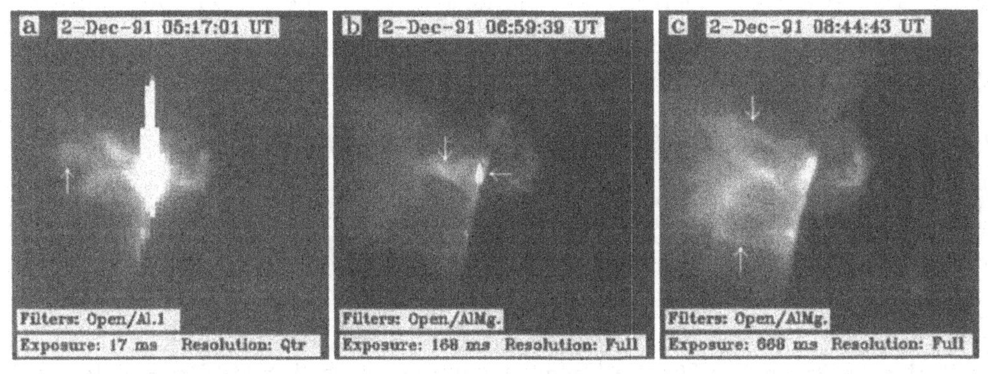

Figure 5: *Same type flare of Dec. 2, 1992 which occurred also at the east limb. Left pannel: Before the rise of the flare, there was an ejected blob which is likely to be the heated part of the rising dark filament seen along its length. Middle pannel: Cusp type structure emerged in the middle, again with a bright feature lying low in the middle of the tunnel. Right pannel: Large-scale structure connecting the top of the cusp-type structure back to the photosphere together with the bright feature low along the axis of the tunnel just like the event of Feb.21, 1992 is clearly seen (from Uchida 1995).*

This flare started around 03UT of Feb. 21, 1992, on the east limb, and the data of the flare as well as the pre-flare state were available for this case. In SXT, it is seen that there appeared a distorted "MacDonald's logo"-shaped (overlapped double arch shape) structure in which the top part of the inner triangular core was connected back to the photosphere on both sides in the pre-flare stage, at around 23:30UT of Feb 20 (Figure 4). The inner part of this structure, the preflare triangular core, which started to become brighter at around 03:10UT of Feb. 21, increased its brightness at around 03:27UT, and the bright core part further brightened as it expands. This developed into a tilted "candle flame"-like high temperature structure with a neat semi-circular "tunnel" below it. There were also characteristic features in the preflare stage having a partition shape vertically dividing the tunnel, with a low-lying array of bright feature appearing *along* the axis of the "tunnel". The bright "candle flame"-shaped extended source,which was the main part of the X-ray source of the flare, gradually faded, but lasted visible until around 09UT. The temperature-emission measure analysis by Tsuneta et al. (1992a) shows that the temperature was the highest, around 1.3 MK, along the outer edges of the candle-flame type shape, and the density was the highest ($4 \times 10^{10}\mathrm{cm}^{-3}$) at the central part right above the "tunnel". Rather soft non-thermal emission existed, but imaging by the hard X-ray telescope (HXT) aboard Yohkoh could not give clear source shape probably due to the wide-spread configuration of the relatively weak sources.

Dec 2, 1992 flare observed at the limb (Figure 5) also had a structure connecting the top of the triangular central part of the flare back to the photosphere on both sides. A rising blob (probably, the heated part of the rising dark filament seen along its length) was seen to be ejected from the core part before the flare (Tsuneta et al.1993). Existence of a brightening feature at the position of the axis of the dark tunnel below the bright triangular core structure is also seen in this event, suggesting that this situation similar to that of the Feb 21, 1992 event is pretty general. The shape of the source near the flare maximum ("candle flame" type cusped structure with a tunnel below, for example) alone would seem, at first sight, to be consistent with the "opening up - reclosing arcade" type models (Sturrock 1966, Hirayama 1974, Kopp and Pneuman 1976, Priest and Forbes 1990, Forbes 1992), *but*, a more careful examination of the preflare state readily tells, as pointed out above, that the relevant magnetic field source in the photosphere can not be a simple dipole array source as supposed in the classical model, because a structure connecting the top of the triangular core part of the flare back to the photosphere on both sides at several times the arcade width indicates that magnetic source relevant to the arcade flares is likely to be a quadruple array in the photosphere. Also, the existed bright feature at the axis of the tunnel, together with a vertical partition type structure inside the tunnel in the preflare stage are difficult to explain by the classical model with the bipolar array source.

These remind us of a model proposed years ago by the present author in the Skylab Workshop (Uchida and Sakurai 1980) in which a quadruple magnetic array configuration was proposed as a most natural configuration providing a magnetic structure suspending a dark filament having a weak field along the length. In the quadruple array model, a magnetic neutral point appears in the magnetic field components in the plane perpendicular to the field polarity reversal line in the photosphere at the center (we call this components as $B_\perp$). This, in three dimensions, is a magnetic neutral line lying parallel to and above the field polarity reversal line in the photosphere. This neutral line will be deformed into a vertical neutral sheet when the footpoints are squeezed from sideways. When the outer pair of the magnetic source has some skewness in position with respect to the polarity reversal line, there will be a $B_{||}$ (parallel to the polarity reversal line in the photosphere) which will dominate in the "neutral sheet" even if it is week, because $B_\perp$ is zero there, separating and stabilizing the structure having antiparallel fields which are otherwise pushed into contact. Dark filaments are very naturally the mass suspended by $B_{||}$ in the neutral sheet of $B_\perp$ by being syphoned in from the footpoints located far . The neutral sheet in that model is produced not by the opening-up of the closed bipolar array field stretched by the rising dark filament, but by the influence of the outside pair of magnetic sources. This allows us to avoid the difficulty that the rise of the dark filament has to have a greater energy than the flare itself, because the dark filament rise in the classical model should be able to cut up the strong magnetic field which so-far has been suspending the dark filament mass, and the "repairing process" of the arcade is to explain the flare energy. It also relieves us from the difficulty of storing mass on the top of the convex arches. (It should be noted that the Kippenhahn-Schluter model, which was somehow considered to be applicable to this situation, was based on different boundary conditions and actually was not applicable to the top of such a convex arch type situation!) In Uchida-Sakurai model, the "neutral" sheet (which is actually a neutral sheet of $B_\perp$, containing relatively weak $B_{||}$ isolating the antiparallel fluxes on both sides) becomes a real neutral sheet that can liberate magnetic energy through reconnection when the dark filament, that is passively held in the relatively weak $B_{||}$ but separating the opposite polarity fields effectively,is squeezed out due to some global destabilization. Uchida and Sakurai (1980) discussed a model of arcade flare after this, and pointed out that an "interleaved book-page" type structure can be formed by the interchange instability along the surface of contact of opposite polarity at the center after the dark filament is squeezed out. The dynamical collapse into this interleaved state occurs in Alfven-transit time of 100 s, and the enhanced annihilation on the extended interface of the opposite polarity fields may explain the observed

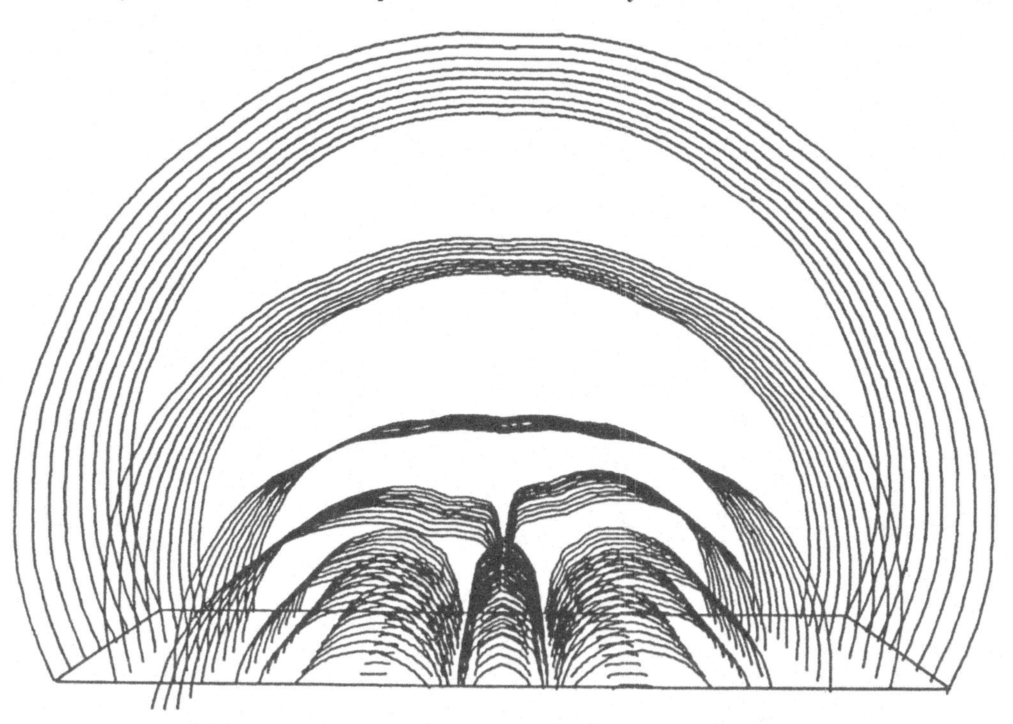

Figure 6: *Magnetohydrostatic solution for a quadruple source in the photosphere (from Uchida et al. 1980, and 1995). There is a neutral sheet in the middle, and the lines of force connecting it back to the photospheric source at several times of the width of the core part on both sides. It is suggested that magnetic reconnection will occur between the field lines on both sides of the vertical neutral sheet of $B_\perp$, when the dark filament separating them is removed from there. The reconnection then produces heated potential field loops created accross the neutral sheet. The process results in similar features as the "open up – reclosing models" if seen locally.*

enhanced rate of energy liberation in arcade flares (the Yohkoh results on arcade flares seem to be compatible with this). We are now reviving this flare model in relation to the quadruple magnetic source model for dark filaments, in the light of the Yohkoh observations (Uchida et al. 1995d).

This model predicted that the upper part of the dark filament may be pushed out upward while some lower part may remain inside the flare arcade, probably observationally corresponding to the bright feature in the dark tunnel, or brightened S-shape when seen top-on.

Since we cannot directly examine magnetograms in these events at the limb, we picked up some of the arcade type flares seen on the disk and examined the Kitt Peak magnetogram of the closest available time. In the case of an arcade type flare observed on the disk, Jun 17, 1992, it is clear that the flare

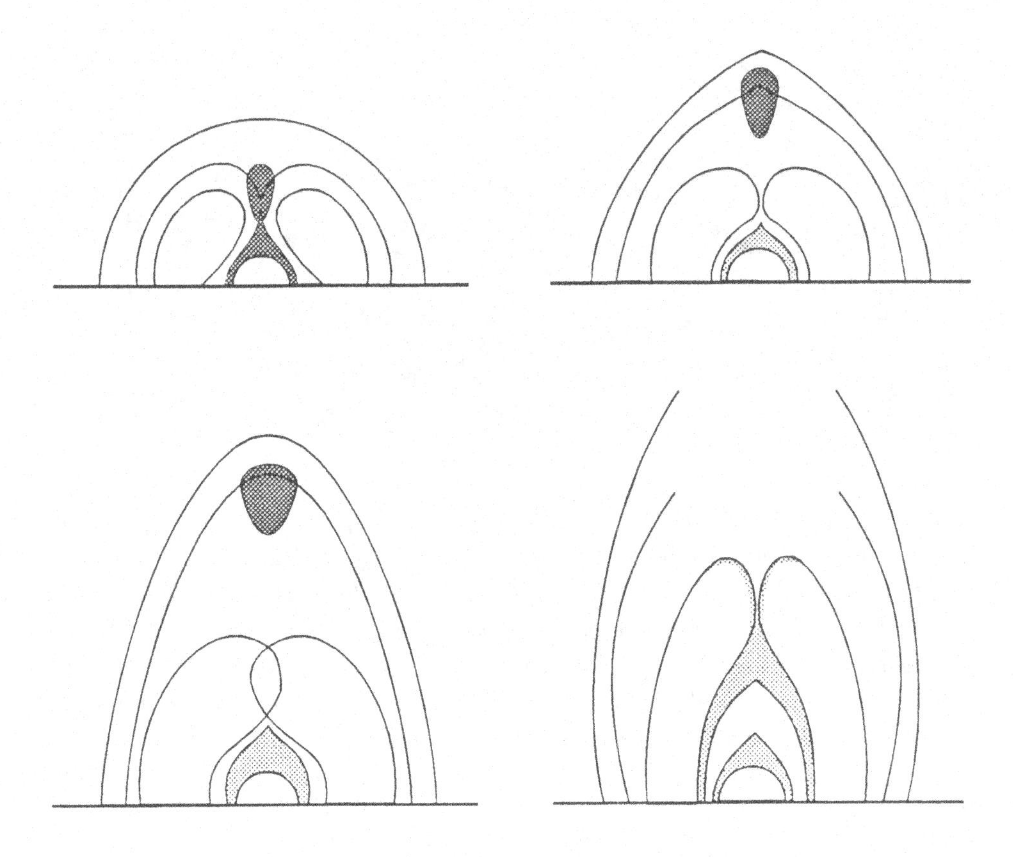

Figure 7: *Illustration for what occurs in the quadruple source model. This is depicting some typical lines of force alone for the sake of explanation. Left pannel: Connectivity of structural threads before the dark filament is squeezed out. Right pannel: After the upper part of the dark filament (the longitudinal threads with* $B_{\parallel}$ *lying in the neutral sheet of* $B_{\perp}$*) rose, and the lower part of it may be pressed down as bright S-shaped structure (probably the axial and partition-like structure seen in Figures 4, and 5 in the dark tunnel), the magnetic reconnection takes place between the opposite polarity fields thus-far separated on both side of the longitudinal field, and a heated loops of potential arcade field emerge. Shaded part is cool dark filament whereas the halftone part represents the heated part by compression or reconnection.*

is associated with the disappearance of a dark filament which existed on the central field-polarity-reversal line in a quadruple array type field distribution (elongated +, -, +, - polarity regions side by side) (McAllister et al. 1995).

The quadruple source model proposed by the present author (Uchida and Jockers 1979, Uchida and Sakurai 1980) did not attract much attention, probably because it was against the main stream of the flare model at that time. It was pointed out then as demerits that it based upon a quadruple magnetic source which was not believed to exist that time. Actually it was found to exist in the case of arcade formation outside active regions to be mentioned in 4.2.4, and it is highly possible that active region field with winding field-polarity-reversal lines can provide such a quadruple-like field effectively, though, of course, it is not a neat quadruple linear array type source. It is worth reviving that model, because some features observed by Yohkoh actually show the features suggesting strongly the quadrupolarity of the field involved as mentioned above (revived version by Uchida et al. 1994a, 1995d). It may be remarked that many years after the model was first proposed, some people came to think of the quadrupolar field, eg., Mandrini et al. (1991).

4 OBSERVED PHENOMENA IN THE BACKGROUND CORONA

4.1 Introduction — Previous Notion of the Corona

As mentioned in Chapter 1, the corona was previously considered as a quiet atmosphere merely having an extraordinarily high temperature. The observations at the limb at eclipses or by coronagraphs gave the side-view of the structures in heavy overlap with other structures, and the detail of the structure was not seen well as mentioned earlier. The X-ray observations, however, permits us to see the structures from their top without overlapping with others, and from their side at the limb, and therefore revealing the structures of the corona with less ambiguity. For example, coronal holes were first seen by the Skylab observations, and made us face the enigma of rigidly rotating coronal holes above the differentially rotating photosphere. There was a confusion based upon a suggestion that the rigid rotation of the coronal hole might reflect the rotation of the deeper layers. But how can a week field in the coronal hole passive to the differentially rotating photosphere insist its rigid rotation. We come back to this problem in Chapter 5.

The unexpected finding that the corona is quite a dynamic entity is one of the most remarkable findings obtained by Yohkoh again due to the merit of wide-dynamic range, high cadence observations mentioned above, together with the video-movie representation of the data. We can perceive much more by making effective use of the correlation in time of the phenomena in the video movie format. It is remarked here that when we talk about the dynamicity of the corona, we are talking about the changes with time scales shorter compared with the time scale like a fraction of the solar cycle which was the generally expected time scale for changes before such observational results were obtained, and not, for example, in terms of the acoustic or Alfven wave traversing time etc., as done in some theoretical treatment. (This is the same way as defining the "dynamicity" of animals. They would be extremely inert if measured in the sound traversing time scales.) We describe the "dynamic behavior" of the corona in such a sense.

4.2 Observation of the Background Corona

4.2.1 Overview of the Observation of the Background Corona

The background corona was found by Yohkoh to be unexpectedly highly dynamic, and many new findings have been reported (Acton et al. 1992b, Tsuneta and Lemen 1992, Uchida 1992). It was revealed that the "background corona" is not a quiet and isolated entity, but forms quite a dynamic system in conjunction with changes in active regions, and many examples of interesting dynamic interplay between them have been observed. The phenomena to be described below are typical ones of such dynamical phenomena occurring quite commonly, but making quite a contrast to the "static" impression of the corona we had before Yohkoh, based upon previous lower cadence observations.

Types of activities in the corona may be preliminarily categorized into three types: (a) Small brightening in active regions influencing a large coronal regions, sometimes the regions with sizes exceeding several hundreds of thousand kilometers, and in some cases releasing the so-called CME's (coronal mass ejections) in the process, (b) Injection of heated mass into pre-existing magnetic structures of the surrounding corona from active regions. In some cases, the injection occurs into a straight open-field like a coronal hole, are called "X-ray jets", and (c) Arcade formations above the field-polarity-reversal lines in the photosphere (the small scale high intensity ones of this type occurring in active regions with strong magnetic field corespond to arcade type X-ray flares), and we call the larger scale lower intensity ones in the background corona as X-ray arcade formation. Sometimes they occur even in the high latitude circumpolar zones with much fainter and larger scale arcade, and we call these as circumpolar arcade formation in the following.

We discuss these in more detail in the following, together with their suggested interplanetary counterparts which are to be investigated further in cooperation with the satellite GEOTAIL, the solar polar mission Ulysses, and the quiet-sun mission SOHO, and other ones to come. In the following, we describe (a), (b) and (c) above in Sections 4.2.2, 4.2.3, and 4.2.4, respectively.

4.2.2 Reconfiguration of the Large Scale Corona due to Small Brightenings in Active Regions

The interplay between the small events in active regions and their surroundings may be most clearly demonstrated in the Mar 27, 1992 event (Figure 8) (Uchida 1992). A brightening in the active region led to a brightening of a large area of the quiet corona to its north-west, and the brightening of that region in turn agitated the thus-far invisible highly sheared S-shaped threads to its north-west through its tipped structure contacting it, and caused a large arcade formation which is very similar to that of the Sep 28, 1991 arcade formation event to be described in 4.2.4. The S-shaped threads were brightened from the contact point, and the brightening developed towards south-west, and an X-ray arcade is formed over the S-shaped threads. A "spine-like" structure existed under the arcade also in this case, developed from the pre-existing S-shaped threads themselves and clearly not the locus of the reconnecting points, exactly similar to that of Sep 28, 1991.

The initial disturbance came from the small brightening in the active region, and that caused a brightening of the adjacent area, and then this secondary effect in turn triggered an arcade formation as a third effect. Such a propagation of influence originating in an active region itself is quite frequently seen, and one of the probable interpretations for this type phenomenon may be a propagation of the effect of the magnetic reconnections; a small fraction of the magnetic flux left out in the process of reconnection with a newly emerged flux in the strong field region may reconnect itself with the weaker field of the surrounding region nearby , and can strongly affect the surrounding weak field structures. The flux disconnected in this process at the far end again exerts strong influence on the outer region of weaker field, and so on. (It should be remarked in this connection that a magnetic reconnection itself does not always necessarily involve appreciable liberation of energy in the form of flares etc.. Substantial energy is liberated only when the part of the magnetic field storing substantial energy in the form of magnetic stress is involved.) A small change in active region, not necessarily a major flare, can affect weaker large scale field as in this case, and

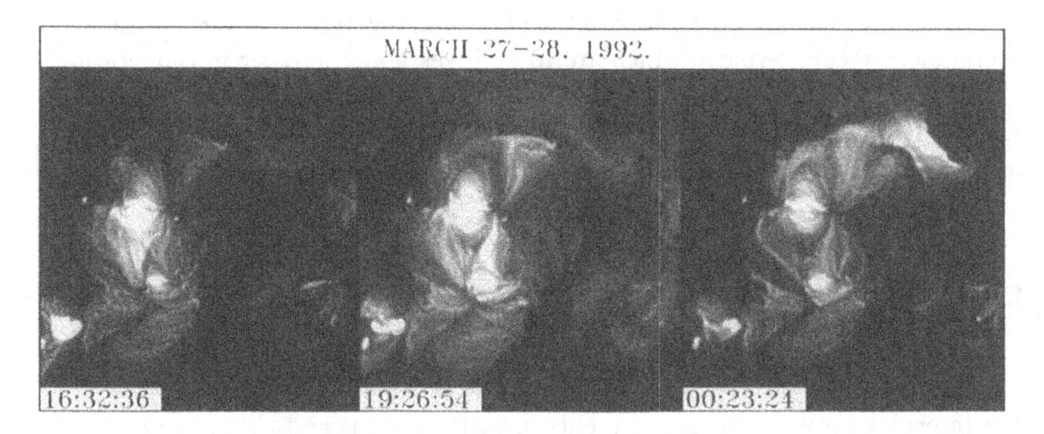

Figure 8: *Propagating effect of a small brightening in active region influencing a vast area of the quiet corona on Mar 27, 1992. A small change in a strong field active region caused a brightening of a neighboring region, and the change of that region triggered through its tipped structure the region lying further in the north-west, and caused a typical arcade formation from a bunch of threads which were not visible before. The feature developped into a typical X-ray arcade having a "spine" which was obviously a preexisting thread brightened in the arcade formation process, not a locus of the reconnecting points of the once opened up arcade reclosing through magnetic reconnection.*

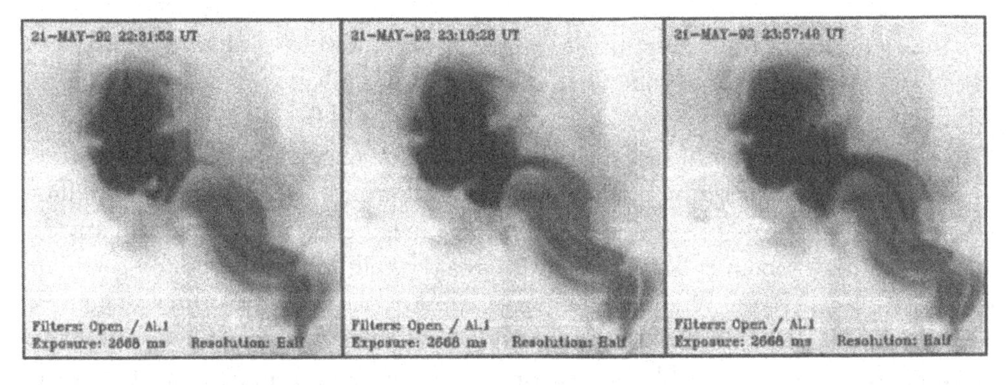

Figure 9: *Repetitive ejection of heated mass from the contact area of an active region and a region showing a helix-like structure on the south-west on May 21-22, 1992. The velocity of the injection of the hot mass is a few hundred km/s, and the process seems to be magnetic field-dominated process since the mass injection did not change the helical loop pattern appreciably (from Uchida et al. 1994).*

can change the stability of the latter, and in suitable situation, a part of the flux may ultmately be released and rise due to its own magnetic buoyancy exerted from the region in which the magnetic field has been squeezed due to horizontal motion of the footpoints in the global evolution in dynamo action, explaining huge CME launching. More discussion about this point will be given in 4.2.4.3.

4.2.3 Mass Ejection from Bright Points into Existing Surrounding Coronal Magnetic Structures

"Active region transient brightenings" mentioned in Chapter 2.2.1 may be one of the examples of the present topic occurring in small closed loops inside the active regions, and possible mechanisms for such heated mas ejection were mentioned there. Larger scale versions going far into the surrounding large scale magnetic structures will be mentioned in the following in 4.2.3.1, and those going into straight magnetic field of open coronal holes will be mentioned in 4.2.3.2. as X-ray jets (Shibata et al. 1992).

4.2.3.1. *Injection of Heated Mass into Adjacent Loop Structures from Active Regions.* There are cases in which hot mass is ejected from an "active" active region into the surrounding structure as a flow along the pre-existing magnetic field patterns. A typical one of this kind was seen repeatedly during the period May 21 through May 27, 1992, as events of heated mass ejection into "helix-like" field pattern adjacent to an agitated active region (Figure 9) (Uchida et al. 1994). An active region which experienced a flare on May 18, 1992 in the north-east quardrant, still maintained activity within it after being rearranged into an "activated" active region on the north, together with a helix-like magnetic structure in contact at its south-west.

The structure of loops in the active region on the north seemed from their connectivity to have one helicity sign and the southern helix-like part seemed to have an opposite, at the contact region of these two sets of fluxes. There occurred brightenings near the boundary of these two regions, and the heated mass is injected into the helix-like region toward south-west from that brightened point, repeatedly during the period of May 21 through 27. In these events the injected mass did not affect very much the shape of the helix-like structure on the south-west, indicating that these processes are magnetic field dominated process. The velocity of injection is estimated to be of the order of a few hundred km/s, and the injection occurred in relation with the brightenings in the region of contact. There is a possibility of estimating the rate of magnetic reconnection if the mass transfer from the hot active region to the helix-like part occurred through the reconnection process can be estimated (Uchida et al. 1994).

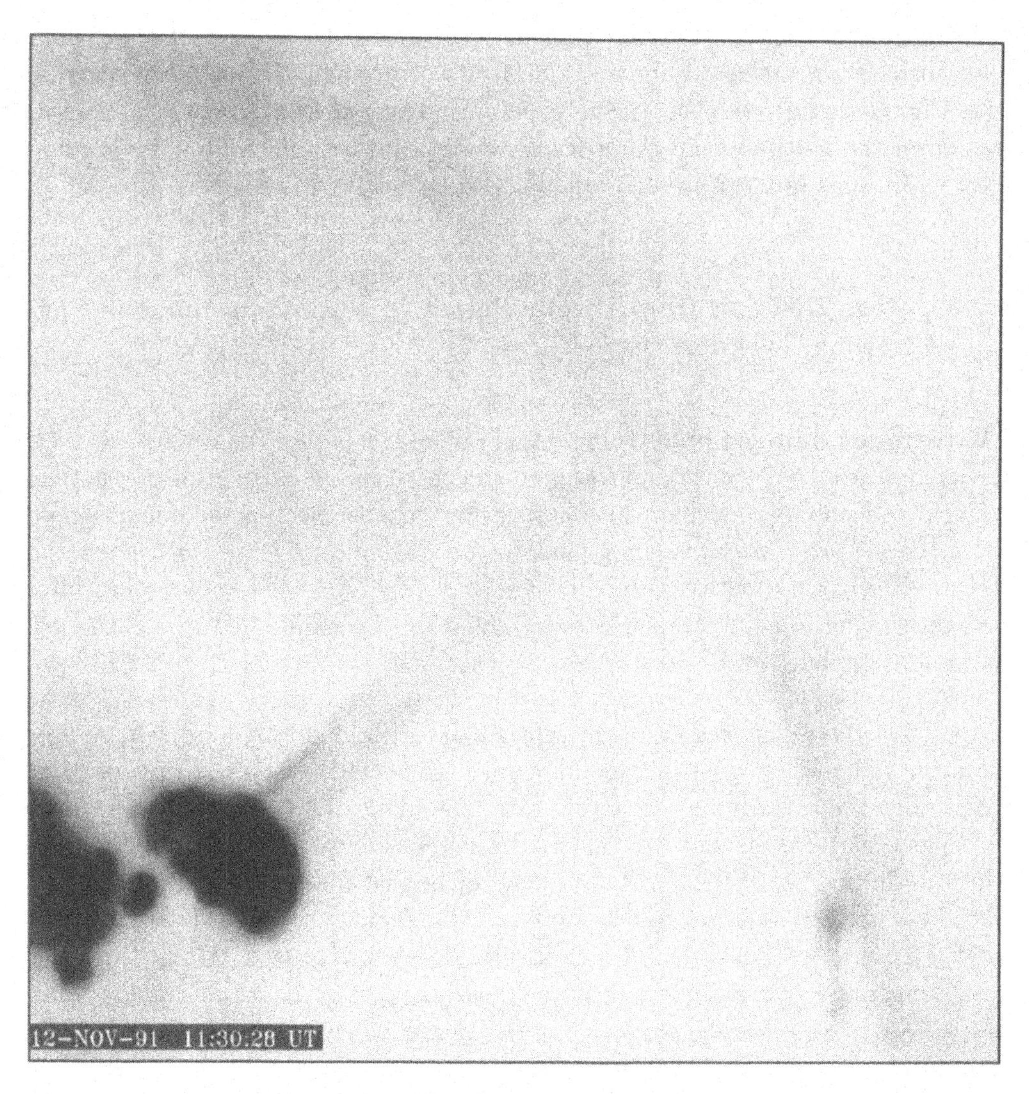

Figure 10: *X-ray jet studied by Shibata et al. (1992). It seems that the magnetic reconnection between the emerging magnetic bipoles with the open magnetic field (like that in the coronal hole regions) causes a jet which is straight. When the same process takes place in other types of coronal field, the mass ejection can take various form like the one given in Figure 9, and in certain cases of a closed field, the hot plasma is injected into two directions, which Shibata et al. (1994) called "two-sided" jets.*

The April 8, 1992 event of similar injection of heated mass into an odd shaped loops, suggesting the shape of the critical lines of force in an X-type reconnecting field, is another example of this kind. We will analyse these examples as cases of mass and energy (and also magnetic helicity) transfer due to the magnetic reconnections.

4.2.3.2. *X-ray Jets.* It may be added that many jets (escape of heated gas confined in a straight magnetic flux tubes) have been observed in soft X-rays with a wide range of velocities in the solar atmosphere (Shibata et al.1992). A most impressive and physically clear ones are jets from the region of emerging magnetic flux region into open field regions like coronal holes. This type occurs when an emerging magnetic flux region appears in the open field regions (Figure 10). Shibata called these as "anemone jets" because the shape of the bright region causing a jet has a shape like a sea-anemones. The interpretation for this proposed by Shibata et al. (1992) is due to the reconnection of the emerged magnetic flux with the surrounding open magnetic flux, and explains the observations well.

4.2.4 Arcade Formation in the Background Corona

4.2.4.1.*Large Arcade Formation outside Active Regions.* It is well-known from the Big Bear H_α observation in the 1970's that H_α dark filament disruption (disappearance) outside active regions sometimes caused an extended low-energy H_α double-ribbon flare (Michalitsanos and Kupferman 1974) showing much less non-thermal phenomena which require high energy particles. On the other hand, it was reported from the result of the Skylab observation that an X-ray arcade was formed in the corona above, connecting these Ha double-ribbons after such a disappearance of dark filament, and lasted more than 20 hours (Svestka 1976). The disappearing dark filament material is often seen as a thready material flying away (usually referred to as eruptive prominences), and it is known that a large "mantle of loops" containing a considerably large amount of mass expands ahead of this (Hundhausen et al. 1988).

The relation of the X-ray arcade formation to the eruptive prominence, or erupting H_α dark filament, and to the double-ribbon H_α flaring below, has been discussed in terms of the "opening up - reclosing magnetic arcade" type models (Sturrock 1966, Hirayama 1974, Kopp and Pneuman 1976) with an elongated bipolar array type magnetic source in the photosphere below.

We here note that the Yohkoh observation indicates that this "classical model" seems to have encountered serious questions in the detailed examination of the observations in the case of arcade flares in active regions (Uchida et al.1995d). This is also true with the off-active region arcades as seen below.

Sep. 28, 1991 event is a typical example of arcade formation outside active regions (McAllister et al.1992). This event of X-ray arcade formation (sub-flare brightening in GOES low-C class) appeared in the north-east quadrant near the disk center at around 11:18UT on Sep. 28, 1991. An activity started when two neibouring Ha dark filaments formed separately merged the day before the event. After an interruption of observation by the satellite night, SXT found that a brightening at the crossing point developed into an increasing brightening of the region centered at that location by 12:27UT, expanding towards north and south along the pre-existed thread structure. Within 30 minutes, the region with a length of 3×10^5 km brightened, and an arcade-like feature appeared perpendicular to the "spine-like" feature. At 15:56UT, the "spine" seemed to have risen, and the arcade was stretched into a tent-like structure by the rising "spine" (Figure 11). This continued and the "spine" ballooned up at the northern part. The satellite changed to other modes of observation without noting the event, since it was in the very initial phase of the instrument testing less than one month after the launch of the satellite.

We found that the "spine-like" structure along the length of the arcade was not the locus of the reconnecting points of the once opened-up overlying magnetic arches as expected in the previous models. The fine structure-enhanced images clearly show that the spine-like structure has more than two bright threads, which is absolutely impossible as the locus of the reclosing points. They turned out to be part of the field structure pre-existed below! This "spine-like" structure, identified with the structure existing below before the event, stretches the initially round X-ray arcade above into a triangular shape like a tent, suggesting that actually it came rising from below the X-ray arcade. This is conspicuously different (McAllister et al. 1992, Uchida 1992) from the previous picture that the "spine-like" structure was expected as the locus of the reconnecting points between the opened up field lines and come shrinking from above. It is clear that some magetic reconnection process is involved, however, and the observed features are in favor of the quadruple source model we discussed above in which the S-shaped bright feature, probably a remaining lower half of the dark filament, comes rising from below the arcade which is formed by the reconnection between the oppositely-directed fields facing each other after the dark filament, that-far separated both, is gone.

4.2.4.2. *Arcade Formation in the Polar Regions.* The process of the X-ray arcade formation in relation to the disappearing H_α dark filament and H_α flaring seems to be a rather general process, and physically similar phenomena exist

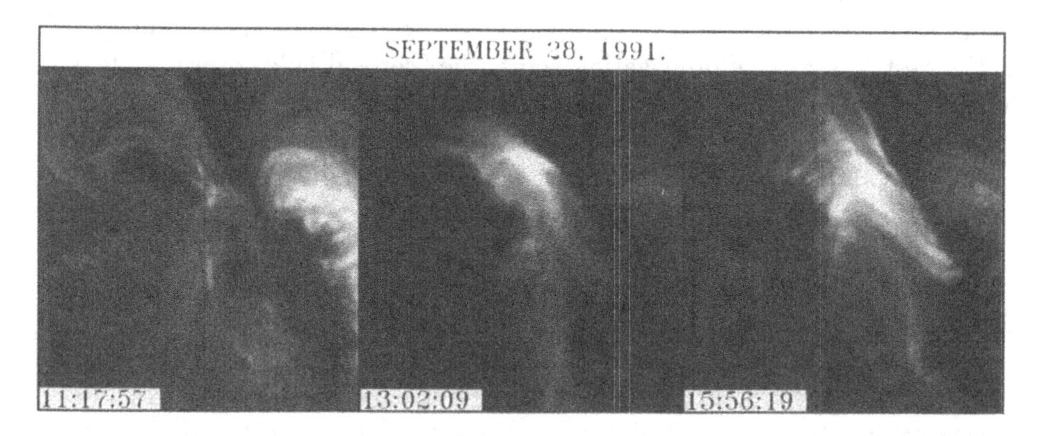

Figure 11: *An example of arcade formation outside active regions occurred on Sep 28, 1991. The initial brightening occurred at a cross point of two magnetic loop systems which occurred when two of the dark filaments merged. The brightening developed and a system of bright loops in the form of arcade appeared. There was a structure corresponding to the "spine" of the arcade, but contrary to the expectation from the classical "opening up – reclosing" picture, the "spine" was not the locus of the reclosing points in the reconnection of the once opened up arcade, but clearly seen to be consisted of the preexisted longitudinal threads heated up and rose. This rising longitudinal threads came up from below the arcade, since as it continued to rise, it seemed to make the arcade into a tent-like shape.*

in a wide range of scales and therefore different magnetic field strengths are involved, so long as the magnetic topology has some similarity; ranging from those occurring in active regions as X-ray arcade type flares (mentioned in 3.2.3.2), quieter X-ray arcade formation associated with the disappearances of dark filament outside active regions (as described in 4.2.4.1), to the formation of still fainter and larger X-ray arcades in the circumpolar region with a disappearance of faint polar region dark filaments without appreciable Ha enhancements. Here we describe this circumpolar arcade formation.

Larger scale fainter version of this type leading to the formation of a faint large arcade in the circumpolar region was first found on November 12, 1991 (Tsuneta et al.1992b), and several similar examples have been observed after that. The November 12 event occurred in an area seemingly a part of the north polar coronal hole and surprised us, but it was confirmed later that the region had a faint polar dark filament and the region was not actually a true unipolar coronal hole region (K. Harvey, private communication). The event started as a weak brightening in a coronal structure on the east side of a "coronal hole" extending in the north-west quadrant, and faint loops successively "popped up" from the coronal structure at the east end of the "coronal hole", and propagated to the west, like a wave propagating across the "coronal hole" over 6×10^5 km in about 1 day's time. The event indicated that a faint large scale version of the arcade formation mentioned above can exist in the polar area, and that there can be a "fake coronal hole" which continued to be a dark polar area for a long time, but is not a magnetically unipolar region.

A more recent sequence of similar events occurred in the April – May, 1994 period (Alexander, et al. 1994, McAllister et al. 1994) and is known to be the sources of large geomagnetic storms that caused damages to the electric power lines in the US. In the case of the event on April 14, 1994, Kitt Peak magnetogram shows a quadruple array type distribution of the photospheric magnetic source as in the case of the on-disk arcade type flare of Jun. 17, 1992 mentioned in 2.3.2. above.

4.2.4.3. *Arcade Formation and Its Relation with CME and "Giant Cusps".* We now give some discussion about the relation of the X-ray arcade formation and the CME launching, and the formation of "giant cusps" found by Yohkoh. The "giant cusp" formation event of Jan 24-25, 1992, which had a good coverage of observation was noted (Strong 1994) as a feature related to the coronal mass ejection (CME). In this event, there existed in soft X-rays various phenomena suggestive of the launch of CME, namely, the rise of X-ray threads suggestive of heated part of the flying dark filament, etc., and finally a very remarkable formation of a "giant cusp" in that region. It was examined in comparison with the Mauna Loa K-coronameter data (Hiei et al. 1993), and it was concluded that the large scale cusp-like structure was formed after a CME pushed through the pre-existing helmet streamer, and the cusp seems to be the heated region in the process of reclosing of the once-opened up arcade of arches. They considered this as evidence for the "opening up - reclosing" process (Hiei et al. 1993).

There are other events of X-ray arcade formation some time after the dark filament disappearance, and Miyazaki et al. (1993) showed that a gigantic Ha dark filament was actually flying away after disappearing from the low corona, almost keeping its shape. This process was also seen in radio with the Nobeyama Radioheliograph, and an X-ray arcade was formed behind low above the field-polarity-reversal line on the solar surface. Therefore, there is no doubt that the dark filament rise and X-ray arcade formation is related that way. The remaining question is how are they related physically.

An example of interest of such large scale arcade formation in the circumpolar region was the Feb. 24-25, 1993 event in the southern polar region (Figure 12) (Uchida et al. 1995e). In this case, there existed a pair of arrays of X-ray

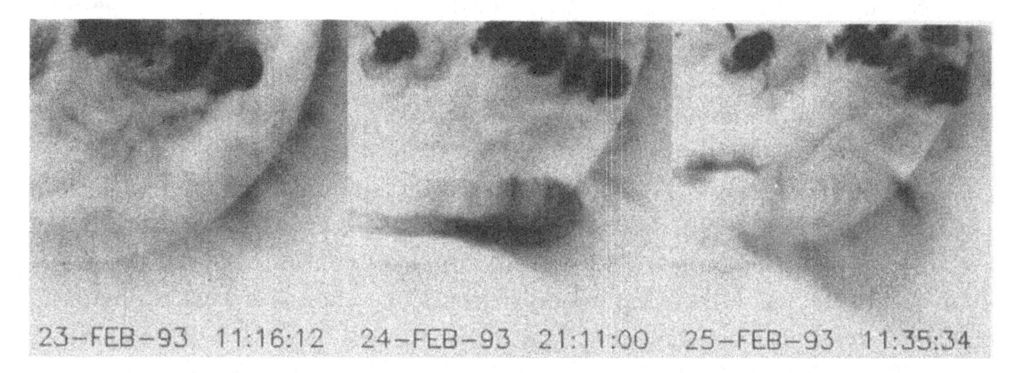

Figure 12: *The time development of the polar arcade formation of Feb 24-25, 1993. There existed two rows of arches side by side, with a dark filament channel between them. A brightening started around 17UT of Feb 24, 1993 in the dark lane, and this progressed toward west along the dark lane, and formed a "spine"-like structure together with an arcade of loops across the original dark lane. This "spine" rose, especially on the western side, pulling up the arcade of loops, and finally ballooned up, and formed a shape that could give an impression of a large cusp if seen from the direction along the initial dark lane.*

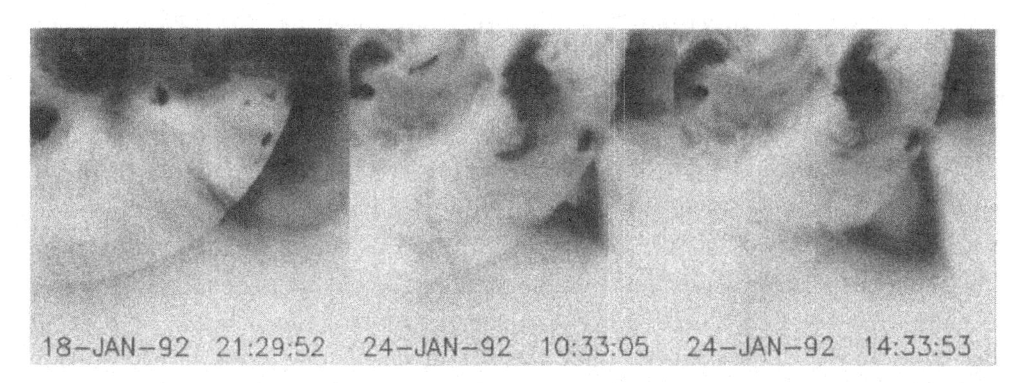

Figure 13: *The time development of the "giant cusp" formation of Jan 25, 1992. Two rows of arches similar to that of the Feb 24-25, 1993 event existed with a dark filament channel in between. In this case, the event occurred when the main part of the structure had rotated beyond the western limb, and the arcade formation with rising feature was seen in projection almost along the axis, or the structure is seen parallel to the dark lane. We were surprised to find that the southern leg of the "giant cusp" remained on the visible hemisphere while the main body of the arcade should be much beyond the limb by that time! This is not expected if the "giant cusp" is the product of the "opened up - reclosing" process.*

arches on both sides of a dark lane lying from east to north-west, and the Kitt Peak magnetogram showed that there existed a field-polarity-reversal line below this X-ray dark lane. At around 17UT of Feb 24, a brightening started along this dark lane, and the brightening progressed again towards west, and the bright legs of a "spine"-structure appeared on the north side of the dark lane. These seem to correspond to the X-ray structure connecting both sides in a skew geometryas seen in Figure 6b. The bright "spine" in this case was again consisted of pre-existed threads along the dark lane which was brightened and rose. It finally ballooned near its west end and, as it continued to balloon up, it began to give a "cusp-like" impression *if* seen along the axis. This is obviously something different from the rising dark filament itself because it came up much later and was rising from below, or at least together with, the "X-ray arcade", which should be formed much after the so-called dark filament eruption itself [cf., Jul 30, 1992 event (Miyazaki et al.1993)]. The "spine" seemed to be one of the longitudinally elongated threads whose legs connecting both sides of the field polarity reversal lines with a skewed geometry. The longitudinal threads having the westmost leg on the north side of the original dark lane seemed to have ballooned up. This situation was entirely similar to the Sep 28, 1991 event described above, and difficult to explain in the classical model, but compatible with the quadruple source model (Uchida and Sakurai1980, Uchida et al.1995d).

A careful examination hinted by this of the Jan 24-25, 1992 event showing the "giant cusp", revealed a surprising fact that a few of the SXT frames at around 14:33UT of Jan 24 indicated that the south-east leg of this "giant cusp" *was seen to stay on the visible hemisphere*, while the main body of it was already rotated beyond the west limb by the solar rotation!. This was a surprise because, if the cusp was the result of reconnection of the once opened-up arcade of loops, then the elementary structure of the "cusp" which was produced through reconnection should lie ideally in the planes perpendicular to the axis. In other words, both footpoints should be already rotated to the same longitude as the "reclosing cusp" itself and *should not* be seen on the visible hemisphere! Figures 13 show that the Jan 25, 1992 event is very similar to the Feb 25, 1993 event discussed above. This suggests that the "giant cusp" may have interpretation quite different from that of the classical "open up - reclosing" picture. It is possible that the "cusp" may be a ballooned up part of the bright S-shaped feature (Uchida 1995, Uchida et al. 1995e), and what we found and described for the Sep 28, 1991 event (McAllister et al. 1992, the "spine" is rising below the X-ray arcade) turns out to be not an exception, but an intrinsic part of the process to be explained by a correct model. We do not repeat citing discussion about the models, and refer the reader to the discussion of 3.2.3.2, by citing our paper to appear soon based on our quadruple source model (Uchida et al. 1995d).

5 BRIEF COMMENTS ON GLOBAL FEATURES AND CYCLE PHASE-DEPENDENT PHENOMENA SUGGESTED BY YOHKOH

The time of launch of Yohkoh was the end of August 1991, probably one year or so after the maximum of the solar activity. Since the satellite is basically healthy about 3.5 years after launch (as of February 1995), we may have some hope for the satellite to be able to cover a period from the near solar maximum to the minimum. Comparison of the SXT pictures of December 1991 and of the most recent date shows a very drastic difference. Much less number of active regions exist, and the massive background corona disappeared. Flares occur but very seldom, and mainly weak ones in recent days. The investigation of the solar cycle variation of various features is obviously what can be done with the Yohkoh data, but not too many works have been accomplished in this area yet. This may be natural because the coverage of the cycle phase by Yohkoh is still only a fraction of it, and therefore, the topics of the investigations thus far taken up have mainly been concentrated on the short time-span, and are mainly individual events or phenomena about flares or the corona. Investigation of the solar cycle phase-dependence of various features as well as of the global characteristics that can be done by using the Yohkoh data, however, is extremely valuable for the study of solar cycle, or dynamo process, and it is expected that the research in this area increases from now on.

Some of the points of interests are:

(i) Systematic study of the location and its change of the active areas in latitude (active latitudes) and in longitude (active longitudes) with the solar cycle-phase is very important. This is somewhat like the traditional investigation of the same problem by using the sunspot groups, but the great difference will be that the information about the connectivity of the weak magnetic field outside the sunspot groups to the active regions and to each other is available in the Yohkoh data. The variation of the weak magnetic field in conjunction with the active region field may be essential in clarifying the global magnetic behavior of the Sun, and therefore vital in investigating the dynamo action of the Sun. The SXT data is very suitable for this. In this context, the investigation of the behavior of the small X-ray bright points will also be of importance. In the Skylab results, the trend of the occurrence of the X-ray bright points had an anti-correlation with that of active regions in the solar cycle phase (Golub et al. 1979), but somehow, the X-ray bright points are not pronounced in the Yohkoh-SXT data. There is a possibility that they were not well visible due to the temperature sensitivity of the Yohkoh-SXT in comparison with that of Skylab, but the comparison of the Yohkoh results with the results of NIXT (Golub 1990) which has its sensitivity in lower temperature range indicates that the X-ray bright points have been actually fewer in the present solar cycle. Whether this changes with the cycle phase towards the activity minimum and the X-ray bright points will increase in number towards the activity minimum or not will be confirmed from now on.

(ii) Behavior of coronal holes will be another important topic for which the Yohkoh data will contribute much. An expectation from the impactful result from Skylab was that the coronal holes are rotating rigidly, as seen in the case of the "American Continent Coronal Hole" which connected the north and south hemispheres across the equator without changing the shape over several rotations. (In the Yohkoh data, the corresponding features are what we call "Coronal Dark Channels" because this type of coronal holes appeared as a thin dark channel connecting the north and south hemispheres. The "American Continent" hole was a specially wide version of the "coronal dark channel" connecting both hemispheres). Preliminary study has been made of the rotation of the coronal hole by using synoptic charts and their stuck plots, and Takahashi et al. (1993, 1994) showed that the part of the coronal hole executing rigid rotation is only the low latitude (active latitude zones) part, and the parts with latitudes higher than the active latitude zones are rotating differentially. The enigmatic conclusion from Skylab (Krieger et al. 1973), that the magnetic field of the corona rotates rigidly while the photospheric material in between rotates differentially, was a large surprise. Some people argued that the large scale coronal magnetic field might reflect the influence of a larger scale magnetic ield in deeper interior of the Sun. But, how does the field escape from being influenced by the current-conducting photospheric layer executing differential rotation in between? What Takahashi et al. found was that the higher latitude zones are executing nicely differential rotation, and only the part below the edge of the active latitude perform rigid rotation. This will be interpreted that what are rotating rigidly are active longitudes, and the connectivity of the weak coronal field is entirely governed by the strong field of active longitude belt, as seen from the behavior of the corona discussed in Chapter 4. Namely, some slight excess of magnetic flux on one side of the active longitude belt (say +) may absorb the − flux of the neighboring region, and so on, and eventually a weak field region with + flux is left open (if the influence of another active longitude belt on the opposite hemisphere happen to produces a + flux region there from other direction), then a coronal hole of + flux may be resulted there. In this case, another hole with − flux may exist on the opposite longitude, 180 degrees apart. This will occur if an imbalance of + and − polarity fluxes exists in the active longitude belts in the opposite hemisphere in the opposite azimuthal directions. This was claimed by Saito et al. (1989) from the variation of interplanetary magnetic field around the polarity reversal process of the solar global field around the solar activity maximum when we saw the coronal dark channels. The north and the south hemisphere may behave differently, but in such a period they may be in phase in longitude. Thus, the coronal dark channel will rotate rigidly as a "shadow" of the active longitude belts as the latter rotate rigidly. Rigid rotation of active longitude belts has explanation for itself (eg., Yoshimura 1974). There are many other topics of interest in the context of global and solar cycle phase-dependent problems, but we do not go into the details before more analyses will be made.

6 CONCLUSION

In the above, we have described some of the new findings about the activity in the active regions, about flares as the conspicuously pronounced activity in them, and about the background corona which shows an unexpectedly active behavior. It was actually a surprize that wide dynamic range and high cadence observations worked so effectively in making the scientific outcome greater. Their role was to allow us to see what is occurring in the darker preflare stages of flares, or what is occurring in the darker background corona. High spatial resolution observation in the hard X-ray range (without contamination from the stronger soft X-ray range) coaligned with the soft X-ray telescope, and together with the context observations by the spectrometers, was likewise scientifically very profitable. All of them combined are expected to contribute toward the solution of the long-standing problems in the solar physics by extending the fronts of observations toward the darker, and more rapidly varying features which, as in most of the observations in astronomy, contain essential clues to the secrets hidden in them.

ACKNOWLEDGEMENT

The author acknowledges the Japanese, US, and UK colleagues whose contribution made the success of the Yohkoh Program possible through the excellent collaboration in the construction of hardware, in the data taking, and in the data reduction. The author aknowledges the colleagues for an exciting discussions in collaborative research.

7 REFERENCES

Acton, L.W., et al., 1992, *Publ.Astron.Soc.Japan*, **44**, L71.

Acton, L.W., Tsuneta, S., Ogawara, Y., Bentley, R., Bruner, M., Canfield, R.C., Culhane, J.L., Hiei, E., Hirayama, T., Hudson, H.S., Kosugi, T., Lang, J., Lemen, J., Nishimura, J., Makishima, K., Uchida, Y., and Watanabe, T.,

1992, *Science*, **258**, 618
Alexander, D., et al., 1994, in *Proceedings of 3rd SOHO Workshop*, in preparation.
Alfven, H., and Carlqvist, K., 1967, *Solar Phys.*, **1**, 220.
Antonucci, E., et al., 1986, *Adv. Space Res.*, **6**, 151.
Culhane, J.L., et al., 1991, *Solar Phys.*, **136**, 89.
Doschek, G., Cheng, C.C., Oran, E., Boris, J., and Mariska, J., 1983,*Astrophys. J.*, **265**, 1103
Doschek, G.A., et al., 1993, *Astrophys. J.*, **416**, 845
Feldman, U., et al., 1994, *Astrophys. J.*, **424**, 444
Forbes, T.G., 1992, in *Eruptive Solar Flares*, eds. Z. Svestka, B.V. Jackson, and M. Machado (Springer Verlag), p79
Galeev, A.A., Rosner, R., Serio, S., and Vaiana, G.S., 1981, *Astrophys. J.* , **243**, 301
Golub, L., Davis, J.M., and Krieger, A. S., 1979, *Astrophys. J. Lett.*, **229**, L145
Golub, L., 1991, in *Flare Physics in Solar Activity Maximum 22*, ed Y.Uchida, R.Canfield, T.Watanabe, and E.Hiei (Springer Verlag), p271.
Hiei, E., Hundhausen, A., and Sime, D., 1993, *Geophys. Res. Letters*, **20**, 2785.
Hirayama, T., 1974, *Solar Phys.*, **34**, 323.
Hundhausen, A., 1988, in *Sixth International Solar Wind Conference*, ed. V.J.Pizzo.
Kopp, R.A. and Pneuman, G.W., 1976, *Solar Phys.*, **50**, 85.
Kosugi, T., 1992, in *Advances in Stellar and Solar Coronal Physics*, ed. J.L. Linsky and S. Serio (Kluwer: Dordrecht), p51.
Kosugi, T., et al., 1991, *Solar Phys.*,**136**, 17.
Krieger, A., et al., 1973, *Solar Phys.*, **29**, 505 .
Mandrini, C., Demoulin, P., Henoux, J-C., and Machado, M.E., 1991, *Astron. Astrophys.*, **250**, 547 .
Masuda, S., 1994, Thesis, University of Tokyo.
McAllister, A.H., Uchida, Y., Tsuneta, S., Strong, K., Acton, L.W., Hiei, E., Bruner, M.E., Watanabe, Ta. and Shibata, K., 1992, *Publ. Astron. Soc. Japan*, **44**, L205.
McAllister, A.H., et al., 1994, in *Proceedings of 3rd SoHO Workshop*, in preparation .
McAllister, A.H., Uchida, Y., and Khan, J., 1995, in preparation Michalitsanos, A., and Kupferman, P., 1974, *Solar Phys.*, **36**, 403.
Miyazaki, H., Miyashita, M., Yamaguchi, A., Ichimoto, K., Kumagai, K., Hirayama, T., and Tsuneta, S., 1993, in *X-ray Solar Physics from Yohkoh*, eds. Y.Uchida, T.Watanabe, K.Shibata and H.Hudson (Universal Academy Press), p277.
Nakagawa, T., Nishida, A., and Saito, T., 1989, *J. Geophys. Res.*, **94**, 11761.
Nakagawa, T., Uchida, Y., 1995, in preparation.
Ogawara, Y., Takano, T., Kato, T., Kosugi, T., Tsuneta, S., Watanabe, T.,

Kondo, I., and Uchida, Y., 1991, *Solar Phys.*, **136**, 1.

Parker, E.N., 1963, *Interplanetary Dynamical Process*, (Interscience: New York).

Parker, E.N., 1963,*Astrophys.J. Suppl. 8.*,**177**, 1963.

Pallavicini, R., et al.,1975, *ESA SP*,**45**, 411.

Petchek, H.E., 1964, *NASA Symposium Physics of Solar Flares*, p 425.

Priest, E., and Forbes, T.G., 1990, *Solar Phys.*, **126**, 319.

Saito, T., Oki, T., Olmsted, C., and Akasofu, S., 1989, *J. Geophys. Res.*, **94**, 14993 .

Sakao, T., 1994, Thesis, University of Tokyo

Sakurai, T., 1991, *Flare Physics in Solar Activity Maximum 22*, ed. Y. Uchida, R.C. Canfield, T. Watanabe, and E. Hiei (Springer Verlag: Berlin), p245.

Sato, T., and Hayashi, T., 1979, *Phys. Fluids*, **22**, 1189.

Shibata, K. and Uchida, Y., 1986, *Solar Phys.*, **103**, 299.

Shibata, K., Ishido, Y., Acton,L.W., Strong, K.T., Hirayama, T., Uchida, Y., McAllister, A.H., Matsumoto, R., Tsuneta, S., Shimizu, T., Hara, H., Sakurai, T., Ichimoto, K., and Ogawara, Y., 1992, *Publ.Astron. Soc. Japan*,**44**, L173.

Shibata, K., Yokoyama, T., and Shimojo, M., 1994, in *New Look at the Sun*, eds. S.Enome and T.Hirayama (Nobeyama Radio Observatory Printing), p75-78.

Shibata, K., 1995, in preparation.

Shimizu, T., Tsuneta, S., Acton, L.W, Lemen, J.R. and Uchida, Y., 1992, *Publ. Astron. Soc. Japan*, **44**, L147.

Strong, K.T., 1994, in *2nd SOHO Workshop "Mass Supply and Flows in the Corona"* eds. B.Fleck, G.Noci, and G.Polleto, p133.

Sturrock, P.A., 1966, *Nature*, **221**, 695.

Svestka, Z., 1976, in *Solar Flares* (Reidel: Dordrecht), p44

Sweet, P.A., 1958, in *Electromagnetic Phenomena in Cosmical Physics*, ed, B.Lehnert (Cambridge University Press), p123.

Takahashi, T., 1992, Master's Dissertation, University of Tokyo.

Takahashi, T., S.Tsuneta, K.Hayashi, and H.Yoshimura, 1994, in*X-ray Solar Physics from Yohkoh*, eds. Y.Uchida, T.Watanabe, K.Shibata, and H.Hudson (Universal Academy Press), pp293-296.

Tanaka, K., Nitta, N., Akita, K., and Watanabe, T., 1983,*Solar Phys.*, **86**, 91.

Tsuneta, S., Acton, L.W., Bruner, M., Lemen, J., Brown, W., Calavalho, R., Catura, R., Freeland, S., Jurcevich, B., Morrison, M., Ogawara, Y., Hirayama, T,, and Owens, J., 1991, *Solar Phys.*, **136**, 37.

Tsuneta, S., et al., 1993, in *The Magnetic and Velocity Fields of Solar Active Regions*, eds. H. Zirin, G-X. Ai, and H. Wang (ASP Conference Series No.46), p239 .

Tsuneta, S., and Lemen, J., 1992, in *Advances in Stellar and Solar Coronal Physics*, ed. J.L. Linsky and S. Serio (Kluwer: Dordrecht), p113.

Tsuneta, S., Hara, H., Shimizu, T., Acton, L.W., Strong, K.T., Hudson, H.S., and Ogawara, Y., 1992a, *Publ. Astron. Soc. Japan*, **44**, L63.

Tsuneta, S., Takahashi, T., Acton, L.W., Harvey, K., 1992b, *Publ. Astron. Soc. Japan*, **44**, L211.

Uchida, Y., and Jockers, K., 1979, Max Planck Institute Preprint .

Uchida, Y., and Sakurai, T., 1980, in *Skylab Workshop, Solar Flares*, ed. P.A. Sturrock (University of Colorado Press), p67, and p110.

Uchida, Y. and Shibata, K., 1985, in *Unstable Current Systems and Plasma Instabilities in Astrophysics*, ed. M. Kundu and G.D. Holman (Reidel: Dordrecht), p287.

Uchida, Y. and Shibata, K., 1988, *Solar Phys.*, **116**, 291.

Uchida, Y., 1992, in *Advances in Stellar and Solar Coronal Physics*, ed .J.L. Linsky and S. Serio (Kluwer: Dordrecht), p97.

Uchida, Y., McAllister, A., Strong, K., Ogawara, Y., Shimizu, T., Matsumoto, R., and Hudson, H.S., 1992, *Publ. Astron. Soc. Japan*, **44**, L155.

Uchida, Y., 1993, in *Advances in Space Res.*. (Proc. COSPAR Washington), **13**, 9 (205).

Uchida, Y., et al., 1994a, in *X-ray Solar Physics from Yohkoh*, eds. Y.Uchida, T.Watanabe, K.Shibata, and H.Hudson (Universal Academy Press), p161.

Uchida, Y., Fludra, A., and Khan, J., 1994 b, in *New Look at the Sun* ed. S.Enome and T.Hirayama (Nobeyama Radio Observatory Prints), p83-88.

Uchida, Y., 1995, *Adv. in Space Res.. (Proc. COSPAR Hamburg)*, in press Uchida, Y., Ono, Iwai, and Hirose, S., 1995a, in preparation .

Uchida, Y., Khan, J., Doschek, G., Masuda, S., McAllister, A., Hirose, S., Feldman, U., and Cheng, C.C.,1995b, in preparation.

Uchida, Y., Yashiro, S., Kohara, N., Watanabe, T., Kosugi, T., Hirose, S., and Cable, S., 1995c in preparation.

Uchida, Y., Jockers. K., Khan, J., and McAllister, A., 1995d, in preparation.

Uchida, Y., Fujisaki, K., Okubo, H., Tsuneta, S., Hirose, S., and Cable, S., 1995e, in preparation.

Yeh and Axford, 1970, *J. Plasma Phys.*, **4**, 161.

Zwaan, C., 1987, *Ann. Rev. Astron. Astrophys.*, **25**, 83

Printed in the United States
By Bookmasters